FUNCTIONING
IN THE
REAL WORLD

A PRECALCULUS EXPERIENCE

SHELDON P. GORDON
Suffolk Community College

FLORENCE S. GORDON
New York Institute of Technology

B.A. FUSARO
Florida State University

MARTHA J. SIEGEL
Towson State University

ALAN C. TUCKER
SUNY at Stony Brook

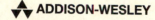 **ADDISON-WESLEY**

An imprint of Addison Wesley Longman

Reading, Massachusetts · Menlo Park, California · New York · Harlow, England
Don Mills, Ontario · Sydney · Mexico City · Madrid · Amsterdam

Sponsoring Editor Bill Poole
Development Editor Elka Block
Senior Production Supervisor Karen Wernholm
Marketing Manager Andy Fisher
Senior Manufacturing Manager Roy Logan
Editorial Production Services Barbara Pendergast
Text Designer Rebecca Lemna
Cover Designers Linda Wade/Susan Carsten
Cover Photo ©Telegraph Colour Library/FPG International Corp.
Art Coordinator Sandra Selfridge
Compositor New England Typographic Services
Illustrators Techsetters, J. A. K. Graphics

Library of Congress Cataloging-in-Publication Data

Functioning in the real world : a precalculus experience / Sheldon P.
   Gordon . . . [et al.].
      p.   cm.
   Includes index.
   ISBN 0-201-84628-4
   1. Functions.   I. Gordon, Sheldon P.
QA331.3.F84 1996
515′ .25—dc20   96-43342
                                CIP

This book is based on a portion of the materials developed with the support of
the Division of Undergraduate Education of the National Science Foundation
under grants #USE-91-50440 and #DUE-9254085 for the Math Modeling/
PreCalculus Reform Project. However, any views expressed are not necessarily
those of the Foundation.

Reprinted with corrections, April 1997.

2 3 4 5 6 7 8 9 10 CRW 999897

CONTENTS

PREFACE   vii

CHAPTER 1

FUNCTIONS IN THE REAL WORLD

1.1   Functions Are All Around Us   1

1.2   Describing the Behavior of Functions   9

1.3   Representing Functions Symbolically   16

1.4   Connecting the Geometric and Symbolic Representations   22

1.5   Mathematical Models   28

1.6   Introduction to Monte Carlo Methods   32

Chapter Summary   37

Review Exercises   38

CHAPTER 2

FAMILIES OF FUNCTIONS

2.1   Introduction   41

2.2   Linear Functions   41

2.3   Exponential Functions   60

2.4   Power Functions   74

2.5   Logarithmic Functions   85

2.6   Comparing Rates of Growth   100

2.7   Inverse Functions   106

Chapter Summary   115

Review Exercises   116

CHAPTER 3

FITTING FUNCTIONS TO DATA

3.1   Introduction to Data Analysis 119

3.2   Linear Regression Analysis 122

3.3   Curve Fitting and Nonlinear Regression Analysis 135

3.4   What the Residuals Tell Us About the Fit 153

3.5   Analyzing Data 163

3.6   Analyzing the *Challenger* Data: A Case Study 172

Chapter Summary 176

Review Exercises 176

CHAPTER 4

EXTENDED FAMILIES
OF FUNCTIONS

4.1   Polynomial Functions   179

4.2   Fitting Polynomials to Data   194

4.3   The Roots of Polynomial Equations: Real or Complex?   199

4.4   Building New Functions from Old   208

4.5   Finding Roots of Equations   224

4.6   Finding Polynomial Patterns   232

Chapter Summary   248

Review Exercises   248

CHAPTER 5

SEQUENCES AND THEIR
APPLICATIONS

5.1   Introduction to Sequences   251

5.2   Applications of Sequences   256

5.3   Mathematical Models and Difference Equations   264

5.4   Eliminating Drugs from the Body   279

5.5   The Logistic or Inhibited Growth Model   293

5.6   Radioactive Decay   306

5.7   Geometric Sequences and Their Sums   312

5.8   Newton's Laws of Cooling and Heating   320

5.9   Sequences and Differences   329

Chapter Summary   337

Review Exercises   338

CHAPTER 6

MODELING WITH DIFFERENCE EQUATIONS

6.1   Solutions of Difference Equations   341

6.2   Constructing Solutions of First Order Difference Equations   354

6.3   Applications of First Order, Nonhomogeneous Difference Equations   363

6.4   Fitting Quadratic Functions to Data   374

6.5   Financing a Car or a Home   385

6.6   Sorting Methods in Computer Science   391

6.7   Modeling the Stock Market   400

6.8   Fitting Logistic Curves to Data   405

6.9   Iteration and Chaos   412

Chapter Summary   422

Review Exercises   423

CHAPTER 7

MODELING PERIODIC BEHAVIOR

7.1   Introduction to the Trigonometric Functions   425

7.2   Trigonometric Functions and Periodic Behavior   436

7.3   Relationships Between Trigonometric Functions   455

7.4   Solving Trigonometric Equations: The Inverse Functions   464

7.5   The Tangent Function   471

7.6   Approximating Periodic Functions with Sine and Cosine Terms   477

7.7   Approximating Sine and Cosine with Polynomials   481

7.8   Properties of Complex Numbers   496

7.9   The Road to Chaos   503

Chapter Summary   512

Review Exercises   513

CHAPTER 8

GEOMETRIC MODELS

8.1 Introduction to Coordinate Systems **515**

8.2 Analytic Geometry **517**

8.3 The Conic Sections **525**

8.4 Parametric Curves **543**

8.5 The Average Value of a Function **553**

8.6 The Polar Coordinate System **561**

8.7 Families of Curves in Polar Coordinates **567**

Chapter Summary **576**

Review Exercises **576**

APPENDIX A   Introduction to Trigonometry **A-1**

APPENDIX B   Arithmetic of Complex Numbers **A-27**

APPENDIX C   Some Mathematical Moments to Remember **A-30**

APPENDIX D   1995 World Population Data **A-37**

SELECTED ANSWERS **A-41**

INDEX **I-1**

PREFACE

To the Student

Picture yourself as a homeowner whose only tools are a set of screwdrivers. You are perfectly capable of driving screws into or out of wood. But what about other jobs around the house? A screwdriver is not useful for banging in a nail or cutting a board in two. Other tools are needed, and for larger jobs, power tools are essential.

In many ways this situation is analogous to standard mathematics courses in which the emphasis has been almost completely on algebraic methods. These methods give you a powerful set of tools—you can collect like terms, factor various expressions, cancel common factors, expand powers of binomial terms, multiply out polynomials, and so forth. But there are many jobs requiring mathematics that cannot be solved at all using only algebraic methods. For such problems there are other mathematical tools, including graphical and numerical methods, that are far more useful.

Suppose a doctor wants to study a patient's heartbeat using an EKG or a patient's brainwaves using an EEG. There are no known formulas to express these quantities algebraically, but the doctor certainly can get critical information about a patient by interpreting the graphs produced by these machines. Suppose an engineer develops a new tread design for automobile tires and wants to test its braking effectiveness for a car going 20, 30, 40, 50, and 60 miles per hour. There is no exact formula for either the braking distance or the time until the car comes to rest—too many unpredictable factors are involved. All the engineer has is a set of measurements from the experimental runs, and he or she must make decisions based on an understanding of what information the data provides.

Both cases illustrate themes that run through this book. We focus on the applications of mathematics to situations all around us and on the *function* concept that allows us to study these phenomena. That's why this book is titled *Functioning in the Real World.* In this course you will learn to use a combination of algebraic, graphical, and numerical methods, depending on which is the most helpful tool in any given context. You will develop your understanding of the mathematical concepts and learn how to apply them to realistic problems, and not merely perform operations mechanically. You will learn to interpret results, not just obtain answers. You will use technology as a tool for answering the kinds of questions that arise naturally. But this tool is not intended merely to give you answers: Technology will help you learn the mathematics. Above all, you will increase your ability to think mathematically so that you can apply mathematics in many other arenas—in other math courses, in courses in other disciplines, in your eventual careers, and in all other aspects of your life.

To do all this, you will look at a much wider variety of topics than you probably have seen in previous math courses. You also will face much more varied types of problems than you might have encountered before. Many of these problems will require you to *think* about the mathematics, not just redo a worked-out example from the text, where the numbers have been changed slightly. To do such nonroutine problems, you will have to understand the mathematical ideas, not merely memorize solutions. To gain that understanding, you will have to pay careful attention in class and read the text thoroughly.

In some ways, this course will be more challenging than others you have taken, but it certainly will be much more rewarding because you will see the value of mathematics all around you. We'd like to give you some suggestions that will make things easier:

- Read the book. It was written for you and is very readable.
- Work in teams outside of class. Students are very good at explaining mathematical ideas to one another in terms that they understand.
- Attacking problems in teams is not only a good learning strategy but also the way people in science, engineering, business, and other fields function in the real world.
- Ask questions in class. Many of the problems are ideal for class discussions, but it is very hard for a teacher to answer questions no one asks.
- Feel free to suggest your own interpretations. Many of the problems can be approached in very different ways, and different students (and instructors) are likely to come up with different solutions depending on their viewpoint.
- Talk to your instructor during office hours if you need help. Your professor understands that this is a demanding course for some students and will be happy to work with you. But you must seek out the help.
- If you use a graphing calculator, carry it to class and use it often to add a graphical dimension to whatever you are studying.
- Some of you already may use more sophisticated computer programs, the mathematical power tools that are widely available today. These might include a software package, such as Derive™, Mathematica™, or Maple™, that can perform virtually any algebraic operation or a spreadsheet, such as Lotus 1-2-3™, that organizes data and produces sophisticated graphs quickly and easily. Become comfortable with this software as soon as you can so that they are familiar tools throughout the course.
- Realize that the most powerful and effective tool you have is your mind. It does things no machine is capable of doing—thinking, understanding, creating, and interpreting.

Remember, a carpenter equipped with *all* the right tools is able to build almost anything. Likewise, a student equipped with all the right mathematical tools and the knowledge and judgment to select the right one is prepared for almost anything. Above all, we hope that you will have a very exciting and rewarding experience as you are *Functioning in the Real World.*

To the Instructor

The reform of the mathematics curriculum is well under way with changes that seek to establish a better balance among geometric, numerical, symbolic, and verbal approaches. There is a much greater emphasis on understanding fundamental mathematical concepts, on realistic applications, on the use of technology, on student projects, and on more active learning environments. These courses tend to make greater intellectual demands on the students compared to traditional courses that place heavy emphasis on rote memorization and manipulation of formulas.

With support from the National Science Foundation, the Math Modeling/PreCalculus Reform Project has developed a new precalculus or college algebra/trigonometry experience with the following goals:

- Extend the common themes in most of the calculus reform projects.
- Focus more on mathematical concepts and mathematical thinking by achieving a balance among geometric, numerical, symbolic, and verbal approaches rather than focusing almost exclusively on algebraic manipulation.
- Provide students with an appreciation of the importance of mathematics in a scientifically oriented society by emphasizing mathematical applications and models.
- Introduce some modern mathematical ideas and applications that usually are not encountered in traditional courses at this level.
- Provide students with the skills and knowledge they will need for subsequent mathematics and related courses.
- Make appropriate use of technology without becoming dominated by the technology to the exclusion of the mathematics.

Philosophy of the Project

To accomplish these goals, we have adopted several basic principles advocated by most leading mathematics educators.

Students should see the power of mathematics.

Most students take precalculus or college algebra/trigonometry courses at the college level because they are prerequisites for future math courses or for courses in one of the client disciplines, not because they are turned on by the elegance of mathematics. When students see interesting and significant applications, they see the power of mathematics in action and so are willing to put in the effort needed to learn the subject.

Such applications are the primary focus of the *Functioning in the Real World* course. For instance, in Chapter 3 we analyze the data related to the *Challenger* disaster showing how mathematics could have been used to decide against the launch that day. In Chapter 4 we analyze the spread of AIDS to determine what kind of mathematical model best describes its growth. In Chapter 7 we see how periodic phenomena in nature, such as temperatures, tides, or hours of daylight over the course of a year, can be modeled mathematically.

Students should focus on mathematical ideas, not mathematical calculations.

Our goal is to achieve mathematical understanding, which too often is lost when students concentrate on routine computations than can, should, and in practice will be done by machine. Calculus reform projects have reduced the level of symbolic manipulation mainly because of the existence of technology. Computer Algebra Systems (CAS) now perform virtually any type of manipulation we could ever expect a student to do. Even though such systems are not necessarily incorporated into the new calculus courses, their existence has major implications about what is important to teach. Further, graphing calculators allow us to study considerably more complicated problems from a geometric perspective than conventional problems that require factoring artificially constructed functions. Consequently, there is less need to develop as high a level of manipulative skills in precalculus courses as in the past. Rather than focusing on producing students who are poor imitations of such computer systems, we should be emphasizing the power of the human mind to inquire, explore, analyze, and interpret.

Technology certainly has a role in the *Functioning* course. We use graphing technology to compare the growth behavior of different members within a family of functions or to compare the behavior among different families of functions; this theme comes up repeatedly throughout the book. Another technological theme is to find the equation of the function (linear, exponential, power, logarithmic, trigonometric, logistic, and so on) that best fits real sets of data. In Chapters 5 and 6 we use technology to generate and display solutions of difference equations and investigate their dependence on initial conditions.

Whether you choose to have your students use graphing calculators, CAS, spreadsheets, or other computer packages, technology should be used to motivate or explain the mathematical ideas, not just to produce answers. *Whatever technology you and your students use, we firmly believe that the overriding focus should remain on the mathematics and not on the machinery.*

Students should DO mathematics, not just passively watch mathematics.

The typical mathematics course involves classroom lectures and homework. But mathematics is all around us, and the best way to appreciate its power and usefulness is to apply it directly. To achieve this, we suggest having students do several mathematical projects either individually or in small groups. Such projects give the students the opportunity to "get their hands dirty" by:

- formulating mathematical questions
- collecting appropriate sample data
- analyzing that data
- drawing conclusions based on the analysis
- preparing reports, which promote the organization of their ideas as well as their writing and communication skills

For instance, when we look at linear functions early in Chapter 2, students can be asked to select a set of data of interest to them and estimate, by eye, the equation of the line that best fits the data. In Chapter 3, students can be asked to fit a variety of functions (linear, exponential, power, and logarithmic) to a set of data and make predictions based on the results. (See the *Instructor's Resource Manual* for additional suggestions). The results of doing projects increase the students' level of enthusiasm for the subject matter and their understanding of the mathematical ideas.

Students should be exposed to a broad view of mathematics.

Traditional precalculus courses focus exclusively on preparing students for calculus, particularly in terms of the algebraic skills they may need. The *Functioning* course is intended to prepare students for mathematics in a more general sense. We have incorporated a variety of nontraditional topics in the course that simultaneously introduce new mathematical ideas while advancing the students toward a calculus experience.

- The laboratory sciences routinely use mathematics to analyze laboratory data, but rarely explain the underlying mathematical foundations—students typically use semi–log paper or log–log paper to produce the appropriate results without understanding why. Throughout the book, we emphasize the concept of fitting functions to data using real-life situations, which motivates the mathematical ideas as well as providing contexts in which to develop important algebraic skills at a high level. For example, in Chapter 3 we analyze the growth of the U.S. population via an exponential function; in Chapter 6 we return to the U.S. population and attempt to fit it with a logistic, or inhibited growth, model.
- In most calculus reform courses the emphasis on differential equations has greatly increased. We extensively discuss difference equations to introduce most of the comparable ideas and models in a a discrete setting. This discussion reinforces critical ideas on functional behavior and the modeling of real-world phenomena while providing opportunities for honing important algebraic skills.
- We have integrated a probability thread throughout. Probalistic reasoning has become increasingly important in recent decades, and we use it in the service of preparing students for calculus.

The Intended Audiences

The materials in this book were developed with different courses in mind:

1. An alternative to the usual one- or two-semester precalculus course designed to prepare students for calculus, one that is in the spirit of the AMATYC *Crossroads* Standards.
2. An alternative to a one- or two- semester course in college algebra and trigonometry.
3. An alternative to traditional high-school precalculus courses, one that is in the spirit of the NCTM Standards and the recommendations of the Pacesetter curriculum project.
4. An alternative to related courses that often are used as a terminal or capstone mathematics course.
5. An alternative to a precalculus-level course for education majors.

We presume that the students taking this course have had a reasonably good mathematical background at the level of intermediate algebra, a previous exposure to some right-angle trigonometry (but likely remember very little of it), and a previous exposure to logarithms (but do not remember anything about them other than a strong sense of aversion). For those students who need to review right-angle trigonometry, we have developed the key ideas in Appendix A.

As an alternative to traditional precalculus and college algebra/trig courses, the *Functioning* course certainly provides a strong preparation for a standard calculus course. However, the primary emphases we have adopted make the approach particularly well-suited as preparation for virtually any of the reform calculus courses. As with those projects, we put a strong emphasis on:

- the applications of the mathematics
- mathematical reasoning and understanding of mathematical concepts, not just symbol manipulation
- the use of the Rule of Three: Topics are approached geometrically, numerically, and symbolically wherever possible
- the use of verbal reasoning and communication skills

Overall, we believe that it is very important to emphasize repeatedly why you are teaching a course with a very different approach. Students need to be reminded of the reasons why you are expecting more of them and asking them to do different things. Pointing out the limitations of purely algebraic methods or the power of modern technology to solve problems that could not be touched just a few years ago helps. Pointing out the type of traditional manipulative operations that now can be done easily by machine also helps. Most importantly, remind students that they now need to develop the thinking skills to know the questions to ask, to decide which tools to use for answering those questions, and to develop the ability to interpret and communicate these solutions.

Suggested Time Frame

The suggested pacing below is for a one-semester course that meets four hours per week. By including the optional sections and Chapter 6, this text can give a two-semester sequence that provides a strong foundation in pre-calculus ideas and in applied mathematics, particularly discrete mathematics.

Ch. 1 Functions in the Real World **1–1.5 weeks**	Much of this chapter can be assigned as independent reading. However, we suggest spending some of the first week talking about and developing the critical ideas on the behavior of functions and introducing students to the function concept from geometric, numerical, symbolic, and verbal points of view.

Ch. 2   Families of Functions
3 weeks
Section: 2.2   Class hours: 2–2.5
        2.3   2–2.5
        2.4   2
        2.5   1.5
        2.6   1–1.5
        2.7   1.5
	This chapter is critical in helping students to reach the same plateau of mathematical background and to make the transition to a new way of looking at and thinking about mathematics. You should give students time to reorient themselves.

Ch. 3   Fitting Functions to Data
2 weeks
Section: 3.1, 3.2   Class hours: 1.5–2
        3.3   2–2.5
        3.4   0.5–1
        3.5   1
        3.6   0.5
	This chapter provides the link between mathematics and the real world. It shows where functions come from, reinforces ideas about the behavior of families of functions, and provides the opportunities to develop important algebraic skills. Section 3.6 may be assigned as reading.

Ch. 4   More about Functions
2.5–3 weeks
Section: 4.1   Class hours: 1.5–2
        4.2   0.5
        4.3   0.5
        4.4   2
        4.5 (optional)   1
        4.6 (optional)   1.5
	This chapter extends the idea of families of functions to polynomials. It also introduces the idea of constructing new functions from old.

Ch. 5   Sequences and Their Applications
1–4 weeks
Section: 5.1–5.3   Class hours: 3.5
        5.4   1.5
        5.5   1.5
        5.6   1
        5.7   1
        5.8   1.5
        5.9 (optional)   1.5
	This chapter develops and analyzes models for describing: • population growth • logistic growth • eliminating drugs from the body • radioactive decay • Newton's laws of heating and cooling • geometric sequences and their sums. If pressed for time, you may wish to select some, but not all, of the models given in Sections 5.5, 5.6, and 5.8.

Ch. 6	Modeling with Difference Equations (Optional)		This chapter explores a variety of difference equation models. Although the chapter is optional, you may wish to cover one or more of the topics, if time permits. Most sections depend on Sections 6.1 and 6.2. Other sections depend on Section 6.3.
	2–3 weeks		
	Section: 6.1	Class hours: 1.5–2	
	6.2	1.5–2	
	6.3	1	
	6.4	1	
	6.5	1	
	6.6	1	
	6.7	1	
	6.8	1	
	6.9	1	
Ch. 7	Modeling Periodic Behavior		This chapter: • uses trigonometric functions to model periodic phenomena • examines the relationship between the trigonometric functions • approximates periodic functions with sine and cosine terms • approximates sine and cosine functions with polynomials • examines the properties of complex numbers • explains chaotic phenomena. Coverage of Section 7.9 may take several class hours, particularly if you wish to include live computer graphic demonstrations.
	3 weeks		
	Section: 7.1	Class hours: 1.5	
	7.2	1.5	
	7.3	1.5	
	7.4	1–1.5	
	7.5	1	
	7.6 (optional)	1.5	
	7.7 (optional)	1.5	
	7.8	1.5	
	7.9 (optional)	1	
Ch. 8	Geometric Models		This chapter includes analytic geometry, the conic sections, parametric curves, the average value of a function, and curves in the polar coordinate system. The two parts of this chapter, Sections 8.1–8.5 (analytic geometry) and Sections 8.6 and 8.7 (polar coordinates) can be considered as minichapters that could be covered independently, if so desired.
	2.5–3 weeks		
	Section: 8.1	Class hours: 1–1.5	
	8.2	1–1.5	
	8.3	2.5	
	8.4 (optional)	1.5	
	8.5 (optional)	1	
	8.6	1	
	8.7	1.5–2	

Detailed suggestions can be found in the *Instructor's Resource Manual*.

Supplements

- The **Instructor's Solutions Manual** contains complete solutions to all problems for instructors.
- The **Student Solutions Manual** contains complete solutions to selected problems, particularly nonroutine problems.
- The **Instructor's Resource Manual** contains
  —detailed suggestions for topic selection and pacing for the course,
  —additional problems and exercises,
  —ideas for classroom activities and suggestions for student projects,
  —suggestions for computer laboratory exercises and assignments, and
  —samples of tests, exams, and project assignments.

Acknowledgments

We gratefully acknowledge the contributions of many people whose advice and assistance immeasurably helped the project and the development of these materials. The project's Advisory Board consisted of Chris Arney (U.S. Military Academy), John Brunsting (Hinsdale Central High School), John Dossey (Illinois State University), Deb Hughes Hallett (Harvard University), Bill Lucas (Claremont Graduate School), Joe Malkevitch (York College of CUNY), Warren Page (New York City Technical College), Henry Pollak (Columbia University), and Karl Smith (Santa Rosa Junior College). The project's evaluation was coordinated by Geoffrey Akst of the Borough of Manhattan Community College. We thank them all for their advice and guidance throughout the project.

We especially appreciate the contributions and suggestions regarding the content of this book from many people. Judy Broadwin (Jericho High School, NY) brought the secondary school perspective into the project, contributed many innovative suggestions for graphing calculator usage, and assisted in project dissemination activities; we greatly appreciate her ongoing efforts. We also want to thank Jim Sandefur (Georgetown University), Walter Meyer (Adelphi University), Walter Yurek (WorWic College), Tony Peressini (University of Illinois at Champaign-Urbana), Paula Maida (American University), I-Lok Chang (American University), Harry Hauser (Suffolk Community College), William Abrams (Longwood College), Anne Landry (Dutchess Community College), William Steger (Essex Community College), Sylvia Sorkin (Essex Community College), and Chuck Laufman (Westbury High School) for their valuable suggestions and contributions to this edition.

We are indebted to Joe Fiedler and Ignacio Alarcón (both of California State University–Bakersfield) for the lovely and thorough job they did in producing both the *Instructor's Solutions Manual* and the *Student's Solutions Manual* that accompany the book. We want to thank Paula Maida for her creative contributions to the *Instructor's Resource Manual*. We also acknowledge the careful and intensive work that Paul Lorczak (MathSoft, Inc.) has done in checking the entire manuscript and all solutions. We also want to thank Steve Ouellette (Winchester High School) for checking the solutions for the *Instructor's Solutions Manual*. Finally, we want to thank Joe Fiedler, Ignacio Alarcón, Judy Fethe (Pellissippi State Technical College), Ann Landry, Gary Lawlor (Adirondack Community College), and Joanne Manville (Bunker Hill Community College) for assisting in a variety of faculty training workshops to prepare instructors to teach the course.

We want to thank the National Science Foundation for their support and for making this project possible. In particular, we are indebted to James Lightbourne, William Haver, and Elizabeth Teles at the NSF Division of Undergraduate Education for their ongoing encouragement and assistance. It is truly appreciated. We also gratefully appreciate the local support provided by K.V. Cheek of New York Institute of Technology.

The project materials were class-tested by instructors from the following institutions whose feedback and suggestions significantly contributed to this edition:

California

California State University—Bakersfield
California State University—Monterey
  Bay Seaside
Loyola High School
Mount San Antonio College

Connecticut

Manchester Technical College

District of Columbia

American University

Florida

Hillsborough Community College
Lyman High School
Miami-Dade Community College
Valencia Community College

Georgia

Columbus College

Illinois

Latin School of Chicago
University of Illinois at Champaign-
  Urbana

Indiana

Wabash College

Iowa

Drake University

Kentucky

Somerset Community College

Maryland

Bowie State University
Essex Community College
Frederick Community College
Towson State University
WorWic Technical Community College

Massachusetts

Bunker Hill Community College
Deerfield Academy
Emerson College
Essex Community College
Massachusetts Bay Community College

Michigan

Eastern Michigan University

Montana

Bainville High School
Billings West High School
Custer County District High School
Carroll College

New Hampshire

Phillips Exeter Academy

New Jersey

Raritan Valley Community College

New York

Adirondack Community College
College of Saint Rose
Dutchess Community College
Empire State College
Finger Lakes Community College
Ithaca College
Jamestown Community College
Jericho High School
Queensborough Community College
St. John Fisher College
Schreiber High School
Suffolk Community College
  Selden Campus
Suffolk Community College
  Western Campus
SUNY College at Cortland

Ohio

Wittenberg University
Xavier University

Oregon

Southwestern Oregon Community College

Pennsylvania

Academy of New Church College
Holy Family College
King's College

Rhode Island

Salve Regina University

South Carolina

The Citadel
Columbia College
University of South Carolina

Tennessee

Pellissippi State Technical Community
 College
Walters State Community College

Texas

College of the Mainland
Collin County Community College
Midwestern State University
Sam Houston University
University of Houston

Vermont

Johnson State College

Virginia

Longwood College
Old Dominion University

Washington

Bush School

Wyoming

Casper College
Northwest College
University of Wyoming

We are indebted to the wonderful team at Addison-Wesley, who came to share our vision for the course and helped bring this book to fruition: Bill Poole, our sponsoring editor, and Elka Block, our development editor, provided tremendous encouragement, assistance, and innumerable helpful suggestions throughout the process. We also appreciate all of the outstanding contributions of Geri Davis, Andy Fisher, Joanne Foster, Doug Fruscione, Christine O'Brien, Barbara Pendergast, Esther Podany, Ben Rivera, Sandra Selfridge, Greg Tobin, Joe Vetere, and Joe Will in many other critical aspects of the project. Finally, we would like to thank Beverly Fusfield and the staff of Techsetters, Inc., and Joanna Koperski of JAK for the excellent artwork they created for this book. They are a great team, and it has been a true pleasure working with each and every one of these professionals.

Sheldon P. Gordon
E. Northport, NY

Florence S. Gordon
E. Northport, NY

B. A. Fusaro
Tallahassee, FL

Martha J. Siegel
Baltimore, MD

Alan C. Tucker
Stony Brook, NY

1

FUNCTIONS IN THE REAL WORLD

1.1 *Functions Are All Around Us*

The notion of *function* is a fundamental idea in mathematics. Functions are the basis for most mathematical applications in nearly all areas of human endeavor. To see how functions can come up in unexpected areas, look at the graph in Figure 1.1.

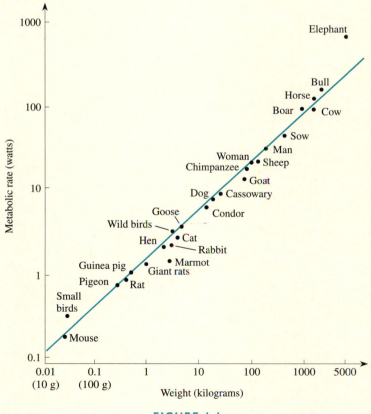

FIGURE 1.1

This graph appears in many introductory biology textbooks. It shows the results of a study comparing the weights of various mammals and birds with their metabolic rates. The biologist who conducted the study first plotted the *data*—the raw measurements on body weight measured in kilograms and metabolic rate measured in watts—on a graph and then drew the line shown that passes very close to all the points. What does this graph tell us? It is clear from the pattern of the points that there must be some relationship between the body weight and the metabolic rate of mammals and birds. If there were no relationship, the points would not fall into such a clear pattern. Thus we conclude that, in some fashion, the metabolic rate of an organism depends on the body weight of that organism. Such a relationship is a *function* and we say that metabolic rate R is a function of body weight W.

In general, a function is a rule that associates one set of values with another. Functions are usually given in words, by formulas (or equations), by tables, or by graphs. For instance, the metabolic rate for different species is *related* to their body weight, as shown in the graph in Figure 1.1. Similarly, pediatricians use a graph to predict both the weight and the height of their patients at different ages since both *depend* on a child's age. Financial planners study tables and graphs on the performance of different stocks over time to give sound financial information to their clients about the behavior of the stocks *as a function of* time. The chief of operations at Walt Disney World must keep track of the average number of guests *corresponding to* each week during the year in order to provide adequate staffing throughout the year.

Let's summarize this idea of a function more formally:

> A **function** is a rule that assigns to each value of one quantity precisely one related value of another quantity.

The rule might be expressed in words, as in "The cost of postage is 32 cents for the first ounce and 23 cents for each additional ounce." The function might be expressed as a mathematical formula such as $A = \pi r^2$ for the area of a circle, or $P = \frac{kT}{V}$, which expresses the pressure P of a gas as a function of its temperature T, where V is the volume of the container that holds the gas and k is a constant. The function might be given via a table of data. For instance, clerks in retail stores often assess sales tax from a table without ever seeing the formula based on the tax rate that leads to the table. Similarly, you compute your income tax for the Internal Revenue Service by using a table—for each level of taxable income, there is an associated tax levied, as shown in Figure 1.2.

For any such situation involving two quantities, we will be interested in several questions:

1. Is there a functional relationship between the two quantities?
2. If there is a relationship, can we find a formula for it?

3. Can we construct a table or graph relating the two quantities, especially if we can't find a formula?

4. If we can find a formula, or if we have a graph of the relationship, or if we have a table of values relating the two quantities, how do we use it?

That is, how can knowledge of the function help us to understand the relationship between the two quantities or allow us to make predictions or informed decisions about one of the variables based on the other?

For example, Figure 1.1 clearly suggests a definite relationship between metabolic rates R in mammals and birds and their body weights W. This same relationship can then be used to predict the metabolic rate of other species, say lions or Kodiak bears, based on our knowledge of their weights. We could even use this relationship to predict the metabolic rate of an extinct pterodactyl based on estimates of its body weight made from its skeletal remains. However, we would not use the relationship to predict the metabolic rate for a crocodile since it is neither a mammal nor a bird; the relationship we observe in Figure 1.1 for mammals and birds may not apply to reptiles. Based on the graph, would you use the relationship to predict the metabolic rate for extinct mammoths, which were slightly larger than today's elephant?

If line 37 (taxable income) is—		And you are—		
At least	But less than	Single	Married filing jointly	Married filing sepa-rately
			Your tax is—	
29,000				
29,000	29,050	5,092	4,354	5,592
29,050	29,100	5,106	4,361	5,606
29,100	29,150	5,120	4,369	5,620
29,150	29,200	5,134	4,376	5,634
29,200	29,250	5,148	4,384	5,648
29,250	29,300	5,162	4,391	5,662
29,300	29,350	5,176	4,399	5,676
29,350	29,400	5,190	4,406	5,690
29,400	29,450	5,204	4,414	5,704
29,450	29,500	5,218	4,421	5,718
29,500	29,550	5,232	4,429	5,732
29,550	29,600	5,246	4,436	5,746
29,600	29,650	5,260	4,444	5,760
29,650	29,700	5,274	4,451	5,774
29,700	29,750	5,288	4,459	5,788
29,750	29,800	5,302	4,466	5,802
29,800	29,850	5,316	4,474	5,816
29,850	29,900	5,330	4,481	5,830
29,900	29,950	5,344	4,489	5,844
29,950	30,000	5,358	4,496	5,858

FIGURE 1.2

Representing Functions with Graphs

Many effective ways are used to display functions graphically in everyday life—in newspapers, magazines, and scientific, business, and government reports. The graph of a function is valuable because it displays accurate information about a quantity while simultaneously giving us an overview of the behavior of that quantity. In particular, a graph can show any trends or patterns in the process being studied.

The graph in Figure 1.3 shows the increase in life expectancy in years since the beginning of this century. This graph is a function of time, t, because, for any given

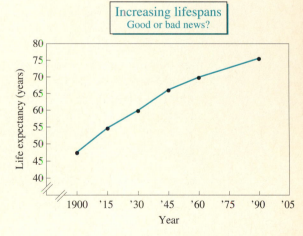

FIGURE 1.3

year, there is a single value for the life expectancy of a child born in that year. From the graph, for example, we can estimate that a child born in 1900 would have had, on the average, a life expectancy of about 47 years, although a child born in 1990 would have a life expectancy of about 75 years. Certainly, the rise in life expectancy is a remarkable achievement due to advances in science and medicine and improvements in our lifestyles. However, there are also some unfortunate aspects connected with living longer. Can you think of any?

From this graph, not only can we observe the rising trend, but we can also look ahead to predict life expectancies in the not-too-distant future. Notice that the level of life expectancy is not merely increasing, but is actually increasing ever more slowly as time goes by. What is the significance of this growth pattern if it continues?

Another approach commonly used to display information on several related quantities is to stack their respective graphs, as in Figure 1.4, which shows trends in new car sales since 1983. The vertical distance from the horizontal axis to the lower graph represents domestic car sales in any given year. The vertical distance between the lower graph and the upper graph gives the sales of imported cars. The vertical distance from the horizontal axis to the upper graph represents the total new car sales; it shows the sum of the other two functions. Thus, we see that in 1990 the total number of passenger cars sold in the United States was approximately 9.3 million, while the total number of domestic cars sold was about 6.9 million. The difference, represented by the height of the line from the lower graph to the upper graph, is about 2.4 million cars; this represents the number of imported cars sold in the United States in 1990. From the figure, it is easy to determine when the U.S. government pressured the Japanese government to reduce its automobile exports to this country. Can you explain what's happening in terms of the number of imports coming into this country around 1990? While total sales are dropping somewhat, which component of the car market is absorbing the decrease?

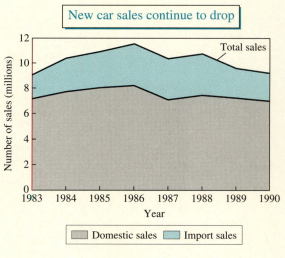

FIGURE 1.4

There are many other ways that functions are displayed graphically in newspapers and magazines. Keep an eye out for them in your daily activities.

Representing Functions with Tables

Consider the following table, which shows the acceleration of a Pontiac Trans Am. The table gives the time in seconds needed to reach different speeds.

Final speed, v (mph)	30	40	50	60	70
Time, t (seconds)	3.00	4.29	5.52	7.38	9.81

Although you may not have thought of something like this as being a function, the time t needed to accelerate to a given speed v *is* a function of the speed. We may not have an explicit formula for this time as a function of the final speed, but it nevertheless satisfies the definition of a function: For each final speed v, there is a unique time t needed to accelerate to that speed.

We can plot these points in a graph and connect them with a series of straight line segments or even by a smooth curve, as shown in Figure 1.5. Notice that the times depend on the final speed. Thus, we plot the speeds on the horizontal axis and the times needed to achieve those speeds on the vertical axis. Also, realize that the values in the table give us only the actual measured points. By drawing a smooth curve through the points, we are making assumptions about what happens between the given points. The curve is just an artist's rendition of what the pattern could be; the actual pattern might have some very minor variations.

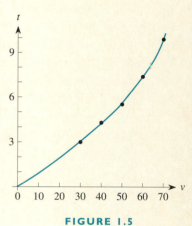

FIGURE 1.5

Think About This

Estimate the time needed for a Trans Am to accelerate from 0 to 45 mph; from 0 to 75. Which estimate do you think is more accurate? Why?

Frequently, when we observe that one quantity is a function of another, we would like to determine an appropriate formula that expresses this relationship. For example, throughout most of human history, people believed that objects fall at a constant speed. Then, in about 1590, Galileo realized that this might not necessarily be the case. He also had the insight to realize that this conjecture could be tested experimentally. Galileo conducted his now-famous experiments of dropping objects from the top of the Tower of Pisa and found that they fell at ever increasing rates and that the weight of the objects did not affect how fast they fell. Galileo's study of the relationship between the distance that an object falls and the time it takes to fall was the key connection for Newton to develop his theories of motion that transformed the physical sciences into what we know them as today. It is now fairly simple to interpret the information on the acceleration of the Trans Am because we have this understanding.

As another example, consider the following data on high temperatures in Phoenix during a severe heat wave in June 1990. The table shows the high temperature corresponding to each day during the heat wave.

Date in June	19	20	21	22	23	24	25	26	27	28	29
Temperature (°F)	109	113	114	113	113	113	120	122	118	118	108

Notice that there is a single high-temperature reading associated with each day. This function only makes sense for the 11 days, June 19 through June 29, and its values consist of the high-temperature readings 108, 109, 113, 114, 118, 120, and 122.

The function that associates the high temperature in Phoenix with the corresponding day of the month can be depicted graphically by plotting the individual points, as shown in Figure 1.6. The points in the figure can be joined by a series of line segments or by a smooth curve to give a sense of an overall trend or pattern, as shown in Figure 1.7. However, doing this requires some careful thought. When the points are so connected, we are *not* indicating that this is the graph of temperature as a function of time; we are just connecting the maximum temperatures recorded each day, and the curve shown gives absolutely no information about the temperature at any intermediate time. In fact, the actual graph of temperature versus time would typically show the kind of oscillatory effect depicted in Figure 1.8.

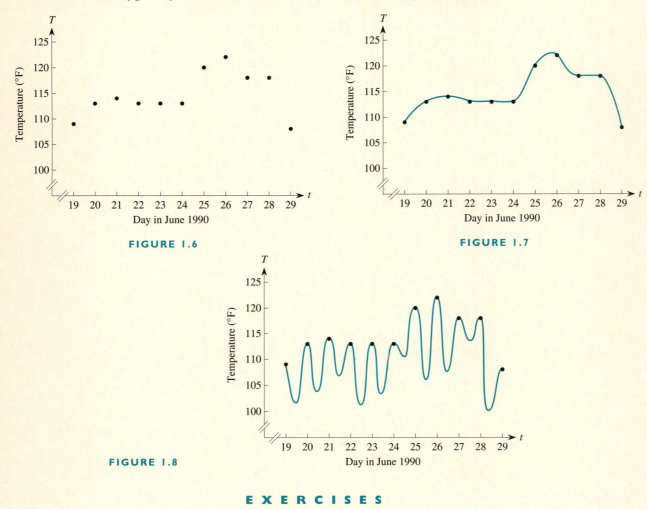

FIGURE 1.6

FIGURE 1.7

FIGURE 1.8

EXERCISES

1. Consider again the graph in Figure 1.3. Write a paragraph or two interpreting what the increase in life expectancy over the past century means. For example, you might consider it in terms of your own expected life span compared to those of your children and grandchildren. Alternatively, you might consider the effects on the overall distribution of people

of different ages in the population at large, or you might discuss the question of whether there is a natural limit to how long the human life span can be extended in the future. Compare the values for life expectancies in the United States in Figure 1.3 with the values for life expectancies of other nations given in Appendix D.

2. Match each of the following functions with a corresponding graph. Explain.

    a. The population of a country as a function of time.
    b. The path of a thrown football as a function of time.
    c. The distance driven at a constant speed as a function of time.
    d. The daily high temperature in a city as a function of time over several years.
    e. The number of cases of a disease as a function of time.
    f. The percentage of families owning VCRs as a function of time.

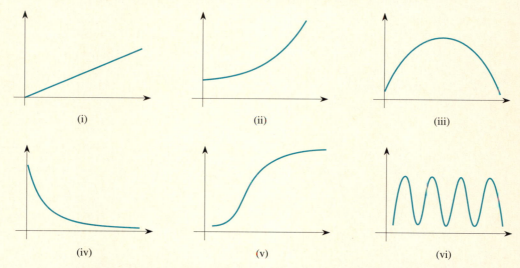

    (i)   (ii)   (iii)

    (iv)   (v)   (vi)

3. The graphs that follow show the noise level of a crowd of college students watching their school's basketball team playing at home in the championship finals for the league title. Match the three graphs with the corresponding scenarios (reactions), and then draw a graph for the remaining scenario.

    a. Our team started slowly, but eventually began to pull away.
    b. It was a disaster from start to finish.
    c. The score kept seesawing, but we finally won on a three-point shot at the buzzer.
    d. Our team started well, then the opposition took the lead, but we finally beat them out.

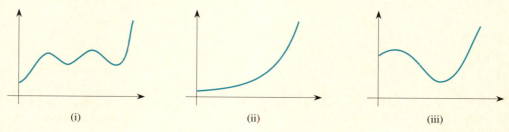

    (i)   (ii)   (iii)

4. Consider the following scenario: "You left home to run to the local gym. You started at a constant rate of speed, but sped up when you realized how energetic you felt. About halfway there, you began to tire, so you started slowing down." Sketch a graph of your distance from home as a function of time.

5. Sketch a graph of your distance from home as a function of time for each of the following situations:

   a. You drove steadily across town, speeding up as traffic diminished until the road turned into a highway.

   b. You drove steadily toward town, but slowed down as the traffic increased. Eventually you inched forward around a car that had broken down before you could resume normal speed.

   c. You drove steadily, but realized you had left something behind, so you returned home and then drove all the way to school without any further trouble.

   d. You drove steadily across town, but then had a flat tire; after changing it, you drove much faster so that you wouldn't be too late for class.

6. For each of the scenarios in Exercise 5, sketch a graph of the total distance you've traveled as a function of time.

7. Which of the following tables represents a function and which does not? Explain your reasoning.

   a.
x	0	3	6	1	5	2	4
y	8	6	2	2	4	5	3

   b.
x	0	2	3	4	1	3	5
y	8	4	7	2	6	10	9

8. The following table lists the number of runs batted in (RBIs) for a major league baseball player over the course of his career. Identify the years in which his run production is a maximum and the years in which it is a minimum. When is it growing most quickly?

Year	1	2	3	4	5	6	7	8	9	10	11
RBIs	36	52	86	73	94	87	77	65	73	32	41

   Write a short biographical note describing the player's RBI performance over the course of his career.

9. The Dow-Jones average of 30 industrial stocks is probably the most closely watched measure of performance of stocks. The following are the values for the Dow at the beginning of the year for the last 17 years.

Year	1980	1981	1982	1983	1984	1985	1986	1987	1988	1989
Dow	839	964	875	1047	1259	1212	1547	1896	1939	2169

Year	1990	1991	1992	1993	1994	1995	1996
Dow	2753	2634	3169	3301	3758	3824	5117

   Write a short note describing the behavior of the stock market over the course of this period. When was it rising? When was it dropping? Which years would have been the best times to buy stocks? Which would have been the worst times to do so?

1.2 *Describing the Behavior of Functions*

We have seen that we use functions to represent quantities in the real world. Most of these quantities are changing, depending either on time or on some other quantity. In order to describe these changes, we need to introduce some terminology to describe the *behavior of the function;* that is, how the function changes. There are two levels at which we can describe the behavior of a function. The first and most immediate level is to observe whether the function is *increasing* or *decreasing.* The graph of a function *f* is **increasing** if, as we look from left to right, the values get bigger and bigger; that is, the graph rises. Similarly, the graph of a function *f* is **decreasing** if, as we look from left to right, the values get smaller and smaller; that is, the graph falls. See Figure 1.9.

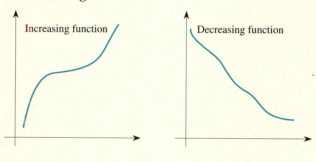

FIGURE 1.9

For instance, we know that the world's population is growing. Therefore, the function that expresses the population over time is an increasing function. We also know that the heavier a car is, the lower its gas mileage will be. Therefore, the function that relates gas mileage to the weight of a car is a decreasing function.

Of course, not every quantity merely increases or decreases. Often, a quantity will rise and fall: for example, the height of a ball tossed up into the air, the value of the Dow-Jones average, or the high temperature recorded in a particular location each day of the year. Thus, we might have a function whose graph looks like the one shown in Figure 1.10. Such a function increases for some values of the variable and decreases for others. For the graph shown, the function rises (increases) up to a maximum or largest value, then falls (decreases) down to a minimum value, and then increases again. We call any point where the behavior of the function changes from increasing to decreasing or from decreasing to increasing a **turning point** of the function. The turning points occur at points where the function reaches a *maximum* (a value where it is larger than any nearby value) or a *minimum* (a value where it is smaller than any nearby value).

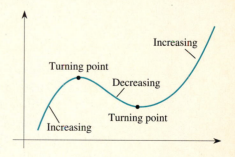

FIGURE 1.10

Notice that a function increases or decreases over an interval; it has a turning point at a particular point. So we might describe a function of t as decreasing from $t = 0$ to $t = 4$, increasing from $t = 4$ to $t = 6$, decreasing from $t = 6$ to $t = 10$, and then increasing after $t = 10$. This function would have turning points at $t = 4$, at $t = 6$, and at $t = 10$. Can you sketch the graph of a function having this behavior?

We describe a function that is only increasing or only decreasing (it has no turning points) as being *strictly increasing* or *strictly decreasing*.

There is a second aspect to a function's behavior. We show two increasing functions in Figure 1.11. How do they differ? In Figure 1.11(a), the function is not merely increasing; it is actually increasing faster and faster as time goes by. Think of the curve as bending upward. For instance, population growth follows this type of growth pattern. The function in Figure 1.11(b) is also increasing, but it is increasing more and more slowly as time goes by. Think of the curve as bending downward. For instance, the increasing human lifespan in Figure 1.3 in Section 1.1 grows in this way.

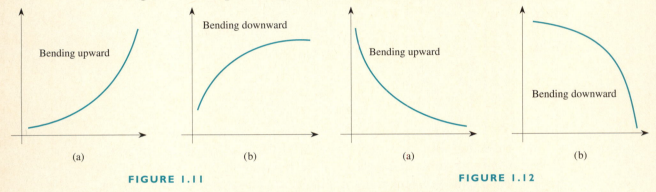

FIGURE 1.11   FIGURE 1.12

Now look at the two decreasing functions in Figure 1.12. The function in Figure 1.12(a) is decreasing very rapidly at first and then more slowly as time passes. For instance, if a pollutant is released into a lake, the level of pollution in the lake will decrease in this manner as time goes by. As in the function in Figure 1.11(a), this curve is also bending upward. The graph in Figure 1.12(b) is also decreasing, but it is decreasing rather slowly at first and then more and more rapidly. For instance, if an object is tossed off the roof of a tall building, its height above ground will decrease in this manner as it speeds up in its descent because of the effects of gravity. Notice that this curve is bending downward, as is the curve in Figure 1.11(b).

We use the term *concavity* to describe the type of bend we see in these behavior patterns. Curves that bend upward, such as those in Figures 1.11(a) and 1.12(a), are said to be **concave up.** Notice that one curve is increasing while the other is decreasing, so concavity is a completely different concept from increasing/decreasing. Similarly, curves that bend downward, such as the ones in Figures 1.11(b) and 1.12(b), are called **concave down.** Again, notice that one is increasing and the other is decreasing. We show the two types of concave up behavior in Figure 1.13(a) and the two types of concave down behavior in Figure 1.13(b).

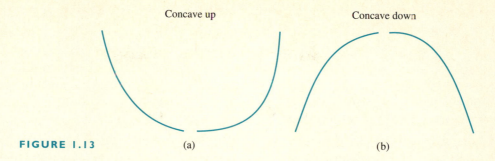

FIGURE 1.13

(a)

(b)

Just as it is possible for a function to be increasing in one region and decreasing in another, it is possible for a function to be concave up in one region and concave down in another. Moreover, the point where the concavity changes from concave up to concave down or vice versa is known as a **point of inflection** or an **inflection point.** In Figure 1.14, we show two curves, one having a point of inflection where the curve changes from concave up to concave down, and the other where the curve changes from concave down to concave up. Observe that neither point of inflection occurs where the curve reaches a maximum or a minimum at the turning points, so turning points are not the same as inflection points.

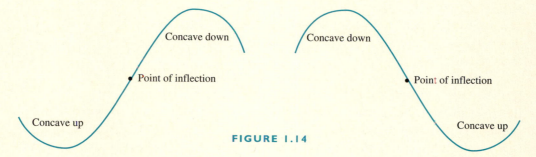

FIGURE 1.14

We summarize the preceding information as follows:

A function f is **increasing** if the values of the function increase as x increases.

A function f is **decreasing** if the values of the function decrease as x increases.

The graph of an *increasing* function *climbs* from left to right.

The graph of a *decreasing* function *descends* from left to right.

The points where a function changes from increasing to decreasing or from decreasing to increasing are the **turning points.**

The graph of a function is **concave up** if it bends upward.

The graph of a function is **concave down** if it bends downward.

The points where the concavity changes from concave up to concave down or from concave down to concave up are the **points of inflection.**

E X A M P L E

Identify all intervals where the function f shown in Figure 1.15 is

a. increasing;　　**b.** decreasing;
c. concave up;　　**d.** concave down.

Then indicate all points where the function has a

e. turning point;　　**f.** maximum;
g. minimum;　　**h.** point of inflection.

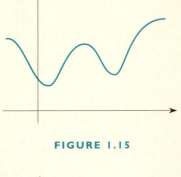

FIGURE 1.15

Solution　It is important that you realize that the question is first asking for intervals of x-values where the different types of behavior occur. We begin by redrawing the graph and introducing all points $x_1, x_2, \ldots, x_9$ where the behavior of the function changes, as shown in Figure 1.16.

(a) We see that the function is increasing for values of x between x_3 and x_5 and again between x_7 and x_9. (b) The function is decreasing between x_1 and x_3 and again between x_5 and x_7. (c) The curve is concave up between x_2 and x_4 and again between x_6 and x_8. (d) The function is concave down between x_1 and x_2, between x_4 and x_6, and again from x_8 to x_9.

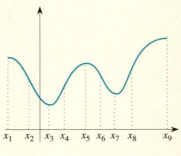

FIGURE 1.16

Next, we look for particular points on the curve. (e) The turning points for this function are at $x = x_3$, at $x = x_5$, and at $x = x_7$. (f) The function has a maximum at $x = x_1$ (when compared to other nearby points); the function also has a maximum at $x = x_5$ and again at $x = x_9$. (g) Similarly, the function reaches a minimum at $x = x_3$ (when compared to other nearby points) and again at $x = x_7$. (h) The points of inflection occur where the concavity changes, and this happens at $x = x_2$, at $x = x_4$, at $x = x_6$, and at $x = x_8$.

Periodic Behavior

There is another pattern of behavior for functions that is extremely important. Many processes in nature have the property of being *periodic*—the pattern repeats over and over again. We see this in the height of tides that rise and fall in the same pattern roughly every 12 hours in most coastal locations. We see it in the pattern of temperature readings in any location from one year to the next. It is very easy to spot a periodic function from its graph: The identical pattern repeats over and over again. For instance, consider the following data based on historical records giving the average number of tornados reported in the United States, per month, in a typical year.

Month	Jan	Feb	Mar	Apr	May	Jun	Jul	Aug	Sep	Oct	Nov	Dec
Tornados	16	24	60	111	191	179	96	66	41	26	31	22

We show a graph of these points in Figure 1.17. Notice, either from the table or the graph, how the values increase from a minimum level in January up to a maximum number of tornado sightings in May and then decrease back down toward the minimum as the year ends. Since these values are based on historical averages, this cycle will repeat yearly with little, if any, change from one year to the next. It is therefore a roughly periodic phenomenon. In Figure 1.18 we show a smooth curve that captures the longer term behavior of this roughly periodic function.

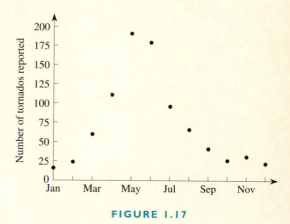

FIGURE 1.17

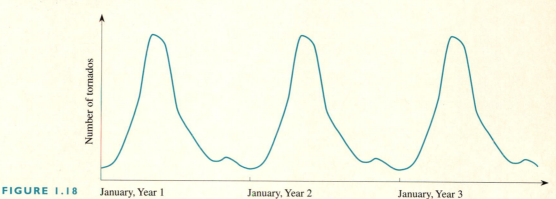

FIGURE 1.18   January, Year 1   January, Year 2   January, Year 3

EXERCISES

1. Consider the function shown in the accompanying graph. Use four different colored pens or pencils. With one, trace all parts of the curve where the function is increasing. With a second color, trace all parts of the curve where the function is decreasing. With a third color, trace all parts of the curve where the function is concave up. With the fourth color, trace all parts of the curve where the function is concave down. Then mark all turning points and points of inflection on the curve.

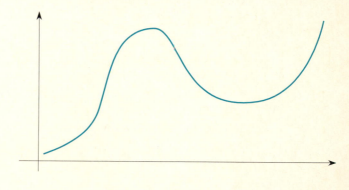

2. Sketch the graph of a single smooth curve that is first increasing and concave up, then increasing and concave down, and finally decreasing and concave down. Mark all turning points and points of inflection on your curve.

3. Sketch the graph of a single smooth curve that is first decreasing and concave up, then increasing and concave up, and finally increasing and concave down. Mark all turning points and points of inflection on your curve.

4. Sketch a possible graph of the temperature in your home town over the course of an entire week as a function of time. On the graph, indicate all the turning points. Where is the temperature function increasing? Where is it decreasing? Where is the temperature function concave down? Where is it concave up?

5. Sales of VCRs grew slowly at first and then increased tremendously as people came to accept them widely. Eventually new sales began to level off as market saturation was neared. Sketch a possible graph of the percentage of U.S. homes owning a VCR as a function of time paying careful attention to the behavior of the function. Indicate any turning points and any points of inflection.

6. Decide which of the following functions are periodic. (Assume that the graph continues indefinitely to the left and right in the same pattern.)

a.
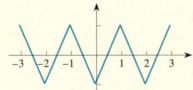

b.
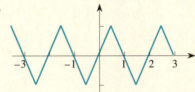

c.
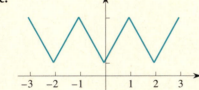

d.
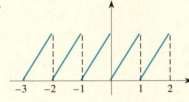

e.
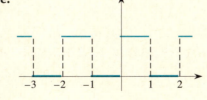

f.
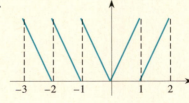

g.
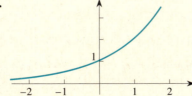

h.
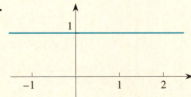

i.   j.

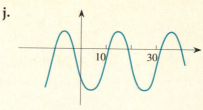

7.   Janis trims her fingernails every Saturday morning. Sketch the graph of the length of her nails as a function of time. Is this a periodic process?

8.   Sam is a perfectly normal individual with a pulse rate of 60 beats per minute and a blood pressure of 120 over 80. Thus his heart is beating 60 times each minute and his blood pressure is oscillating between a low (diastolic) reading of 80 and a high (systolic) reading of 120. Sketch the graph of his blood pressure as a function of time. Be sure to indicate appropriate scales on each axis.

9.   a.   The thermostat in Sylvia's home in Baltimore is set at 66°F during the winter. Whenever the temperature drops to 66° (roughly every half-hour), the furnace comes on and stays on until the temperature reaches 70°. Sketch the graph of the temperature in her house as a function of time. Be sure to indicate appropriate scales on each axis.

     b.   Gary, who lives in upstate New York, also has his thermostat set to come on at 66°. How will a sketch of the temperature in his house differ from the one you drew in part (a) for Sylvia's house?

     c.   Jodi, who lives in central Florida, likewise has her thermostat set to come on at 66°. How will a sketch of the temperature in her house differ from the other two?

10. Astronomers have been observing sunspots on the face of the sun for centuries. These dark spots on the sun, which are accompanied by the release of bursts of electromagnetic radiation that disrupt radio and TV signals, occur in periodic cycles. The accompanying figure is a graph of the number of sunspots observed each year.

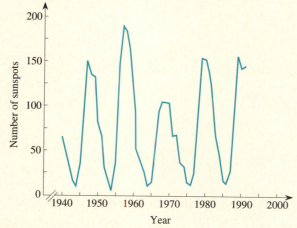

     a.   Estimate the period of the sunspot cycle.
     b.   Estimate when the next two peaks will occur in the cycle.
     c.   How might you estimate the maximum number of sunspots that will occur during the next peak in the cycle?

1.3   *Representing Functions Symbolically*

We will use a single letter such as f, g, or h as the name for a function. The particular formula for the function, if it is known, is usually written as $y = f(x)$. We read this as "y equals f of x" or possibly "y is a function of x." For example, the function f that takes any real number x and squares it can be written as

$$y = f(x) = x^2.$$

Some particular values of this function are

$$f(3) = 3^2 = 9 \qquad\qquad f(8) = 8^2 = 64$$
$$f(-5) = (-5)^2 = 25 \qquad\qquad f(0.02) = (0.02)^2 = 0.0004$$
$$f(-0.01) = (-0.01)^2 = 0.0001 \qquad\qquad f(\pi) = \pi^2 \approx 9.8696.$$

Similarly, the function g that takes the square root of any nonnegative real number x can be written as

$$y = g(x) = \sqrt{x}.$$

For instance, some values of the function g are

$$g(16) = \sqrt{16} = 4$$
$$g(400) = \sqrt{400} = 20$$
$$g\left(\frac{1}{4}\right) = \sqrt{\frac{1}{4}} = \frac{1}{2}.$$

But $g(-25) = \sqrt{-25}$ does not make sense because it is not possible to take the square root of a negative number. That is, the function g is not defined for $x = -25$.

The function h that gives the reciprocal of any nonzero number x can be written as

$$y = h(x) = \frac{1}{x}.$$

For example, some values of the function h are:

$$h(5) = \frac{1}{5} = 0.2$$

$$h(200) = \frac{1}{200} = 0.005$$

$$h(-0.125) = -\frac{1}{0.125} = -8.$$

But $h(0)$ does not make sense because we cannot divide by 0. That is, the function h is not defined at $x = 0$.

To work with functions, we will need some terminology. In the form $y = f(x)$, we call x the ***independent* variable** because it can assume any appropriate value. We call y the ***dependent* variable** because its value depends on the choice of x.

We can use letters other than f, g, or h to represent functions; other common choices are F, G, or f_1, f_2, f_3, and so on. We can use letters other than x to represent the independent variable; other common choices are t (for time), θ (for an angle), r (for radius), and so forth. Similarly, we often use letters other than y to represent the dependent variable; for instance, we can use A for area, D for distance, P for population, or C for cost.

Thus, since we know that the area A of a circle is a function of its radius r, we write this as $A = f(r) = \pi r^2$. Here r is the independent variable and A is the dependent variable because its values depend on the choice of r. The distance D that a car moves in t hours at a steady speed of 50 miles per hour (mph) is given by $D = g(t) = 50t$. Here t is the independent variable and D is the dependent variable.

Suppose you toss a ball straight up into the air with an initial velocity of 64 ft/sec. The function

$$y = f(t) = 64t - 16t^2$$

gives the height in feet of the ball above ground level after t seconds. Picture what happens. As the ball rises, it slows down due to the effect of gravity. Eventually it reaches a maximum height and then begins to fall back to the ground. As the ball falls, its speed increases, again because of gravity.

Now let's see how the function f gives the height of the ball above ground at any time t. For instance, after half a second, when $t = \frac{1}{2}$, the ball is 28 feet above ground level since

$$y = f\left(\frac{1}{2}\right) = 64\left(\frac{1}{2}\right) - 16\left(\frac{1}{2}\right)^2 = 32 - 4 = 28.$$

After one second, it is at a height of

$$y = f(1) = 64(1) - 16(1)^2 = 48 \text{ feet.}$$

After two seconds, it is at a height of

$$y = f(2) = 64(2) - 16(2)^2 = 64 \text{ feet,}$$

which happens to be the maximum height the ball reaches. After three seconds, the height is

$$y = f(3) = 64(3) - 16(3)^2 = 48 \text{ feet,}$$

and the ball is on its way down to earth. After four seconds, the ball is back at ground level since $f(4) = 0$.

In each of these examples of functions, there were some natural limitations on the possible values for both the independent variable and the dependent variable. The ball is released at time $t = 0$ and returns to ground level at $t = 4$ seconds. It therefore makes no sense in this problem to think about what happens before time $t = 0$ or after time $t = 4$. Thus, the permissible values for t are between 0 and 4 seconds. Furthermore, the ball rises to its maximum height of 64 feet and then falls back to ground level.

Therefore, the only meaningful values for the height of the ball are between $y = 0$ and $y = 64$ feet. (Of course, you should realize that it is unrealistic to think of throwing a ball upward from ground level—we have
done this here just to simplify the mathematics.)

In a similar way, the function $y = g(x) = \sqrt{x}$ makes sense only if the
independent variable x is not negative. Thus, the possible corresponding
values for y must be positive or zero. The function $y = h(x) = \frac{1}{x}$ makes
sense only if x is not zero. Thus, the possible corresponding y-values of
this function can be any number other than 0 since there is no value of x
such that $y = \frac{1}{x} = 0$. Finally, for the function $y = f(x) = x^2$, there is no limitation on the possible values of x, but there certainly is a limitation on the
corresponding values for $y = x^2$ because they can never be negative.

For any function f, the set of *all possible values for the independent variable*
is called the **domain** of f; the set of *all possible values for the dependent variable* is called the **range** of f.

E X A M P L E   I

Consider the function $y = g(x) = \sqrt{x}$ at the values $x = 0, \frac{1}{4}, 1, 2, \pi$, and 4 in its
domain. The corresponding y-values in its range are

$$y = g(0) = \sqrt{0} = 0$$

$$y = g\left(\frac{1}{4}\right) = \sqrt{\frac{1}{4}} = \frac{1}{2}$$

$$y = g(1) = \sqrt{1} = 1$$

$$y = g(2) = \sqrt{2} \approx 1.41421\ldots$$

$$y = g(\pi) = \sqrt{\pi} \approx \sqrt{3.14159\ldots} \approx 1.77245$$

$$y = g(4) = \sqrt{4} = 2$$

The domain and range of this function consist of all nonnegative numbers.

E X A M P L E   2

Consider the function $y = F(x) = x + \frac{1}{x}$. Suppose the domain of F consists of
all positive numbers. The range of F then turns out to be all values $y \geq 2$. Try
different values for x with your calculator to see that this is indeed the case. The
smallest possible value for y, which is $y = 2$, corresponds to $x = 1$; for any
other value of x, the value for y will be larger.

You can visualize a function f as an operation that transforms each value from its domain into the corresponding value in its range. See Figure 1.19.

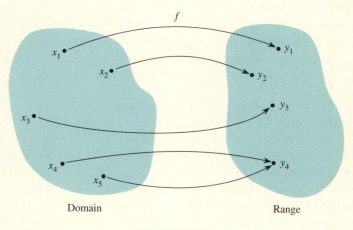

FIGURE 1.19

Each point x in the domain is transformed into a single point in the range. Thus x_1 is transformed into y_1. We also can say that x_1 is carried into y_1 or x_1 is mapped into y_1. Similarly, x_2 is carried into y_2 and x_3 is mapped into y_3. Note that x_4 and x_5 are both transformed into y_4, which is perfectly legitimate for a function. Each x-value must be mapped into a single y-value, although it is certainly possible for several different x's to be mapped into the same y. Think about the function $y = f(x) = x^2$, where both $x = 2$ and $x = -2$ are transformed into $y = 4$.

We now summarize the preceding ideas in a formal definition of a function.

A **function** f is a rule that assigns to each permissible value of the independent variable x a single value of the dependent variable y.

The **domain** of f is the set of all possible values for the independent variable.

The **range** of f is the set of all possible values for the dependent variable.

Consider the relationship between people and their telephone numbers. Is this a function? If there is even one person who has two different telephone numbers, then the relationship does not satisfy the definition and so *is not* a function. On the other hand, a person's height *is* a function of the person—each individual has one and only one height.

Throughout this book, unless some restriction is indicated, we will assume that all functions discussed are defined (either mathematically or practically) on the largest possible domain that makes sense.

EXERCISES

1. Which of the following relationships are functions and which are not? For those that are functions, identify the independent and dependent variables. For those that are not functions, explain why.

   a. The cost of first-class postage on January first of each year since 1900.
   b. The weight of letters you can mail with $n = 1, 2, 3, \ldots$ postage stamps.
   c. The time of sunrise associated with each day of the year.
   d. The time of high tide associated with each day of the year.
   e. The high temperature associated with each day of the year.
   f. The closing price of one share of IBM stock each trading day on the stock exchange.
   g. The area of a rectangle whose base is b.
   h. The area of an equilateral triangle whose base is b.
   i. The height of a bungee jumper t seconds after leaping off a bridge.
   j. The time it takes the bungee jumper to reach a height H above ground level.
   k. The number of baseball players who have n home runs in a full season.
   l. The height of liquid in a 55-gallon tank h hours after a leak develops.
   m. The daily cost to a family of heating their home versus the average temperature that day.

2. Kim has a peanut butter sandwich on white bread each day. The number of calories, C, in the sandwich, as a function of the number of grams, P, of peanut butter, is $C = f(P) = 150 + 6P$.

   a. What is $f(1)$? What does it mean?
   b. What is $f(10)$? $f(15.5)$? $f(20)$? $f(30)$?
   c. How many calories come from the bread alone?
   d. Explain why using $P = -1$ makes no sense.
   e. What is a reasonable domain and range for this function?

3. Suppose that Jim wants his peanut butter sandwich on rye bread instead of white. Rye bread contains 85 calories per slice. What would be the corresponding formula for the number of calories in Jim's sandwich?

4. The number of calories in a peanut butter and jelly sandwich on white bread is $C = 150 + 6P + 2.7J$, where P and J are the number of grams of peanut butter and jelly, respectively.

   a. How many calories are in a sandwich with 24 grams of peanut butter and 20 grams of jelly?
   b. Suppose that Adam is on a diet and wants to limit his calorie intake from a peanut butter and jelly sandwich to a maximum of 300 calories. Find two reasonable combinations of amounts of peanut butter and jelly that produce a sandwich with exactly 300 calories.
   c. Which is more caloric, a gram of peanut butter or a gram of jelly?

5. A car rental company charges a fixed daily rate for a mid-size car plus a charge for each mile above 100 miles that the car is driven per day. A formula for the rental cost of a car driven over 100 miles is $c = f(m) = 35 + 0.25(m - 100)$.

   a. Find $f(100)$. What does it mean?
   b. Find $f(150)$, $f(200)$, and $f(500)$.
   c. What is a reasonable domain and range for this function?

6. Suppose that you throw a ball upward, with initial velocity of 60 feet per second, from the roof of a 120-foot-high building.

   a. Sketch a possible graph of the height of the ball as a function of time, as you visualize it.
   b. Suppose you are told that the height of the ball as a function of time is given by

$$H(t) = 120 + 60t - 16t^2 .$$

Find the height of the ball when $t = 1$; when $t = 4$.

   c. Find $H(2)$ and $H(3)$. What do they represent?
   d. Use your function grapher to estimate how long it takes for the ball to reach its maximum height. What is the maximum height?
   e. How long does it take until the ball first hits the street below?
   f. What are the domain and range for this function?

7. For the function $f(t) = t^2 - 5$, find the values corresponding to $t = 2, 4, 6, 10$.

8. For the function[1] $F(x) = \frac{1}{x^2 - 4}$, find $F(0)$, $F(1)$, $F(3)$, $F(4)$, $F(5)$. Why did we skip $x = 2$? Are there any other values that should be skipped? What is the domain of this function?

9. For the function $g(x) = \frac{x^2 + 4}{x^2 - 9}$, find $g(0)$, $g(1)$, $g(2)$, $g(4)$, $g(-1)$. Why did we skip $x = 3$? Are there any other values that should be skipped? What is the domain of this function?

10. For the function $g(s) = s + \sqrt{s}$, find the values corresponding to $s = 4, 16, 25, 100$. Are there any values for s that will make the function come out negative? What does this tell you about the range of g? What is its domain?

11. For the function $z = f(q) = q^3 + 5$, find the value of the dependent variable that corresponds to a value of the independent variable of 4; find the value of the independent variable that corresponds to a value of the dependent variable of 6.

12. For the function $f(x) = x^3 - 8x^2 + 15x - 1$, find three different values of x between 1 and 8 for which $f(x) < 0$. Then find at least two noninteger values of x for which $f(x) < 0$.

---

[1] Note that when you enter this expression into a calculator or most computer programs, you must key the expression in as $1/(x^2 - 4)$. Pay careful attention to when you need to use parentheses in any such expression.

1.4 *Connecting the Geometric and Symbolic Representations*

One of the most significant advances in mathematics is based on the idea of connecting geometric and symbolic representations of functions. It allows us to think of functions from a visual rather than an exclusively symbolic perspective.

We begin by drawing two lines, called *axes*, which are perpendicular to each other, as shown in Figure 1.20. We label their point of intersection O for *origin*; it represents our point of reference. The *horizontal axis* represents values of the *independent variable*, (which in this case is x). By convention, these values increase from left to right, as indicated by the arrow. The *vertical axis* represents values of the *dependent variable* (which in this case is y); we typically think of these values as increasing upward, as indicated by the vertical arrow. The two axes divide the entire plane into four parts, called *quadrants*. They are named the first quadrant (I), the second quadrant (II), and so on (see Figure 1.20). Further, whenever appropriate, we indicate the units used for each variable and label the axes accordingly.

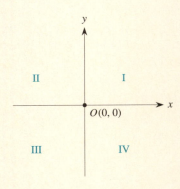

FIGURE 1.20

This representation is called a *rectangular* or *Cartesian coordinate system*, and it is a way of associating points in the plane with ordered pairs of numbers. Thus, every point P in the plane can be represented by an ordered pair of numbers, (x, y). Alternatively, every ordered pair, such as $(2, 5)$ or $(-7, 1)$ or $(-3.84, -1.02)$, represents a point in the plane. We call (x, y) the *coordinates* of the point P. In mathematics, we typically use the letters x and y in a generic sense to represent the independent and dependent variables, respectively. However, in any given context, we urge you to use letters that suggest the quantities being studied.

As an example, let us again consider the function

$$y = f(t) = 64t - 16t^2,$$

which represents the height, at any time t, of a ball tossed upward with initial velocity of 64 feet per second. The formula for the function f gives the vertical height y of the ball at any instant t. When $t = 0$, we have $y = 0$, and the corresponding point $(0, 0)$ is the origin. Further, as we found before, $f\left(\frac{1}{2}\right) = 28$, $f(1) = 48$, $f(2) = 64$, $f(3) = 48$, and $f(4) = 0$, which give rise to the points $\left(\frac{1}{2}, 28\right)$, $(1, 48)$, $(2, 64)$, $(3, 48)$, and $(4, 0)$ in the coordinate system. These six points are plotted in Figure 1.21.

We can determine many other ordered pairs (t, y) satisfying the equation $y = 64t - 16t^2$. (Simply pick any other value for t between 0 and 4 and calculate the associated value of y from the equation.) Each such ordered pair can be plotted as a point in the coordinate system. When all possible points are plotted, they form the curve shown in Figure 1.22. This curve is the *graph of the function f*. It consists of *all* points in the plane whose coordinates (t, y) satisfy the given equation. Thus we have a direct connection between the graph of a function and its algebraic equation. The graph of a function is therefore also a representation of the function even if no formula is given or known. In fact, functions are often given by graphs instead of by algebraic formulas. Notice that the graph in Figure 1.22 represents the *height* of the ball at any time t; it does not show the *path* of the ball, which goes straight up and then down.

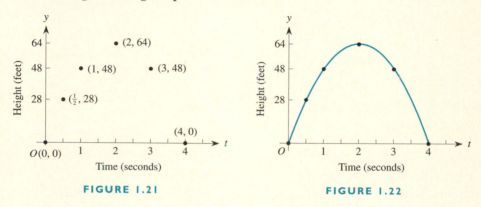

FIGURE 1.21   **FIGURE 1.22**

> The **graph** of a function $y = f(x)$ consists of *all* points $P(x, y)$ in the plane whose coordinates satisfy the equation of the function.

A table of values for a function is also useful when creating a hand-drawn graph of the function f from the formula $y = f(x)$. It provides a simple method to organize the values of the independent variable x and the associated values of the dependent variable y that produce each point to be plotted. The number of points that you need to calculate for a table to draw a reasonable graph of a function depends on how complicated the behavior of the function is. For a line, all you need is two points, since two points completely determine a line. We used six points to produce the graph of the height of the thrown ball in Figure 1.22. For comparison, a graphing calculator uses about 100 points to construct a curve and a computer uses about 600 points.

When drawing the graph of a function, it usually is important to determine several key points. One point is where the graph crosses the vertical axis. It is very simple to find this point: Just set the independent variable x equal to zero in the algebraic formula for the function (if it is known) and calculate the corresponding y-value. It is often desirable, though usually considerably more complicated, to find the points where the curve crosses the horizontal axis; to find them, set the dependent variable y equal to zero and then attempt to solve the resulting equation. For the function representing the height of the ball, we set $y = 0$ and find

$$y = 0 = 64t - 16t^2 = 4t(16 - 4t).$$

When we solve this equation for t, we get $t = 0$ and $t = 4$. The time $t = 0$ is the instant when the ball is first released into the air, so that $y = 0$. At the instant when $t = 4$, the corresponding value for y, which represents the height of the ball, is also zero. That is, at time $t = 4$, the ball has come back down to Earth. You can see the pattern for the values of this function (and thus the pattern for the height of the ball) from the following table.

Time t (seconds)	0	0.5	1.0	1.5	2.0	2.5	3.0	3.5	4.0
Height y (feet)	0	28	48	60	64	60	48	28	0

E X A M P L E   I

Determine the domain and range of the function shown in Figure 1.23.

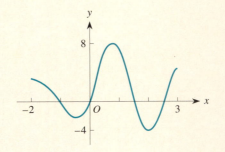

FIGURE 1.23

Solution   Notice that the axes shown are labeled x and y; x is the independent variable and y is the dependent variable. Further, observe that the graph extends from $x = -2$ at the left to $x = 3$ at the right; so the domain of this function is from -2 to 3. You can write this domain in terms of inequalities as $-2 \le x \le 3$. Similarly, the graph extends vertically from a low of $y = -4$ to a high of $y = 8$, so the range is the interval $-4 \le y \le 8$.

In many situations, we typically start with a set of data collected from some experiment or from measurements taken on some process. We then graph the data to get a feel for the behavior of the quantity being studied. Often, we may try to connect the points in the graph with a smooth curve to give a better indication of the behavior of the quantity. Finally, we would like to obtain an equation for a function that fits these data values since there are many questions that can be answered far more easily and accurately when an equation is available. We illustrate this methodology in the following examples.

E X A M P L E   2

The snow tree cricket, which lives in the Colorado Rockies, has been studied by field biologists who have gathered the following measurements on how the chirp rate depends on the air temperature.

Temperature T (° Fahrenheit)	50	55	60	65	70	75	80
Rate R (chirps per minute)	40	60	80	100	120	140	160

We start by plotting these data points to give us a visual dimension, as shown in Figure 1.24. We see that the chirp rate is growing at a constant rate as the temperature increases. Moreover, when we look at the corresponding points in the figure, we see that they seem to fall into a linear pattern, as shown by the fact that we can draw a line through them. In the next chapter, we will discuss how to find the equation of this line and how to predict the chirp rate R of the cricket based on the temperature T, or vice versa.

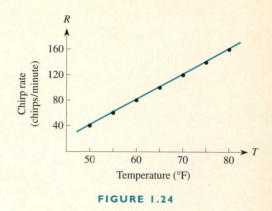

FIGURE 1.24

E X A M P L E   3

The following table gives the population, in millions, of the state of Florida since 1980. We would like to use these values to predict future population growth there.

Year	1980	1981	1982	1983	1984	1985	1986	1987	1988	1989	1990	?
Population	9.75	10.03	10.32	10.62	10.93	11.24	11.56	11.90	12.23	12.58	12.94	?

The graph of this set of data is shown in Figure 1.25. We see that the growth pattern is clearly not a linear one; rather, the population grows ever faster. The function is both increasing and concave up. We will determine what the equation of this function is in the next chapter so that we can make some intelligent predictions about the future growth in the Florida population.

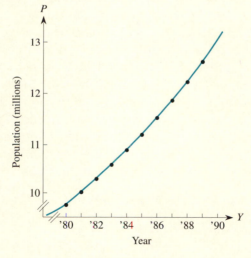

FIGURE 1.25

EXAMPLE 4

The following table shows measurements, at different times, of the height of an object dropped from the top of the Empire State Building, 1250 feet high.

Time (seconds)	0	1	2	3	4	5	6	7	8
Height (feet)	1250	1234	1186	1106	994	850	674	466	226

From the associated graph in Figure 1.26, we see that the object is falling ever faster as time goes by. The function is decreasing and concave down. Although we could estimate from either the table or the graph how long it will take until the object either hits the ground or passes, say, the thirtieth floor, we could answer such questions more precisely if we knew the formula for the function shown in the figure.

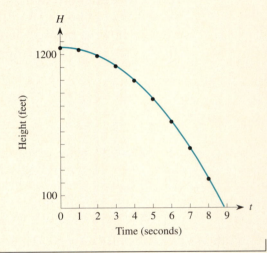

FIGURE 1.26

In Examples 2–4, we simply connected the points to construct the smooth curve that seemed to fit the pattern. Doing so can sometimes lead to serious errors. Suppose we had some data on the turkey population of the United States taken on January 1 of every year. It would likely show a growth trend similar to that in Example 3 on the population of Florida. However, a little thought will convince you that this population will change quite drastically around the middle of each November. The smooth curve drawn using the January 1 turkey census data would therefore be a rather poor description of the actual population over all intermediate times.

Nevertheless, this idea of connecting a series of points to form a curve is precisely how a computer or graphing calculator produces the graph of a function. We strongly urge you to become very comfortable with using a graphing calculator or a computer program in order to investigate the graph of any desired function. The visual dimension invariably provides a wealth of information about the behavior of the function, and we will continually turn to graphical images throughout this book.

Does Every Curve Represent a Function?

Let's consider two related questions:

1. Is every curve the graph of some function $y = f(x)$?
2. If a curve does represent a function, can we identify a formula for that particular function?

Consider the five curves shown in Figure 1.27.

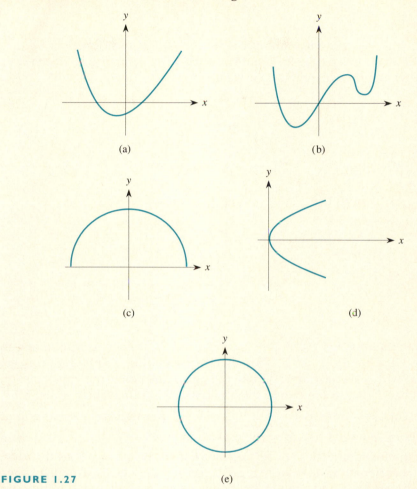

FIGURE 1.27

Let us see if they are the graphs of functions. That is, does every value of the independent variable x give rise to one and only one value of the dependent variable y? We can test a curve in the following simple way to see if it passes this test. Picture a vertical line moving across the curve from left to right, so that it passes through every possible value of x in the domain. If, for each x, the line crosses the curve at only one point, then there is exactly one y-value for that x and so the curve represents a function. If the vertical line crosses the curve at more than one point for *any* value of x, then the curve does not represent a function. This criterion, called the *vertical line test*, shows us that the curves (a), (b), and (c) are all graphs of functions. However, when the vertical line test is applied to curves (d) and (e), we encounter values of x for which the line crosses the curve at more than one y-value. We thus conclude that neither of the curves (d) and (e) are graphs of functions.

In general, identifying the particular algebraic formula for a function from its graph is usually more difficult.

EXERCISES

1. Which of the following graphs are functions? For each function, (a) give its domain and range, (b) identify where it is increasing or decreasing, and (c) identify where it is concave up or concave down.

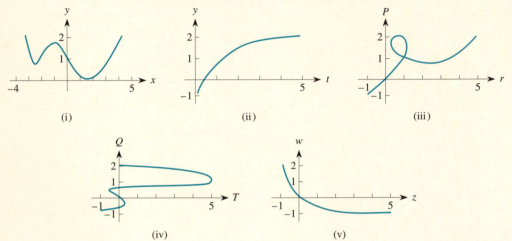

(i)   (ii)   (iii)

(iv)   (v)

2. Given the function $f(x) = x^2 - 3x + 2$, find the values corresponding to $x = -3, -2, -1, \ldots, 4, 5$. Plot the corresponding points and connect them with a smooth curve. Graph the given function using your function grapher. How do the two graphs compare? Find the values of the function corresponding to $x = \frac{1}{2}$, $x = \frac{3}{2}$, $x = -\frac{5}{2}$ and indicate the location of the corresponding points on the curve you drew.

3. For the function $g(t) = 9 - t^2$, use an appropriate set of values for t and the corresponding y values to get a feel for the behavior of the curve when you draw and connect the points. How does your sketch compare to what you see when using your function grapher?

4. Repeat Exercise 3 using $h(s) = s^3 - 7s + 5$.

1.5 *Mathematical Models*

A *model* is an image or representation of an object or process. A diagram of the human circulatory system is a model of the veins and arteries in the human body; this picture can be used to help us understand how blood circulates throughout the body. Similarly, an architect's sketch of a proposed shopping center is a model of the actual center; a wooden model built to scale is a still more realistic representation or model for that shopping center.

> A **model** is a representation that highlights the most important characteristics of an object or process.

Models can be found all around us: the tide tables used by fishermen; a computer scientist's flowchart for a new program; plastic replicas of jet fighters; or even magazine covers featuring human models wearing the newest fashions. Because our focus is on mathematics, the models we will study are *mathematical models*. A **mathematical model** is a representation of a process that is expressed in a mathematical form by a formula, an equation, a graph, a sequence of numbers, or a table of values. Once we have developed any such mathematical representation, we can use it to examine the behavior of the actual process.

For example, the equation for the motion of the ball thrown vertically upward,

$$y = f(t) = 64t - 16t^2,$$

is a mathematical model that describes one critical aspect of the motion of that ball—its height at any time t. There may be other aspects of the motion that may not be captured in this mathematical model (such as the effects of air resistance or how fast the ball is moving at any instant). These other factors may likely require our developing a more sophisticated model, but often we will be satisfied to have a simple model that gives a reasonably accurate first approximation.

How do we use mathematics to describe the real world via a mathematical model? We begin by looking at some process in the real world, such as the motion of a ball thrown upward, the growth of a population, or a person's reaction to a drug. Typically, in the process of trying to explain what is happening, we have to make some simplifying assumptions. For example, in modeling the motion of a ball, we assume that the only force acting on the ball is the force of gravity and ignore the hopefully negligible effects of air resistance. (Of course, if the ball were replaced with a balloon, a feather, or a piece of paper, this assumption would be invalid.)

After making reasonable assumptions, we then express the process in a mathematical form, which leads to a mathematical representation of the process in terms of a formula, an equation, a graph, or a table. Having produced this mathematical model, we then have to interpret it. Does it truly seem to reflect what happens in the real world? Does the behavior of the function mirror the behavior of the process? If so, we can use the mathematical model as an explanation of the process under study and as the basis for predictions about the process. If our model does not adequately reflect the actual process, then we have done something wrong: We have overlooked some important aspect of the situation or ignored some critical factor; we have made some erroneous assumptions or we have made an error in our work. We show this interplay between mathematics and the real world via mathematical modeling in Figure 1.28.

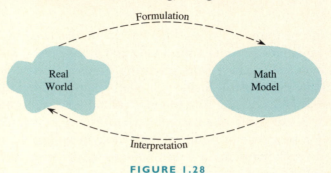

FIGURE 1.28

The concept of function is closely connected to the idea of a mathematical model, and most mathematical models we construct are expressed as functions. Let's look at an example of this process. Researchers have studied the correlation between the level of animal fat in women's diets and the death rate from breast cancer in different countries. Some of their data are shown in the following table, which gives the average daily intake of animal fats in grams per day and the age-adjusted death rate from breast cancer per 100,000 women.

Country	Japan	Spain	Austria	U.S.	U.K.
Daily fat intake (grams)	20	40	90	100	120
Death rate per 100,000	3	7	17	19	23

We begin by looking at this data, first as a set of numerical values in the given table and then visually on a graph, as shown in Figure 1.29. Notice that the death rate D increases as the daily fat intake F increases, so it is an increasing function. There is obviously a relationship between the death rate and the daily fat intake. Moreover, when we look at the corresponding points on the graph, we see that they clearly fall into a linear pattern.

It turns out that the equation for this line is

$$D = f(F) = 0.2F - 1.$$

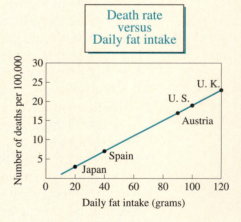

FIGURE 1.29

What does this equation mean in terms of the actual phenomenon we are trying to model? Let's look at various cases for nations not included in the study. The average daily fat intake in Mexico is $F = 23$ grams, so the equation predicts that the death rate from breast cancer in Mexico will be $0.2(23) - 1 = 3.6$ per 100,000 Mexican women.

This type of prediction is called *interpolation* because we are predicting the value of a quantity using a measurement *within* the set of data. Similarly, the average daily animal fat intake in Denmark is $F = 135$ grams, so the equation predicts a death rate of $D = 0.2(135) - 1 = 26$ per 100,000 Danish women. This type of prediction is called *extrapolation* because we are predicting a quantity for a value of the variable *beyond* the range of the data.

In Section 2.2, we will show how to find such an equation. Once we have such an equation as our mathematical model, we can make some informed judgments based on it. This model is based on the *average* daily intake in each country, and there can be tremendous variation among individuals. Even so, there is obviously a link between consumption of animal fat and the incidence of breast cancer. Thus the mathematical relationship indicates that women should drastically reduce their daily animal fat intake to reduce their chance of breast cancer. Furthermore, knowing that such a link exists, researchers have since been conducting follow-up studies to determine why the link exists. They have also found links to other items in the diet as well as in the environment. Thus, it turns out that the incidence of breast cancer depends not just on a single variable, but actually on a number of different variables. Although the study of functions of several variables is somewhat beyond the scope of this course, we note that many situations that you encounter in real life are examples of such functions.

EXERCISES

1. An uncooked chicken (temperature of 70° F) is placed into a hot oven at a temperature of 350° to cook. The chicken is removed when its internal temperature reaches 180°. Sketch a possible graph for the temperature T of the chicken as a function of time t. What would be appropriate values for the domain and range of this function? Describe the behavior (increasing/decreasing, concavity) for the graph.

2. A warm can of soda (80° F) is placed in a refrigerator at a temperature of 36° and left there to cool. Sketch the graph of the temperature T of the soda as a function of time t. Identify appropriate intervals for the domain and range of this temperature function. Describe the behavior (increasing/decreasing, concavity) of this function.

3. An Olympic diver jumps off the 10-meter platform, performs a clean entry into the water, and rises slowly back to the surface. Sketch a possible graph for the height of the diver above water level as a function of time. What might be appropriate values for the domain and range of this function? (Estimate how long it will probably take the diver to reach the water from the platform.) Describe the behavior of this function.

4. Repeat Exercise 3 by sketching the graph of the diver's height above the level of the diving platform as a function of time. How does the shape of this graph compare to the one you drew in Exercise 3?

5. Police sometimes use the formula $s = f(d) = \sqrt{24d}$ as a model to estimate the speed s in miles per hour that a car was going on dry concrete pavement if it left a set of skid marks d feet long. Using this model, estimate the speed of a car whose skid marks stretched:

   a. 60 feet   b. 100 feet   c. 140 feet   d. 200 feet

   e. Suppose you are driving at 60 mph and slam on your brakes. How long will your skid marks be?

1.6   *Introduction to Monte Carlo Methods*

When statisticians conduct a political poll, they typically survey about 800 or 1000 randomly selected registered voters to find out their preferences between the candidates. If 43% of a sample group prefer Harold Smith, then the statistician would conclude that *about* 43% of *all* voters prefer Smith.

Obviously, there is a certain degree of uncertainty in drawing this conclusion about the entire electorate based on the results of just a relatively small group. A different group might have produced slightly different results, perhaps 45% or 40% in favor of Smith. However, it is very unlikely that there would be a substantial difference between the percentages in different groups, assuming they are representative of the population, and all of these percentages should be reasonably close to the true value for the percentage of all voters who prefer Smith. The uncertainty could be reduced by taking a much larger sample group, say 10,000 voters, but that becomes much more expensive and time consuming. Statisticians and all users of statistics must therefore accept a certain degree of uncertainty.

A similar philosophy occurs in business and industry. Consider a company that produces computer chips for use in cars, television sets, CD players, and other devices. It is not possible to test fully each of the millions of chips produced; rather, a sample group of chips is randomly selected and tested. The percentage of defective chips from that sample is then used to assess, with some degree of uncertainty, the percentage of all chips produced that are defective. If this percentage is very small, the company might be willing to accept it and allow a few consumers to buy defective units that would then be replaced. If the percentage of defects is unacceptably high, the company would likely demand higher standards in its production process or impose tighter specifications on its suppliers to reduce the percentage of defective chips.

Let's look at these two scenarios from a slightly more mathematical point of view. In the case of a computer chip manufacturer, we would likely expect the percentage of defective chips to be constant from day to day or even from year to year. Thus the results from the samples are used to estimate the value of this constant function. In the case of the political poll, we would expect that, as time passes, there will be some variation in the percentage of voters who prefer Smith; this function is not constant, but will vary in quite unexpected ways depending on many different factors. Thus each poll conducted estimates the value of a function on a particular day.

Statistical examples such as these can be used to develop a very useful tool for providing information on many different processes, including mathematical ones. The method involves creating or simulating a random sample that is representative of the process under study; studying this sample provides insight into what happens in reality. Since we generate the random samples ourselves, usually on a computer, we can make them as large as we want and so reduce the degree of uncertainty associated with any such approach. The act of generating such a random sample is called a *simulation*. We illustrate one type of simulation: the Monte Carlo method.

The Monte Carlo Method

One of the fundamental numbers in mathematics is the number π, which arises most naturally in terms of properties of a circle, namely

$$\text{Circumference} = 2\pi r$$
$$\text{Area} = \pi r^2,$$

where r is the radius of the circle. Because the decimal representation of π is a non-terminating, non-repeating decimal, it cannot be written exactly as a fraction or rational number. So π is an irrational number. We often approximate the value of π using $\pi \approx 3.14$ or $\pi \approx 22/7 = 3.142857\ 142857\ldots$. A somewhat more accurate approximation is $\pi \approx 3.14159$ or even $\pi \approx 3.141592654$. In fact, two mathematicians at Columbia University have recently calculated π to several billion decimal places. Have you ever wondered how approximations to the value of π are actually obtained?

In principle, the simplest way to obtain a value for π would be to use the following facts. For any circle of radius r, the circumference is $2\pi r$ while the diameter is $2r$. Thus, in any circle, the ratio of the circumference to the diameter is always π. Therefore, we can measure both the circumference and the diameter of any circle as accurately as possible, and use their ratio to estimate the value of π. The problem with this approach is that it is hard to achieve a sufficiently high level of accuracy in the measurements. In biblical times, this method produced an approximation to π of 3, which is quite inaccurate. Even with sophisticated scientific equipment available today, the accuracy possible for the measurements is not adequate to achieve the number of decimal places stated above.

Instead, we will demonstrate a rather surprising way to obtain an accurate approximation to π. It is known as the *Monte Carlo method* (named after the European gambling capital in Monaco) because it is based on a random process. We begin by considering a simpler problem. Assume we have a square that is centered at the origin and whose sides have length 2, as shown in Figure 1.30. Suppose we have a mechanism for generating a random point somewhere in this square. For instance, suppose we can throw a dart that is guaranteed to land inside the square. How likely is it that the dart lands in the smaller square in the first quadrant? The dart is just as likely to land in any of the four quadrants as in any other. The *chance*, or *probability*, that the dart lands in the first quadrant is $\frac{1}{4}$. Now suppose that we continue this process repeatedly—we generate many different random points in the large square. If we generate 1000 such points, then we would expect that *approximately* $\frac{1}{4}$

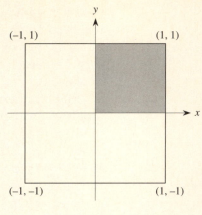

FIGURE 1.30

of them, about 250, will fall in the smaller square in the first quadrant. If we generate 1 million random points in the entire square, we would expect about 250,000 to fall in this shaded square.

We can conclude that about $\frac{1}{4}$ of all points should fall in the shaded square since the area of this square is $\frac{1}{4}$ the area of the entire square. In other words, the ratio of the areas of the two squares should be very close to the ratio of the number of points that fall in the shaded square to the number that fall in the entire square. We express this mathematically as

$$\frac{\text{Area of shaded square}}{\text{Area of entire square}} \approx \frac{\text{Number of points that land in shaded square}}{\text{Number of points that land in entire square}}.$$

We now use this same line of reasoning to estimate the value for π. Consider the circle of radius 1 centered at the origin. It is a *unit circle* whose equation is $x^2 + y^2 = 1$. (See Appendix C.) We inscribe this circle in the same large 2×2 square we used above (see Figure 1.31). The area of the square is $(2)^2 = 4$ and the area of the unit circle is $\pi r^2 = \pi \cdot (1)^2 = \pi$ square units. Suppose we now generate large numbers of random points inside the square. To test whether such a point $P(x, y)$ lies on, inside, or outside the unit circle, we use the equation of the unit circle:

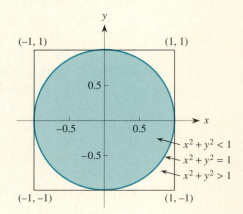

FIGURE 1.31

$$x^2 + y^2 = 1.$$

For instance, the point $P(0.6, 0.8)$ lies *on* the circle because $(0.6)^2 + (0.8)^2 = 1$; the point $Q(0.5, 0.7)$ lies *inside* the circle because $(0.5)^2 + (0.7)^2 = 0.74 < 1$; the point $R(0.8, 0.9)$ lies *outside* the circle because $(0.8)^2 + (0.9)^2 = 1.45 > 1$. Plot these points to verify their locations.

In general, a random point $P(x, y)$ lies

> *on* the circle if $x^2 + y^2 = 1$;
>
> *inside* the circle if $x^2 + y^2 < 1$;
>
> *outside* the circle if $x^2 + y^2 > 1$.

The chance that any single random point lands inside the circle is given by the ratio of the area of the circle, π, to the area of the square, 4, or $\frac{\pi}{4}$. Suppose that we repeat this process and keep track of the number of random points that land in the circle. Then

$$\frac{\text{Area of circle}}{\text{Area of square}} \approx \frac{\text{Number of points that land in circle}}{\text{Number of points that land in square}}.$$

Since the first ratio is $\frac{\pi}{4}$, the second ratio provides a way of estimating the value of $\frac{\pi}{4}$ and hence of π. Typically, such processes are implemented with computer programs that generate large numbers of random points. For instance, if we run such a program with $n = 10{,}000$ points, then a typical outcome might show that 7842 points fall inside or on the circle. Therefore,

$$\frac{\pi}{4} \approx \frac{7842}{10{,}000} = 0.7842$$

$$\pi \approx 4(0.7842) = 3.1368,$$

which is reasonably accurate. A different run of the same program might produce 7893 points out of 10,000 that fall inside the circle, so that the corresponding approximation to π is $4(7893/10{,}000) = 3.1572$, which is also reasonably accurate.

Note that the Monte Carlo method is a random process. That is, each time you repeat it, even with the same number of random points, the specific results will likely be different because there are an unlimited number of random points that can be chosen. The same would be true if you threw 1000 darts—you would expect different outcomes on each set of throws. Nevertheless, the results of different runs are quite consistent, indicating that the approach provides surprisingly accurate values. The key idea is that the Monte Carlo method, while based on random points that are individually unpredictable, becomes highly accurate and highly predictable in the long run when large numbers of points are used. More importantly, the larger the number of random points generated, the more accurate the results are likely to be. Thus if we generate 1 million random points, then we would expect to obtain a better approximation for π. If we generate 1 billion random points, then the result should be highly accurate.

We note that one trade-off in using such a method is that it can be very time consuming, even by computer. Thus, if better methods are available, they should be employed. There are considerably more effective methods for calculating π, and you will see some of them in calculus. However, there are many situations where a Monte Carlo method is the only method that can be used to study certain processes, and we will use it throughout this book.

EXERCISES

1. Suppose you keep track of the colors of M&M candies as you nibble them (purely in the interests of mathematics, of course). You find that 33 out of 312 are red. What would be your estimate of the percentage of all M&Ms that are red?

2. While waiting at a street corner for your best friend to pick you up, you observe that of 120 cars that come to the corner, 70 have to stop for a red light. Estimate the percentage of all cars that arrive at the intersection during the green portion of the cycle.

3. The equation

$$\frac{x^2}{9} + \frac{y^2}{4} = 1$$

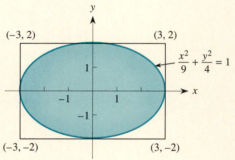

represents an ellipse (see Appendix C) that is inscribed in a rectangle extending horizontally from -3 to 3 and vertically from -2 to 2. Out of 10,000 random points generated inside this rectangle, suppose 7820 land inside the ellipse. Estimate the area of the ellipse.

4. All sophisticated calculators have a random number generator function, often indicated as RAND, which produces a number between 0 and 1. Use the RAND function to generate 20 random numbers and, based on them, estimate the probability that the random number produced by your calculator is between 0 and 0.5.

5. Consider $f(x) = 2x - x^2$ for values of x between 0 and 2. The accompanying figure shows that the corresponding graph lies inside a rectangle of height 1 and width 2. The following set of 10 points inside this rectangle were randomly generated: $(1.82, 0.94)$, $(1.90, 0.15)$, $(0.44, 0.51)$, $(0.74, 0.41)$, $(0.02, 0.73)$, $(1.87, 0.04)$, $(0.22, 0.34)$, $(0.01, 0.99)$, $(1.10, 0.20)$, $(1.71, 0.80)$. Plot these points carefully and use them to estimate the area of the region under the curve and above the x-axis.

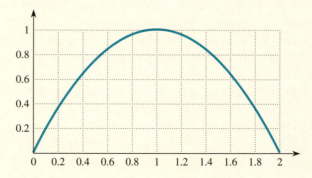

6. The function RAND generates a random number between 0 and 1. To produce a random number between 0 and 2, you need to consider $2 \times$ RAND. How would you produce a random number

   a. between 0 and 5?
   b. between 0 and 100?
   c. between 1 and 6?
   d. between 20 and 50?
   e. between any two numbers a and b?

7. Use the random number generator of your calculator or computer to generate 10 random points inside the rectangle given in Exercise 5. Plot these points on the same graph used in Exercise 5. Does this change the estimate you previously obtained for the area?

8. Consider $f(x) = \sqrt{1 + x}$ for x between 0 and 3. The corresponding graph lies inside a rectangle of height 2. The following set of 10 points inside that rectangle were randomly generated: $(2.93, 0.06)$, $(0.83, 1.67)$, $(0.82, 1.24)$, $(0.37, 0.41)$, $(0.16, 1.97)$, $(2.17, 1.45)$, $(0.04, 0.60)$, $(1.26, 0.50)$, $(0.92, 1.89)$, $(2.92, 0.09)$. Calculate the value of the function that corresponds to the x-value of each generated pair (x, y), and decide whether $y > f(x)$ or $y \leq f(x)$. Use these results to estimate the area of the region under the curve and above the x-axis without graphing the region.

9. The function $y = f(x) = x^2 - 4x + 2$ assumes positive and negative values of y for x between 0 and 1. Generate 10 random numbers between 0 and 1 for x and use them to estimate the percentage of all x-values between 0 and 1 for which the function f is positive.

CHAPTER SUMMARY

In this chapter, you have learned the following:

- How functions arise all around you in the real world.
- The important characteristics about the behavior of functions—where they increase and decrease, where their turning points are, where they are concave up and concave down, where their points of inflection are, and whether they are periodic.
- How to interpret concavity—whether the growth (increase) or decay (decrease) in a function is speeding up or slowing down.
- How functions can be represented in different ways—by graphs, by tables, by formulas, and in words—and how to move from one representation to another.
- How mathematics is used to model phenomena in the real world.
- How probabilistic simulations can be used to obtain information on quantities for making estimates and decisions.

REVIEW EXERCISES

1. In determining the amount of radiation to apply to a tumor site, doctors take into account the depth of the tumor within the body. What is the independent variable and what is the dependent variable in such a relationship? Give reasons for your answer.

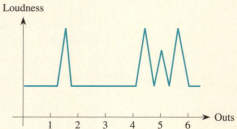

2. The accompanying graph describes the loudness of a crowd watching a baseball game during the ninth inning. Write a scenario that might explain what was happening on the field as the inning progressed.

3. An experimental form of insulin is being administered every 4 hours to a person with diabetes. The body uses or excretes about 40% of the drug over the 4-hour period. Draw a graph that shows the amount of the drug in the body as a function of time over a 24-hour period.

4. Populations tend to grow steadily until there are too many members for the space and resources available. Then the population size levels off. Sketch a function that gives population size as a function of time.

5. Determine if the following tables could represent functions. If not, explain why.

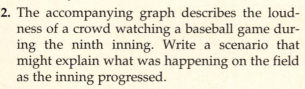

a.

x	1	2	3	4	5	6
$f(x)$	10	10	12	14	18	25

b.

x	11	15	9	20	15	8
$g(x)$	12	13	13	15	16	17

6. The table below shows the budget and the attendance at 15 U.S. zoological parks. Write a short description of how attendance and budget are related.

Budget ($ millions)	10.0	3.4	27.0	6.2	9.7	7.0	4.8	18.0	6.5
Attendance (millions)	1.0	0.5	2.0	0.6	1.3	1.0	1.1	4.0	0.6

13.0	9.0	15.7	7.0	3.2	14.7
3.0	0.5	1.3	1.0	0.5	2.7

7. Social Security Administration figures show the contribution and benefit base (in thousands of dollars) for Old Age and Survivors Disability Insurance (OASDI) over the years from 1983 to 1992. Draw a graph of the benefit base as a function of the year since 1983. In what years did this rise most rapidly? Most slowly?

Year	1983	1984	1985	1986	1987	1988	1989	1990	1991	1992
OASDI Base	35.7	37.8	39.6	42.0	43.8	45.0	48.0	51.3	53.4	55.5

8. For the function $f(x) = 3x^2 - 2x + 1$, find $f(0)$, $f(1)$, $f(1.1)$, $f(1.01)$, $f(-3)$, and $f(a)$.

9. For 1993, governors' salaries as a function of the average salary in his or her state are approximated by the function $g(S) = 2.56S + 30.6$, where salaries, S, are given in thousands of dollars.

   a. Determine the salary of the governor of a state in which the average salary is $20,000.

   b. The governor of West Virginia earned $90,000 per year. What was the average salary in West Virginia, as predicted by the function g?

   c. The average salary in the 50 states ranged from a minimum of $16,430 in South Dakota to a maximum of $29,946 in Alaska. What was the range of values for the salaries of the governors as predicted by the function?

10. A study of the relationship between the average longevity (in years) and the gestation period (in days) for a sample of animals shows that the animals' average longevity can be predicted reasonably well as a function of the gestation period by the function: $f(t) = 1.04\, t^{0.49}$, where t is the gestation period in days.

   a. Estimate the lifetime of a chipmunk whose gestation period is 31 days.

   b. The gestation times in the study extend from 15 days (opossum) to 645 days (elephant). What would the range of the average longevity be if the given function were a good predictor?

   c. Use your function grapher to graph the function. Is it increasing or decreasing? Is it concave up or concave down?

   d. Use the graph to estimate the gestation period of an animal whose average longevity is 15 years.

   e. The gestation time for humans is nine months, or about 270 days. What does the formula predict for the average longevity of human beings? Can you think of any reasons why the value you obtained is so inaccurate?

11. Give the domain of the following functions.

   a. $f(x) = x^3 - 9$ **b.** $f(x) = 4x^2 + 3x - 5$

   c. $f(x) = \sqrt{x + 5}$ **d.** $f(x) = \sqrt{x^2 - 16}$

   e. $g(x) = \dfrac{x^2 + 4}{x^2 - 9}$ **f.** $h(x) = \dfrac{x^2 - 4}{x^2 + 9}$

What is the range for the functions in parts (a) to (d)?

12. The U.S. Postal Service rates for first-class mail in 1996 were 32 cents for the first ounce and 23 cents for every additional ounce.

   a. Construct a table showing the cost of postage for mail weighing (in ounces) for the following intervals:

$$0\text{–}1,\ 1\text{–}2,\ 2\text{–}3,\ 3\text{–}4,\ 4\text{–}5$$

   b. Sketch a graph of this function $F(w)$, showing the cost of sending a first-class letter as a function of the weight w of the letter in ounces.

13. Consider the roller coaster shown in the accompanying figure. The points A, B, C, D, E, and F divide the curve representing the track into portions that are increasing and concave up, increasing and concave down, decreasing and concave up, and decreasing and concave down. For each of the five portions of the track, A to B, B to C, and so on, (a) identify the mathematical behavior of the curve, and (b) describe whether the speed of the cars is increasing at an increasing rate, increasing at a decreasing rate, decreasing at a decreasing rate or decreasing at an increasing rate. In each case, explain your answer.

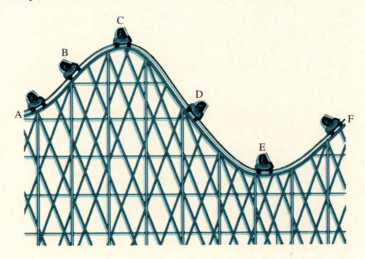

2

FAMILIES OF FUNCTIONS[1]

2.1 Introduction

Functions are fundamental to mathematics and its applications, as you saw in Chapter 1. There are many different kinds of functions. Most of the work we do focuses on a few types: functions that are simple and yet sufficiently powerful to meet our needs. These classes of functions can be thought of as different *families of functions* since the members of each family are closely related to one another in terms of their essential properties. In this chapter, we will study the family of *linear functions,* the family of *exponential functions,* the family of *power functions,* as well as several other useful families. In later chapters, we will consider other families of functions including *polynomial functions* and *trigonometric functions.*

As we discussed in Section 1.3, we will use the letters x and y in a generic sense for the independent and dependent variables, respectively. However, in any specific context, we will try to use letters that more directly suggest the quantities under discussion.

2.2 Linear Functions

The simplest and probably most useful family of functions is the class of *linear functions.* These functions model any quantity that increases steadily or decreases steadily. The graph of such a function is always a straight line.

The most elementary type of linear function is of the form $y = mx$, which can be read as *y is proportional to x.* For example, suppose you go to the deli counter to buy some roast beef that is selling at \$5.99 per pound. If you purchase 1 pound, the roast beef costs $C = 5.99 \times 1 = \$5.99$; if you

---

[1]This chapter owes much to the spirit of Chapter 1 of *Calculus* by Hughes-Hallett, Gleason, et al. (New York: Wiley, 1994). Many of the examples and exercises included here are based on similar ones in that book, of which Sheldon Gordon is also a co-author.

purchase 2 pounds, it costs $C = 5.99 \times 2 = \$11.98$. In general, if you buy R pounds of roast beef, the cost is $C = 5.99R$, and so the cost C of the roast beef is proportional to the number of pounds R that you buy. The multiple 5.99 is the *constant of proportionality*. We also say that the cost of the roast beef is a linear function of the number of pounds of roast beef purchased.

Similarly, the distance D a car travels at a constant speed of 50 miles per hour is proportional to the number of hours t driven, so $D = 50t$. Here's another example: It is reasonable to assume that the quantity Q of garbage produced in a city is proportional to the number of people P living there, so that $Q = kP$, for some constant multiple k. Consequently, both of these are instances of linear functions.

Let's look at another example. The following table of values shows some entries from a recent IRS Income Tax Table.

Income (\$)	3000	4000	5000	6000	7000	8000	9000	10,000	11,000	12,000
Tax due (\$)	450	600	750	900	1050	1200	1350	1500	1650	1800

When you inspect the entries, you should observe that every \$1000 increase in taxable income carries with it an additional \$150 in taxes. What does this mean mathematically? Clearly, the tax you owe is proportional to your taxable income. The fact that each \$1000 earned costs you \$150 in taxes means that your effective tax rate is $\frac{150}{1000} = 0.15$, or 15%. That is,

$$\text{Tax due} = 15\% \text{ of income}$$

or
$$T = 0.15I.$$

The relationship between taxes and income is a linear function with a constant of proportionality equal to 0.15. What would be the tax due if your taxable income were \$16,000?

The Graph of a Linear Function

The graph of any linear function of the form $y = mx$ is a straight line that passes through the origin, as shown in Figure 2.1. What distinguishes one line from another is the constant m, which represents how quickly y changes as x changes. A large value for m means that y changes by a large amount for a fixed change in the variable x. A small m means that y changes relatively little for a fixed change in x. A negative value for m means that y gets smaller as x gets larger. The quantity

$$m = \frac{\text{Change in } y}{\text{Change in } x} = \frac{\Delta y}{\Delta x} = \frac{\text{rise}}{\text{run}}$$

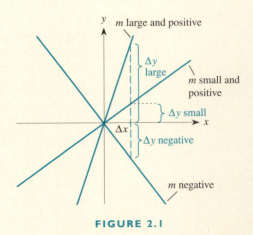

FIGURE 2.1

is called the **slope** of the line. More generally, the slope of a line is

$$m = \frac{\text{Change in dependent variable}}{\text{Change in independent variable}},$$

since the letters used for the independent and dependent variables will reflect what those quantities are and so will often be different from x and y. For instance, based on the data in the tax table, where the independent variable I is income and the dependent variable T is the tax due, the slope of the line is

$$m = \frac{\Delta T}{\Delta I} = \frac{600 - 450}{4000 - 3000} = \frac{150}{1000} = 0.15.$$

Next let's consider lines that do not pass through the origin. The equation of any such line is of the form

$$y = mx + b,$$

where

m is the **slope** of the line   and

b is the **vertical intercept,** or the value of y when x is zero. The vertical intercept is sometimes called the **y-intercept.**

The special case when $b = 0$ corresponds to the simplest form of a line $y = mx$ which passes through the origin as discussed previously.

The graphs associated with the equations $y = f(x) = 2x - 1$, $y = g(x) = 2x + 1$, and $y = h(x) = 2x + 2$ are shown in Figure 2.2. Notice that the three lines are parallel because they all have the same slope, $m = 2$, but their vertical intercepts are different.

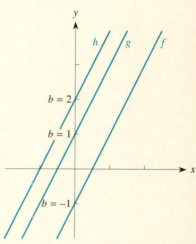

FIGURE 2.2

The graphs associated with the equations $y = f(x) = 2x + 1$, $y = g(x) = 3x + 1$, and $y = h(x) = -2x + 1$ are shown in Figure 2.3. Notice that because all three lines cross the y-axis at the point $y = 1$, they have the same y-intercept. However, the three lines have different slopes and so they behave differently. The functions f and g are growing as you read from left to right (as x increases) while the function h is decaying as x increases. We therefore again see that the slope of the line determines whether a line rises or falls and how rapidly it does so. When m is positive, the line rises as x increases from left to right; when m is negative, the line falls as x increases from left to right. Thus, when the slope is positive, the linear function is increasing. When the slope is negative, the linear function is decreasing. (However, because a line is straight and doesn't bend, either up or down, it is neither concave up nor concave down.)

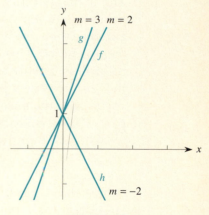

FIGURE 2.3

Furthermore, the more positive the slope is, the faster the line rises. For example, the line whose equation is $y = 1.5x + 1$ passes through the same point $b = 1$ on the y-axis, as do $y = 2x + 1$ and $y = 3x + 1$. All three lines are increasing, but $y = 1.5x + 1$ slopes upward more gradually than the other two lines. The line whose equation is $y = -2.5x + 1$ also passes through the same point $b = 1$ on the y-axis, decreasing as x increases, but slopes downward more sharply than the line $y = -2x + 1$.

To summarize, the **slope** of a line is given by the following definition:

$$m = \frac{\text{change in } y}{\text{change in } x} = \frac{\Delta y}{\Delta x} = \frac{\text{rise}}{\text{run}}$$

If $P(x_1, y_1)$ and $Q(x_2, y_2)$ are two points on the line, then

$$\Delta x = x_2 - x_1 \quad \text{and} \quad \Delta y = y_2 - y_1,$$

where the order of the coordinates must be the same in both Δx and Δy. Therefore, the slope definition can be written as

$$m = \frac{\Delta y}{\Delta x} = \frac{y_2 - y_1}{x_2 - x_1}$$

Now let's see what these ideas really mean.

E X A M P L E   I

The manager of a telemarketing firm needs to analyze the weekly cost of the firm's operation as a function of the number of telephone operators used. Suppose the weekly cost of running the facilities (rent, utilities, etc.) and equipment is $1500. This amount is called the *fixed cost* because the company pays it regardless of how many operators are used. The manager also must consider the cost of wages paid to the telephone operators. Because the cost of wages varies with the number of operators, it is called a *variable cost*. If each operator is paid $300 weekly, then the variable cost for n operators is $300n$. The total weekly cost, C, of running this business with n operators is

$$C = f(n) = 1500 + 300n.$$

Notice that $C = f(n)$ is an *increasing function* of n because, the larger n is, the larger C is. You can see this clearly in the following table. Observe that every time n goes up by one, C increases by 300.

n	0	1	2	3	4	5	6
C	1500	1800	2100	2400	2700	3000	3300

The 300 represents the variable cost; that is, every additional operator hired costs the company an extra $300. We therefore say that the *rate* at which the cost is increasing is $300 per operator. This rate is also the *slope* of the line $f(n) = 1500 + 300n$ In Figure 2.4 you can picture the slope as the ratio

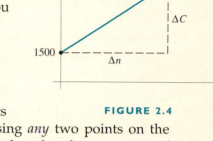

$$m = \frac{\Delta C}{\Delta n} = \frac{\text{rise}}{\text{run}} = \frac{300}{1},$$

or simply $m = \$300$ per operator.

The value for $m = 300$ was based on the two points $(0, 1500)$ and $(1, 1800)$. If you calculate the slope using *any* two points on the line, you will get the same value. It is this fact—that the slope, or rate of change, is the same at every point—that makes a line straight. If the rate of change varies from one point to another, then the function is not linear.

FIGURE 2.4

What is the significance of the 1500? For $C = f(n) = 1500 + 300n$, the intercept on the vertical axis is 1500 because

$$C = f(0) = 1500 + 300(0) = 1500.$$

Because 1500 is the value of C when $n = 0$, it represents only the expenditures for facilities and equipment, which is the fixed cost.

The form for the equation of the line that we found above,

$$C = 300n + 1500,$$

is known as the **slope-intercept form** because it highlights the *slope m* of the line and the *vertical intercept b*. The slope-intercept form is very useful for *displaying* the equation of a line. However, it is almost always a poor choice for *determining* the equation of a line because the vertical intercept is often difficult to find. Even when we do find the vertical intercept, it may have little to do with the quantities we are studying.

A much better method for determining an equation of a line is to use an equivalent form known as the **point-slope form.** This form is based on the idea that a line is determined once we know its slope m and one point $P(x_0, y_0)$ on it. Suppose $Q(x, y)$ is any other point on the line, as shown in Figure 2.5. Since

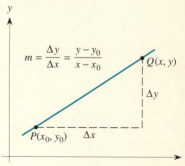

$$m = \frac{\Delta y}{\Delta x} = \frac{y - y_0}{x - x_0},$$

we can multiply both sides by $(x - x_0)$ to get

$$y - y_0 = m(x - x_0),$$

for an equation of the line with slope m that passes through the known point $P(x_0, y_0)$.

FIGURE 2.5

You are almost always better off using the point-slope form. To illustrate how it works, let's consider the following example.

E X A M P L E   2

Consider our old friend the snow tree cricket from Example 2 in Section 1.4. We have the following set of measurements relating its rate of chirping, in chirps per minute, to the temperature, in Fahrenheit:

Temperature, T (°F)	50	55	60	65	70	75	80
Rate, R (chirps per minute)	40	60	80	100	120	140	160

As you can see from the table, the chirp-rate is increasing steadily; it goes up 20 chirps per minute for every 5°F increase in temperature, so the chirp-rate is an increasing function of the temperature T. Equivalently, we can say that the chirp-rate goes up 4 chirps per minute for every 1°F increase in temperature. This fact is reinforced when we look at the corresponding points in Figure 2.6, and we also see that the points clearly fall into a linear pattern. The function giving the chirp-rate R as a function of temperature T is a linear function.

How do we find the equation for this line? Any two points determine a line, so we can use any two of the given points, say (55, 60) and (75, 140). Using these two points, as shown in Figure 2.7, we see that the slope of the line is

$$m = \frac{\Delta R}{\Delta T} = \frac{140 - 60}{75 - 55} = \frac{80}{20} = 4.$$

This value for the slope means that for each 1°F increase in temperature, the cricket chirps 4 more times per minute. If the temperature goes up 5°F, the cricket chirps 20 more times per minute; if it goes up 10°F, the cricket chirps 40 more times per minute, and so forth.

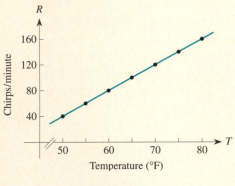

FIGURE 2.6

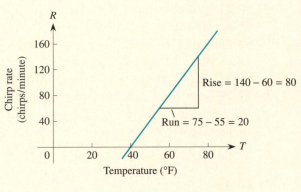

FIGURE 2.7

Next we use the point-slope formula to find the equation of the line using any point on the line. If we pick the point (55, 60) that we used earlier, then we obtain

$$R - 60 = 4(T - 55)$$

or, when simplified,

$$R = f(T) = 4T - 160.$$

This equation tells us that the vertical intercept is $R = -160$ (when $T = 0$). Realize, though, that a chirp-rate of $R = -160$ is meaningless! Does this mean that the formula is wrong? No. It just means that it makes sense to describe the snow tree cricket's chirp-rate only for temperatures between, or possibly near, the given set of readings—that is, from 50°F to 80°F. It does not make real-world sense to use this linear relationship far outside of this interval, such as at 0°. The formula does not sensibly represent chirp-rates for temperatures less than 40°F, since R becomes negative. It doesn't hold at temperatures high enough to cook the crickets either. It is unlikely for temperatures in the Colorado Rockies to be above 100°F, so there is a natural domain for this function:

Domain of f = all T values between 40°F and 100°F,   or $40 \leq T \leq 100$.

Since this function is strictly increasing, the corresponding range is

Range of f = all R values from 0 to 240,   or $0 \leq R \leq 240$.

How reasonable are these results? At 100°F, the equation predicts that a snow tree cricket will chirp 240 times per minute, or 4 times per second. If you stop and think about this, you might decide that this seems a bit unreasonable. Thus, even though the linear model does predict this value, we might want to rethink whether it makes sense to extend the linear model as far as $T = 100°F$ when the upper limit on our data values is $T = 80°F$. It is often misleading to extrapolate too far beyond the actual data values.

So far we have used the temperature to predict the chirp-rate, and we thought of the temperature as the *independent variable* and the chirp-rate as the *dependent variable*. However, we could reverse this and think of the temperature as a function of the chirp-rate. How we view a relationship determines which variable is dependent and which is independent. Thinking of temperature as a function of chirp-rate would enable us to approximate temperature for given chirp-rates. To do so, we start with the formula we had before,

$$R = f(T) = 4T - 160,$$

and solve it algebraically for T as a function of R. That is,

$$4T = R + 160$$

$$T = \frac{1}{4}(R + 160) = \frac{1}{4}R + 40 = g(R).$$

We note that this is also a function, except now the independent variable is R. So, if you ever encounter a snow tree cricket who is chirping merrily away, knowing this equation can help you determine the local temperature just by using your watch. Count the number of chirps in a one-minute interval and apply the formula to calculate the temperature.

We summarize the important information about linear functions.

A **linear function** has the form

$$y = mx + b,$$

where

m is the **slope**, or rate of change of y with x,

$$m = \frac{\Delta y}{\Delta x} = \frac{\text{rise}}{\text{run}}$$

b is the **vertical intercept**, or value of y when $x = 0$.

The **point-slope form** for the equation of a line with slope m that passes through the point $P(x_0, y_0)$ is

$$y - y_0 = m(x - x_0).$$

E X A M P L E   3

During the early years of the Indianapolis 500 race held annually on Memorial Day, the average winning speed increased, as shown in the following table. Find a formula to model these values.

Year	1919	1922	1925
Average speed (mph)	88	94.5	101

Solution   Because the average winning speed S increased consistently by 6.5 mph every three years, we see that S is a linear function of time over the period 1919 to 1925. The average winning speed starts at 88 mph and increases at the rate of one-third of 6.5, or 2.17 mph, each year. Therefore, if t is the number of years since 1919, we have the linear model

$$S = f(t) = 88 + 2.17t.$$

The slope of this line is 2.17; it shows the rate at which the winning speed increases each year. You can visualize the slope in Figure 2.8 as a ratio:

$$\text{Slope} = \frac{\text{rise}}{\text{run}} = \frac{\Delta S}{\Delta t} = \frac{6.5}{3} = 2.17.$$

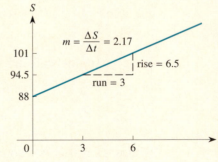

FIGURE 2.8

As we said before, if you calculate this ratio using any two points on the line, you will obtain the identical value for the slope. Since $S = f(t)$ increases as t increases, we see that f is an increasing function.

What is the significance of the number 88? This number represents the initial winning speed in 1919, when $t = 0$. Geometrically, 88 is the vertical intercept.

You may wonder whether this linear trend continues beyond 1925. Let's compare what it predicts with what actually happened. The fastest average winning speed in the Indy 500 was 186 mph in 1990. This winning speed corresponds to $t = 71$ years (after 1919). Using the linear equation $S = 88 + 2.17t$, we obtain a prediction of 242 mph in 1990. Clearly, although speeds have increased dramatically, they have not quite kept up with the linear function we constructed based on just a few carefully chosen data points. Further, this model again illustrates the danger in extrapolating too far from the given data.

Think About This

What does this information tell us about the length of the race? How much longer did it take the winning car to drive the 500 miles in 1919 than it took in 1990?

Because the data in the table is given only at specific points (every three years), we say that the data is *discrete*. However, because the function $y = 88 + 2.17t$ makes sense for *all* possible values of t, we treat the variable t as though it were *continuous* (or defined for all points). The graph in Figure 2.8 is of the continuous function because it is a solid line rather than three separate points representing the winning speeds in the race in three particular years.

EXAMPLE 4

Search and Rescue teams[2] are often called upon to find lost hikers in remote areas in the Southwest. Members of the search team walk through the search area parallel to each other at a fixed distance d between searchers. Experience has shown that the team's chance of finding lost persons is related to the distance of separation, d. The closer together the searchers are, the better the chances of success. Based on a number of simulated missions, the percentage of lost persons found was used to assess the probability of finding someone based on various separation distances, as shown in the following table. Find a formula to model this.

Distance d (feet)	20	40	60	80	100
Probability of success $P(\%)$	90	80	70	60	50

(These values correspond to searches conducted in the relatively open terrain of the Southwest; searchers in other regions where there is dense forest or undergrowth would have to use much narrower separation distances to achieve comparable levels of success.)

---

[2] This example is adapted from *Calculus* by Hughes-Hallett, Gleason, et al. (New York: Wiley, 1994).

Solution   Since the value for the probability of success P decreases as distance d (the independent variable) increases, we see that the function $P = f(d)$ is a decreasing function of d. From the data, we can also see that each 20-foot increase in distance causes the probability of success P to decrease by 10. Since this fact is true for any successive pair of points, we see that P is a linear function of d, so the graph of the probability of success versus distance is a straight line. See Figure 2.9. Further, notice that based on the first two data points, the slope of this line is

$$m = \frac{80 - 90}{40 - 20} = \frac{-10}{20} = -\frac{1}{2}.$$

The negative sign reinforces the fact that P decreases as d increases. The slope is the rate at which P is decreasing as d increases.

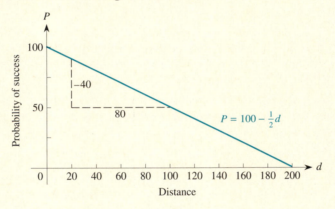

FIGURE 2.9

To find the equation of the line, we use the point-slope formula. We choose any one of the given points, say $(20, 90)$, and obtain

$$P - 90 = -\frac{1}{2}(d - 20).$$

If we simplify this expression, we get

$$P = f(d) = 100 - \frac{1}{2}d.$$

What is the meaning of the vertical intercept, $P = 100$? Suppose that $d = 0$, so the searchers are walking shoulder to shoulder; we would expect everyone to be found, or $P = 100$. What is the horizontal intercept? When $P = 0$, we have $0 = 100 - \frac{1}{2}d$, or $d = 200$. According to the model, the value $d = 200$ represents the separation distance at which no one is found. This outcome is unreasonable, because even when the searchers are far apart, the search will sometimes be successful. What this situation suggests is that, somewhere outside the data given, the linear relationship ceases to hold. As in the Indy 500 example, extrapolating too far beyond the given data may not make sense.

How Do We Know a Set of Data Is Linear?   Being able to tell that there is a linear relationship between the two variables by looking at a table of values is often important. We could plot the data points to see if the points fall into a linear pattern, but this approach is very imprecise. Alternatively, we can decide if a function $y = f(x)$ given by a table of values is linear by examining the data. If the data fall into a linear pattern, then we should obtain the same slope no matter which pair of points we use. This reasoning gives a very simple criterion for determining linearity: See whether the differences in y-values are constant for equally spaced x-values. If the x-values are uniformly spaced and there is a constant difference among the y-values then the data fall into a linear pattern.

E X A M P L E   5

The following two sets of data represent values for a linear function and a non-linear function. Identify which is the linear function and find the equation of the line.

x	$f(x)$		x	$g(x)$
1.0	7.0		1	2
1.2	7.8		2	3
1.4	8.6		3	6
1.6	9.4		4	11
1.8	10.2		5	18
2.0	11.0		6	27
2.2	11.8		7	38

Solution   Notice that in both sets of values, the x-values are evenly spaced, so we can proceed by examining the differences $\Delta f(x)$ and $\Delta g(x)$ in the values of the functions. For the values of function f, we find the following:

x	$f(x)$	$\Delta f(x)$
1.0	7.0	$0.8 = 7.8 - 7.0$
1.2	7.8	0.8
1.4	8.6	0.8
1.6	9.4	0.8
1.8	10.2	0.8
2.0	11.0	0.8
2.2	11.8	

Because there is a constant difference of 0.8 between the values of the function f, we conclude that this set of data is indeed linear and so the slope of the line through these points is

$$m = \frac{\Delta y}{\Delta x} = \frac{\Delta f(x)}{\Delta x} = \frac{0.8}{0.2} = 4.$$

Further, using the first point $P(1, 7)$ and the point-slope form for the equation of a line, we find that the equation of the line is

$$y - 7 = 4(x - 1).$$

When we simplify this expression and solve for y, we get

$$y = 4x + 3 = f(x).$$

Either expression represents the equation of the linear function. Verify both equations by substituting in values from the table.

Suppose we try the same analysis on the values for the function g. We find

x	$g(x)$	$\Delta g(x)$
1	2	
		1
2	3	
		3
3	6	
		5
4	11	
		7
5	18	
		9
6	27	
		11
7	38	

Since the differences are not constant, we conclude that these points do not fall into a linear pattern, and hence there is no line that passes through them. Consequently, we see that the function g *cannot* be a linear function. In fact, since the differences are successively larger, we conclude that the function is growing faster than a linear function grows. You may want to graph the points to see that this increasing function is, in fact, concave upward.

So far, we have been given information on some process or quantity that clearly is a linear function. In practice, however, we often face data that appear to be linear in nature, but the particular data points known do not precisely fall onto a line. We illustrate such a situation in Example 6.

E X A M P L E   6

The following table gives some measurements for the rate of chirping (in chirps per minute) of the striped ground cricket as a function of the temperature.

$T(°F)$	89	72	93	84	81	75	70	82	69	83	80	83	81	84	76
Chirps	78	60	79	73	68	62	59	68	61	65	60	69	64	68	57

Even though the measurements for the snow tree cricket from Chapter 1 fell exactly onto a straight line, we see in Figure 2.10 that the comparable measurements for the striped ground cricket clearly do not. The difference between the two species may be due to errors in measurement; it may be that the striped ground cricket is less sensitive to temperature; or perhaps the snow tree cricket has more mathematical aptitude to get the situation right. Even though the points for the striped ground cricket do not fall precisely on a line, they do fall into a *linear pattern*. Let's find an equation that represents this linear pattern.

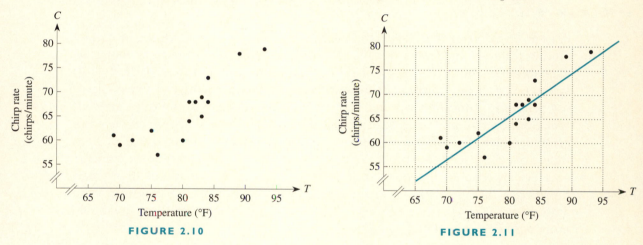

FIGURE 2.10   FIGURE 2.11

Suppose you take a piece of black thread, hold it taut, and move it around over the points in Figure 2.10. Each possible orientation for the thread represents a different line. It is possible to select an orientation that seems, by eye, to give the best match or *fit* to the linear pattern in the data. Such a line is superimposed over the data points in Figure 2.11. (Obviously, different people will come up with slightly different lines.) We now estimate the equation of this line that captures the overall trend of the chirp-rate function. From Figure 2.11, we see that this line passes through the point $(85, 70)$ and passes approximately through $(80, 65.5)$. Therefore the slope of the line is approximately

$$m = \frac{70 - 65.5}{85 - 80} = \frac{4.5}{5} = 0.9.$$

This means that the chirp-rate increases about 0.9 chirp per minute for each 1°F increase in temperature. Further, since the line apparently passes through the point $P(85, 70)$, we conclude that the equation of the line is

$$C - 70 = 0.9(T - 85),$$

or, when simplified,

$$C = 0.9T - 6.5.$$

Note that this result is just an *estimate* for the equation of the line that visually best fits the linear trend in the data. In Chapter 3, we will introduce methods for finding the equation of the one line that is the *best fit* to a set of data.

Implicit Linear Functions

We now consider a somewhat different type of example, in which two quantities are related to each other, but in such a way that it is not necessarily obvious which variable is independent and which one is dependent.

E X A M P L E   7

An ongoing debate at all levels of government concerns the allocation of money among different programs. Because typically only a fixed amount of money is available, the more that is spent on one program, the less money there is to spend on other programs. Let's look at a simple case involving just two competing programs, funding road and highway repairs versus funding day-care centers. Suppose we have a total of $100,000 available to divide between day-care centers, which cost $20,000 per center, and road repaving, which costs $5000 per mile. Let c represent the number of day-care centers and r represent the number of miles of road to be repaved. Then the amount of money spent on road repaving is $5000r$ dollars (because it costs $5000 to repave each mile), and the amount spent on day-care centers is $20,000c$ dollars. Assuming all the available money is spent, we get

$$\text{Amount spent on centers} + \text{Amount spent on repaving} = \$100,000$$
$$20,000c \qquad + \qquad 5000r \qquad = \$100,000$$
$$r + 4c = 20.$$

This equation is called the *budget constraint*. To graph this equation, we first find the points at which the graph crosses the axes, as shown in Figure 2.12. If $c = 0$, then $r + 4(0) = 20$ so that $r = 20$. At the other extreme, if $r = 0$, then we have $0 + 4c = 20$, or $c = 5$.

Since all the money that is not spent on road work is used for day-care centers, the number of centers funded is a function of the number of miles of roads repaved. That is, c is a function of r, and we can solve the budget constraint equation $r + 4c = 20$ for c to get

$$c = f(r) = \frac{20 - r}{4} = 5 - \frac{1}{4}r.$$

Similarly, the number of miles repaved is a function of the number of centers funded, so r is a function of c. We can

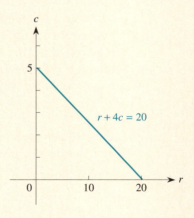

FIGURE 2.12

solve the budget constraint equation for r to get

$$r = g(c) = 20 - 4c.$$

Any actual situation typically determines the applicable domain and range. In the situation in Example 7, r makes sense only for values between 0 and 20, whereas c makes sense only for values between 0 and 5. Which of these is the domain and which is the range depends on which variable we think of as the independent variable and which is the dependent variable.

Note that the budget constraint equation

$$r + 4c = 20$$

is called an *implicit function*, since neither quantity is given explicitly in terms of the other.

Some Useful Facts

Several facts about lines are useful to remember:

1. *Parallel lines* have the same slope. This means that the quantities they represent are growing at the same rate. For example, the lines $y = 4x + 3$, $y = 4x - 15$, and $y - 4x = 11$ are all parallel. What is their common slope?

2. *Perpendicular lines* have slopes that are negative reciprocals. For example, the lines $y = 2x - 9$ and $y = -\frac{1}{2}x + 3$, having slopes of 2 and $-\frac{1}{2}$, respectively, are perpendicular to each other. Sketch their graphs to convince yourself of this fact. Similarly, the lines $y = 0.162x + 7.4$ and $y = -6.173x + 1.03$, which have slopes of 0.162 and $-6.173 = -\frac{1}{0.162}$, respectively, are perpendicular to each other. Write the equation of a line that is perpendicular to $y = \frac{5}{4}x - 7$. (Of course, your answer will likely be different from your classmates' choices.)

3. The point where any two lines cross is known as their *point of intersection*. The x- and y-coordinates of this point must satisfy both equations simultaneously. You find the point of intersection by solving the system of simultaneous equations either algebraically or graphically.

EXERCISES

1. Match each of the following equations with its graph. (Note that the scales of the graphs are different.)

   a. $y = x + 2$ **b.** $y = x - 3$ **c.** $y = -2x + 4$

   d. $y = -3x - 4$ **e.** $y = \frac{1}{2}x$ **f.** $y = 3$

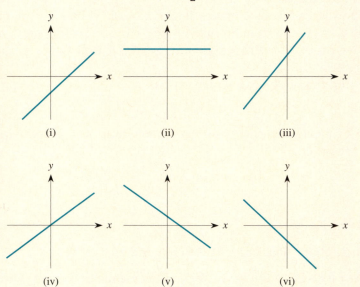

2. Estimate the slope of each line. Then use the slope to find an equation of the line.

   a.
   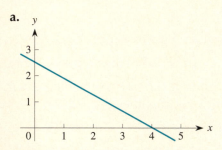

   b.
   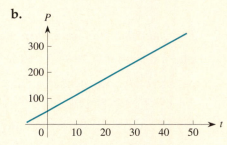

3. Find the equation of the line passing through points A and B.

   a. $A(1, -2)$, $B(2, 5)$ **b.** $A(1, -2)$, $B(3, -2)$

   c. $A(3.52, 4.96)$, $B(-1.91, 8.36)$

4. The data in each table lie along a line. For each set of data, carefully plot the points on graph paper, estimate by eye the slope and vertical intercept, and use these values to approximate the equation of the line. Then find the equation of the line algebraically. How close was your estimate?

   a.

x	1	2	3	4
y	1.81	3.34	4.87	6.40

   b.

x	1	2	3	4
y	1.08	0.69	0.30	-0.09

5. Find the equation of a linear function that fits the following set of values.

x	4	5	6	7
y	1.557	1.614	1.671	1.728

6. Inspector Clueless, while investigating the murder of Mr. Jones, has found the murderer's footprint from a size $11\frac{1}{2}$ shoe in a flower bed. The inspector mutters something about the killer being "a man who is %#$#$&&# tall." If the equation of the best-fit line relating shoe size to height in inches is $S = 0.51H - 25.2$, decipher Clueless's muttering.

7. The graph of Fahrenheit temperature F versus Celsius temperature C is a line. We know that water boils at 212°F and 100°C and freezes at 32°F and 0°C.

   a. Sketch the graph of the line.
   b. Find the slope of the line relating the two temperature scales.
   c. Find the equation of this line.
   d. Use the equation to find the Fahrenheit temperature that corresponds to 30°C.
   e. Use the equation to find the Celsius temperature that corresponds to 98.6°F.
   f. When is the Fahrenheit temperature the same numerical value as the Celsius temperature?

8. In 1986, 1048 women and 12,099 men died of AIDS. In 1992, 6312 women and 39,160 men died of AIDS. Assuming (incorrectly) that the pattern of growth in AIDS for both groups is linear,

   a. Find the equation for the number of AIDS-related deaths among men as a linear function of time.
   b. Find the equation for the number of AIDS-related deaths among women as a linear function of time.
   c. What is the practical significance of the slope in part (a) and part (b)?
   d. If the trends in AIDS-related deaths for men and women were indeed linear and continued indefinitely, explain why there will never be a time when the same number of men and women will die from AIDS.
   e. Use the fact that 4883 women and 36,770 men died of AIDS in 1990 to explain why it is incorrect to assume that the growth trends are linear.

9. A disk jockey (DJ) charges a flat fee of $120 per party plus $60 for each hour of the party. A second DJ charges $100 per party plus $75 for each hour.

   a. For each DJ find a formula that gives the cost of hiring the DJ as a function of the number of hours the party lasts.
   b. Sketch the graphs of both functions on the same set of axes.
   c. How should you decide which DJ costs less?

10. The net income of the Apex Company was $240 million in 1980 and has been increasing $30 million per year ever since. Over the same period, the net income of its chief competitor, the Best Corporation, has been growing $20 million per year starting with $300 million in 1980. Which company earned more in 1990? When did Apex surpass Best?

11. According to the IRS, the formula $T = 0.15I$, which gives income tax as a function of taxable income, applies only for single taxpayers with taxable incomes up to $21,450. The IRS tax table states: "If the taxable income is over $21,450, But not over $51,900, Enter on Form 1040: $3,217.50 + 28% of the amount over 21,450."

     a. Rewrite this statement as an equation that can be used to calculate your taxes. What are the domain and range of the resulting function?

     b. What is the practical meaning of the value you get for the slope? In particular, compare its significance to the slope of 0.15 we found for lower income levels.

     c. Sketch a single graph showing both tax formulas. Is there any discrepancy?

12. When filing income tax returns, many people can claim deductions for depreciation on items such as cars and computers used for business purposes. The idea is that the value of such an asset decreases, or depreciates, over time. The simplest method used to find the depreciated value is called *straight-line depreciation*, which assumes that the item's value decreases as a linear function of time. If an $1800 computer system depreciates completely in five years, find a formula for its value as a function of time. What is it worth after three years?

13. The following table gives data on the IQ and grade point averages of 10 students.

IQ	100	120	110	105	85	95	130	100	105	90
GPA	3.0	3.8	3.1	2.9	2.6	2.9	3.6	2.8	3.1	2.4

     a. Plot these points carefully on a sheet of graph paper and use the black thread method described in Example 6 to locate and draw the line that seems to best fit the data points.

     b. Estimate the equation of this line.

     c. Using your answer to part (b), what is your best estimate of the GPA for a student who has an IQ of 80? An IQ of 112?

     d. What is your best estimate of the IQ of a student whose GPA is 3.3?

14. The following table gives data relating a car's gas mileage to its weight.

Weight (lb)	2100	2200	2400	2500	2800	3000	3200
Mileage (mpg)	37	34	29	27	26	25	23

     a. Plot these points carefully on a sheet of graph paper and use the black thread method to locate and draw the line that seems to best fit the data points.

     b. Estimate the equation of this line.

     c. Using your answer to part (b), what is your best estimate of a car's gas mileage if it weighs 2350 pounds? 3100 pounds? 1950 pounds?

     d. What is your best estimate of the weight of a car that gets 32 mpg?

15. A student who works as a waiter in a restaurant records the cost C of meals and the tip T left by couples. His data recorded one evening are

Cost ($)	18.55	21.04	22.76	23.38	26.10	28.54
Tip ($)	2.75	3.00	3.50	3.75	4.00	4.50

    a. Plot these points on a sheet of graph paper and draw the best line you can to fit the points. Explain your choices of the independent and the dependent variable.

    b. Suppose the equation for this function is $T = 0.18C - 0.60$. In terms of this mathematical model, what is the increment in the tip for each $1 increment in the cost of the meal?

    c. What does the slope of the line in part (b) represent? What significance does the vertical intercept have?

    d. Suggest possible values for the domain and range of this function.

16. You have a fixed budget of $30 to spend on nuts and Gummi Bear™ candy for a party. The nuts cost $3 per pound, and the candy costs $2 per pound.

    a. Write an equation expressing the relationship between the number of pounds of nuts and of Gummi Bears that you can buy if you completely spend your budget. This equation is your budget constraint.

    b. Graph the budget constraint, assuming that you can buy any fractional amount of a pound. Label the intercepts.

    c. What are the domain and range for this function?

    d. Suppose your roommate chips in an additional $30 for the party. Graph the new budget constraint on the same set of axes used for the budget constraint graphed in part (b).

    e. Keeping the original budget at $30, suppose the Gummi Bears go on sale for half the price. Sketch the new budget constraint on the same axes used in part (d).

    f. Keeping the original budget at $30, suppose the price of nuts suddenly doubles because of a frost in the Southeast. Sketch the new budget constraint on the same axes used in part (d).

17. Jen is typing her term paper for Psych 101. She types the body of the paper at the rate of 35 words per minute for 30 minutes, then takes a 5-minute break, and comes back to do the references at a rate of 20 words per minute for 12 minutes.

    a. Sketch the graph of Jen's typing rate as a function of time.

    b. Sketch the graph of the total number of words she types as a function of time.

    c. Find the equations of the different line segments you drew in part (b).

18. A bicyclist pedals at the rate of 1000 feet per minute for 20 minutes, then slows to 500 feet per minute for 6 minutes, then races at 1200 feet per minute for 4 minutes, and cools down at 500 feet per minute for 5 minutes.

    a. Sketch the graph of the bicyclist's rate as a function of time.

    b. Use the graph from part (a) to determine the total distance biked.

    c. Sketch the graph of the distance traveled as a function of time.

    d. Find the equations of the different line segments you drew in part (c).

19. Find the equation of the line that passes through the point $(6, 4)$ and is

    a. parallel to the line $y = 5x - 3$.

    b. perpendicular to this line.

20. Find the equation of the line that passes through the point $(6, 4)$ and also passes through the point of intersection of $y = -2x + 1$ and $y = 3x + 6$.

21. An equation of a line is $4x + 3y = 24$. Find the length of the portion of this line that lies in the first quadrant. What is the area of the triangle formed by the line and the two axes?

22. Which of the following functions are strictly increasing, strictly decreasing, or neither?

    a. The cost of first-class postage on January first of each year.
    b. The time of sunrise associated with each day of the year.
    c. The high temperature associated with each day of the year.
    d. The closing price of one share of IBM stock for each trading day on the stock exchange.
    e. The area of an equilateral triangle in terms of its base b.
    f. The height of a bungee jumper t seconds after leaping off a bridge.
    g. The height of liquid in a 55-gallon tank h hours after a leak develops.
    h. The daily cost of heating a home as a function of the day's average temperature.

2.3 Exponential Functions

The population of Florida was 9.75 million in 1980 and has been growing steadily ever since, as you can see from the following table of values. How is this population growing? If the population is growing in a linear pattern, then ΔP, the changes or increases in population from one year to the next, would all be the same. Let's check these differences in the following table.

Year	Population (millions)	ΔP
1980	9.75	
		0.28
1981	10.03	
		0.29
1982	10.32	
		0.30
1983	10.62	
		0.31
1984	10.93	
		0.32
1985	11.25	
		0.32
1986	11.57	
		0.34
1987	11.91	

We see that the successive differences themselves are increasing. This pattern makes sense because as the population grows, there are more people around to have babies. Consequently, the population of Florida has been growing at a rate that is faster than a linear rate. Let's see if we can find out what type of growth pattern is involved here.

Suppose we divide the population in any year by the population in the previous year. This quotient gives us the following approximate values.

$$\frac{\text{Population in 1981}}{\text{Population in 1980}} = \frac{10.03 \text{ million}}{9.75 \text{ million}} \approx 1.029$$

$$\frac{\text{Population in 1982}}{\text{Population in 1981}} = \frac{10.32 \text{ million}}{10.03 \text{ million}} \approx 1.029$$

$$\frac{\text{Population in 1983}}{\text{Population in 1982}} = \frac{10.62 \text{ million}}{10.32 \text{ million}} \approx 1.029.$$

Thus, we see that the population of Florida in any given year is 1.029 times the population in the previous year. Equivalently, we say that between 1980 and 1983 the population grew by about 2.9% from one year to the next. If you check the population figures for the subsequent years through 1987, you will find that each year the population grew by the same factor of about 1.029, or 2.9%. Whenever the growth factor is constant (here it is 1.029), we have *exponential growth*.

Let's find an equation for this function. If t is the number of years since 1980,

when $t = 0$, population $= 9.75 = 9.75(1.029)^0$

when $t = 1$, population $= 10.03 = 9.75(1.029)^1$

when $t = 2$, population $= 10.32 = 10.03(1.029) = 9.75(1.029)^2$

when $t = 3$, population $= 10.62 = 10.32(1.029) = 9.75(1.029)^3$

and so on. More generally, after t years, the population of Florida is

$$P(t) = 9.75(1.029)^t.$$

This equation is called an *exponential function* with *base* 1.029. The name *exponential* is used because the variable (in this case, t) occurs in the exponent. The base (in this case, 1.029) is the *growth factor* by which the population increases from one year to the next.

Assuming that this relationship continues for the next 80 years, we can graph this population function as shown in Figure 2.13. The function obviously is increasing. Moreover, the graph grows faster and faster as time goes on, and so the curve bends upward. Thus, the graph of the exponential function is concave up (see Section 1.2). This behavior is typical of an exponential growth function. Compare this function's behavior with that of an increasing linear function. Because a linear function grows at the same rate everywhere, its graph is a straight line. However, exponential functions, such as this one, that climb slowly at first eventually climb extremely rapidly. This type of behavior explains why there is widespread concern about the exponential growth of the world's population.

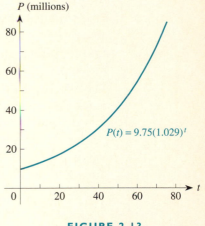

P (millions)

$P(t) = 9.75(1.029)^t$

FIGURE 2.13

The graph shown in Figure 2.13 is only an approximation to the actual graph of Florida's population. Since we can't have a fraction of a person, the graph theoretically should be jagged with small steps up or down each time someone is born, dies, or moves to or from Florida. However, on the scale we used for Figure 2.13, such changes would be insignificant with so large a population. So our smooth curve actually is a good approximation to the population.

E X A M P L E   I

Estimate the population of Florida in the year

a. 2004, when $t = 24$ **b.** 2028, when $t = 48$ **c.** 2052, when $t = 72$

Solution Extrapolating into the future is based on our assumption that the population continues to grow exponentially at the same rate of 2.9% per year. The further we project into the future, the riskier our prediction becomes because other factors can affect the growth rate. (Can you think of any?) Nevertheless, we will use the model we found to predict the following values.

a. $P(24) = 9.75(1.029)^{24} = 19.36 \approx 2 \times 9.75$ million.

b. $P(48) = 9.75(1.029)^{48} = 38.45 \approx 4 \times 9.75$ million.

c. $P(72) = 9.75(1.029)^{72} = 76.37 \approx 8 \times 9.75$ million.

Let's look at what the population values we found tell us. After 24 years, the population has doubled. After roughly another 24 years (that is, $t = 48$ years), it has doubled again. After roughly another 24 years ($t = 72$), the population has doubled once again. Therefore, we say that the *doubling time* of Florida's population is about 24 years: If you take the population in any given year and compare it to the population 24 years later, you would find that it has doubled.

Every population that grows exponentially has a fixed doubling time. The world's population currently has a doubling time of about 38 years. Since the current population is about 5.5 billion, there will be about 11 billion people in 38 years and roughly 22 billion people in 76 years, all competing for an ever diminishing amount of resources. As another way of looking at it, if you live to be 76, the world's population will quadruple during your lifetime.

A Decaying Exponential Function

Our next example demonstrates a quantity that *decreases exponentially*.

E X A M P L E   2

The strength of any signal in a fiber-optic cable, such as the type used for telephone and other communication lines, diminishes 15% for every 10 miles. Find an expression for the strength of a signal remaining after a given number of miles. (a) How much of the signal is left after 100 miles? (b) How far does a signal go until its strength is down to 1% of the original level?

Solution   a. If the signal diminishes by 15% every 10 miles, then after each ten-mile stretch, only 85% of the original signal strength remains. Let S_0 be the initial strength of some signal in a fiber-optic cable and let $S(n)$ be the strength of the signal remaining after n ten-mile lengths. Therefore, after the first ten-mile length ($n = 1$), 85% of S_0 is left. Similarly, after the second ten-mile length of cable ($n = 2$), 85% of the signal strength remaining after the first ten-mile length is left. That is,

$$S(2) = 85\% \text{ of } S(1).$$

Continuing this pattern, we get

$$S(0) = S_0$$
$$S(1) = (0.85)S_0$$
$$S(2) = (0.85)S(1) = (0.85)(0.85)S_0 = (0.85)^2 S_0$$
$$S(3) = (0.85)S(2) = (0.85)(0.85)^2 S_0 = (0.85)^3 S_0.$$

After n ten-mile lengths of a cable,

$$S(n) = S_0 (0.85)^n.$$

Thus, after 100 miles, when $n = 10$ ten-mile lengths, the amount of signal strength remaining is

$$S(10) = S_0 (0.85)^{10} = S_0 (0.1969),$$

so just under 20% of the original signal strength is left.

b. To find out how far it takes until only 1% of the signal strength is left, we must find when the strength remaining is $0.01 S_0$, or

$$S(n) = S_0 (0.85)^n = 0.01 S_0.$$

If we divide both sides of this equation by the initial signal strength S_0, we get

$$(0.85)^n = 0.01.$$

If you use your calculator to solve this equation by trial and error, you will find that $n \approx 28$. Therefore, the signal deteriorates by 99% after about 28 ten-mile lengths, or about 280 miles.

In practice, this model means that such fiber-optic signals may need to be boosted if they are to go any great distance. For instance, if a booster station can clearly detect a signal at 1% of its original strength level, then such stations would have to be located every 280 miles. Suppose that the equipment used can clearly detect a signal at 0.1% of its original level. How far apart would the booster stations have to be?

For the signal strength model to make sense, n must be nonnegative. However, in general, the *exponential decay function* $y = S(x) = S_0(0.85)^x$ can be defined for any real x. In particular, for the case with $S_0 = 1$, we list some values of the function in the following table.

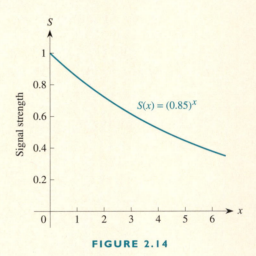

x	$S(x) = (0.85)^x$	x	$S(x) = (0.85)^x$
0	1	6	0.3771
1	0.85	7	0.3206
2	0.7225	8	0.2725
3	0.6141	9	0.2316
4	0.5220	10	0.1969
5	0.4437		

FIGURE 2.14

The graph of this function is shown in Figure 2.14. Notice how this function is decreasing or *decaying*. Each step down is smaller than the previous one. In the signal strength model as the signal gets weaker, there is less of the signal left to diminish, and so the amount of decrease in signal strength diminishes every successive ten-mile length. Recall that for exponential growth, each step up is greater than the previous one. In exponential decay, each step down is less than the previous one. Notice that the graph for exponential decay is concave up, as was the graph for exponential growth.

Furthermore, as the process continues, the strength of the signal in the fiber-optic cable obviously gets smaller and smaller and hence closer and closer to 0. We say that the strength level approaches 0 *asymptotically* in the sense that it never reaches 0 in any finite time interval. In Figure 2.14, we call the horizontal axis a *horizontal asymptote* for the graph of the decaying exponential function.

Formula for an Exponential Function

In general, P is an **exponential function** of t with base a, if

$$P(t) = P_0 a^t,$$

where P_0 is the initial quantity (when $t = 0$) and a is the **growth** or **decay factor** by which P changes when t increases by 1 unit. We always assume that $a > 0$ and $a \neq 1$.

If $a = 1 + r$, then P is increasing by $100 \cdot r\%$ each time period.

If $a = 1 - r$, then P is decreasing by $100 \cdot r\%$ each time period.

r is the *growth rate* or the *decay rate*.

$a = 1 + r$ is the *growth factor*.

$a = 1 - r$ is the *decay factor*.

For example, if a quantity (e.g., the balance in your bank account) is growing 5% per year, then $r = 0.05$ is the growth rate and $a = 1 + r = 1.05$ is the growth factor. If a quantity is decreasing at the rate of 25% an hour (e.g., the effectiveness of a medication in the body), then $r = 0.25$ is the decay rate and $a = 1 - r = 0.75$ is the decay factor; this reflects the fact that if a quantity decreases at 25% an hour, then 75% remains after the hour.

We can recognize that a function $P = f(t)$ given by a table of data values is either growing or decaying exponentially by looking at the successive ratios of the P values, as we did with the population of Florida. If the ratios are constant for equally spaced t values, then we conclude that the values follow an exponential pattern. Further, the common ratio is precisely the growth or decay factor for the process if the given values of t increase by 1 unit. With Florida's population values, we saw that the common ratio was 1.029, which is the growth factor, and the associated growth rate is 0.029, or 2.9% per year.

Doubling Time and Half-life

The **doubling time** for an exponentially increasing quantity is the time needed for it to double. The **half-life** for an exponentially decaying quantity is the time needed for it to be reduced by half. You can visualize what this means in Figure 2.15.

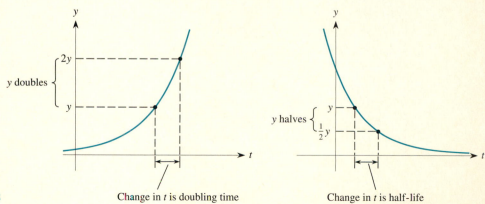

FIGURE 2.15

Change in t is doubling time   Change in t is half-life

Note that the doubling time T for an exponential growth process is the same at any quantity level; that is, for any point (t, y) you pick on the curve, the quantity always will increase to $2y$ after T time-units. Similarly, the half-life is also the same at any quantity level; no matter which point (t, y) you select, the quantity will decrease to $\frac{1}{2}y$ after T time-units.

E X A M P L E   3

Find the half-life for the strength of a signal being transmitted along the fiber-optic cable given in Example 2.

Solution   We know the signal strength after n ten-mile lengths is

$$S(n) = S_0(0.85)^n.$$

We find that after $n = 5$ ten-mile lengths, the strength of the signal is down to $0.4437S_0$, or about 44% of S_0, so the half-life must be somewhat less than 5. We also know that $S(4) = S_0(0.85)^4 = 0.522S_0$, so the half-life must be somewhat greater than 4. Using a trial-and-error approach, you can find that the actual value is about $n = 4.265$ ten-mile lengths or 42.65 miles. That is, the strength of the signal drops by half approximately every $42\frac{2}{3}$ miles. (See Figure 2.16.)

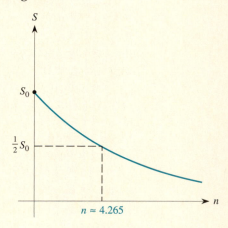

FIGURE 2.16

Note that we were lucky in Example 3 to be able to determine the half-life approximately by trial and error. In general, either the doubling time for an exponential growth process or the half-life for an exponential decay process can be found exactly by using logarithms, as you will see in Section 2.5.

It is important to realize that questions asking "where" a certain behavior occurs relate to the independent variable. Questions asking something about the values of the function ("what") relate to the dependent variable.

Since exponential functions involve working with exponents, all the usual algebraic rules for manipulating exponents apply. As a reminder, we list some definitions and rules for exponents.

Definitions and Rules for Exponents

Property	Example
1. $a^x \cdot a^y = a^{x+y}$	$10^3 \cdot 10^2 = (10 \cdot 10 \cdot 10) \cdot (10 \cdot 10) = 10^5$
2. $\dfrac{a^x}{a^y} = a^{x-y}$	$\dfrac{10^5}{10^2} = \dfrac{10 \cdot 10 \cdot 10 \cdot 10 \cdot 10}{10 \cdot 10} = 10^3$
3. $(a^x)^y = a^{xy}$	$(10^3)^2 = 10^3 \cdot 10^3 = 10^6$
4. $a^0 = 1$	$10^0 = 1$
5. $a^{-1} = \dfrac{1}{a}$	$10^{-1} = \dfrac{1}{10}$
6. $a^{-n} = \dfrac{1}{a^n}$	$10^{-3} = \dfrac{1}{10^3} = \dfrac{1}{1000}$
7. $a^{1/n} = \sqrt[n]{a}$	$10^{1/2} = \sqrt{10}, \ 10^{1/3} = \sqrt[3]{10}.$

Just as two points determine a line (that is, there is one and only one line that passes through the points), two points also determine an exponential function. If we have any two points (x_1, y_1) and (x_2, y_2), where y_1 and y_2 are either both positive or both negative and $y_1 \neq y_2$, then there is one and only one exponential curve that passes through the two points. We illustrate how to apply this fact in Example 4.

EXAMPLE 4

The circulation of the *USA Today* newspaper grew from 1.4 million in 1985 to 2.0 million in 1993. Assuming that the growth pattern has been exponential, find the equation of the exponential function that models the paper's circulation.

Solution   Let t represent the number of years since 1985. Then we have the two points $(0, 1.4)$ and $(8, 2.0)$. The exponential function is of the form

$$C(t) = C_0 \cdot a^t,$$

where the constants C_0 and a must be determined. Substituting the first point $(0, 1.4)$ into the function—that is, $t = 0$ and $C(0) = 1.4$—we find

$$C(0) = C_0 \cdot a^0 = 1.4.$$

Because $a^0 = 1$, we have $C_0 = 1.4$ million. Using the second point $(8, 2.0)$ gives

$$C(8) = 1.4a^8 = 2.0.$$

Solving for a^8 gives

$$a^8 = \frac{2.0}{1.4} = 1.429.$$

Just as we solve $x^2 = 10$ for x by taking the square root of 10 or solve $x^3 = 10$ for x by taking the cube root of 10, we solve $a^8 = 1.429$ for a by taking the eighth root of 1.429. (We discuss the details more formally in the next section.) Thus,

$$a = \sqrt[8]{1.429} \approx 1.0456.$$

(Check that $\sqrt[8]{1.429} \approx 1.0456$ by taking the eighth power of 1.0456.) Thus, the exponential function that models the growth in the circulation of *USA Today* is

$$C(t) = 1.4(1.0456)^t.$$

EXERCISES

1. Each year, the world's annual consumption of water rises. Also, the amount of increase in water consumption rises each year. Sketch a graph of the annual world consumption of water as a function of time.

2. A human fetus grows rapidly at first and then grows with decreasing rapidity. Draw a graph showing the size of a fetus as a function of time.

3. Sales of microwave ovens grew slowly when they were first introduced and then this growth increased dramatically as more people appreciated their usefulness. Eventually, sales began to slow as the marketplace neared saturation. Sketch the graph of microwave oven sales as a function of time. Indicate the location of the point of inflection.

4. For the function shown in the given figure, indicate the following:

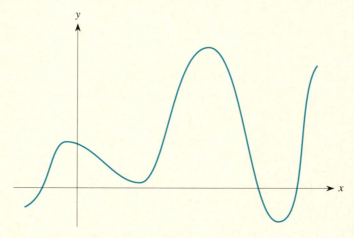

   a. the intervals of x-values where the function is increasing.
   b. the intervals where the function is decreasing.
   c. all points x where the function has a turning point.
   d. all points x where the function has a maximum.
   e. all points x where the function has a minimum.
   f. approximately where the function is increasing most rapidly.
   g. approximately where the function is decreasing most rapidly.
   h. the intervals of x-values where the function is concave up.
   i. the intervals where the function is concave down.
   j. where the function has points of inflection.
   k. the location of any *zeros* of the function (points where the curve crosses the x-axis).

5. Consider the data in the following table. Assume that these values represent a sample of values for a smooth or continuous function.

x	−2.5	−2.0	−1.5	−1.0	−0.5	0	0.5	1.0	1.5	2.0	2.5	3.0
$f(x)$	62.3	28.4	6.8	4.3	11.9	33.2	14.7	2.3	−12.5	−38.8	−5.2	11.7

    a. Over what intervals of x-values is the function increasing?
    b. Over what intervals is the function decreasing?
    c. Near what x-values is the function at a maximum?
    d. Near what x-values is the function at a minimum?
    e. Between what pair of successive x-values is the function increasing most rapidly?
    f. Between what pair of successive x-values is the function decreasing most rapidly?
    g. Over what intervals is the function concave up?
    h. Over what intervals is the function concave down?
    i. Near what x-values does the function have points of inflection?
    j. Estimate the location of any *zeros* of the function (points where the curve crosses the x-axis).

6. The accompanying graph shows population growth curves for four different nations. Which nation(s)

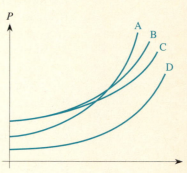

    a. has the greatest growth rate?
    b. has the smallest growth rate?
    c. have the same growth rates?
    d. has the largest initial population?
    e. has the smallest initial population?

7. Andy opens a bank account with $1200 at 4% annual interest. Bill opens an account with $1000 at 4.5% annual interest. Christine opens an account with $1500 at 3.8% annual interest. Doug opens an account with $1200 at 4.5% annual interest. Elka opens an account with $1300 at 4.25% annual interest. Sketch a graph showing the balances in the five accounts over time on the same set of axes. Be sure to label which account belongs to which person.

8. Determine which of the following pairs of points determine an exponential function of the form $y = C \cdot a^x$ and which do not. For those that do, sketch the graph of the exponential function and indicate the sign of C and whether the growth/decay factor a is greater or less than 1.

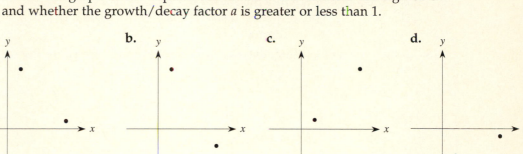
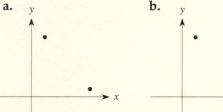
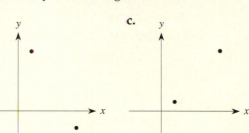

9. The population shown in the accompanying figure is growing exponentially.

   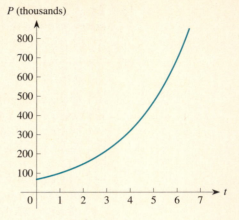

   P (thousands)

   a. Use the graph to estimate the doubling time of the population.
   b. Verify graphically that the doubling time does not depend on where you start on the graph.

10. A certain country has a population of 10 million and an annual growth rate of 2%. Estimate the doubling time by trial and error.

11. The population of a certain country doubles every 20 years. Estimate its annual growth rate by trial and error.

12. A radioactive substance decays exponentially so that after 10 years, 40% of the initial amount remains. Find an expression for the quantity remaining after t years. How much will be present after 25 years? What is the half-life of the substance? How long will it be before only 2% of the original amount is left? (Use trial and error where necessary.)

13. In 1986, 1048 women and 12,099 men died of AIDS. In 1992, 6312 women and 39,160 men died of AIDS. Assume that the pattern of growth in AIDS-related deaths for both groups is exponential.

   a. Find the equation of the exponential function for the number of AIDS-related deaths among men as a function of time.
   b. Find the equation of the exponential function for the number of AIDS-related deaths among women as a function of time.
   c. What is the practical significance of the growth factors and growth rates in part (a) and part (b)?
   d. Since the growth rate of the function for women is greater than that for men, determine when the same number of men and women will die of AIDS if the trends were indeed exponential and continued indefinitely.

14. Let $f(x)$ be an exponential function of x. If $f(7) = 25.6$ and $f(8) = 28.8$, find the following:

   a. the growth factor;
   b. the growth rate;
   c. the value of the function when $x = 10$.
   d. Suppose you are told instead that $f(7) = 25.6$ and $f(7.5) = 27.2$. Repeat parts (a) through (c).

15. Match up each of the following formulas with the corresponding table of values.

   a. $y = a(1.1)^s$ b. $y = b(1.05)^s$ c. $y = c(1.03)^s$

   (i)
s	2	3	4	5	6
$f(s)$	1.06	1.09	1.13	1.16	1.19

   (ii)
s	1	2	3	4	5
$g(s)$	2.20	2.42	2.66	2.93	3.22

   (iii)
s	3	4	5	6	7
$h(s)$	3.47	3.65	3.83	4.02	4.22

16. The filter in a swimming pool removes 30% of all impurities in the water every hour it operates. Find an expression for the level of impurities left in the pool after n hours. How much is left after five hours?

17. The Dow-Jones average of 30 industrial stocks is the most famous measure of performance of the New York Stock Exchange. At the start of 1980 the Dow was 839, and at the start of 1996 it was about 5771. Assuming (incorrectly) that the Dow increased continuously over this time frame and that the pattern is exponential, find the exponential function that models the behavior of the Dow between 1980 and 1996. What would you predict as the value for the Dow at the beginning of the year 2000? (We will return to this situation in more detail in Section 6.8.)

18. A study completed in 1994 showed that a four-year private college education costs an average of approximately $70,000 and has been growing at an annual rate of about 7%. Assuming that this rate continues indefinitely, complete the table below to estimate the cost of a college education at each school starting in the year 2019, which is 25 years after the study.

College	1994–95 Cost	Projected Cost
American University	21,500	?
Amherst College	24,152	?
Baylor University	10,940	?
Boston University	24,130	?
Morehouse College	13,224	?
Oberlin College	24,485	?
Stanford University	24,310	?
University of Michigan	8,650	?
University of Notre Dame	20,072	?
University of Tulsa	13,043	?

19. Functions f, g, and h given in the table are increasing functions of x, but each function increases according to a different behavior pattern. Which of the following graphs best fits each function?

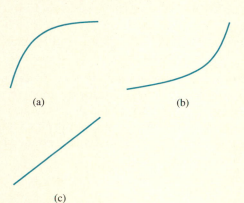

(a)

(b)

(c)

x	$f(x)$	$g(x)$	$h(x)$
1	11	30	5.4
2	12	40	5.8
3	14	49	6.2
4	17	57	6.6
5	21	64	7.0
6	26	70	7.4

20. Functions f, g, and h given in the table are decreasing functions of t, but each function decreases according to a different behavior pattern. Which of the following graphs best fits each function?

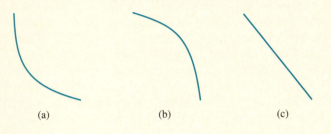

(a)   (b)   (c)

t	$f(t)$	$g(t)$	$h(t)$
1	200	30	5.4
2	180	27.6	5.2
3	164	25.2	4.8
4	151	22.8	4.1
5	139	20.4	3.1
6	129	18.0	1.8

21. The net income of the Acme Company was $240 million in 1980 and has been increasing at an annual rate of 10% per year ever since. Over the same period, the net income of its chief competitor, the Finest Corporation, has been growing 8% annually from an income of $300 million in 1980. Which was the richer company in 1990? Does Acme surpass Finest? If so, when?

22. (Extension of Exercise 21) Suppose Finest grew by a fixed amount of $25 million per year since 1980 while Acme grew exponentially at an annual rate of 10%. By using trial and error, estimate when Acme surpassed Finest.

23. **a.** Use the three points P, Q, and R shown on the accompanying graph of $y = f(x)$ to determine three line segments, PQ, QR, and PR. List these line segments in the order of *increasing* slope (smallest to largest).

b. Repeat part (a) if the function is increasing and concave down instead.

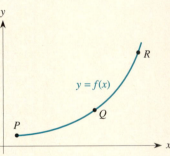

24. Sketch the graph of a function that passes through the point $(0, 1)$ and is

a. increasing and concave up for $x < 0$ and increasing and concave down for $x > 0$.

b. decreasing and concave up for $x < 0$ and increasing and concave up for $x > 0$.

c. decreasing and concave up for $x < 0$ and decreasing and concave down for $x > 0$.

d. increasing and concave up for $x < 0$ and decreasing and concave up for $x > 0$.

25. When Steven was five years old, his grandmother decided to set up a trust account to pay for his college education. Granny wanted the account to grow to $80,000 by Steven's eighteenth birthday. If she was able to invest her money at 6% per year, how much did Granny have to put into this trust account? (This amount is known as the *present value* of the investment. The $80,000 is known as the *future value*.)

26. Find possible equations for the exponential functions graphed in (a)–(c).

a.

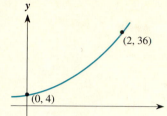

b.

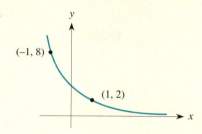

c.

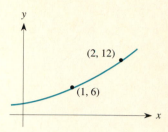

27. The function shown is a modified exponential function of the form $y = A + B \cdot C^x$. Find appropriate values for the three constants A, B, and C.

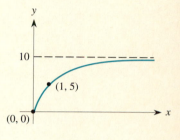

28. Simplify the following:

a. $x^5 \cdot x^3$ **b.** $x^4 \cdot x^2$ **c.** $a^8 \cdot a^4$ **d.** $\dfrac{a^{15}}{a^6}$

e. $x^{-5} \cdot x^3$ **f.** $a^5 \cdot a^{-3}$ **g.** $\dfrac{r^8}{r^{-4}}$ **h.** $\dfrac{b^{15}}{b^{-6}}$

29. Show that $x^{5/3} \neq \dfrac{x^5}{x^3}$

a. numerically: by finding at least one value of x for which the two expressions are different; and

b. graphically: by comparing the graphs of the two functions.

30. The Apollo-12 mission involved a flight to the moon (250,000 miles from Earth), five circular orbits about the moon, and a return to Earth.

a. Assume (incorrectly) that the spacecraft traveled at a constant speed between the Earth and the moon. Sketch a rough graph of the distance from Earth as a function of time.

b. Assume (correctly) that the spacecraft's speed diminished the farther it got from Earth's gravity until it neared the moon and then increased due to the moon's gravitational force. The behavior of the spacecraft's speed reversed on the return trip. Sketch a rough graph of the spacecraft's distance from the Earth as a function of time. (Think concavity!)

31. You have been asked to design a slide at a water amusement park that extends vertically from point A to point B. As a person slides down, he or she will speed up due to the force of gravity. For the three possible shapes of the slide shown, along which will a person make the trip from A to B most rapidly? Give reasons for your answer. (The specific curve along which an object will slide without friction from A to B in the shortest possible time is known as the *brachistochrone* and was first solved by Jacques Bernoulli around 1700.)

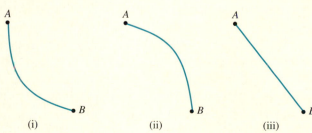

(i)   (ii)   (iii)

2.4 *Power Functions*

The area A of a circle with radius r is

$$A = f(r) = \pi r^2.$$

The surface area S of a sphere with radius r is

$$S = g(r) = 4\pi r^2.$$

(Picture a tennis ball whose surface is made up of four roughly circular regions, as shown in Figure 2.17). The volume V of the sphere is

$$V = h(r) = \frac{4}{3}\pi r^3.$$

Similarly, the Inverse Square Law of Gravitation describes how the force of gravity of one object on any other object in the universe varies with distance. The gravitational force F on a unit mass at a distance d from the center of the Earth is

$$F = \frac{k}{d^2} \qquad \text{or} \qquad F = k \cdot d^{-2},$$

where k is a positive constant.

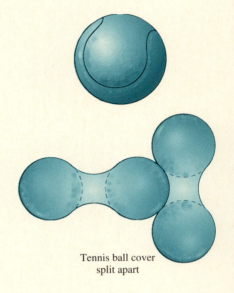

Tennis ball cover
split apart

FIGURE 2.17

All four of these functions are examples of *power functions* because the independent variable is raised to a constant power. In each case, the dependent variable is a constant multiple of some power of the independent variable. In general, a **power function** is any function of the form

$$y = f(x) = k \cdot x^p,$$

where k and p are any constants. (Compare this kind of expression with an exponential function of the form $y = k \cdot a^x$, where the independent variable x is the exponent, or power, and the base a is a constant.) You will see in the next chapter that power functions are used very frequently to model many different processes.

In this section we investigate the effect on the power function of different values for the power p. In the special case $p = 1$, the power function reduces to $f(x) = kx$ (where k is a constant), which is a linear function passing through the origin. Thus, such linear functions are also included in the comparison. We investigate where power functions are increasing or decreasing and where they are concave up or concave down.

Another important characteristic of some functions in general is that their graphs show **symmetry**—one portion of the graph is a mirror image of another portion. (See Appendix C.) Exponential functions do not have this property, but some power functions do, as we will see. Further, we investigate how power functions behave near the origin when x is close to 0 and how they behave in the long run as x approaches either $+\infty$ or $-\infty$

(that is, as x becomes very large positively or very large negatively). We're often interested in the relative growth rates of functions in order to tell which function grows most rapidly for large values of x. The reason that it is important to focus on the behavior pattern is that the same description applies to the process that the function represents.

Positive Integer Powers

For simplicity, we begin by letting $k = 1$ in the general power function $f(x) = k \cdot x^n$ to consider power functions of the form $f(x) = x^n$, where n is a positive integer. These functions are $y = x, y = x^2, y = x^3, \ldots$. Notice, from Figure 2.18, that the graphs of these functions fall into two groups: functions with odd powers and functions with even powers.

Let's first examine the power functions with odd powers. All of these functions, $y = x, y = x^3, y = x^5, \ldots$ are increasing everywhere as x increases from left to right. Also, each graph passes through the origin because $0^n = 0$ for any n. Further, the portion of each curve in the third quadrant is the mirror image of the corresponding portion in the first quadrant, so each function is symmetric about the origin. Also notice that all the odd power functions for $n > 1$ are concave down for $x < 0$ and are concave up for $x > 0$. Thus the concavity in each curve changes at $x = 0$. This change in concavity means that every odd power function except $y = x^1$ has a point of inflection at the origin. You should examine some of these functions on your own using your function grapher.

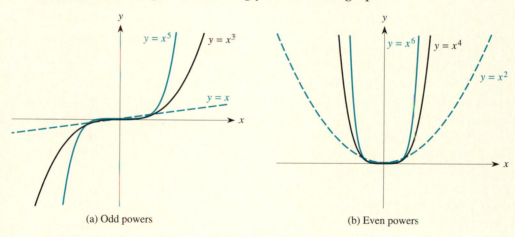

(a) Odd powers   (b) Even powers

FIGURE 2.18

Now let's look at the even power functions $y = x^2, y = x^4, y = x^6, \ldots$. These functions all first decrease (until $x = 0$) and then increase as x increases from left to right. Also, they all pass through the origin and are symmetric about the y-axis (the left and right halves of the curves are mirror images). Another characteristic is that they are all U-shaped. Thus all the even powers are concave up everywhere, so even power functions do not have a point of inflection.

You have seen that all positive integer power functions, whether odd or even, pass through the origin. Notice that they also all pass through the point $(1, 1)$ because $1^n = 1$ for any power n. In addition, all even power functions pass through the point $(-1, 1)$; consider $y = x^2$. All odd power functions pass through the point $(-1, -1)$; consider $y = x^3$. Thus, these particular points serve to "pin down" the power functions. They also serve as key points where different characteristics of the functions come into play.

The origin $(0, 0)$ represents the place where an even power function achieves its minimum value; it also represents the place where an odd power function has its point of inflection. Further, the point $(1, 1)$ represents the demarcation between two different types of behavior for power functions—the behavior for "small" values of x near the origin and the behavior for "large" values of x away from the origin.

Because all power functions pass through the origin, we can think of their relative behavior for x near 0 as a race to see which power function approaches 0 faster. That is, to compare power functions, we want to decide which function has y-values that are closest to 0 when x is close to 0.

Similarly, because all power functions of the form $y = x^n$ grow indefinitely large as the value of x increases, we also can think of their relative behavior for large x-values as a different race to see which power function approaches infinity faster.

Let's first examine the race toward infinity: In Figure 2.19, we see that the higher the power of x is, the faster the function increases. For large values of x (in fact, for all values of $x > 1$), $y = x^5$ is above $y = x^4$, which is above $y = x^3$, and so on. Check this graphically using your function grapher. You can check this numerically by considering $x = 10$: For example, 10^5 is greater than 10^4, which is greater than 10^3, and so on. Thus not only do the higher powers of x get larger, but they get *larger much faster*.

This comparison gets more pronounced when we look at larger and larger values of x. Repeat the numerical check mentioned above using, for example, $x = 20$ instead of $x = 10$. As x gets ever larger (that is, as x approaches infinity, denoted by $x \to \infty$), any positive power completely overwhelms, or dominates, any smaller power.

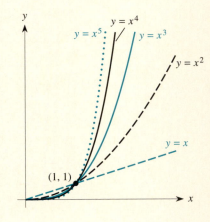

FIGURE 2.19

Now let's examine the race toward 0: We want to check the behavior near the origin. As illustrated in Figure 2.20, we see that when x is between 0 and 1, the size order is reversed: $y = x^5$ is smaller than $y = x^4$, which is smaller than $y = x^3$. (You can check this graphically or numerically by considering $x = \frac{1}{10}$: for example, $\left(\frac{1}{10}\right)^5 = \frac{1}{10^5} = 0.00001$, which is smaller than $\frac{1}{10^4} = 0.0001$, which is smaller than $\frac{1}{10^3} = 0.001$, and so on.) Thus we see that for small values of x near zero, higher powers of x get *smaller much*

faster
x, th
wa

in the race toward 0. In comparison, we saw that for large values of
... he higher powers of x get larger faster, so they dominate in the race to-
... rd infinity.

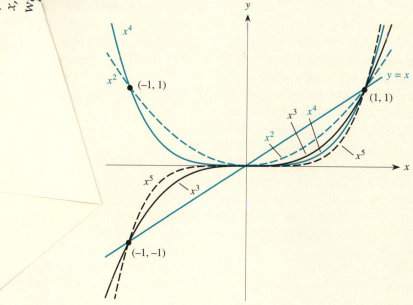

FIGURE 2.20

As a special case, consider the power function with $p = 0$: $y = x^0$. Since
we define the 0th power of any nonzero number to be 1 ($a^0 = 1$ for $a \neq 0$),

$$y = x^0 = 1,$$

for all values of $x \neq 0$. Therefore, the graph of $y = x^0$ is a horizontal line at
height 1. It is neither increasing nor decreasing, and it is neither concave
up nor concave down.

Fractional Powers

We now turn to the case of power functions with fractional powers such as
$y = x^{1/2}$, $\quad y = x^{1/3}$, $\quad y = x^{3/2}$, ...

In algebra, fractional exponents are usually introduced purely as a means
for simplifying operations with terms involving radicals. By definition,

$$x^{1/2} = \sqrt{x} \qquad\qquad x^{1/3} = \sqrt[3]{x} \qquad\qquad x^{5/8} = \sqrt[8]{x^5}$$

and, in general,

$$x^{m/n} = \sqrt[n]{x^m}.$$

As you will see, functions of the form $y = f(x) = x^{m/n}$ arise naturally in
many applications. For instance, biologists have found a relationship
between the height H of an animal and its weight W:

$$W = k \cdot H^{3/2},$$

for some constant multiple k. (Recall that $H^{3/2}$ means either first cu[...]
and then take the square root or first take the square root of H and [...]
cube the result.) For example, if the height H increases 100-fold, th[...]
$W = k(100H)^{3/2} = 1000kH^{3/2}$, so the weight increases 1000-fold. This rel[...]
tionship is known as the *square-cube law* because it can be rewritten a[...]
$W^2 = k^2 \cdot H^3 = m \cdot H^3$, for some new constant of proportionality $m = k^2$.

One consequence of this law is that none of those old horror movies involving the attack of 100-foot spiders makes sense biologically; if a creature's size increased 100-fold, its weight would increase 1000-fold and its relatively thin legs could not support it. Another implication is that an ant can carry many times its own weight whereas an elephant can carry only a small fraction of its weight.

To understand better how this process behaves, we examine the behavior of the underlying class of mathematical functions of the form $y = x^{m/n} = \sqrt[n]{x^m}$, which model the process. Since fractional powers such as $x^{1/2} = \sqrt{x}$ and $x^{1/4} = \sqrt[4]{x}$ are defined only for nonnegative values of x, we cannot extract even roots of negative numbers. For example, $\sqrt{-4}$ and $\sqrt[4]{-10}$ are undefined, but $\sqrt[3]{-8} = -2$. Therefore, we often restrict the domain of all power functions with fractional exponents to $x \geq 0$.

Furthermore, for any fractional power, if $x = 0$, then $y = x^p$ is zero also, so every power function with a positive fractional exponent passes through the origin. (We will discuss negative powers later.) Also, as x increases, every power function of the form $y = x^p$ with a positive fractional exponent approaches infinity. Therefore, we again can think of the two races: Which fractional power functions decay to zero faster and which grow to infinity faster?

Figure 2.21 shows that for large values of x (in fact, for all $x > 1$), the graph of $y = x$ is above the graph of $y = x^{1/2}$, which in turn is above the graph of $y = x^{1/3}$. You can see that this makes sense by considering what happens numerically, for example, when $x = 10$:

$$10^{1/2} = \sqrt{10} \approx 3.162$$

$$10^{1/3} = \sqrt[3]{10} \approx 2.154,$$

and so, $10 > 10^{1/2} > 10^{1/3}$. Where would you expect to see the graph of $y = x^{3/4}$? What about $y = x^{0.99} = x^{(99/100)}$ and $y = x^{1.01}$? How do they behave compared to the line $y = x$? In general, the higher the fractional power, the larger the power function is for $x > 1$.

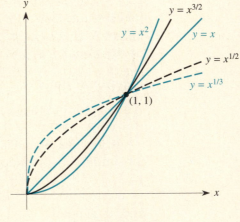

FIGURE 2.21

What happens near the origin when x is between 0 and 1? For example, suppose $x = \frac{1}{10} = 0.1$, so that $\left(\frac{1}{10}\right)^{1/2} = \sqrt{0.1} \approx 0.316$ and $\left(\frac{1}{10}\right)^{1/3} = \sqrt[3]{0.1}$ ≈ 0.464. Therefore, we conclude that near the origin, the situation is reversed: $y = x$ is below $y = x^{1/2}$, which in turn is below $y = x^{1/3}$. Further,

$y = x^{3/2}$ is between $y = x$ and $y = x^2$ for all x. Therefore the higher the fractional power is, the smaller the power function is for x near 0. Again, the higher power wins the race toward zero. Be sure to verify these ideas graphically and numerically using your function grapher.

The other important thing to notice about the graphs of $y = x^{1/2}$ and $y = x^{1/3}$ is their concavity. For $x > 0$, the graphs of $y = x^2$ and $y = x^3$ are concave up because they are growing faster and faster as x increases. However, the graphs of $y = x^{1/2}$ and $y = x^{1/3}$ are concave down because they are growing ever more slowly as x increases. Nevertheless, all fractional power functions become infinitely large as x increases.

Negative Powers

We next examine the behavior of power functions involving negative exponents. For instance, we look at functions such as $y = x^{-1}, y = x^{-2}$, and $y = x^{-1/2}$, and in general, $y = x^{-p}$, for any $p > 0$.

To begin, one of the basic definitions for exponents recalled in Section 2.3 is

$$x^{-p} = \frac{1}{x^p}$$

Using this definition, we can rewrite these functions as

$$y = x^{-1} = \frac{1}{x}, \qquad y = x^{-2} = \frac{1}{x^2}, \qquad y = x^{-1/2} = \frac{1}{x^{1/2}}, \text{ and so forth.}$$

Again, we are concerned with two questions: What happens to these power functions as x increases? What happens when x is close to 0?

Let's first look at what happens to the function $y = \frac{1}{x^p}$ when x is positive. We show the graphs of $y = \frac{1}{x^2}, y = \frac{1}{x^{3/2}}, y = \frac{1}{x}$, and $y = \frac{1}{x^{1/2}}$ in Figure 2.22. From the graphs, it is clear that all these curves pass through the point (1, 1) and have the x-axis as a horizontal asymptote. To see why, consider what happens when x increases in size. Whenever the power p in $\frac{1}{x^p}$ is greater than 1, the denominator increases ever faster as x increases, and so the function must decrease toward 0. Whenever the power p in $\frac{1}{x^p}$ is between 0 and 1, the denominator increases ever more slowly, but it nonetheless increases as x increases, and so the function must decrease toward 0. In general, any function of the form $y = x^{-p} = \frac{1}{x^p}$ decreases toward 0 as $x \to \infty$. Moreover, we see that the larger the power p is, the more rapidly the function decreases, so that $y = x^{-2}$ is below $y = x^{-1.5}$, which in turn is below

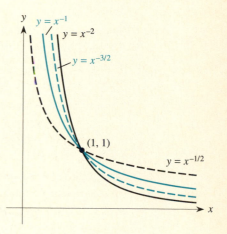

FIGURE 2.22

$y = x^{-1}$, and so forth. Thus the power function $y = \frac{1}{x^p}$ with the higher power p always wins the race toward 0 as x increases. Check this by using some reasonably large value of x, say $x = 10$, and then by viewing various curves on your function grapher to convince yourself.

Now let's see what happens near the origin when x is positive. From Figure 2.22, we see that all the curves appear to have the positive y-axis as a *vertical asymptote*; that is, as x gets closer and closer to 0, the values of these functions get larger and larger and the corresponding curves get closer and closer to the y-axis. To see why, notice that the closer x is to 0, the smaller x^p is for any positive power p. The smaller that x^p is, the larger $\frac{1}{x^p}$ becomes, and so, as x approaches 0, all power functions of the form $y = \frac{1}{x^p}$ must increase toward positive infinity. Moreover, from the graph in Figure 2.22, we see that the larger the power p is, the more rapidly the curve is increasing as x approaches 0. That is, for a given value of x close to 0, the graph of $y = \frac{1}{x^2}$ is higher than the graph of $y = \frac{1}{x^{3/2}}$, which in turn is higher than the graph of $y = \frac{1}{x}$, which in turn is higher than the graph of $y = \frac{1}{x^{1/2}}$. Check this numerically with $x = 0.1$, say, and then graphically by viewing some curves on your function grapher. Alternatively, you can think of what it takes to reach a given height as x approaches 0. We reach any height faster along $y = \frac{1}{x^2}$ than we do along $y = \frac{1}{x^{3/2}}$, which requires getting closer to 0. Similarly, we reach the same height even more slowly along $y = \frac{1}{x}$, because we have to get even closer to 0 before y reaches that height. Thus the power function $y = \frac{1}{x^p}$ with the higher power p wins the race toward $+\infty$ as x approaches 0, for $x > 0$.

Finally, we can consider what happens to the power function $y = x^{-p}$ when $x < 0$ for cases where the power is a negative integer. Each of the power functions with an even power $y = x^{-2} = \frac{1}{x^2}$, $y = x^{-4} = \frac{1}{x^4}$, and so on is positive while each of the power functions with an odd power $y = x^{-1} = \frac{1}{x}$, $y = x^{-3} = \frac{1}{x^3}$, and so on is negative. Thus to the left of the y-axis, the curves for all power functions with even negative integer powers are above the x-axis and the curves for all power functions with odd negative integer powers are below the x-axis. In either case, the more negative x is, the closer the function is to 0. For instance, what is the value of the function $f(x) = \frac{1}{x^4}$ when $x = -10$,

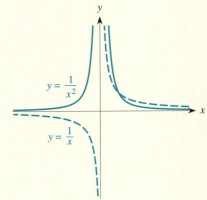

FIGURE 2.23

$x = -50$, and $x = -100$? Each of the curves has the x-axis as a horizontal asymptote. The only difference is whether the graph is above or below the x-axis. See Figure 2.23.

Think About This

Compare the rate at which different power functions of the form $y = x^{-p}$ approach $\pm\infty$ as $x \to 0$. Which ones win the race?

We summarize the behavior patterns of the power functions $y = x^p$ when $x > 0$ in Figure 2.24. Whenever the power $p < 0$, the curve is decreasing and concave up as it decays toward 0. Also, whenever p is between 0 and 1, the curves are all increasing and concave down as they slowly rise toward infinity; observe that all of these power functions are below the line $y = x$ if $x > 1$. Whenever the power $p > 1$, the curves are all increasing and concave up as they grow fairly rapidly toward infinity; observe also that all of these power functions are above the line $y = x$ for $x > 1$.

Finally, note that while each of the power functions $y = x^p$ could be multiplied by a constant (that is, $y = k \cdot x^p$), the constant multiple does not affect the overall behavior of the function. A positive multiple retains the same overall shape; a negative multiple flips the curve over about the x-axis, but the overall shape remains the same. Explore this fact on your function grapher by comparing, say, $y = x^2$ and $y = -x^2$ or $y = 5x^2$ and $y = -5x^2$. Now consider why this "flip" occurs. Look at the following two tables corresponding to the functions $y = x^2$ and $y = -x^2 = -(x)^2$. What does the negative multiple do to the y-values? Can you relate this behavior to the two graphs?

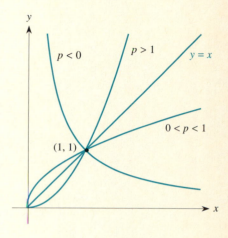

FIGURE 2.24

x	$y = x^2$
-2	4
-1	1
0	0
1	1
2	4

x	$y = -(x)^2$
-2	-4
-1	-1
0	0
1	-1
2	-4

EXERCISES

1. Identify which of the following functions in parts (a)–(n) are exponential functions, which are power functions, and which are neither.

    a. $f(x) = 40x^{1.05}$ **b.** $f(x) = 40(1.05)^x$ **c.** $f(x) = \dfrac{1}{(1.4)^x}$

    d. $f(x) = -\dfrac{3}{x^{2.4}}$ **e.** $f(t) = 5t^{-3.7}$ **f.** $f(q) = 1.09q - 4.37$

    g. $f(t) = 12(0.35)^{-t}$ **h.** $f(t) = 5\sqrt{t}$ **i.** $f(s) = \sqrt{s^2 + 3}$

     j. $f(r) = \frac{4}{3}\pi r^3$ **k.** $f(z) = z \cdot z^{3/5}$ **l.** $f(x) = x^x$

     m. $f(w) = w^2 \cdot 3^w$ **n.** $f(u) = 7(1.62)^{-u}$

2.   Match each of the following formulas with the corresponding table of values:

     a. $y = 4x^{1.2}$ **b.** $y = 5x^{0.8}$ **c.** $y = 4(1.2)^x$.

     (i)

x	2	3	4	5	6
$f(x)$	8.71	12.04	15.16	18.12	20.96

     (ii)

x	1	2	3	4	5
$g(x)$	4.80	5.76	6.91	8.29	9.95

     (iii)

x	2	4	6	8	10
$h(x)$	9.19	21.11	34.34	48.50	63.40

3.   Consider the sequence of values 10^5, 10^4, 10^3, 10^2, and 10^1 and use it to provide a reason for defining $10^0 = 1$. What about 10^{-1}?

4.   By trial and error, determine the largest power of 10 that your calculator can handle. What is the smallest positive number?

5.   Data from three different functions are shown in the accompanying table. One function is exponential, one is of the form $y = ax^2$, and one is of the form $y = bx^3$. Which function is which?

     (i)

x	3	3.5	4	4.5	5
$f(x)$	28.8	39.2	51.2	64.8	80.0

     (ii)

x	3	3.5	4	4.5	5
$g(x)$	4.39	5.01	5.71	6.51	7.42

     (iii)

x	3	3.5	4	4.5	5
$h(x)$	10.80	17.15	25.60	36.45	50.00

6.   Just as two points determine a line or an exponential function, two points also determine a power function. Find a power function that passes through the following pairs of points:

     a. $P(1, 3)$ and $Q(4, 6)$ **b.** $P(1, 3)$ and $Q(4, 8)$

     c. $P(1, 3)$ and $Q(4, 10)$ **d.** $P(5, 20)$ and $Q(6, 30)$

     e. $P(1, 10)$ and $Q(4, 5)$ **f.** $P(2, 20)$ and $Q(5, 8)$

7.   Police sometimes use the formula $s = \sqrt{30kd}$ to estimate the speed s in miles per hour that a car was going if it left a set of skid marks d feet long. The coefficient k depends on the road conditions (dry or wet) and on the type of pavement. For instance, $k = 0.8$ for dry concrete; $k = 0.4$ for wet concrete; $k = 1.0$ for dry tar; and $k = 0.5$ for wet tar.

     a.   A car left a set of skid marks 120 feet long. How fast was it going on dry concrete?

     b.   Suppose the concrete pavement in part (a) was wet. How fast was the car going?

   c. If the car in part (a) left a set of skid marks 240 feet long, how fast was it going?

   d. Suppose a car is going 50 mph on a dry tar surface when the driver slams on the brakes. How far will it skid?

   e. Suppose that the tar pavement in part (d) was wet. How far will the car skid?

8. Scientists are actively investigating the potential of using windmills to generate electricity. They have found that, for moderate wind speeds, the power P in watts generated by a windmill is related to the windspeed v in miles per hour according to the equation

$$P = 0.015v^3.$$

   a. How much power is generated by a steady wind at 10 mph?

   b. How much power is generated by a steady wind at 20 mph?

   c. Based on your results in parts (a) and (b), by what factor does doubling the windspeed increase the power generated?

   d. Compare the power generated by a steady wind at 5 mph to that of a steady wind at 10 mph. Does doubling of the wind speed increase the power generated by the same factor found in part (c)?

   e. Suppose a certain community has power needs for an additional 250 kilowatts of electricity and can anticipate winds on the average of 8 mph. How many windmills would be needed to meet the added electric demand?

   f. What wind speed would be needed to light up a 100-watt light bulb?

9. **a.** Use your function grapher to plot on the same screen the graphs of the power functions x^2, x^5, and x^8 for the interval $-0.2 \leq x \leq 0.2$. Determine an appropriate range for y so that all powers will be distinguishable in the viewing rectangle.

   b. Plot the same graphs for $-2 \leq x \leq 2$ and determine an appropriate range for y.

   c. Plot the same graphs for $-20 \leq x \leq 20$ and determine an appropriate range for y.

10. **a.** Use your function grapher to plot on the same screen the graphs of the power functions $x^{1/2}$, $x^{1/3}$, and $x^{1/4}$ for the interval $0 \leq x \leq 0.2$. Determine an appropriate range for y so that all powers will be distinguishable in the viewing rectangle.

   b. Plot the same graphs for $0 \leq x \leq 2$ and determine an appropriate range for y.

   c. Plot the same graphs for $0 \leq x \leq 20$ and determine an appropriate range for y.

   d. What happens if you use the interval $-2 \leq x \leq 2$?

11. What happens to

   a. x^3 as $x \to \infty$? As $x \to -\infty$?   **b.** $-x^3$ as $x \to \infty$? As $x \to -\infty$?

   c. $x^{1/3}$ as $x \to \infty$? As $x \to -\infty$?   **d.** $-x^{1/3}$ as $x \to \infty$? As $x \to -\infty$?

   e. x^{-3} as $x \to \infty$? As $x \to -\infty$?   **f.** x^{-3} as $x \to 0$?

12. In 1986, 1048 women and 12,099 men died of AIDS. In 1992, 6312 women and 39,160 men died of AIDS. Assume that the growth in AIDS-related deaths for both groups is growing according to a power function pattern.

  a. Find the equation of the power function for the number of AIDS-related deaths among men as a function of time.

  b. Find the equation of the power function for the number of AIDS-related deaths among women as a function of time.

  c. Since the number of deaths among women is growing faster than that for men, determine when the same number of men and women will die of AIDS if the trends were indeed power functions and continued indefinitely.

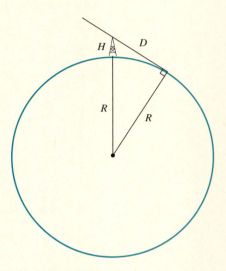

13. In the accompanying figure let R be the radius of the Earth (about 3960 miles). Find an expression for the distance D to the horizon from a point at a height of H miles above the Earth's surface. (*Hint:* Recall that in a circle, any tangent line is perpendicular to a radius.)

14. The observation deck on the top of the World Trade Center in New York is 1377 feet high. If you are standing up there, complete the following phrase: "On a clear day, you can see . . ."

15. UHF (ultra high frequency) TV transmissions travel along a line of sight from a transmitter as far as the horizon. In the Chicago area, the UHF stations broadcast from a transmitter atop the 1454-foot ($= 0.275$ mile $\approx \frac{1}{4}$ mile) high Sears tower. What is the greatest distance that someone could receive a UHF signal from the tower?

16. Suppose a mast 250 feet (about $\frac{1}{20}$ of a mile) high is being planned for the Sears tower to extend the broadcast range of UHF stations. How much farther would the signal extend? How much larger a receiving area would be covered?

17. NASA's space shuttles orbit the Earth at altitudes of about 200 miles. Find the maximum line-of-sight transmission distance from the shuttle to the surface of the Earth. Approximately how large a receiving area on the Earth is in range of this shuttle?

18. Communications satellites orbit the Earth in geosynchronous orbits (carefully chosen heights and velocities so that they appear to be permanently above a fixed point on the surface of the Earth as the Earth rotates). Suppose such a satellite is in orbit at a height of 23,000 miles above a point on the equator. Since the radius of the Earth is about $R = 3960$ miles, the distance around the equator is approximately $2\pi R = 24,900$ miles. Consequently, a point on the equator is rotating at a velocity of about 1037 miles per hour. Find the orbital velocity of such a communication satellite in a geosynchronous orbit.

19. Explain why it is not possible to have a communications satellite whose signals are able to cover a full half of the Earth's surface.

20. Using $R = 3960$ miles for the radius of the Earth, the formula you found in Exercise 13 for the line-of-sight distance to the horizon from a height of H miles is $D(H) = \sqrt{H^2 + 2RH} = \sqrt{H^2 + 7920H}$. When H is small, the term H^2 seemingly has little effect on the value of D, so we might be tempted to approximate the distance D using the simpler formula $D \approx \sqrt{7920H} \approx 89\sqrt{H}$. To see if using this approximation formula is reasonable, complete the following table comparing the estimated value for this distance with the actual value:

H	$D \approx 89\sqrt{H}$	$D(H) = \sqrt{H^2 + 7920H}$
0.1 mile		
1 mile		
10 miles		
100 miles		

21. Consider the function $f(x) = x^2$ and let P be the point on the curve where $x = 0$, R be the point where $x = 2$, and Q be the midpoint where $x = 1$. Find the slopes of the three line segments PQ, QR, and PR. How does the slope of PR compare to the slopes of the other two segments?

22. Repeat Exercise 21 using the function $g(x) = x^3$. Does the relationship among the three slopes you found in Exercise 21 also hold for g?

23. Consider the function $f(x) = x^2$ and let P be the point where $x = a$, Q be the point where $x = a + h$, and R be the point where $x = a + 2h$, for any quantity h. Find the slopes of the three line segments PQ, QR, and PR. Show that the slope of PR is the average of the other two slopes.

24. Use the properties of exponents to evaluate the following (do not use calculators):

   a. $9^{1/2}$ b. $9^{-1/2}$ c. $8^{4/3}$ d. $8^{-4/3}$

25. Simplify the following:

   a. $x^4 \cdot x^3$ b. $a^6 \cdot a^{-8}$ c. $\dfrac{r^8}{r^4}$ d. $\dfrac{z^{12}}{z^{-9}}$

2.5 *Logarithmic Functions*

In Section 2.3, we constructed an exponential function to approximate (in millions) the population of Florida as

$$P = f(t) = 9.75(1.029)^t,$$

where t is the number of years since 1980. This expression gives the population as a function of time based on the 1980 population being 9.75 million people with an annual growth rate of 2.9%. Using this model, we can predict Florida's population at any given time, assuming that the growth rate doesn't change.

Suppose that instead of calculating the size of the population at a given time, we want to predict when the population will reach 20 million so that adequate services can be provided for the population. That is, what is the value of t for which

$$f(t) = 9.75(1.029)^t = 20?$$

Since we know that this exponential function is always increasing, we know that there must be exactly one value of t when $P = 20$. We could try to find it by trial and error:

Try $t = 10$: $P = f(10) = 9.75(1.029)^{10} \approx 12.98$ (so $t = 10$ is too small)
Try $t = 20$: $P = f(20) = 9.75(1.029)^{20} \approx 17.27$ (so $t = 20$ is too small)
Try $t = 30$: $P = f(30) = 9.75(1.029)^{30} \approx 22.99$ (so $t = 30$ is too big)

Thus the desired time t is clearly between 20 and 30 years. Narrowing down still further, we find

$$P = f(25) = 9.75(1.029)^{25} \approx 19.92$$
$$P = f(26) = 9.75(1.029)^{26} \approx 20.50,$$

so that t is between 25 and 26 years. Thus we predict that Florida's population will reach 20 million people sometime during the year 2005.

Even though it is always possible to find t by trial and error, it clearly would be better to have a function that gives t in terms of P. We seek a process that extracts the variable t from the exponent in $P = a^t$. This process involves a new function called the *logarithm*. As with exponentials, logarithms involve knowing a base. Although it is possible to have logarithms to any base b (denoted by $\log_b$), we will work primarily with logarithms to base 10.

Definition of Logarithms to Base 10

$$\log_{10} x = y \quad \text{means} \quad 10^y = x.$$

The logarithm to the base 10 of x is the power of 10 that produces x.

For example, $\log_{10} 100 = \log_{10} 10^2 = 2$ because 2 is the power of 10 needed to produce 100: $10^2 = 100$. Also $\log_{10} 1000 = \log_{10} 10^3 = 3$ because 3 is the power of 10 needed to produce 1000: $10^3 = 1000$. Similarly, $\log_{10}(0.1) = \log_{10}\frac{1}{10} = \log_{10} 10^{-1} = -1$ because -1 is the power to which 10 must be raised to get 0.1: $10^{-1} = 0.1$.

The logarithm to the base 10 of x, written $\log_{10} x$, is usually written as simply $\log x$. Since the logarithm is a function, it would actually be preferable to write $\log(x)$ rather than just $\log x$. Because $\log(x)$ is not standard usage, we will avoid it. However, we do use parentheses for expressions such as $\log(5x)$.

Behavior of the Logarithmic Function

Let's now consider the behavior of the log function. Keep in mind that the logarithm represents that power of 10 needed to produce a given number x. Because no power of 10 ever produces 0 (10 raised to what power is 0?), log 0 is undefined. Similarly, because no power of 10 ever produces a negative number, log x is not defined for negative values of x. Consequently, the domain of the log function is $x > 0$. On the other hand, because it is possible to have a negative power of 10, log x can be negative. For example, $10^{-0.25} = 0.56234$ means that log $0.56234 = -0.25$. Thus, the range of the log function includes both positive and negative values. In fact, you can check that the log of any number between 0 and 1 is negative and the log of any number larger than 1 is positive. (See Figure 2.25.) Finally, log $1 = 0$. Therefore, the range of the log function consists of all real numbers.

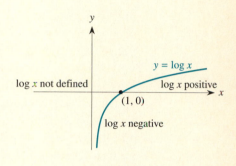

FIGURE 2.25

Let's use these ideas to examine the behavior of the log function $f(x) = \log x$ by comparing it to the related exponential function $g(x) = 10^x$. We can compare these functions in two ways: numerically by calculating the values for the two functions, as shown in the following table or graphically, as shown in Figure 2.26.

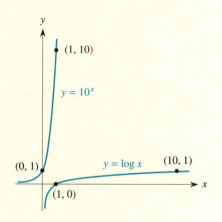

FIGURE 2.26

x	$\log x$
0	undefined
0.01	-2
0.1	-1
1	0
2	0.301
3	0.477
4	0.602
⋮	⋮
10	1

x	10^x
-2	0.01
-1	0.1
0	1
1	10
2	100
3	1000
4	10^4
⋮	⋮
10	10^{10}

Let's consider the comparative growth patterns of the two functions. Both are clearly increasing functions. The exponential function is concave up while the log function is concave down. However, the most fundamental difference in their growth patterns is that the exponential function $g(x) = 10^x$ grows extremely rapidly while the logarithm function $f(x) = \log x$ grows extremely slowly. These two facts are directly related.

You know that

$$\log 10 = 1$$
$$\log 100 = 2$$
$$\log 1000 = 3$$
$$\vdots$$
$$\log 1{,}000{,}000 = 6,$$

and so forth. With the log function, to gain one unit vertically it is necessary to go 10 times as far horizontally. Thus you need an extremely large value of x to make log x large. For instance, what value of x do you think you need to make log $x = 100$? By definition, x must be 10^{100} because $\log(10^{100}) = 100$; don't bother trying to evaluate 10^{100} with a calculator because it likely exceeds your calculator's capacity. Knowing that 10^{100} is this large tells us that it takes an incredibly long time for the log curve to reach a height of 100—the log function grows very slowly. The log function does go to infinity as x increases, even though it does so exceedingly slowly.

We know that the log function is not defined at $x = 0$ or at negative x-values. But what happens for small values of x? Consider the following:

$$\log 1 = 0$$
$$\log 0.1 = \log(10^{-1}) = -1$$
$$\log 0.01 = \log(10^{-2}) = -2$$
$$\log 0.001 = \log(10^{-3}) = -3$$
$$\vdots$$
$$\log 0.000001 = \log(10^{-6}) = -6,$$

and so forth. We therefore see that as x gets closer and closer to 0, log x becomes more and more negative. Thus the line $x = 0$ (which is the y-axis) is a *vertical asymptote* for the graph of $y = \log x$ since the curve gets closer and closer to this vertical line but never reaches it. This vertical asymptote reinforces the fact that the log function is not defined at $x = 0$, and so the graph of $y = \log x$ has no y-intercept. It does, however, have an x-intercept at $x = 1$ because log $1 = 0$. (See Figure 2.26 on the previous page.)

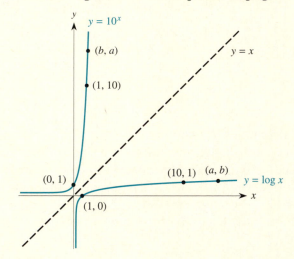

FIGURE 2.27

From Figure 2.27 you can see something rather striking about the graphs of the two functions $y = 10^x$ and $y = \log x$. They are reflections of one another about the diagonal line $y = x$, which means that the two curves are *symmetric* about this line (see Appendix C). We know that

$$\log 10 = 1,$$

so the point $(10, 1)$ is on the log graph. By the definition of the logarithm,

$$\log 10 = 1 \qquad \text{means} \qquad 10^1 = 10.$$

But $10^1 = 10$ tells us that the point $(1, 10)$ satisfies the equation $10^x = y$, and so the point $(1, 10)$ is on the exponential graph. The points $(10, 1)$ and $(1, 10)$ are reflections of one another about the line $y = x$. In general, if the point (a, b) is on the graph of $y = \log x$, then

$$\log a = b.$$

This expression is equivalent to saying

$$10^b = a,$$

which means the point (b, a) is on the graph of the exponential function $y = 10^x$. Hence, the log graph and the exponential graph are reflections of each other about the line $y = x$.

Properties of Logarithms

To use logarithms, you need to know their basic properties.

Properties of Logarithms

1. $\log (A^p) = p \cdot \log A$
2. $\log (A \cdot B) = \log A + \log B$
3. $\log \dfrac{A}{B} = \log A - \log B$

The first property is precisely the tool we need for extracting a variable from the exponent. For example,

$$\log (3^x) = x \cdot \log 3.$$

The second property allows us to simplify the *log of a product* by writing it as the *sum of the individual logs*. For example,

$$\log (5 \cdot 12) = \log 5 + \log 12 \qquad \text{or} \qquad \log (100x) = \log 100 + \log x$$
$$= 2 + \log x.$$

Note that you can easily find the log of any power of 10 by using the definition: $\log_{10} 100 = \log_{10} 10^2 = 2$. However, all logarithms can be evaluated quickly using a calculator.

The third property allows us to simplify the *log of a quotient* by writing it as the *difference of the individual logs*. For example,

$$\log \tfrac{9}{4} = \log 9 - \log 4 \qquad \text{or} \qquad \log \tfrac{1000}{x} = \log 1000 - \log x$$
$$= 3 - \log x$$

Do you think that $\log \tfrac{9}{4}$ is the same as $\tfrac{\log 9}{\log 4}$? Try it on your calculator to see. Now check to see that

$$\log \tfrac{9}{4} = \log 9 - \log 4.$$

Similarly, do you think that $\dfrac{\log 1000}{\log x} = \dfrac{3}{\log x}$ is the same as $\log 1000 - \log x$? Graph both $y = \dfrac{\log 1000}{\log x}$ and $y = \log 1000 - \log x$ to see whether or not this is true.

All these formulas apply to logarithms with any base b, not just the base 10.

E X A M P L E   1

Solve for x in the equation $3^x = 8$.

Solution   If $3^x = 8$, then we extract the x from the exponent by taking the log of both sides and using Property 1:

$$3^x = 8$$
$$\log (3^x) = \log 8$$
$$x \cdot \log 3 = \log 8$$
$$x = \frac{\log 8}{\log 3} \approx 1.893$$

E X A M P L E   2

Determine when the population of Florida will reach 20 million.

Solution   We begin with the equation

$$P = f(t) = 9.75(1.029)^t = 20.$$

Dividing both sides of the equation by 9.75, we get

$$(1.029)^t = \frac{20}{9.75} = 2.05128$$

We now take logs of both sides and use the fact that $\log (A^t) = t \cdot \log A$ to get

$$\log (1.029^t) = t \log (1.029) = \log (2.05128).$$

So
$$t = \frac{\log 2.05128}{\log 1.029} = 25.13,$$

which is between $t = 25$ and $t = 26$, as expected.

Note that if you round differently in the various steps, your answer could change considerably. See what happens, for instance, if you were to use $\frac{20}{9.75} = 2.051$, or 2.05 instead of 2.05128.

Alternatively, we could solve this problem somewhat differently by not first dividing 20 by 9.75. Instead, consider

$$9.75 \, (1.029)^t = 20.$$

If we take logs of both sides and apply both the first and the second property of logs, then we get

$$\log [9.75 \, (1.029)^t] = \log 20$$
$$\log 9.75 + \log (1.029)^t = \log 20$$
$$\log 9.75 + t \cdot \log 1.029 = \log 20.$$

So

$$t \cdot \log (1.029) = \log 20 - \log 9.75 = 0.3120,$$

and therefore

$$t = \frac{0.3120}{\log 1.029} = 25.13,$$

which is the same result we found above.

The definition of the logarithm also gives us the following two formulas.

Fundamental Logarithmic–Exponential Identities

$$\log 10^x = x \qquad \text{for all real } x$$
$$10^{\log x} = x \qquad \text{for all } x > 0$$

Because these formulas hold for all appropriate values of x, they are called **identities**. Think about the two results to be sure you understand them thoroughly. For the first, $\log 10^x$ is that power of 10 needed to produce 10^x. Clearly, that power must be x itself. For the second identity, the exponent in $10^{\log x}$ is $\log x$. But $\log x$ is the power of 10 that gives the number x. So 10 raised to the $\log x$ power gives simply x.

Unfortunately, logarithms cannot always be used to find exponents. An equation such as

$$3^x = x^4$$

involves both a power function and an exponential function. This equation can be solved numerically or graphically (that is, finding an approximate solution), to any desired degree of accuracy by using a graphing calculator or a computer. But the equation cannot be solved algebraically (that is, finding an exact solution).

Applications of Logarithmic Functions

Logarithms have many applications. For instance, chemists use a quantity known as the pH to measure how acidic a water solution is. The pH is based on the concentration of hydrogen-ions (measured in moles per liter) in the solution. The hydrogen–ion concentration of pure water is 10^{-7} moles per liter. Thus the pH of pure water is

$$pH = -\log \text{(concentration)} = -\log (10^{-7}) = -(-7) = 7,$$

which is used as the reference point for a neutral solution. Water solutions whose pH values are less than 7 are said to be acidic whereas water solutions with pH values greater than 7 are basic, or alkaline. In fact, the lower the pH is, the more acidic the solution; the higher the pH is, the more basic the solution. For example, orange juice, which is somewhat acidic, has a hydrogen–ion concentration of 2×10^{-4} moles per liter and so its pH is

$$-\log (2 \times 10^{-4}) = -[\log 2 + \log (10^{-4})] = -[0.301 - 4] \approx 3.7.$$

Hydrochloric acid, with a hydrogen-ion concentration of 10^{-1} moles per liter, has a pH of 1, which indicates that it is extremely acidic. On the other hand, human blood, with a concentration of 4×10^{-8} moles per liter, has a pH of $-\log (4 \times 10^{-8}) = 7.4$ and is slightly basic; household ammonia, with a pH of 11.5, is extremely basic. By now you should realize that each one-point decrease in the pH value represents a tenfold increase in the hydrogen–ion concentration.

EXAMPLE 3

The crust of the Earth is composed of about 20 rigid plates that "float" on the liquid magma (the molten material beneath the Earth's crust). The study of this phenomenon is called "plate tectonics." A geologic fault, such as the famous San Andreas Fault in California, is the "space" between two plates. As the plates move, they bump into one another and sometimes one plate passes slightly under another, causing the upper plate to shift and heave. The result is an earthquake on the Earth's surface. There are about a million earthquakes, mostly very minor, each year. The American seismologist Charles Richter developed a way of measuring the intensity of an earthquake. The *Richter scale* is based on the idea that there is a minimum level of earthquake intensity, denoted by I_0, which is noticeable. I_0 is the threshold level below which we would not be aware of a quake. Any stronger quake has an intensity denoted by I. The Richter scale relates the magnitude R of an earthquake to its intensity:

$$R = \log \frac{I}{I_0}.$$

That is, the magnitude given by the Richter scale measurement is the logarithm of the ratio of the actual intensity to the threshold level.

How are different measurements on the Richter scale related? For instance, if the measurement for one earthquake is double that of another, how much greater is it? How extreme is the largest recorded earthquake whose Richter scale reading was 8.9? For comparison, the energy involved in a threshold level earthquake is approximately equal to the energy released by 10,000 atomic bombs.

Suppose that an earthquake measures $R = 5$ on the Richter scale, so that

$$\log \frac{I}{I_0} = 5$$

and therefore

$$\frac{I}{I_0} = 10^5 = 100,000.$$

Since $I = 100,000 I_0$, the intensity of such a quake is 100,000 times the threshold level: This quake's energy is equivalent to roughly $100,000 \times 10,000 = 10^9$, or one billion atomic bombs exploding simultaneously.

How does this compare to an earthquake measuring $R = 6$ on the Richter scale? We now get

$$\frac{I}{I_0} = 10^6 = 1,000,000.$$

The intensity of this quake is one million times the threshold level. Thus an increase of 1 Richter unit corresponds to a tenfold increase in the intensity of the earthquake.

Similarly, for the largest recorded quake, with $R = 8.9$, we have

$$8.9 = \log \frac{I}{I_0},$$

so that

$$\frac{I}{I_0} = 10^{8.9} = 794,328,234.$$

This quake had an intensity almost 800 million times greater than the threshold level!

Suppose one quake has a reading that is twice another on the Richter scale. How much stronger is it? Is it four times as strong? Is the relative intensity the same? Does it depend on the value for R? Let's compare $R = 4$ to $R = 2$ to see what happens. (We ask you to compare $R = 6$ to $R = 3$ in the exercises at the end of this section.) With $R = 4$, we have

$$R = 4 = \log \frac{I}{I_0}$$

so that

$$I = 10^4 \cdot I_0.$$

For $R = 2$, we have

$$R = 2 = \log \frac{I}{I_0},$$

so that

$$I = 10^2 \cdot I_0.$$

Therefore a magnitude 4 quake is actually $\frac{10^4}{10^2} = 100$ times stronger than a magnitude 2 quake.

Changing Bases

Throughout this book, we will be working with logarithms to the base 10 to undo exponential functions of the form $y = k \cdot 10^x$. However, it is possible to have bases other than 10, say $a = 2$ or $a = 1.045$, as the base for an exponential function $y = k \cdot a^x$. Each possible base gives rise to a corresponding logarithmic function. For instance, we could work with logarithms to the base 2, written $\log_2 x$.

Definition of Logarithms to Base a

$$\log_a x = y \quad \text{means} \quad a^y = x.$$

The logarithm to the base a of x is that power of a needed to produce x.

In practice, there is one particular base besides 10 that is widely used. This base is the number $e = 2.71828\ldots$, and the corresponding logarithm is called the **natural logarithm.** You will see why e is important when you study calculus. Even though we could write log to the base e as $\log_e x$, it is customary to write $\ln x$, which is often read as "lin of x."

Although $\log_{10} 10 = 1$, we have $\ln 10 = 2.3026$ because $e^{2.3026} \approx 2.71828^{2.3026} = 10.0001$. Similarly, while $\log_{10} 100 = 2$, with the natural logarithm, $\ln 100 = 4.6052$ because $e^{4.6052} \approx 2.71828^{4.6052} = 100.003$.

We previously said that all of the properties of logarithms apply no matter what base is used. Thus if we work with base e, we have the following properties.

---

1. $\ln (A^p) = p \cdot \ln A$
2. $\ln (A \cdot B) = \ln A + \ln B$
3. $\ln \dfrac{A}{B} = \ln A - \ln B$
4. $\ln e^x = \log_e e^x = x$
5. $e^{\ln x} = x, \quad \text{if } x > 0$

Furthermore, it is sometimes convenient to have a way of converting either an exponential function or logarithm in one base into an exponential function or logarithm in a different base. That is, for any x, how do we convert a^x to 10^x or convert $\log x$ to $\ln x$, and vice versa? Let's first look at the question of converting bases of exponential functions.

We saw that the population of Florida can be modeled by the exponential function $P(t) = 9.75(1.029)^t$. How do we convert this into an equivalent expression that involves base 10 or base e? Suppose we try to find the appropriate power q so that

$$(1.029)^t = 10^q.$$

If we take logs of both sides, we find

$$t \log (1.029) = q \log 10 = q \cdot 1 = q,$$

so the expression for the population becomes

$$P(t) = 9.75(1.029)^t = 9.75(10^{t \log (1.029)}) = 9.75(10^{0.0124t}).$$

Alternatively, you might think of this result as coming from

$$10^{t \log (1.029)} = (10^{\log (1.029)})^t = (1.029)^t.$$

Suppose we now want to convert the expression for the Florida population into base e. Using the property $e^{\ln x} = x$, we see that

$$(1.029)^t = (e^{\ln 1.029})^t.$$

Therefore,

$$P(t) = 9.75(1.029)^t = 9.75(e^{\ln 1.029})^t = 9.75(e^{0.0286t}).$$

These three expressions for the population of Florida are mathematically equivalent—just the bases are different. Graph the three functions given for $P(t)$ using your function grapher and convince yourself that they are truly identical.

Now let's consider the problem of converting a log in one base to a log in another base. To see how to proceed, we begin by looking at some typical values of $\log x$ and $\ln x$, rounded to four decimal places, as shown in the following table. To see if there is any clear relationship between the two sets of logarithmic values, we plot the values of $\ln x$ versus $\log x$, as shown in Figure 2.28.

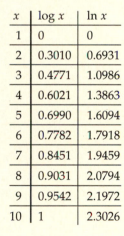

x	$\log x$	$\ln x$
1	0	0
2	0.3010	0.6931
3	0.4771	1.0986
4	0.6021	1.3863
5	0.6990	1.6094
6	0.7782	1.7918
7	0.8451	1.9459
8	0.9031	2.0794
9	0.9542	2.1972
10	1	2.3026

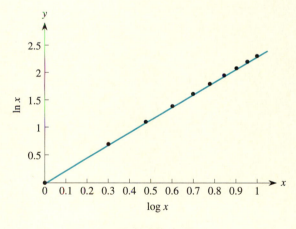

FIGURE 2.28

From the graph, we see there is a linear pattern relating them. Since the line passes through the origin, its vertical intercept is 0, and so we can write

$$\ln x = m \cdot \log x,$$

for some constant of proportionality m. You also can see this fact by looking at the ratio of ln x and log x for any value of x—in every instance, the ratio always will be approximately equal to 2.3026. Check this ratio on your calculator using several different values of x. Realize that Figure 2.28 shows the values of ln x plotted against those of log x. If we plotted either ln x against x or log x against x, we would get the graph of a *logarithmic function*—only ln x versus log x results in a straight line.

The value of the constant of proportionality, $m = 2.3026$, is also the slope of the line through the points shown in Figure 2.28. Thus, we can write

$$\ln x = 2.3026 \log x$$

for any x. Moreover, notice that 2.3026 appears in the last row of the preceding table as precisely the value of ln 10, so we can rewrite this relationship as

$$\ln x = (\ln 10) \cdot \log x,$$

or equivalently,

$$\log x = \frac{\ln x}{\ln 10}.$$

Rewriting this equation to highlight the base of the logarithm, we get

$$\log_{10} x = \frac{\log_e x}{\log_e 10}.$$

In fact, if we perform the identical analysis with any other base, say a, instead of base e, then we obtain the comparable result for changing between base 10 and base a, for any a:

$$\log_{10} x = \frac{\log_a x}{\log_a 10}.$$

EXERCISES

1. The graphs of the following functions may be surprising to you. Use your function grapher to graph each function and explain what you see.

   a. $y = \log 10^{2x}$
   b. $y = \log (2x) - \log (x)$
   c. $y = \log 10^{x^2}$
   d. $y = 10^{\log (x^2)}$
   e. $y = \log 3^x$
   f. $y = \log \left(\frac{10}{6^x}\right)$

2. Simplify each of the following expressions.

   a. $\log x + \log x^2 + \log x^3$

   b. $\log x^2 + \log y^3 - \log x - \log y^2$

   c. $\log x + \log \sqrt{x}$

   d. $\log \dfrac{x}{y} - \log \dfrac{y}{x}$

   e. $\log 10^{x^2}$

   f. $10^{\log (x^2)}$

3. Use your function grapher to draw simultaneously the graphs of $y = \log (2^x)$, $y = \log (3^x)$, and $y = \log (5^x)$. For each function use the properties of logarithms to explain why you get the graph you see.

4. In computer science the efficiency of algorithms (methods for accomplishing a certain task) are often analyzed by how long it takes to perform the operation with n objects. Typically, as n increases, the time involved for the algorithm increases significantly. Two different algorithms used to put a set of names in alphabetical order are compared. For one algorithm, the time needed to order n names, as a function of n, is $B(n) = \frac{1}{2}n^2$.

   The time for the other algorithm, as a function of n, is $S(n) = n \cdot \log n$. Which algorithm is faster?

5. The population of Argentina was 32 million in 1988 and was growing exponentially at an annual rate of 1.4%.

   a. Find an expression for Argentina's population at any time t.
   b. What population would you predict for the year 2000 if the present trend continues?
   c. What is the doubling time?

6. The population of Kenya is growing exponentially. Its population was 23.3 million people in 1988 ($t = 0$) and 27.0 million in 1993.

   a. Find an expression for the population at any time t.
   b. What would the population be in the year 2000?
   c. What is the doubling time?

7. The Best Company earned $50 million in the year 1995 and its income is growing at a rate of 2% per year. The Acme Corporation earned $30 million that year and its income is growing at a rate of 6.5% a year. When will Acme overtake Best in annual income?

8. Due to ardent fishermen during the summer months, the population of fish in a lake is reduced by 10% each week. Find the half-life of this dwindling fish population.

9. a. Find the doubling time for annual growth rates of 3%, 4%, 5%, 6%, and 7%.
   b. Consider the doubling time d as a function of growth rate r. Plot your results from part (a) and decide what type of function seems to fit the behavior pattern you observe.

10. Bankers use a technique called the *Rule of 70* to estimate the doubling time for money invested at different interest rates. They divide 70 by 100 times the interest rate. Thus for an interest rate of 10% = 0.10, bankers would estimate the doubling time to be

$$\frac{70}{100 \cdot 0.10} = \frac{70}{10} = 7 \text{ years.}$$

Use your results from Exercise 9 to test how accurate this method actually is.

11. Assuming that inflation continues at a rate of 3% per year, determine when the cost of first-class postage for a letter will reach $1. (First-class postage rose to 29¢ in 1990 and to 32¢ in 1995.)

12. How much stronger is a magnitude 6 earthquake than a magnitude 3 earthquake?

13. How much stronger is

a. a magnitude 7 quake than a magnitude 5 quake?
b. a magnitude 7 quake than a magnitude 4 quake?

14. Let I_0 be the minimum (or threshold) level of sound that can be heard by humans. If the intensity of a particular sound is I, then we measure the magnitude of the sound by the number of decibels d, given by

$$d = 10 \log \left(\frac{I}{I_0}\right).$$

a. Normal conversation measures about 60 decibels. How much more intense is this level than the threshold level?
b. A loud noise of about 150 decibels will cause deafness. How much more intense is this level than the threshold level?
c. An aircraft taking off has a loudness level of about 120 decibels. How much more intense is this level than the threshold level?
d. How loud (that is, how many decibels) is a sound whose intensity is one million times the threshold level?
e. The noise level from a rock band is about 100 billion times higher than the threshold level. What is the decibel value of this noise level?

15. Solve for x:

a. $7^x = 11$
b. $1.05^x = 2$
c. $3 \cdot (1.04)^x = 5$
d. $4 \cdot (1.05)^x = 5 \cdot (1.04)^x$
e. $12 \cdot (0.86)^x = 3$
f. $9 \cdot (0.17)^x = 0.25.$

16. The points P, Q, and R lie in order from left to right on the graph of a function f that is increasing. If the slope of line segment PQ is less than that of line segment QR, is the curve concave up or concave down? Explain your reasoning.

17. Picture a water slide at an amusement park. The slide starts at an initial height H_0 above the pool and smoothly drops to water level. The slide is first concave down, then concave up, then concave down, and finally concave up.

    a. Sketch a graph of the slide's height H above water level as a function of horizontal distance x.
    b. Suppose you go down the slide in a sitting position. Sketch a graph of the height of your eye level above the water line as a function of horizontal distance x.
    c. Sketch the graph of the height of your eye level above the water line as a function of time t.
    d. Sketch the graph of your speed as a function of time t.

18. Consider the graph of the function $f(x) = 10^x$ and the line L shown in the accompanying figure.

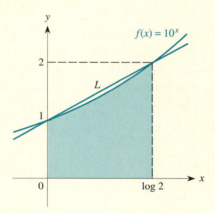

    a. Find the equation of the line L.
    b. Since the line L looks close to the curve for values of x between 0 and $\log 2$, use the equation of L to estimate the values of $10^{0.03}$, $10^{0.25}$, and $10^{0.33}$. How close are these estimates to the true values?
    c. Since the line L looks close to the curve for $0 < x < \log 2$, use it to estimate the area of the shaded region.

2.6 *Comparing Rates of Growth*

Most of the families of functions we have studied in this chapter—linear, exponential, and logarithmic functions—are either strictly increasing or strictly decreasing. If we restrict our attention to nonnegative values of x, then power functions also are either strictly increasing or strictly decreasing. Before going on, let's summarize what we know so far.

Function	Equation	Behavior	Graph
Linear	$y = mx + b$	Strictly increasing when $m > 0$. The larger the slope, the faster the rate of increase. Strictly decreasing when $m < 0$. The more negative the slope, the greater the rate of decrease.	
Exponential	$y = a^x$ (for $a > 0$)	Strictly increasing when growth factor $a > 1$. The larger a is, the faster the function grows. Strictly decreasing when growth factor $a < 1$. The smaller a is, the faster the function decays toward zero. Exponential graphs are always concave up.	
Power	$y = x^p$ (for $x \geq 0$)	Strictly increasing when $p > 0$. The larger p is, the faster the function grows. If $p > 1$, the graph is concave up—it grows more and more rapidly. If $0 < p < 1$, the graph is concave down—it grows more and more slowly. Strictly decreasing when $p < 0$. The more negative p is, the faster the function decays toward zero. If $p < 0$, the graph is concave up.	

Function	Equation	Behavior	Graph
Logarithmic	$y = \log_a x$ (for $x > 0$ and $a > 0$)	Strictly increasing. Logarithmic graphs are always concave down.	

This summary of information involves comparing the growth or decay rate of one function in a family to that of another function in the same family. In this section, we look at how the growth or decay rate of one function in a family compares to the growth or decay rate of a function in a different family. In particular, we want to answer two questions: First, which family of functions grows fastest? Second, which family of functions decays to zero fastest?

Exponential Versus Power Functions: Which Wins the Race?

Power functions ($y = x^p$, $p > 1$) and exponential growth functions ($y = a^x$, $a > 1$) both grow rapidly as x increases. But, do they grow at roughly the same rate, or does one grow much faster than the other?

Suppose we consider $y = 2^x$ and $y = x^4$ for $x \geq 0$. We know that every power function passes through the origin and that every exponential curve of the form $y = a^x$ crosses the vertical axis at $y = 1$ (since $a^0 = 1$). So let's begin by comparing these two functions for small values of x. The *local*, or close-up, view, in Figure 2.29 shows that between $x = 0$ and $x = 1$, the graph of $y = 2^x$ is above the graph of $y = x^4$, but the power function seems to be growing more rapidly. In fact, if you extend the interval somewhat, you find that by the time $x = 2$, the power function has surpassed the exponential function. (Where does this happen?) Using a somewhat more *global*, or large-scale, view such as that shown in Figure 2.30, you see that the power function continues to pull away from the exponential function.

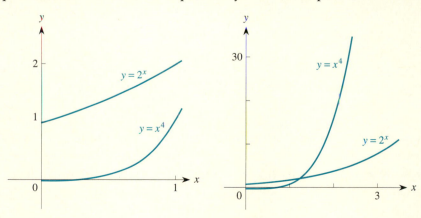

FIGURE 2.29   FIGURE 2.30

However, in Figure 2.31, which shows the interval from $x = 0$ to $x = 20$, we see that the exponential curve has again overtaken the power curve. (Where does this happen?) In a truly global view such as that shown in Figure 2.32 over the interval from $x = 0$ to $x = 25$, we see that for large x-values, $y = x^4$ is quite insignificant compared to $y = 2^x$. In fact, $y = 2^x$ is growing so much faster than $y = x^4$ that its graph appears almost vertical in comparison to the relatively slow growth of $y = x^4$. Verify this comparison numerically by trying several different values of x, say $x = 1$, $x = 10$, and $x = 50$ (but don't go too far because you could exceed your calculator's capacity).

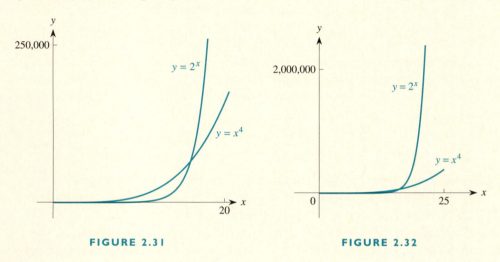

FIGURE 2.31   FIGURE 2.32

Think About This

Use your function grapher to find to two decimal places the points where $y = 2^x$ and $y = x^4$ cross.

The behavior pattern you saw here is typical of any power function $y = x^p$ compared to any exponential function $y = a^x$, for $a > 1$. Although the exponential function starts out more slowly than the power function for small values of x, the exponential function will eventually dominate $y = x^p$ for any value of p, and so it always wins the race toward infinity.

Think About This

Plot $y = 3^x$ and $y = x^5$ for $0 \le x \le 20$ with $0 \le y \le 300,000$ to see where the exponential function overtakes the power function.

You have already seen that a positive constant multiple does not change the overall shape or behavior of a function. For instance, the power function $y = 5000x^3$ has the same shape as the power function $y = x^3$, but it grows much more rapidly. We know that the exponential function $y = 2^x$ eventually overtakes the power function $y = x^3$. It also eventually overtakes the power function $y = 5000x^3$; it just takes longer. The only question is: Where does it happen? In the long run, the exponential function invariably winds the race.

E X A M P L E

Estimate the point x where $f(x) = 1.05^x$ finally overtakes $g(x) = x^{10}$.

Solution   We know that the power function $g(x) = x^{10}$ grows very rapidly and that the exponential function $f(x) = 1.05^x$ has a fairly small growth factor of 1.05. Let's look at their respective function values for different values of x.

x	$g(x) = x^{10}$	$f(x) = (1.05)^x$
10	10^{10}	1.62889
100	10^{20}	131.501
1000	10^{30}	1.5463×10^{21}
10,000	10^{40}	7.8161×10^{211}

From this comparison, it is evident that the exponential function has overtaken the power function sometime after $x = 1000$ but long before $x = 10,000$, where $f(x) = (1.05)^x$ has far exceeded the value of $g(x) = x^{10}$. Suppose we try to narrow our search by trying a few additional values of x, as displayed in the following table.

x	$g(x) = x^{10}$	$f(x) = (1.05)^x$
2000	1.024×10^{33}	2.3911×10^{42}
1500	5.7665×10^{31}	6.0806×10^{31}
1499	5.7282×10^{31}	5.7911×10^{31}
1498	5.6901×10^{31}	5.5153×10^{31}

Therefore, we conclude that the exponential function $f(x) = (1.05)^x$ finally overtakes the power function $g(x) = x^{10}$ just before $x = 1499$.

What do you think happens in the case of a decaying exponential function and a power function with a negative exponent? For example, consider the functions $y = \frac{1}{2^x}$ and $y = x^{-2} = \frac{1}{x^2}$. We know that both their graphs eventually approach the x-axis as a horizontal asymptote, but which one approaches the x-axis faster? You could compare them graphically to see the result. You also could compare them numerically by trying several x-values, such as $x = 100$ or $x = 1000$. Alternatively, we can reason as follows: because 2^x is eventually larger than x^2, we know that $\frac{1}{2^x}$ is eventually smaller than $\frac{1}{x^2} = x^{-2}$. So the graph of $y = \frac{1}{2^x}$ is eventually below the graph of $y = x^{-2}$. See Figure 2.33.

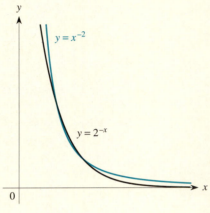

This behavior pattern is typical. All decreasing exponential functions invariably will approach 0 faster than any power function with a negative exponent as $x \to \infty$. A power function could begin dropping at a faster rate when compared to an exponential function; for example, compare $y = x^{-100}$ with $y = 0.9^x$. However, as $x \to \infty$, the exponential function eventually will decay faster than the power function to win the race toward 0. The only question is: When does the decaying exponential function overtake the decaying power function on the way to zero?

FIGURE 2.33

Logarithms and Power Functions: Which Wins the Race?

When you examine a table of values for $f(x) = \log x$, it is evident that the logarithm grows very slowly as x increases beyond $x = 1$. In fact, the logarithm grows more slowly than any positive power of x. Figure 2.34 shows the graphs of $y = \log x$ and $y = x^{1/3}$. Notice that, in the long run, the power function beats the log function in the race toward infinity.

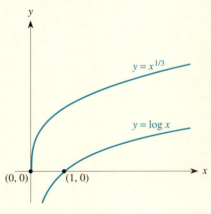

We encourage you to compare the behavior of different power functions, say $y = x^{1/2}$ or $y = x^{1/10}$ to $y = \log x$ on your function grapher to convince yourself how slowly the log function grows.

Furthermore, both logarithmic functions and power functions with power $0 < p < 1$ grow more slowly than linear functions with slope $m > 0$. This fact follows because we can think of the linear function $y = x$ as a power function with $p = 1$, which grows faster than any power function with a smaller

FIGURE 2.34

power p. Also you know that both logarithmic functions and power functions with power $0 < p < 1$ are concave down, and therefore they grow more slowly than linear functions.

In summary, we have the following facts:

Concave Up Growth Functions

Power functions with power $p > 1$ grow faster than linear functions with slope $m > 0$.

Exponential functions with growth factor $a > 1$ grow faster than power functions with power $p > 1$.

Concave Down Growth Functions

> Power functions with power $0 < p < 1$ grow more slowly than linear functions with slope $m > 0$.
> Logarithmic functions grow more slowly than power functions with power $0 < p < 1$.

Decay Functions

> Exponential functions with decay factor $a < 1$ decay more rapidly than power functions with power $p < 0$.

E X E R C I S E S

1. Use your function grapher to graph $y = x^3$ and $y = 2^x$. Determine appropriate x- and y-scales to obtain the diagrams shown.

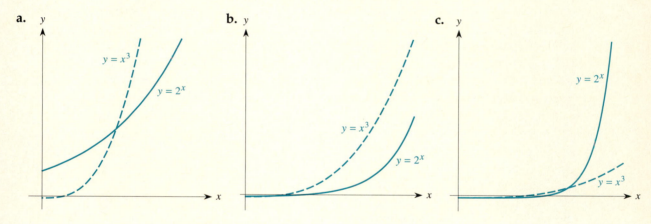

a. y $y = x^3$ $y = 2^x$ b. y $y = x^3$ $y = 2^x$ c. y $y = 2^x$ $y = x^3$

2. a. For what values of x is $2^x > x^2$?

   b. For what values of x is $3^x > x^3$?

   c. For what values of x is $4^x > x^4$?

3. Use your function grapher to estimate when $y = (0.6)^x$ overtakes $y = x^{-6}$ as they both decay to zero.

4. Estimate where $f(x) = x^{0.15}$ finally overtakes $g(x) = \log x$ on their "turtle-versus-snail" race toward infinity.

5. The functions $y = x^2$ and $y = 2^x$ intersect at $x = 2$ and at $x = 4$; $y = x^3$ and $y = 3^x$ intersect at $x = 3$ and at $x \approx 2.478$. In general, for $x \geq 0$ the graphs of $f(x) = x^p$ and $g(x) = p^x$ intersect at two points except for one specific value of p (with $p > 1$) for which the curves intersect at only one point. Use your function grapher and trial and error to locate the one special value of p (accurate to two decimal places) for which the curves $y = x^p$ and $y = p^x$ intersect at only one point.

2.7 *Inverse Functions*

Countries using the metric system report all temperatures in degrees Celsius. Thus an American visiting Canada who wants to know the temperature in degrees Fahrenheit must be able to convert the Celsius readings via the formula

$$F = \frac{9}{5}C + 32.$$

This expression indicates that the Fahrenheit measurement is a function of the Celsius measurement. Canadian visitors to the United States face the reverse problem: They must convert Fahrenheit readings into Celsius using the related function

$$C = \frac{5}{9}(F - 32).$$

These two functions have the effect of undoing each other. For that reason, they are called **inverse functions**.

In general, the *inverse* f^{-1} of a function f is a function that "reverses" or "undoes" f. The two functions F and C, which give temperature conversions between the two systems of measurements, represent a pair of inverse functions because they undo each other. For relatively simple cases such as this one, it is straightforward to determine the inverse function f^{-1} (read as "f inverse") for a given function f. We just solve the original expression algebraically for the independent variable in terms of the dependent variable. For instance, if $F = \frac{9}{5}C + 32$, then

$$\frac{9}{5}C = F - 32$$

$$C = \frac{5}{9}(F - 32).$$

So, $C = \frac{5}{9}(F - 32)$ is the inverse function.

Similarly, suppose we have the function $f(x) = x^3$, which transforms any real number x into the cube of that number, x^3. To undo this function, we want to transform any number that has been cubed into the number that it came from; that is, we want to extract the cube root. Therefore, if

$$y = f(x) = x^3, \text{ then } x = \sqrt[3]{y} = f^{-1}(y)$$

is the inverse function. Consequently, if we need to find the value of x for which $x^3 = 472$, say, then we extract the cube root of 472:

$$x = \sqrt[3]{472} = 472^{1/3} \approx 7.786.$$

You can check that this result is correct by cubing 7.786 to get roughly 472. In an analogous way, if we need to find the value of x for which

$$x^{1175} = 2, \quad \text{then} \quad x = \sqrt[1175]{2} = 2^{1/1175} = 1.00059.$$

If we need to determine the value of the base b for which

$$b^{365} = \frac{1}{2}, \quad \text{then} \quad b = \sqrt[365]{\frac{1}{2}} = \left(\frac{1}{2}\right)^{1/365} = 0.99810.$$

To verify that these results are correct, just calculate $(1.00059)^{1175}$ and $(0.99810)^{365}$. Although these two examples may seem bizarre to you, we will be doing such operations routinely in later chapters because they allow us to answer interesting questions.

Extracting the appropriate root from the preceding functions was quite simple. Unfortunately, complications arise if the power is even. For example, consider the problem of solving $y = f(x) = x^2 = 25$. To find x, we take the square root of both sides to get

$$x = \pm\sqrt{25} = \pm 5.$$

Notice that we get two different answers. We say that the function $f(x) = x^2$ does not have an inverse because we cannot undo its effects uniquely. Thus not all functions have inverses. Later in the section we discuss this situation in more detail.

Suppose you want to find the power of 10 that produces 1024. You may realize immediately from the definition of logarithm that this is precisely log 1024. Alternatively, you could solve the equation $1024 = 10^x$ for x. Taking the log of both sides, you get

$$\log (10^x) = \log 1024$$

$$x \cdot \log 10 = x = \log 1024$$

$$x = \log 1024 \approx 3.0103.$$

Thus we see that the logarithm to the base 10 undoes the effect of the exponential function 10^x. Therefore, the logarithm function is the inverse function of the exponential function.

Be sure you understand the difference between these two situations. To extract an unknown variable that appears as the base in a power function $y = x^p$, take the corresponding pth root. To extract an unknown variable that appears in the exponent of an exponential function $y = a^x$, take the logarithm.

We have said that a function f and its inverse f^{-1} undo each other. Let's see what this means. Suppose we start with a number x. The function f carries x into the corresponding value of $y = f(x)$. The inverse function f^{-1} carries y back into the original x. That is,

$$y = f(x) \quad \text{and} \quad x = f^{-1}(y).$$

Similarly, if we start with any value of y, then f^{-1} maps y into the value of x associated with it, so that $f^{-1}(y) = x$. The function f carries x back into the original y. That is,

$$x = f^{-1}(y) \quad \text{and} \quad y = f(x).$$

Graph of the Inverse Function

It is sometimes desirable to express both f and f^{-1} as functions of the same variable. With the cube function, we previously wrote

$$y = f(x) = x^3 \quad \text{and} \quad x = f^{-1}(y) = \sqrt[3]{y}.$$

Instead, we can write these two functions as

$$y = f(x) = x^3 \quad \text{and} \quad y = f^{-1}(x) = \sqrt[3]{x}.$$

We want the same independent variable so we can draw the graphs of the two functions on the same set of axes to compare their behavior easily. In Figure 2.35, we display both the original function $f(x) = x^3$ and the inverse function $f^{-1}(x) = \sqrt[3]{x}$. Note that if we did not interchange x and y for the inverse function, then the two formulas $y = x^3$ and $x = \sqrt[3]{y}$ would represent the identical curve, so that we would see only one curve. By interchanging the two variables to express the inverse function also as a function of x, namely $y = \sqrt[3]{x}$, we can

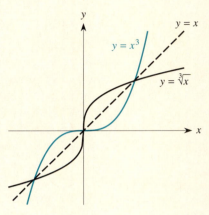

FIGURE 2.35

compare the two related graphs, as shown in Figure 2.35. Notice that the graph of the function $f(x) = x^3$ and its inverse $f^{-1}(x) = \sqrt[3]{x}$ are mirror images of each other about the line $y = x$.

As we pointed out above, the exponential function $y = 10^x$ and the logarithmic function $y = \log x$ are inverses of each other. If $y = f(x) = 10^x$, then we can solve for x by taking the log of both sides to get

$$\log y = \log (10^x) = x = f^{-1}(y).$$

We now interchange the variables so that x is the consistent independent variable and write the inverse function as $f^{-1}(x) = \log x$. Recall that in Section 2.5, we saw that the graphs of these two functions were also mirror images of each other about the line $y = x$, as shown in Figure 2.27.

It is important to realize that $f^{-1}(x)$ is not the same as $\frac{1}{f(x)}$. For instance, if $f(x) = x^3$, then $f^{-1}(x) = x^{1/3}$, while $\frac{1}{f(x)} = \frac{1}{x^3} = x^{-3}$. Similarly, if $g(x) = 10^x$, then $g^{-1}(x) = \log x$, but $\frac{1}{g(x)} = \frac{1}{10^x} = 10^{-x}$, which is not the same as $\log x$. In Figure 2.36 we show the graphs of $g(x) = 10^x$ and $\frac{1}{g(x)} = 10^{-x}$. Notice that they are *not* mirror images of each other about the line $y = x$. Using the symmetry condition, describe where the graph of the inverse function of g would be in Figure 2.36.

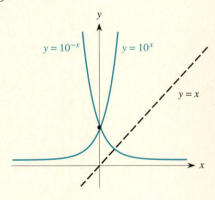

FIGURE 2.36

The following uses of inverse functions will occur repeatedly throughout the rest of this book. Be sure you understand and can apply them:

To solve for the variable x from a power function $y = x^n$, we extract the nth root. That is,

$$\text{if } y = f(x) = x^n, \text{ then } x = f^{-1}(y) = y^{1/n}.$$

In other words, $y = f^{-1}(x) = x^{1/n}$ is the inverse of $y = f(x) = x^n$. If n is even, we must have $x \geq 0$.

To solve for the variable x from an exponential function $y = a^x$, we take logarithms. That is, for $a = 10$,

$$\text{if } y = f(x) = 10^x, \text{ then } x = f^{-1}(y) = \log y.$$

In other words, $y = f^{-1}(x) = \log x$ is the inverse of $y = f(x) = 10^x$.

Not Every Function Has an Inverse

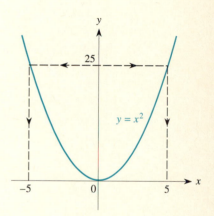

FIGURE 2.37

Using $f(x) = x^2$ as an example, we saw that not every function f has an inverse f^{-1}. In fact, no power function with an even power can have an inverse. Now let's see why. Consider the function $f(x) = x^2$ graphically to see what actually is happening. The graph of $f(x) = x^2$ is the parabola shown in Figure 2.37. When we take $x = -5$ and square it, we get a point on the parabola that is at a height of 25 units. The inverse function, if one exists, should undo this operation. That is, if we start at a height of $y = 25$, we should be able to work backward to end up at the original value $x = -5$. However, there are two different points on the parabola at a height of 25, one corresponding to $x = -5$ and the other corresponding to $x = +5$. Since we cannot reverse the process uniquely to get only the value $x = -5$, the function $f(x) = x^2$ does not have an inverse.

However, we can restrict the domain of this function to make it possible to produce a partial inverse. Suppose we limit our attention to only nonpositive values of x. Let's consider the function $g(x) = x^2$ for $x \leq 0$. In this case, if we start with any nonpositive value for x and square it to get x^2, we can undo the result of the squaring operation by taking the square root and accepting only the nonpositive value. Thus the function $g(x) = x^2$ *with domain restricted to $x \leq 0$* has an inverse, $g^{-1}(x) = -\sqrt{x}$. Alternatively, if we restrict our domain to values of $x \geq 0$, we could also uniquely undo the results of squaring.

Can we come up with a simple criterion to determine if a function f has an inverse? Definitely! In fact, there are several simple criteria we can use. Again, let's illustrate with the function $f(x) = x^2$. We know that it has an inverse if we restrict its domain to either $x \geq 0$ or $x \leq 0$. It does not have an inverse if we allow the domain to include both positive and negative values for x. When we consider only $x \geq 0$, we are considering only the right-hand side of the parabola and that is where the function is *strictly increasing*. When we consider only $x \leq 0$, then we are considering only the left-hand side of the parabola and that is where the function is *strictly decreasing*. In both instances, the restricted function has an inverse. When we allow both positive and negative values for x, the function first decreases and then increases, and so it does not have an inverse. On the other hand, we know that the function $h(x) = x^3$ has an inverse $h^{-1}(x) = \sqrt[3]{x}$ without any restrictions on x. We also know that this function is strictly increasing for all values of x, as seen in Figure 2.35.

These observations suggest a simple criterion for functions to have an inverse: The function must be either strictly increasing or strictly decreasing. We call such a function **monotonic**. Picture the two functions shown in Figure 2.38. The first is strictly increasing and, for any desired height y, it is obvious that we can undo the effect of the function and come back to a unique x that produced that particular y. In contrast, the second function increases and decreases over different intervals. There are some heights, such as y_3 and y_4, where the curve is bending up and down, so different x-values correspond to the same y-value. Thus it is not possible to find the unique x that produced a particular y-value.

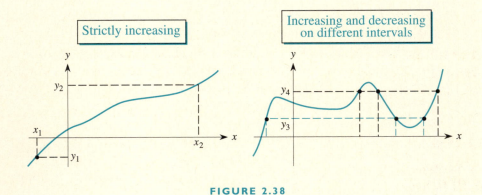

FIGURE 2.38

An alternative criterion is analogous to the vertical line test we discussed in Section 1.4 for determining if a curve represents a function. Recall that a curve represents a function if every vertical line crosses the curve at most once. This means that given any value of x, there is one and only one value of y corresponding to it. In a totally analogous way, we can use the *horizontal line test* to determine if a function has an inverse: If every horizontal line crosses the curve at most once, then a function f has an inverse f^{-1}. This means that given any height y, there is one and only one value of x corresponding to that height.

Now that we have criteria for knowing if a function f has an inverse, there is one important question to address: How does f^{-1} behave, especially in relationship to f? Let us interpret this question graphically for three pairs of inverse functions we already have seen to get an insight. The pairs of functions are

1. The conversions between Fahrenheit and Celsius scales shown in Figure 2.39 where

$$f(x) = \frac{9}{5}x + 32 \quad \text{and} \quad f^{-1}(x) = \frac{5}{9}(x - 32);$$

2. The exponential and log functions shown in Figure 2.40 where

$$g(x) = 10^x \quad \text{and} \quad g^{-1}(x) = \log x, \quad x > 0;$$

3. The squaring and square root functions shown in Figure 2.41 where

$$h(x) = x^2 \quad \text{and} \quad h^{-1}(x) = \sqrt{x}, \quad x \geq 0.$$

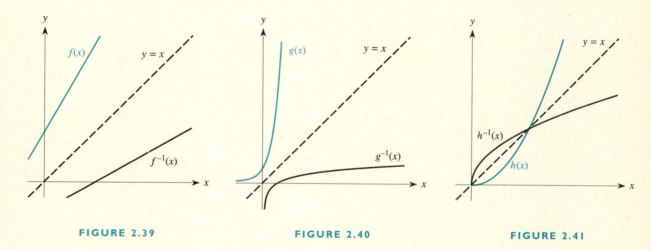

FIGURE 2.39   FIGURE 2.40   FIGURE 2.41

Notice that each pair of curves can be interpreted as reflections, or mirror images, of each other about the line $y = x$.

In general, the graphs of a function f and its inverse f^{-1} are always mirror images of each other about the line $y = x$. Further, in order for the inverse to exist, the function f must be either strictly increasing or strictly decreasing. Consequently, the graph of the inverse function f^{-1} also is monotonic. In particular, both functions increase or both decrease. However, there are no clear patterns for concavity. It is possible for both f and f^{-1} to be concave up, for both to be concave down, for each to have opposite concavity, or for both to have no concavity (if their graphs are straight lines).

Suppose we know that a function f has an inverse because it is either strictly increasing or strictly decreasing. Can we always find the inverse function f^{-1}? If the formula for the function is quite simple, then it is sometimes possible to undo the equation algebraically to obtain a formula for f^{-1}, as we illustrate in the next example. However, it is usually not possible to do this algebraically, and so we must resort to numerical or graphical methods to estimate values for the inverse function; that is, given a particular value for y, we can determine the value of x that corresponds to it by examining either the graph of the original function or successive numerical estimates.

E X A M P L E

Find g^{-1} if g is given by $y = g(x) = 1 + \dfrac{1}{x}$ for $x > 0$. Analyze the behavior of g and g^{-1}.

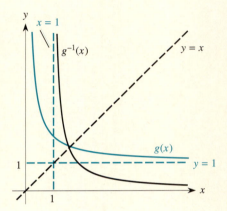

FIGURE 2.42

Solution From its graph (see Figure 2.42), we see that the function g is strictly decreasing and so its inverse exists. Further, since the domain of g is $x > 0$, we also know that $\dfrac{1}{x} > 0$. So $1 + \dfrac{1}{x}$ must be larger than 1, and therefore the range of g must be $y > 1$.

Next, let's find a formula for the inverse function. If $y = 1 + \dfrac{1}{x}$, then

$$\frac{1}{x} = y - 1.$$

Upon taking the reciprocal of both sides, we get

$$x = \frac{1}{y - 1} = g^{-1}(y), \quad y > 1.$$

If we interchange the roles of x and y to use the same independent variable, then

$$y = \frac{1}{x - 1} = g^{-1}(x), \quad x > 1.$$

The graphs of the function g and its inverse g^{-1} are shown in Figure 2.42. Notice how each is the mirror image of the other about the line $y = x$.

Further, the original function g is a decreasing function; it decays from a vertical asymptote at $x = 0$ toward a horizontal asymptote of $y = 1$. The graph of g^{-1} also decreases from a vertical asymptote at $x = 1$ toward a horizontal asymptote of $y = 0$. (Notice that the vertical and horizontal asymptotes are interchanged.) Also note that both curves are concave up.

In summary, suppose we have a function $y = f(x)$ that represents some quantity or process of interest to us. Typically, there are two questions we want to answer. The first question, which is simple, is to determine *the value of that quantity* corresponding to a particular value of x: Just substitute the value of x into the expression for the function. The second question is to determine *when* the quantity achieves a given level; that is, to find the value of the independent variable x that produces a given value for y. Answering the second question requires the existence of an inverse function and the ability either to find its equation algebraically or to estimate its values numerically or graphically.

EXERCISES

1. Which of the following functions have inverses? Explain why or why not. For those functions having an inverse, describe what the inverse function tells you.

   a. $f(t)$ = the height of water after t minutes in a child's pool that you are filling at a steady rate using a garden hose.

   b. $f(t)$ = your distance from New York on an airplane flight from New York to San Francisco as a function of the time t since take-off.

   c. $f(n)$ = the height of the student who is numbered n on your instructor's class roster.

   d. $f(n)$ = the amount that the nth customer in line at Burger Heaven pays for lunch.

   e. $f(t)$ = the length of the fingernail on your right index finger if t is the number of hours since you last clipped your nails.

   f. $f(T)$ = the amount spent by a family to heat their home if T is the temperature at which they keep the thermostat set.

   g. $f(t)$ = the depth of the snow on a person's front lawn in Buffalo as a function of the time t elapsed from October 1 to the following March 1.

   h. $f(t)$ = the total amount of snow that falls on the person's lawn in part (g).

2. For the function f shown at the right, estimate the value for x that corresponds to the given values of y:

   a. $y = 0$ **b.** $y = 2$

   c. $y = 5$ **d.** $y = -1$

   Then plot the resulting points and use them to sketch the graph of the inverse function f^{-1}.

3. Consider the function f with values given in the following table:

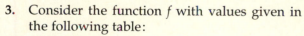

x	0	1	2	3	4	5
$f(x)$	2.94	2.48	2.05	1.84	1.44	1.12

   a. What is the domain of f? What is the range?

   b. Create a table of values for f^{-1}. What are its domain and range?

4. Use your function grapher to decide which of the following functions have inverses. For those functions that do, estimate the value for $f^{-1}(10)$.

   a. $f(x) = x^3 - 9x^2 + 5x - 5$ **b.** $f(x) = x^3 - 2x^2 + 5x - 5$

   c. $f(x) = 2x + x^2$ **d.** $f(x) = 2x - x^2$

   e. $f(x) = 2x + x^3$ **f.** $f(x) = 2x - x^3$

5. The following table gives the time T needed for a Trans Am to accelerate from zero to the indicated final speed v.

Final speed, v (mph)	30	40	50	60	70
Time, T (seconds)	3.00	4.29	5.52	7.38	9.81

 a. Explain why this is a function and why it has an inverse.
 b. Explain what the inverse function tells us. What is $f^{-1}(5.52)$? Estimate the value of $f^{-1}(7)$.

6. We know that 1 inch is equivalent to about 2.54 centimeters.

 a. Write a formula for the function f that gives an object's length C in centimeters as a function of its length I in inches.
 b. Find a formula for the inverse function f^{-1} and explain what f^{-1} tells you, in practical terms.

7. Find the inverse function of $p(t) = (1.04)^t$.

8. Find the inverse function of $f(t) = 50(10)^{0.1t}$.

9. Solve for the unknown in each of the following equations.

 a. $c^{25} = 14$ b. $1.07^t = 3$

 c. $0.84^k = 0.20$ d. $m^{1995} = 4$

10. Suppose the temperature of an object is being measured to the nearest degree in both Fahrenheit and Celsius. In general, which reading would you expect to be more accurate? Why?

11. For each of the functions f shown below, sketch the graph of the inverse function f^{-1} on the same set of axes.

a.
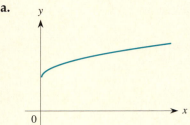

b.
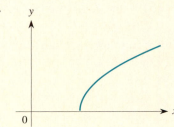

c.
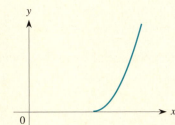

d.
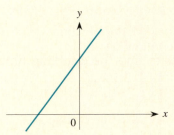

12. Suppose that a function f is increasing and concave up. By thinking of its inverse as the reflection about the line $y = x$, explain why f^{-1} is also increasing. Is it concave up or concave down? What happens if f is increasing and concave down?

13. Repeat Exercise 12 if the function f is decreasing and concave up; the function f is decreasing and concave down.

14. Given the graph of the following function f, sketch the graph of f^{-1}.

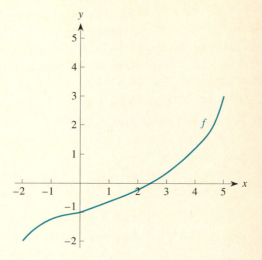

CHAPTER SUMMARY

In this chapter, you have learned the following:

- Important behavior characteristics of four important families of functions—linear functions, exponential functions, power functions, and logarithmic functions.
- How to find the slope and the equation of a line.
- What the slope of a line means.
- A criterion for knowing when a set of data follows a linear pattern.
- How to estimate the equation of a line that captures the linear pattern in a set of data.
- How to set up and solve problems involving linear processes.
- The behavior of exponential growth and exponential decay functions.
- What the growth/decay rate and the growth/decay factor mean.
- A criterion for knowing when a set of data follows an exponential pattern.
- How to find the doubling time for an exponential growth process and the half-life for an exponential decay process.
- How exponential behavior compares to linear behavior.
- How to find the exponential function that passes through two points.
- How to set up and solve problems involving exponential processes.
- The behavior of power functions when the power is a positive integer, a negative integer, or a noninteger.
- How to find the power function that passes through two points.
- How to set up and solve problems involving power function processes.
- The behavior of logarithmic functions.

- How to set up and solve problems involving logarithmic function processes.
- How to use logarithms with bases other than 10.
- How power function behavior compares to exponential behavior.
- How log function behavior compares to power function behavior.
- How to determine if a function has an inverse.
- What the inverse function tells you.

REVIEW EXERCISES

1. For each of the following curves, suggest any types of functions that might have the indicated behavior pattern. If you suggest an exponential function, indicate whether the base a is greater than 1 or less than 1. If you suggest a power function, indicate whether the power p is positive or negative and whether p is greater than 1 or less than 1.

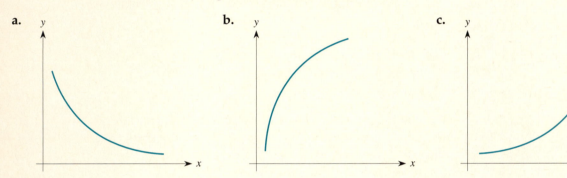

a.   b.   c.

2. A function $F(t)$ is exponential with known values $F(0) = 5$ and $F(3) = 8.5$. Determine the function and give the growth factor.

3. a. The profits of Alamo Paper Company grow $100,000 each year. In 1980, its profits were $1.5 million. Determine the profit function $P(t)$, where t = number of years since 1980. Draw a graph of $y = P(t)$ and determine the year in which profits first exceed $2 million.

   b. Ord Paper Company had profits of $950,000 in 1980, but its profits are growing at the rate of 10% each year. Determine when its profits first exceed $2 million.

   c. Use your function grapher to determine when the profits of Ord first exceed the profits of Alamo.

4. The current population of an endangered species is 20% of its population 40 years ago.

   a. Determine the half-life of the population.

   b. When can we expect that only 10 animals of this species will be left? Assume that, 50 years ago, about 10,000 animals existed in the world.

5. In preparing a holiday cranberry mold, boiling water at 212°F is added to the fruit and gelatin mixture, which is then poured into the mold and put into a 40° refrigerator. After 30 minutes, the temperature of the mixture

is 148°. The temperature $F(t)$ at time t (in minutes) is given by $F(t) = 172(1 + a)^t + 40$. What is the temperature of the mixture 3 hours after the mold was put in the refrigerator?

6. Information from 1970 to 1991 for the Social Security Supplementary Medical Insurance Trust Fund shows that the total of the fund can be approximated by a linear function. Use the following data and the black thread method to find the best line that fits the data.

Year	1970	1975	1980	1985	1990	1991
Income ($ billions)	1.876	4.322	10.275	24.577	46.138	48.166

Use this linear function to estimate the income of the trust fund in 1996.

7. Housing affordability depends on the percentage of income paid in mortgage payments. The National Association of Realtors reports mortgage payments as a percentage of median income from 1981 to 1992, as shown in the following table.

Year	1981	1982	1983	1984	1985	1986	1987	1988	1989	1990	1991	1992
Percentage	36.3	35.9	30.1	28.2	26.2	23.0	21.9	22.0	23.1	22.7	22.3	20.0

Use the black thread method to find the linear function that fits the data best. What is the slope of the line? What does this slope represent?

8. For the data given in Exercise 7, it is possible to determine that the function $f(t) = 38.8t^{-0.25}$, where $t =$ the number of years since 1980, gives a better fit to the data than a linear function. Using this function f, find the percentage of median income spent on mortgage payments in 1996. When can we expect the percentage to drop below 18%?

9. Further investigation shows that the data in Exercise 7 can be better approximated by the logarithmic function $G(t) = 37.69 - 7.03 \ln t$, where $t =$ the number of years since 1980. (Recall that ln is $\log_e$). Draw the graph of the function on the same set of axes as the data. Determine the percentage of median income spent on the mortgage in 1996. When can we expect the percentage to drop below 18%?

10. Find possible equations for the function represented by each of the following graphs.

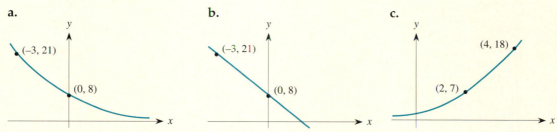

a.
y
(−3, 21)
(0, 8)
x

b.
y
(−3, 21)
(0, 8)
x

c.
y
(4, 18)
(2, 7)
x

11. Solve for the unknown in each equation.

a. $c^{10} = 1.7$ b. $1.2^x = 5$ c. $\log x = 4.8$ d. $\log_5 a = 6.8$

e. $\log_a 40 = 7$ f. $0.72^w = 0.15$ g. $60 \log (3 + x) = 85.$

12. For each of the functions whose graph is shown, draw the inverse function, if one exists. If the function has no inverse, explain why.

a.   b.   c.

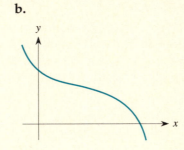
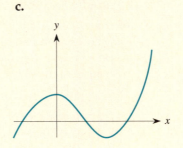

13. For each function, give its domain and find the inverse function.

    a. $f(t) = 0.5 \log (2t - 4)$ b. $g(x) = x^3 + 6$

14. The function $F(x)$ is either linear or exponential. From the values given in the table, decide which is the correct type and find a formula for F.

x	1	2	3	4	5
$F(x)$	6	9	13.5	20.25	30.375

15. Match each of the functions f, g, and h in the table below to the behavior described in the following categories.

    a. increasing, straight line
    b. increasing, concave down
    c. increasing, concave up.

x	1	2	3	4	5
$f(x)$	2.70	3.64	4.92	6.34	8.96
$g(x)$	1.4	4.6	7.8	11.0	14.2
$h(x)$	5.10	5.19	5.27	5.34	5.40

16. a. Since 1960, the price of an ice cream cone in one southern city has been growing approximately exponentially according to the function $f(t) = C a^t$. The price of a one-scoop cone was 20 cents in 1960 and was $1.00 in 1990. Determine the function f.

    b. The average price of a ticket to a first-run movie was $2.00 in 1960. This price also has been growing exponentially. In 1990, the price was $6.00. Which of the prices, ice cream or movies, is growing faster?

    c. When can you expect the ice cream and movie prices to be the same if they continue to grow as they did until 1990?

    d. How much would a ticket to the movies cost at the time you found in part (c)?

3

FITTING FUNCTIONS TO DATA

3.1 *Introduction to Data Analysis*

When the space shuttle *Challenger* exploded some 90 seconds after launch on January 28, 1986, it left an entire nation in a state of shock. The shuttle accident was especially traumatic because many school children were watching the launch live on television since Christa McAuliffe, a teacher, was a member of the crew.

The *Challenger* explosion also left behind a host of unanswered questions. Among the most important questions are:

- What went wrong?
- Could the problem have been anticipated?
- If it could have been anticipated, why wasn't it?

The first question is technical, the second is mathematical, and the third is political. In this section, we address the first two questions.

The *Challenger* disaster involved two factors: a component known as an O-ring and the air temperature at launch. The O-ring is a very thin ring (37.5 feet in diameter but only 0.28 inches thick) that seals the connections, or joints, between different sections of the shuttle engines. The locations of the O-rings are indicated by the arrows in Figure 3.1. On the morning of the *Challenger* launch, the air temperature was 31°F, which was considerably colder than the temperature at any previous launch. In fact, the coldest temperature recorded at any previous launch was 53°F.

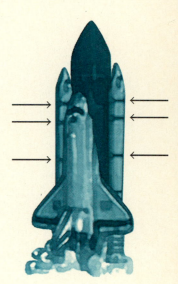

FIGURE 3.1

The Rogers Commission, which studied the *Challenger* disaster, focused on the O-rings as a possible cause of the explosion because there had been problems with O-rings on previous flights. In fact, the night before the launch, some of the project engineers, as part of the standard prelaunch routine, had thought about the O-rings and questioned whether the flight should happen because of the predicted overnight temperatures. The reasoning that went into the flight decision is worth considering because it demonstrates the important role that mathematical analysis can play in making informed decisions.

Prior to the *Challenger* flight, there were 24 shuttle flights. On seven of them, relatively minor problems or "distress incidents" had developed with the O-rings. To review the data on the previous problems, the engineers examined the data shown in Figure 3.2. Notice that the horizontal axis indicates the air temperature at launch and the vertical axis gives the number of O-rings affected. Each black dot represents a particular shuttle launch. Thus the shuttle launched at 53°F experienced problems with three different O-rings. Even though the graph shows one dot for 70°, there were actually two shuttles launched at this temperature that had problems with one O-ring. Nevertheless, when you examine this graph (as the shuttle engineers probably did), you will likely see no consistent pattern regarding the relationship between the number of O-ring problems and the temperature at launch. In fact, notice that the shuttle launched at 75°F had problems with two different O-rings. (Actually, this was the previous launch of the *Challenger*.) Consequently, this data did not give the engineers any solid reason for canceling the *Challenger* launch the following morning.

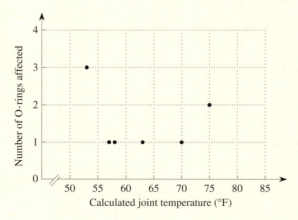

FIGURE 3.2

Unfortunately, the engineers did not realize that the data presented in Figure 3.2 is just part of the story. It reflects only those instances when there were problems with the O-rings. What is missing from the data set are those launches that had no trouble. If we look at the full set of data for all previous shuttle flights shown in Figure 3.3, a very striking pattern emerges: Almost all of the problems occurred at the low-temperature end of the graph. All of the no-problem launches occurred when the air temperature was above 65°F. Thus there is a clear pattern that suggests that it is unlikely to have a problem with O-rings on warm days, but there may likely be a problem on a cool day. And, the predicted temperature of 31°F on the morning when the *Challenger* was due to go up was far colder than the temperature at any previous launch. Had the engineers looked at all the data shown in Figure 3.3, there is no way that they could have allowed the *Challenger* to go up.

Moreover, we can go beyond just an eyeball examination of the data in looking for trends. In Chapter 2, we discussed the idea of linear regression as a technique for finding the straight line that is the best fit to all points in a *scatterplot*, which is the graph of the data points. Figure 3.3 is the scatterplot for the number of O-ring problems as a function of air temperature. However, a straight line is not a particularly good fit to this set of data. Instead, we need to extend the notion of linear regression to determine a curve that is the best fit to a set of nonlinear data. In doing so with the O-ring data, we might obtain a curve such as the one shown superimposed over the data points in Figure 3.4. The resulting curve suggests a decaying exponential function or power function, which means that the colder the temperature, the more likely there would be problems with the O-rings. Although extrapolating beyond the region of the data points (in this case from 53°F to 80°F) is usually risky, there is little doubt that launching the *Challenger* at 31°F was even riskier!

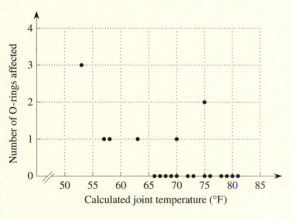

FIGURE 3.3

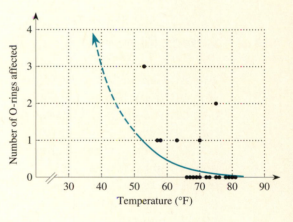

FIGURE 3.4

Most scientific, engineering, and technical work involves collecting and working with data. More important, the resulting decisions made are based on what can be inferred from the data. Thus in order to make intelligent decisions, we must understand all the information that a set of data imparts. To understand this information, it is necessary to have techniques for displaying the data in an intelligible form that enhances the ability to extract the relevant information. Unfortunately, these techniques have not been a major focus in past mathematical and technical training.

This situation illustrates the changing role of mathematics and how people use it. It is not a matter of "Here's an expression. Factor it." or "Here's an equation. Solve it." Rather, one is faced with a situation about which a decision must be made. You have to see how that situation can be viewed mathematically (set up an appropriate mathematical model), identify the appropriate question to ask, obtain the solution (often with some electronic tool), interpret the solution in terms of the original problem, and communicate that solution effectively to others. The emphasis is on reasoning and judgment, not on mechanical operations.

In this chapter, we consider a variety of ways to get key information, both graphically and computationally, from sets of data. To do so, we develop simple yet effective methods to analyze sets of data and determine the nature and equation of the function that best fits the data. Then we use the results to draw appropriate conclusions from the pattern of the data. In a very real sense, these techniques are the methods by which we construct functions. By using this approach, we hope preventable disasters such as the *Challenger* explosion will not recur.

3.2 *Linear Regression Analysis*

The concept of linear regression was introduced in Chapter 2 in a very simplified manner using the black thread method for finding the line that best fits a set of data. We now consider these ideas again from a somewhat more sophisticated viewpoint.

Suppose we have a set of n measurements on two presumably related quantities, x and y: $(x_1, y_1), (x_2, y_2), \ldots, (x_n, y_n)$, where x_1 and y_1 are the coordinates of the first point, x_2 and y_2 are the coordinates of the second point, and so forth. For instance, the coordinates might represent peoples' heights and weights; students' high school averages and their college GPAs; or the year and the new world record time for the mile run, as given in Table 3.1, shown on the next page.

We begin by plotting these points to obtain a *scatterplot* of the data, as shown in the computer-generated plot in Figure 3.5. Notice that this set of data appears to fall into a roughly linear pattern with a negative slope. Our objective is to determine the *best linear fit* to this set of data: the one line that comes closest to all the data points. Notice that we can draw many different lines that all seem to fit the overall pattern of the points in the scatterplot, as shown in Figure 3.6. However, none of them can possibly pass through all of the points. In fact, a *good fit* line may not necessarily pass through any of the points.

Year	Record time	Winner
1911	4:15.4	John Paul Jones, United States
1913	4:14.6	John Paul Jones, United States
1915	4:12.6	Norman Taber, United States
1923	4:10.4	Paavo Nurmi, Finland
1931	4:09.2	Jules Ladoumegue, France
1933	4:07.6	Jack Lovelock, New Zealand
1934	4:06.8	Glen Cunningham, United States
1937	4:06.4	Sidney Wooderson, Great Britain
1942	4:06.2	Gunder Haegg, Sweden
1942	4:06.2	Arne Andersson, Sweden
1942	4:04.6	Gunder Haegg, Sweden
1943	4:02.6	Arne Andersson, Sweden
1944	4:01.6	Arne Andersson, Sweden
1945	4:01.4	Gunder Haegg, Sweden
1954	3:59.4	Roger Bannister, Great Britain
1954	3:58.0	John Landy, Australia
1957	3:57.2	Derek Ibbotson, Great Britain
1958	3:54.5	Herb Elliott, Australia
1962	3:54.4	Peter Snell, New Zealand
1964	3:54.1	Peter Snell, New Zealand
1965	3:53.6	Michel Jazy, France
1966	3:51.3	Jim Ryun, United States
1967	3:51.1	Jim Ryun, United States
1975	3:51.0	Filbert Bayi, Tanzania
1975	3:49.4	John Walker, New Zealand
1979	3:49.0	Sebastian Coe, Great Britain
1980	3:48.9	Steve Ovett, Great Britain
1981	3:48.8	Sebastian Coe, Great Britain
1981	3:48.7	Steve Ovett, Great Britain
1981	3:47.6	Sebastian Coe, Great Britain
1985	3:46.5	Steve Cram, Great Britain

TABLE 3.1: *World Records in the Mile Run*

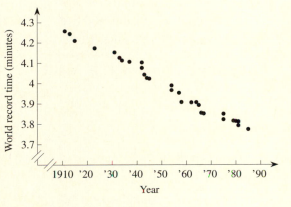

FIGURE 3.5

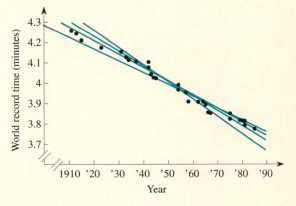

FIGURE 3.6

In the following development, we write the equation of a line in the form $y = a + bx$ instead of the more usual form $y = mx + b$. This form better reflects the display on most calculators. As always, the coefficient of the independent variable represents the slope and the constant gives the vertical intercept no matter what letters are used.

The key to knowing what line to use lies in deciding what is meant by the phrase *comes closest to all the points*. The most common interpretation of this phrase is that the sum of the squares of all the vertical distances from the points to the line should be a minimum, as shown in Figure 3.7. (If we used only the actual vertical distances, rather than their squares, some would be positive, others negative, and they would tend to cancel each other out when finding the sum.) These vertical distances, called *residuals*, will be discussed in more detail in Section 3.4.

Suppose the equation of the best-fit line is $y = a + bx$, where a and b are, for now, unknown constants. For each of the n points (x_1, y_1), (x_2, y_2), . . . , (x_n, y_n), we measure the vertical distance using the square of the distance from each point to the line, namely $[y_1 - (a + bx_1)]^2$, $[y_2 - (a + bx_2)]^2$, ... See Figure 3.8.

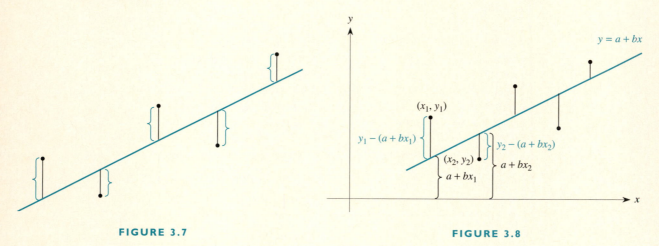

FIGURE 3.7   **FIGURE 3.8**

It turns out that there is precisely one line that gives a *minimum* for this sum of squares. It is known as the *least squares line*, or more commonly, the *regression line*. The derivation of the coefficients[1] a and b in the equation of this line is studied in calculus. The formulas for a and b are built into most sophisticated calculators, usually under the STAT (statistics) menu and may be

---

[1]The regression coefficients are

$$a = \frac{(\Sigma x^2)(\Sigma y) - (\Sigma xy)(\Sigma x)}{n(\Sigma x^2) - (\Sigma x)^2} \qquad b = \frac{n\Sigma xy - (\Sigma x)(\Sigma y)}{n(\Sigma x^2) - (\Sigma x)^2},$$

where n = number of data pairs (x, y), Σx^2 = the sum of the squares of the x's, $(\Sigma x)^2$ = the square of the sum of the x's, Σxy = the sum of the products of x and y in each data pair.

marked as LinReg. Routines for doing these calculations also are widely available in many computer packages. Typically, all you need to do is enter the set of data pairs and the calculator or computer will print out the equation of the regression line and often will draw the scatterplot of the data and the associated regression line. For instance, using the data on the world record time in the mile run, we get the equation of the regression line as

$$y = -0.00669x + 17.05.$$

With your calculator, you typically have to access the STAT (statistics) menu, select the data option, clear any old data, enter the new data, and then go back to the main STAT menu and select linear regression. Most calculators will give you values for the coefficients, a and b, for the regression equation

$$y = a + bx,$$

and for a third quantity we will discuss in a moment. (See the detailed instructions pertaining to your calculator in its manual.)

Once we have the equation of the regression line for the mile-run data, we can use it to predict or extrapolate what the future world record time is likely to be. For example, if we take $x = 2000$, then this equation predicts that the record in the year 2000 will be a 3.67 minute (which is equivalent to 3 minutes, 40.2 second) mile. Is this reasonable for a world record in the mile?

With quantities such as these, it usually helps greatly to reduce the four-digit numbers for the years to a smaller scale. If we let x be the number of years since 1900, then 1911 is entered as 11, 1945 as 45, and so forth. If you then want to predict the record in the year 2000, you will have to convert 2000 into 100 when you put it into the regression equation you obtain.

Think About This

What is the extrapolated value for the world record in the mile run in the year 2500? Is the value reasonable? Similarly, what happens if you extrapolate back to the year 1492? Do you get a reasonable value?

If you use the equation of the regression line to predict values of y corresponding to values of x within the interval of data values (called *interpolating*), the results are usually quite reasonable. However, if you try to use the regression equation for *extrapolating* well beyond the interval of data values, then the results become extremely questionable. Thus you should not extrapolate too far into either the future or the past.

You should be aware of another major concern regarding the use of the regression equation. If you take any set of measurements relating two variables, you can always construct the regression equation based on the data. However, the results are completely meaningless if the two variables are totally unrelated. For example, you could collect data on students' telephone numbers and their social security numbers, construct a scatterplot, and calculate the corresponding regression equation. But, since the two variables are unrelated, the results of predicting students' social security numbers from their phone numbers would be of no value whatsoever.

Therefore we need a way to determine if there is, in fact, a linear relationship between two quantities. The most common way of detecting such a relationship is by using a quantity[2] known as the *linear correlation coefficient*, or simply the *correlation coefficient r*. This quantity is always a number between -1 and $+1$.

- Values of r close to $+1$ indicate a high degree of positive correlation between x and y; that is, they are likely related via a linear relationship and the regression line will have positive slope. For example, we would expect a high positive correlation between a company's profits and its sales: As the sales go up, usually the profits will go up also.
- Values of r close to -1 indicate a high degree of negative correlation between x and y; they are likely related by a linear relationship and the regression line will have negative slope. For instance, there is a high negative correlation between a car's gas mileage and its weight—as weight goes up, gas mileage goes down, and vice versa. Similarly, there is a high negative correlation between the literacy rate and the infant mortality rate in any nation.
- Values of r close to 0 indicate little or no correlation, and thus we would conclude that there is no linear relationship between the variables. As we mentioned earlier, there is no correlation between students' social security numbers and their telephone numbers.

You can visualize the different cases by looking at the scatterplots in Figure 3.9. In the first case, the data points lie more or less along a rising line and the value of the correlation coefficient will be positive and relatively close to 1. In the second case, the points are scattered about a downward-sloping line, which means that the correlation coefficient will be negative and relatively close to -1. In the third case, there is no clear pattern for the points and so the correlation coefficient will be relatively close to 0 to indicate that there is no linear relationship between the variables.

As with the calculation of the regression coefficients, the correlation coefficient usually is obtained either with a computer program or a calculator. It is rare that anyone would have to perform the calculations by hand. In particular, you can obtain the value of the correlation coefficient from your graphing calculator or from any program that performs linear regression analysis. Of course, the key to linear regression is knowing how to interpret the result, not how to calculate it.

---

[2]The correlation coefficient is

$$r = \frac{n\Sigma(xy) - (\Sigma x)(\Sigma y)}{\sqrt{n(\Sigma x^2) - (\Sigma x)^2}\ \sqrt{n(\Sigma y^2) - (\Sigma y)^2}}.$$

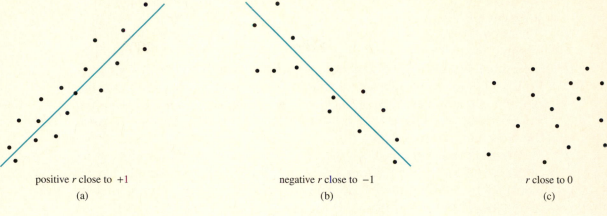

positive r close to $+1$

(a)

negative r close to -1

(b)

r close to 0

(c)

FIGURE 3.9

To use the correlation coefficient, we must be able to distinguish between a high degree of correlation and no correlation. To do this, we have to consider the entire question from a somewhat more global point of view. Typically, when we collect a set of data, it represents just a sample from a much larger population of possible data values. For instance, if we are conducting a study regarding the relationship between gas mileage and vehicle weight, we certainly would not collect data on every car on the road but only on a random sample of all cars. Presumably, the sample we select will be representative of all cars, but there is no guarantee of that. It is conceivable that we could choose an extremely unrepresentative sample. What's worse, there usually is no way of telling that a sample is unrepresentative just by examining the sample results.

To account for this possibility, statisticians have developed a set of techniques that allow us to draw conclusions that are correct with 95% certainty. No statistical conclusion is ever 100% certain because that would require encompassing *every* conceivable sample from a population. Rather, such conclusions are designed to be correct 95% of the time to account for the possibility of a sample containing several very unrepresentative observations. (Statisticians also work with 90% and 99% levels of certainty, but we consider only 95% certainty here.)

The size of the sample also comes into play here. If we take a relatively small sample, say $n = 5$ values, then this sample could contain a nonrepresentative data point and such a "bad" observation may have a significant impact on a group of only five points. If the sample contains $n = 50$ random values, then the effect of a single nonrepresentative data point will likely be diluted. If the sample consists of $n = 500$ random values, then the effect of any single nonrepresentative data point is likely to be negligible.

We apply these ideas in the following way. First collect a random sample consisting of n data points. Use either a computer program or a calculator to analyze the data by graphing the scatterplot, calculating and drawing the regression line, and calculating the value of the correlation coefficient r. If the points on the scatterplot clearly do not appear to fall into a linear pattern, then you should expect that there is little or no linear correlation between the two variables. Even if the pattern seems to be linear, you must still use the information provided by the correlation coefficient. Most graphing calculators determine the equation of the regression line and the value for the correlation coefficient r. Either way, once you have r for the data, you must decide if it indicates a significant level of correlation between the two variables.

Table 3.2 contains a set of so-called *critical values* for the correlation coefficient r. These critical values are the boundary values separating what is considered to indicate correlation from what we interpret as having no correlation. Notice that these critical values change depending on the size of the sample, n.

n	r	n	r	n	r
3	0.997	13	0.553	27	0.381
4	0.950	14	0.532	32	0.349
5	0.878	15	0.514	37	0.325
6	0.811	16	0.497	42	0.304
7	0.754	17	0.482	47	0.288
8	0.707	18	0.468	52	0.273
9	0.666	19	0.456	62	0.250
10	0.632	20	0.444	72	0.232
11	0.602	21	0.433	82	0.217
12	0.576	22	0.423	92	0.205

TABLE 3.2:   *Critical Values of r*

To use the table, we compare the value of r we get from the sample data to the corresponding critical value shown in the table. For example, if we have a sample of size $n = 10$, then the critical value for r is 0.632. If the value for the correlation coefficient for the data is *greater than* 0.632, say $r = 0.758$, then we can conclude with 95% certainty that there is positive correlation between the two variables. If the calculated correlation coefficient for our data is *less than* -0.632, say $r = -0.685$, then we can conclude with 95% certainty that there is negative correlation between the two quantities being studied. However, if the value for r is between -0.632 and $+0.632$, say $r = 0.446$ or $r = -0.583$, then we cannot conclude that there is any linear correlation between the two variables. This means that there does not appear to be a *linear* relationship between x and y. See Figure 3.10. (Of course, there still may be a nonlinear functional relationship between x and y.)

Notice that the critical value for a sample of size $n = 20$ is 0.444 while the critical value for a sample of size $n = 10$ is 0.632. As we

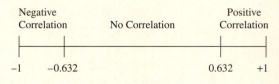

FIGURE 3.10

pointed out, a larger sample is more likely to be representative of the population, so the evidence for correlation does not have to be quite as great. With a small sample, the value for r must be very close to 1 or -1 in order for us to conclude with any certainty that there is correlation.

In the following examples, we consistently use X as the independent variable and Y as the dependent variable for the output from either the computer or calculator. It will then be necessary to interpret what the X- and Y-variables represent in the context of each individual situation.

EXAMPLE

A study is conducted to determine the relationship between a person's height in inches and his or her shoe size. The following set of data pairs is obtained:

$(66, 9)$, $(63, 7)$, $\left(67, 8\frac{1}{2}\right)$, $(71, 10)$, $(62, 6)$, $(65, 8\frac{1}{2})$, $(72, 12)$, $\left(68, 10\frac{1}{2}\right)$, $\left(60, 5\frac{1}{2}\right)$, $(66, 8)$.

(a) Determine the value of the correlation coefficient and find the equation of the regression line relating height to shoe size based on this sample. Use the equation to predict the most likely shoe size for a person who is (b) 70 inches tall and (c) 61 inches tall.

Solution   **a.** We show the scatterplot for the given data and the graph of the associated regression line in Figure 3.11. The equation of the regression line, as given by either a calculator or computer, is

$$Y = 0.51X - 25.016,$$

which is equivalent to

$$S = 0.51H - 25.016$$

for a person's shoe size S as a function of his or her height H in inches. Further, the value of the correlation coefficient is $r = 0.951$, which suggests a very high de-

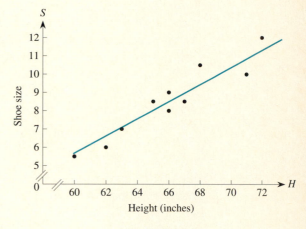

FIGURE 3.11

gree of positive correlation between a person's height and his or her shoe size. Because there are $n = 10$ data points, the critical value for r is 0.632 from Table 3.2. Since $0.951 > 0.632$, we can conclude with 95% certainty that there is a positive correlation between a person's height and shoe size.

   b. Using this regression equation for a person who is 70 inches tall, we estimate that the shoe size corresponding to $H = 70$ is

$$S = 0.51(70) - 25.016$$

$$= 10.684 \approx 10\frac{1}{2}.$$

c. Similarly, if $H = 61$, then we estimate that the corresponding shoe size is

$$S = 0.51(61) - 25.016$$
$$= 6.094 \approx 6.$$

In the context of this example, it seems sensible to think of shoe size as a function of a person's height. As such, we chose height H as the independent variable. However, it is also appropriate to interchange the roles of S and H and think of H as a function of S.

Suppose that you have determined, with 95% certainty, that there is a high degree of linear correlation between two variables. What does this tell you? In the last example, a high correlation coefficient means that there is a linear relationship between H and S. That is, the two variables can be related using a linear equation, namely the regression equation. However, the fact that two quantities have correlation and that a linear relationship exists does not mean that there is a cause-and-effect relationship between them. For example, several studies have found that there is a high degree of positive correlation between teacher salaries in a school district and the amount of alcohol consumed by the students. That is, if teacher salaries are low, then the level of student drinking is also low; if teacher salaries are high, then there is considerable drinking going on. Is it reasonable to conclude that one causes the other? Or, do you think that the level of drinking could be diminished by lowering teacher salaries in a district?

The fact that there is positive correlation between the two quantities says nothing more than that there is a linear relationship between them. There may be other factors present that could contribute to both. For instance, high salaries typically reflect a school district in a relatively affluent community where the students are likely to have relatively large amounts of their own money to spend on alcohol.

In all of the previous discussion, we have concerned ourselves only with the possibility that there is a linear relationship between two variables. The correlation coefficient is then used to detect and measure only a linear relationship. In the next section, we will consider ways of detecting, measuring, and calculating a nonlinear (exponential, power, or logarithmic) relationship between two quantities.

EXERCISES

1. The following table gives some measurements on the rate of chirping (per minute) of the striped ground cricket as a function of the temperature.

$T(°F)$	89	72	93	84	81	75	70	82	69	83	80	83	81	84	76
Chirps	78	60	79	73	68	62	59	68	61	65	60	69	64	68	57

Determine the equation of the best fit line to this set of data. How does it compare to the equation we estimated by eye in Example 6 of Section 2.2? Does the value of the correlation coefficient indicate a high degree of correlation between chirp rate and air temperature?

2. According to a leading road and track magazine, the following information gives the time in seconds for a Mercedes to accelerate from zero to the indicated speed in miles per hour:

Speed (mph)	30	40	50	60	70	80
Time (seconds)	3.6	5.0	7.0	9.1	11.9	15.2

Does the corresponding correlation coefficient indicate a significant level of linear correlation between the two variables? If so, determine the equation of the regression line that best fits the data. Estimate how long it would take a Mercedes to accelerate to 45 mph and to 90 mph. Which is more likely to be accurate?

3. The following table represents the percentage of our Gross Domestic Product (GDP) that was spent on health care over the years.

Year	1960	1965	1970	1975	1980	1985	1990
Percentage	4.4%	5.7%	7.7%	8.3%	10%	10.2%	12.2%

Is there a significant level of linear correlation between these two variables? If so, what is the regression line? Estimate the percentage of our GDP that will be spent on health care in the year 2000. If current trends continue, when will we spend 20% of our GDP on health care?

4. Repeat Exercise 3 using 60, 65, 70, . . . , 90 for the years instead of 1960, 1965, ... How do the results for the equation of the line of best fit and for the correlation coefficient change? What do you expect would happen if you used 0, 5, 10, . . . , 30 for the years instead?

5. The electrical resistance R of a piece of metal depends on the temperature T of the metal. An experiment is conducted by measuring the resistance in ohms in a piece of wire at different temperatures in degrees Celsius. The results are as follows:

T	33.2	40.6	45.3	51.8	58.4	63.8	71.0	76.9	80.6	90.1
R	4.71	4.80	4.93	5.02	5.17	5.34	5.39	5.52	5.55	5.75

Determine the correlation coefficient and the equation of the regression line that best fits the data. Does the value for the correlation coefficient suggest that there is a linear relationship between the two quantities?

6. Repeat the example given in this section on shoe size versus height by interchanging the roles of H and S to consider S as the independent variable. How do the results differ? How are the two values obtained for the correlation coefficient related?

7. Match each of the following values for the correlation coefficient r with its scatterplot:

    a. $r = -0.978$ **b.** $r = 0.447$ **c.** $r = 0.832$

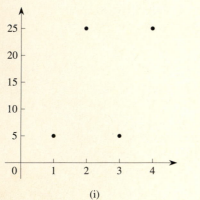
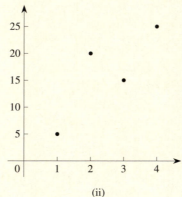
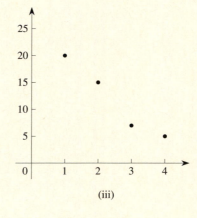

        (i)   (ii)   (iii)

8. Consider the four points $(0, 0)$, $(0, 1)$, $(1, 0)$, and $(1, 1)$. Two reasonable guesses for the best-fit line are (a) the diagonal line passing through $(0, 0)$ and $(1, 1)$ and (b) the horizontal line $y = \frac{1}{2}$. Using the criterion that the best-fit line is the one with the minimal value for the sum of the squares of the vertical distances from each of the points to the line, decide which of these two lines is a better fit. What is the actual best-fit line found by using your graphing calculator? How high is the value for the correlation coefficient?

9. The accompanying graph shows the scatterplot for the points $(1, 48)$, $(2, 68)$, $(3, 93)$ and $(4, 114)$ along with the line $y = 20x + 30$.

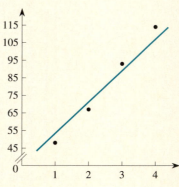

    a. Find the sum of the squares associated with this line.

    b. How would you change the line (slope and/or intercept) to make a better fit?

10. The median family income I in the United States was about \$10,000 in 1970 and rose to about \$27,000 in 1985. Let t be the number of years since 1970.

t	Linear Model	Exponential Model
0	10	10
5	?	?
10	?	?
15	27	27
20	?	?
25	?	?

Income in Thousands of Dollars

a. If you assume that the increase in family income has been linear, find an equation for the line representing income I in terms of t. Use this equation to complete the first column of the preceding table.

b. If you assume that the increase in family income has been exponential, find an equation of the form $I = I_0 a^t$ to represent family income levels between 1970 and 1985 and complete the second column of the table.

c. On the same set of axes, sketch the graphs of the functions you obtained in parts (a) and (b).

d. Use the equations from parts (a) and (b) to predict the median family income in the year 2000 for both types of growth.

e. Suppose that both predictions from part (d) seem unreasonable. Can you suggest any other types of functions that might be a better fit?

11. (Continuation of Exercise 10) Suppose you now learn that the median family income in 1990 was about $34,000.

a. Which of the two models in Exercise 10 now seems more accurate?

b. If you plot the three data points corresponding to 1970, 1985, and 1990, how would you describe the shape of the likely graph of median family income as a function of time? What is the significance of this shape?

12. The Athabasca glacier in southern Alberta, Canada, is part of the largest mass of ice in the Rocky Mountains. (Tourists who visit the Jasper and Banff National Parks can take a side trip out onto the actual glacier.) Over the last 120 years, the glacier has been steadily "withdrawing" at a rate of about 15 meters per year, as it slowly melts.

a. Express the approximate position of the southernmost extent of Athabasca as a function of time, measured in years from 1900. Measure position northward from the U.S.–Canadian Border, which was about 300 kilometers south of the glacier in 1900.

b. If the current rate of withdrawal has been in effect indefinitely, how long ago did the toe of the glacier extend over the border?

c. Can the function in part (a) continue to apply for the next million years? Why or why not?

13. A long-distance telephone company charges $1.32 for the first minute of a call from Los Angeles to London and $1.08 for each additional minute.

a. Write the equation of a linear function that models this situation.

b. What is the practical significance of the slope? Of the vertical intercept?

c. What is the cost of a 26-minute call?

d. Suppose there is a 30% discount on the above rates for calls made in off-peak hours. Repeat parts (a)–(c).

14. (Continuation of Exercise 13) A competing long-distance company claims it is cheaper because its rates on the Los Angeles to London call are $1.24 for the first minute and $1.12 for each additional minute.

a. For the 26-minute call in Exercise 13(c), which carrier is actually cheaper?

b. Graph both lines. What does the point where they intersect signify?

c. Find the length of call at which the second company becomes more expensive than the first.

15. The double bar graph shown associates the percentage of residents who recycle in Manhattan's twelve districts with their median income. Which variable should be considered independent and which dependent? Is there a significant level of correlation between the two? Find the equation of the regression line relating these quantities.

16. The table below gives the world record times for the 100-meter freestyle for men and women. For each data set, find the best-fit line. What is the practical significance of each line's slope? Since the slopes are different, determine the point where the two lines intersect and tell what this point means. Is it reasonable?

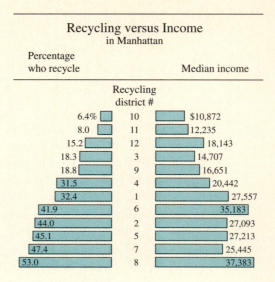

Recycling versus Income
in Manhattan

Men's Records			Women's Records		
Year	Swimmer	Time	Year	Swimmer	Time
1905	Zoltan Halmay (HUN)	65.8	1908	Martha Gerstung (GER)	95.0
1910	Charles Daniels (US)	62.8	1910	C. Guttenstein (BEL)	86.6
1912	Duke Kahanamoku (US)	61.6	1911	Daisy Curwen (GB)	84.6
1918	Duke Kahanamoku (US)	61.4	1912	Fanny Durack (AUS)	78.8
1920	Duke Kahanamoku (US)	60.4	1915	Fanny Durack (AUS)	76.2
1922	Johnny Weissmuller (US)	58.6	1920	Ethelda Bleibtrey (US)	73.6
1924	Johnny Weissmuller (US)	57.4	1923	Gertrude Ederle (US)	72.8
1934	Peter Fick (US)	56.8	1924	Mariechen Wehselau (US)	72.2
1935	Peter Fick (US)	56.6	1926	Ethel Lackie (US)	70.0
1936	Peter Fick (US)	56.4	1929	Albina Osipowich (US)	69.4
1944	Alan Ford (US)	55.9	1930	Helene Madison (US)	68.0
1947	Alex Jany (FR)	55.8	1931	Helene Madison (US)	66.6
1948	Alan Ford (US)	55.4	1933	Willy Den Ouden (NETH)	66.0
1955	Dick Cleveland (US)	54.8	1934	Willy Den Ouden (NETH)	65.4
1957	John Devitt (AUS)	54.6	1936	Willy Den Ouden (NETH)	64.6
1961	Manual Dos Santos (BRA)	53.6	1956	Dawn Fraser (AUST)	62.0
1964	Alain Gottvales (FR)	52.9	1958	Dawn Fraser (AUST)	61.2
1967	Ken Walsh (US)	52.6	1960	Dawn Fraser (AUST)	60.2
1968	Mike Wenden (AUS)	52.2	1962	Dawn Fraser (AUST)	59.5
1970	Mark Spitz (US)	51.9	1964	Dawn Fraser (AUST)	58.9
1972	Mark Spitz (US)	51.22	1972	Shane Gould (AUST)	58.5
1975	Jim Montgomery (US)	50.59	1973	Kornelia Ender (E GER)	57.54
1976	Jonty Skinner (S. AFR)	49.44	1974	Kornelia Ender (E GER)	56.96
1981	Rowdy Gaines (US)	49.36	1976	Kornelia Ender (E GER)	55.65
1985	Matt Biondi (US)	49.24	1978	Barbara Krause (E GER)	55.41
1986	Matt Biondi (US)	48.74	1980	Barbara Krause (E GER)	54.79
1988	Matt Biondi (US)	48.42	1986	Kristin Otto (E GER)	54.73

World Records in Swimming

3.3 *Curve Fitting and Nonlinear Regression Analysis*

In the last section, we developed the ideas of linear regression and correlation that allow us to determine whether a linear relationship exists between two variables, and if so, what the equation of the regression line that best fits the data is. We then can use this equation to make predictions—either interpolations or extrapolations not too far from the range of data values.

Although these methods are extremely powerful, not all relationships between two quantities are linear. Two variables can be related via an exponential function, a power function, or some other function. In such a case, the linear regression methods we previously developed do not apply, at least not directly. For example, if the pattern relating the variables is not linear, then the correlation coefficient r will not detect the relationship; it only measures a linear pattern. Even though we can construct the linear regression equation based on the data points, it may be a very poor fit, and so any predictions based on the linear model will be of no value.

In this section, we consider ways in which a set of data that is not linearly related can be transformed in such a way that the resulting transformed values may, in fact, fall into a linear pattern. This approach is known as **linearizing** the data. Once this is done, we can apply our previous techniques to find the regression line and correlation coefficient for the transformed data. Finally, we undo the original transformation by using the appropriate inverse function to produce the equation of the nonlinear function that best fits the original set of data points.

Fitting Exponential Functions to Data

To illustrate these ideas, consider the population of the United States, in millions, from 1780 to 1900, as shown in Table 3.3 on the next page. The corresponding scatterplot showing population vertically versus time horizontally is displayed in Figure 3.12. In the following discussion, we let the independent variable t represent the number of decades since the year 1780. That is, 1780 corresponds to $t = 0$, 1790 corresponds to $t = 1$, and so forth to 1900, which corresponds to $t = 12$. The reason for this choice will be discussed in a moment.

It is evident that the growth pattern is not linear. Based on our discussions about population growth in Chapter 2, we expect that the pattern for population growth is likely exponential. Let us apply the ideas from Chapter 2 for determining the growth constant a. By taking the ratio of successive terms, we find that the ratios are more or less constant. (Real data do not necessarily precisely fit the mathematical patterns we would expect.) This discrepancy may be due to other factors (political or economic, say), which could give the population a spurt in one year or decade while slowing down the population growth during another time period.

Decade	Year	Population		Ratio
0	1780	2.8		
				1.39
1	1790	3.9		
				1.36
2	1800	5.3		
				1.36
3	1810	7.2		
				1.33
4	1820	9.6		
				1.34
5	1830	12.9		
				1.33
6	1840	17.1		
				1.36
7	1850	23.2		
				1.35
8	1860	31.4		
				1.27
9	1870	39.8		
				1.26
10	1880	50.2		
				1.25
11	1890	62.9		
				1.21
12	1900	76.0		

TABLE 3.3: U.S. Population (1780–1900)

Decade	Population	log (Pop.)
0	2.8	0.447
1	3.9	0.591
2	5.3	0.724
3	7.2	0.857
4	9.6	0.982
5	12.9	1.111
6	17.1	1.233
7	23.2	1.365
8	31.4	1.497
9	39.8	1.600
10	50.2	1.701
11	62.9	1.799
12	76.0	1.881

TABLE 3.4: Logarithms of Population Values

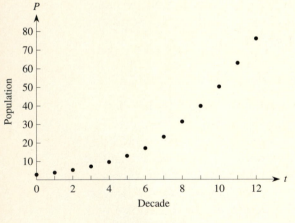

FIGURE 3.12

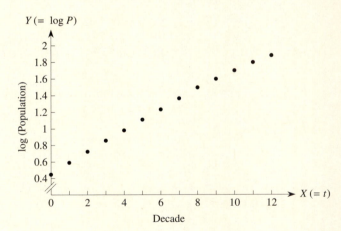

FIGURE 3.13

Alternatively, there is always an error in counting the number of people in any area; for example, consider the debate that rages around the U.S. Census every 10 years when different cities and states claim that large numbers of their people were not counted.

To determine the exponential function that best fits the values for the U.S. population in Table 3.3, we first transform the population data in the following way: We take the logarithm of each of the population values, as shown in Table 3.4.

The resulting scatterplot of log(Population) versus the year, shown in Figure 3.13, now clearly indicates a linear pattern. We find the best linear fit to this transformed set of data (since it is apparently linear) and obtain a correlation coefficient $r = 0.998$, based on the $n = 13$ data points. Comparing this value to the critical value of 0.553 in Table 3.2 in Section 3.2, we see that, with 95% certainty, there is indeed positive correlation between the year (or equivalently, the decade) and the logarithm of the population from 1780 through 1900. The equation of the regression line that best fits this set of transformed data is

$$Y = 0.121X + 0.487,$$

where X is the number of decades since 1780. However, we must interpret this as

$$\log P = 0.121t + 0.487.$$

We now undo the original transformation (the logarithm) of the data by applying the inverse function, which is exponential. Therefore, we obtain

$$P = 10^{\log P} = 10^{0.121t + 0.487}$$
$$= 10^{0.121t} \cdot 10^{0.487}$$
$$= (10^{0.121})^t \cdot (10^{0.487})$$
$$= 3.069(1.321)^t,$$

where t is the number of decades since 1780. This new equation is of the form $P_0 \cdot a^t$ for an exponential function. Notice that the estimated best value for the base $a = 1.321$ is very close to most of the ratios of successive population values we calculated in Table 3.3. Furthermore, in Figure 3.14, we show the original population data with this particular exponential function superimposed. We see that this curve is a very good fit to the data. Thus we have found the best possible exponential curve to fit the set of data. Moreover, because the correlation coefficient 0.998 for the transformed data indicates a high degree of positive correlation, we have further evidence that this curve is a good fit.

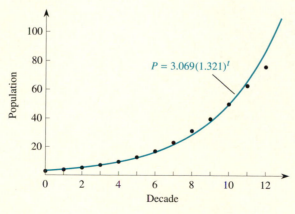

FIGURE 3.14

We could also analyze this set of population data by letting t represent either the year itself, $t = 1780, 1790, \ldots$ or the number of years since 1780, $t = 0, 10, 20, \ldots$. This will produce exponential functions with different bases, but each one will be an appropriate model for the situation. You must be careful that you keep track of what the variables used represent. In general, we suggest that you not use the entire year because that usually leads to extremely large or extremely small numbers when you take, for instance, the 1997th power of a base.

Many of you may wonder why we have only considered the U.S. population up to 1900 and not beyond. It turns out that the population does not follow an exponential pattern quite as closely thereafter. Various factors, such as limitations on immigration, changes in lifestyle to reflect smaller families, and the end of westward expansion have come into play during the twentieth century to slow the rate of population growth. We will introduce a more sophisticated mathematical model for population growth that takes such factors into account in Chapter 5, and we will come back to the question of fitting a better mathematical function to the data on the U.S. population up to the present day in Section 6.9.

Let us see why we took the logarithm of the population values. Suppose that the scatterplot for a set of data appears to follow an exponential pattern, so we hope to fit an exponential function of the form $y = f(x) = Ca^x$ to the data for some constants a and C. If we take logarithms of both sides of this equation, we get

$$\log y = \log (Ca^x) = \log C + \log (a^x) = \log C + x \cdot \log a,$$

which is of the form

$$Y = b + mX.$$

We can write

$$b = \log C \quad \text{and} \quad m = \log a$$

to emphasize the linear relationship between $Y = \log y$ and $X = x$ (see Figure 3.15).

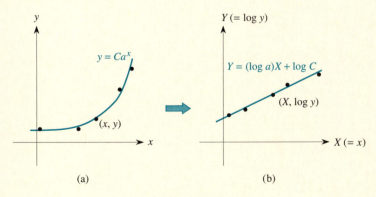

(a)   (b)

FIGURE 3.15

Thus if y is an exponential function of x, log y is a linear function of x, and this transformation serves to linearize the data. We then find the coefficients m and b using the least squares, or linear regression, technique. Finally, we undo the transformation using the inverse function to obtain:

$$b = \log C \qquad \text{so that} \qquad C = 10^b$$
$$m = \log a \qquad \text{so that} \qquad a = 10^m$$

to get the desired exponential function.

Fitting Power Functions to Data

What happens if we have a growth pattern in the data that seems to be less extreme than exponential growth? We might suspect this situation by either looking at the original scatterplot or trying to fit an exponential function to the data and finding that it is not a good fit—the correlation coefficient for the transformed data does not indicate a very high degree of linear correlation.

In this case let's try to model the data with a power function of the form $y = f(x) = Cx^p$, where C and p are two constants to be determined. We begin by taking logarithms of both sides of the equation:

$$\log y = \log (Cx^p) = \log C + \log (x^p) = \log C + p \log x.$$

When we examine this equation, we see that log y is a linear function of $\log x$ because both log C and p are constants. That is,

$$\log y = m \cdot \log x + b$$

is of the form

$$Y = mX + b,$$

where

$$m = p \qquad \text{and} \qquad b = \log C.$$

Based on this observation, we can linearize the original data set by taking the logarithms of *both* the x- and y-values. Presumably we get a linear relationship between the transformed data points. We can then apply the linear regression technique to determine m and b and finally undo the transformation to get the desired values for p and C: $p = m$ and $C = 10^b$.

To verify that this works, suppose we start with a simple power relationship, $y = 5x^2$, and select a few points that are on the curve (see Figure 3.16).

x	y
1	5
2	20
3	45
4	80

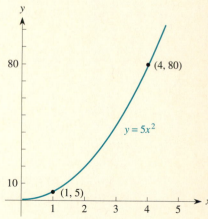

FIGURE 3.16

When we transform this set of data by taking the logarithms of both the x- and y-values, we get

x	y	$\log x$	$\log y$
1	5	0	0.699
2	20	0.301	1.301
3	45	0.477	1.653
4	80	0.602	1.903

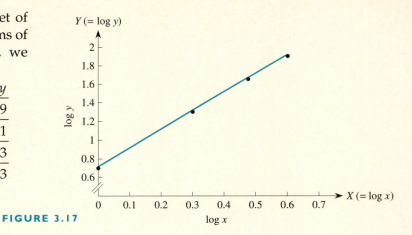

FIGURE 3.17

We plot the transformed data points $(\log x, \log y)$ in Figure 3.17 and see that they appear to lie exactly on a straight line. If we take any two of the transformed points and calculate the slope, we find $m = 2$. Furthermore, the y-intercept is $b = 0.699$. Thus we get $p = m = 2$ and $C = 10^b = 10^{0.699}$. So we see that the original data points do indeed lie on the curve $y = 5x^2$ because

$$y = Cx^p = 10^{0.699}x^2 = 5x^2.$$

To verify the accuracy of the regression equation, we compare the value $mX + b$ predicted by the linear regression equation to the transformed value $Y = \log y$ based on the original measurement y for each data point. If these two sets of values are close, then the fit is a good one. If some of these values are not close, then the fit may not be particularly good. If many of the values are extremely different, then the fit is a very poor one. The difference between the predicted values $mX + b$ and the observed values Y associated with y are called the **residuals** for the linear regression model. Geometrically, the residuals are the vertical distances from the data points to the regression line. In general, if all the residuals are small, then the fit is a good one. However, if several of the residuals are large, then we judge the fit to be poor and usually seek a better fit using a

E X A M P L E   I

In 1619, Johannes Kepler published his third law concerning the motion of the planets around the sun. In particular, this law relates the distance D of a planet from the sun to the period t (the duration of a year) for that planet. His work was based on the best experimental data available at the time. (For instance, astronomers then were not aware of the existence of the three outer planets, Uranus, Neptune, and Pluto.) Use the following current data to determine the re-

lationship between D and t. (D, the average distance of each planet from the sun, is measured in millions of miles; t, the length of its year, is measured in days.)

Planet	Period t	Distance D
Mercury	88	36.0
Venus	225	67.2
Earth	365	92.9
Mars	687	141.5
Jupiter	4329	483.3
Saturn	10,753	886.2
Uranus	30,660	1782.3
Neptune	60,150	2792.6
Pluto	90,670	3668.2

Solution We begin by drawing the scatterplot for this data, as shown in Figure 3.18, where the period t is plotted along the horizontal axis and the distance D from the sun is plotted along the vertical axis.

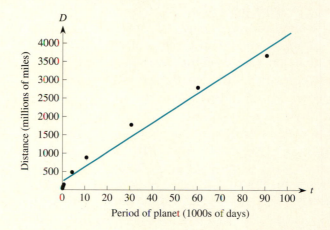

FIGURE 3.18

Although at first glance it might appear that the data points fall into a linear pattern, they actually do not. Consider the scale for the scatterplot. The vertical range extends from 36 million miles out to 3668 million (or 3.668 billion) miles. Furthermore, all of the data points fall rather far from the regression line. While these discrepancies seem small to the eye, they actually are enormous considering the size of the quantities involved. In turn, this suggests that the residuals may be quite large. In particular, the linear regression equation giving the distance from the sun as a function of the length of the year is

$$Y = 0.041X + 211.2.$$

Because the independent variable X is the length of the year t and the dependent variable Y is the distance D from the sun, we can write the linear regression equation as

$$D = 0.041t + 211.2.$$

Thus, the predicted values for the distance from the sun and the corresponding residuals are:

Planet	Period t	Observed values Distance D	Predicted values $D = 0.041t + 211.2$	Residual = Observed − Predicted
Mercury	88	36	214.8	−178.8
Venus	225	67.2	220.4	−153.2
Earth	365	92.9	226.2	−133.3
Mars	687	141.5	239.4	−97.9
Jupiter	4329	483.3	388.7	94.6
Saturn	10,753	886.2	652.1	234.1
Uranus	30,660	1782.3	1468.3	314.0
Neptune	60,150	2792.6	2677.4	115.2
Pluto	90,670	3668.2	3928.7	−260.5

Notice that the most accurate prediction based on the linear regression model is off by 94.6 million miles! As we said previously, the apparently small discrepancies between the data points on the scatterplot and the regression line actually are immense when we take the scale into account.

Next, we might try to fit an exponential curve, $D = C \cdot a^t$, to the original data; we do this by plotting $\log D$ versus t, as shown in Figure 3.19. This scatterplot clearly does not indicate a linear pattern. Using a calculator or computer

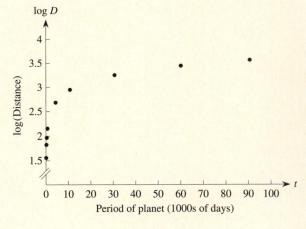

FIGURE 3.19

program and making the appropriate conversions to get the coefficients, the corresponding regression equation is $D = 152.8(1.000043)^t$. To check its accuracy, look at the *errors* we get when using this exponential formula; the **errors** are the differences between the actual values and the predicted values. The smallest error for Mars is

$$\text{error} = \text{actual} - \text{predicted}$$
$$= 141.5 - 152.8(1.000043)^{687}$$
$$\approx 141.5 - 157.4$$
$$= -15.9 \text{ million miles.}$$

The largest error, corresponding to Pluto, is a whopping 3871 million miles!

We now attempt to fit a power function, $D = Ct^p$, to the original data by plotting $\log D$ versus $\log t$, as shown in Figure 3.20. In this case, note that the points seemingly lie along a straight line. But, as you saw when we tried an exponential fit, you always should examine the residuals before jumping to any conclusions.

The equation of the regression line for the transformed data is

$$Y = 0.6667X + 0.2596.$$

In terms of our notation here, this equation is equivalent to

$$\log D = 0.6667 \log t + 0.2596$$
$$= \log (t^{0.6667}) + 0.2596.$$

Using properties of exponents and logarithms, we get

$$D = 10^{\log D} = 10^{\log (t^{0.6667}) + 0.2596}$$
$$= 10^{\log (t^{0.6667})} \cdot 10^{0.2596}$$
$$= t^{0.6667}(1.818).$$

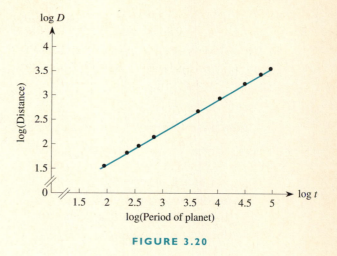

FIGURE 3.20

To verify the accuracy of this fit, we again look at the errors we get when using this formula to predict the power function of the length of each planet's year:

Planet	Period (t)	Distance (D)	Predicted values	Error
Mercury	88	36	36.0	0
Venus	225	67.2	67.3	−0.1
Earth	365	92.9	92.9	0
Mars	687	141.5	141.6	−0.1
Jupiter	4329	483.3	483.0	0.3
Saturn	10,753	886.2	886.0	0.2
Uranus	30,660	1782.3	1781.5	0.8
Neptune	60,150	2792.6	2792.0	0.6
Pluto	90,670	3668.2	3670.6	−2.4

For the inner planets (Mercury, Venus, Earth, and Mars) whose distances from the sun are relatively small, the discrepancy between the actual value and the predicted value is extremely small. As the actual distance increases for the outer planets, the error in the prediction also increases somewhat but remains reasonable, particularly if we consider the discrepancies on a percentage basis. So, the power function is quite a good fit to the data. Furthermore, the correlation coefficient for the transformed data is $r = 0.99999994$.

Incidentally, the relationship obtained here is the form that we typically would want. It is relatively easy, in practice, to measure the period for a planet simply by determining how long it takes to make a complete revolution about the sun, at least for the inner planets whose year is reasonably short. It is much harder to measure the distance from the sun. However, it turns out that there is a more interesting form for the relationship between the period and the distance of the planets from the sun. Since, the best-fit function is given by $D = 1.818t^{0.6667}$ and $0.6667 \approx \frac{2}{3}$, we can write

$$D = 1.818t^{2/3}.$$

We now raise both sides of this equation to the third power to get

$$D^3 = 6.0087t^2.$$

Alternatively, if we divide both sides of the equation by 6.0087, we get

$$t^2 = 0.1664D^3,$$

which is the form in which Kepler's Third Law usually is expressed.

Astronomers discovered the planet Pluto in 1930 after observing some minor perturbations in the orbit of Neptune. They hypothesized that these discrepancies could be accounted for by the existence of a previously unknown outer planet. Knowing the timing of the perturbation, the astronomers knew approximately where to look for this unknown planet by using predictions based on Kepler's Third Law.

E X A M P L E   2

An experiment is conducted in which kernels of corn are heated in an open container until they pop and fly out of the container and land some horizontal distance away. The distance in feet that each piece of popcorn lands away from the container is then measured, giving the following results:

Distance (feet)	0 to 1	1 to 2	2 to 3	3 to 4	4 to 5
Number of popcorn	121	50	18	9	2

Find the function, from among the families we have studied, that best fits this data.

Solution We first recognize that the independent variable is the horizontal distance d in feet that a popped kernel flies and that the dependent variable is the number of pieces of popcorn N that flew a given distance. To make things simple, we take any distance from 0 to 1 feet to be $d = 0.5$; any distance from 1 to 2 feet to be $d = 1.5$, and so forth.

We begin by displaying the data in a scatterplot, as shown in Figure 3.21. From the scatterplot, we observe that the pattern is clearly nonlinear. Moreover, from the shape of the data points, we would expect the best-fit function to be decreasing and concave up; we know

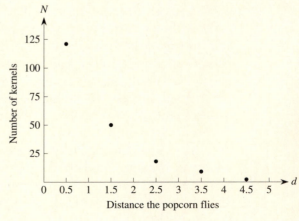

FIGURE 3.21

that both an exponential decay function and a power function with a negative power have that behavior pattern.

We first construct the best exponential fit. To do so, we must transform the data to compare $\log N$ to d, as follows:

d	N	$\log N$
0.5	121	2.083
1.5	50	1.699
2.5	18	1.255
3.5	9	0.954
4.5	2	0.301

The resulting linear regression equation relating the transformed values $\log N$ to d is

$$Y = -0.431X + 2.336,$$

or equivalently,

$$\log N = -0.431d + 2.336$$

with correlation coefficient $r = -0.993$. Note that r is negative because the function is decreasing. As the distance d that the popcorn flies increases, the number of kernels that fly this distance decreases. We undo the transformation using properties of logarithms and exponents:

$$N = 10^{\log N} = 10^{-0.431d + 2.336}$$
$$= 10^{-0.431d} \cdot 10^{2.336}$$
$$= (10^{-0.431})^d (10^{2.336})$$
$$= 216.8\,(0.371)^d,$$

which is the best exponential function to fit this data. Notice that, as expected, the base for this function is less than one because the function decays: The decay factor is $0.371 < 1$. We show this exponential curve superimposed over the scatterplot in Figure 3.22. Note how well it fits the data.

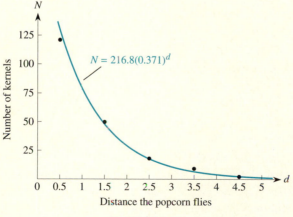

FIGURE 3.22

We next construct the best power function fit to this data. Here we relate $\log N$ to $\log d$, extending the previous table as follows:

d	N	$\log d$	$\log N$
0.5	121	-0.301	2.083
1.5	50	0.176	1.699
2.5	18	0.398	1.255
3.5	9	0.544	0.954
4.5	2	0.653	0.301

The resulting linear regression equation relating the transformed values of $\log N$ and $\log d$ is

$$Y = -1.689\,X + 1.755.$$

This equation is equivalent to

$$\log N = -1.689 \log d + 1.755,$$

with correlation coefficient $r = -0.929$. Again note that r is negative because the function is decreasing. We now undo the transformation.

$$
\begin{aligned}
N = 10^{\log N} &= 10^{-1.689 \log d + 1.755} \\
&= 10^{\log (d^{-1.689}) + 1.755} \\
&= 10^{\log (d^{-1.689})} \cdot 10^{1.755} \\
&= 56.89\, d^{-1.689}.
\end{aligned}
$$

As expected, the power in this power function is negative because the function is decreasing for d greater than zero. We show this power function superimposed over the data points in Figure 3.23.

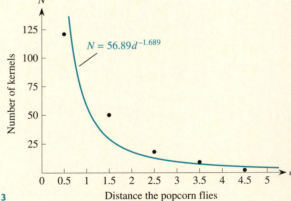

FIGURE 3.23   Distance the popcorn flies

Looking at Figure 3.23, we see that this power curve is not as good a fit to the data as the exponential function shown in Figure 3.22. Moreover, we conclude from the values for r that the exponential function is a better fit than the power function: The correlation coefficient for the exponential function, $r = -0.993$, is closer to -1 than the correlation coefficient for the power function, $r = -0.929$.

Fitting Logarithmic Functions to Data

Finally, suppose that the data we are studying falls into a pattern such as that shown in Figure 3.24; it is increasing and concave down, but does not appear to be leveling off to a horizontal asymptote. Such a shape might suggest a logarithmic function, especially if the interval of values for the independent variable is large compared to the interval of values for the dependent variable. Therefore, we attempt to fit a logarithmic function of the form

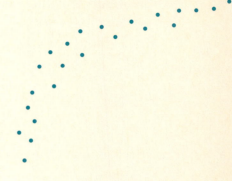

$$y = f(x) = C \log x + b$$

FIGURE 3.24

to the data. This equation suggests that we should transform the original data set by comparing y to $\log x$ rather than comparing y to x. However, most calculators and computer packages that perform this calculation use natural logarithms with base e rather than logarithms with base 10. Thus we actually fit functions of the form

$$y = f(x) = C \ln x + b$$

to the data.

In Section 1.1, we discussed the growth in life expectancy over the years since the beginning of the century. The graph showing this trend is in Figure 3.25. Notice that the data values are growing more and more slowly over time and that the span of years from 1900 to 1990 is considerably larger than the interval of values for life expectancy. So, we could expect that a logarithmic function would be a good fit to this data. We investigate this situation in Example 3.

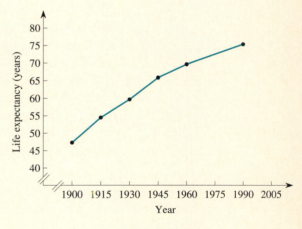

FIGURE 3.25

E X A M P L E   3

The following table shows the trend in life expectancy for a child born in the given year since the beginning of this century. Find the best logarithmic function to fit this data.

Year, t	1900	1915	1930	1945	1960	1990
Life Expectancy, L	47.3	54.5	59.7	65.9	69.7	75.4

Solution   To fit a logarithmic function to this data, we need to compare life expectancy to the natural logarithm of the year, as shown in the following table.

ln t	7.5496	7.5575	7.5653	7.5730	7.5807	7.5959
L	47.3	54.5	59.7	65.9	69.7	75.4

The best linear fit to this transformed data is

$$Y = 609.0X - 4548.5,$$

which is equivalent to

$$L = 609.0 \ln t - 4548.5.$$

Because this expression already gives L as a function of t, there is no need to detransform the equation in order to solve for L. The corresponding correlation coefficient is $r = 0.984$, which indicates a very high level of positive correlation.

Summary of Curve-Fitting Procedures

- If the data appear to follow an exponential pattern, $y = Ca^x$, plot log y versus x to linearize the data and find the best linear fit, log $y = mx + b$, to the transformed data.
- If the data appear to follow a power function pattern, $y = Cx^p$, plot log y versus log x to linearize the data and find the best linear fit, log $y = m \log x + b$, to the transformed data.
- If the data appear to follow a logarithmic pattern, $y = C \log x + b$, plot y versus log x to linearize the data and find the best linear fit, $y = m \log x + b$, to the transformed data.

There is one problem that you should be aware of when doing any of these curve-fitting procedures. Recall that the function $y = \log x$ is defined only for values of x greater than 0. Thus if any of your data values are 0 or negative, you cannot take their logarithm. Often you can circumvent this difficulty by redefining the independent variable. For example, suppose that the data represent values of a quantity versus time starting in 1950. You could count the years since 1950, but then 1950 corresponds to $t = 0$, which causes a problem if you had to take logs. Alternatively, you could count the years since 1900 because 1950 then corresponds to $t = 50$, which circumvents the problem of log 0 being undefined. For that matter, you simply could count the years from 1949 so that 1950 corresponds to $t = 1$. You even could use the year 1950 itself, although having $t = 1950$ creates potential round-off errors due to the size of the numbers, as we previously discussed. Whatever you do, just be careful to keep track of what your independent variable represents.

If the values of the dependent variable are 0 or negative, things become a little more complicated. One way to resolve the problem is to shift all the values up by a fixed amount to make them all positive, perform the analysis on these transformed values, and then undo the vertical shift. We will illustrate this approach in Section 3.6 when we perform a complete analysis of the *Challenger* data given in Section 3.1.

EXERCISES

1. NASA sent seven Apollo missions to the moon from 1969 to 1972; only six made it because the Apollo 13 moon mission was aborted in flight without a moon landing. The following table lists the number of hours that each mission spent on the surface of the moon compared to the date of the mission measured in the number of months from June 1960.

Mission No.	Date of Mission	Number of Months	Hours on Moon
11	July 1969	1	22
12	November 1969	5	32
14	January 1971	19	34
15	July 1971	25	67
16	April 1972	34	71
17	December 1972	42	75

a. Determine the best fit (linear, exponential, or power function) to this set of data.
b. Had the Apollo space program continued as planned with additional Apollo moon missions, estimate the length of stay on the moon for a launch in November of 1973 using each of the three best-fit functions.
c. Which function seems most reasonable? least reasonable?

2. The table at the right, based on data from the U.S. Department of Education, gives the total number in thousands of high school graduates in the indicated years since 1900.

a. Determine the best fit (linear, exponential, or power function) for the number of high school graduates as a function of the year.
b. Use each function to predict the number of high school graduates in the year 2000. Which prediction seems the most reasonable? the least reasonable?

Year	High School Grads
1900	364
1910	502
1920	665
1930	746
1940	848
1950	1461
1960	2029
1970	2694
1980	3621
1990	4863

  c. Use each function to predict the year in which there will be 10 million high school graduates. (Be careful not to let $t = 0$ correspond to the year 1900 or you will not be able to take its log.)

  d. Do you think that the same type of function that turned out to be the best fit to this data for the nation will apply as the best fit to the comparable data for any particular state? Why or why not?

3. Repeat the analysis of the growth of the U.S. population given earlier in this section, but concentrate on the period from 1780 to 1890 instead of 1780 to 1900. Does this exponential function give a better fit? How do you know it is a better fit?

4. The accompanying graph shows the number of deaths per 100,000 women in the United States from both stomach cancer and lung cancer over the last 60 years. Both sets of data appear to be exponential functions.

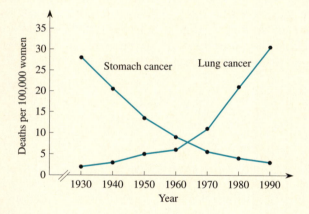

  a. Find the best fit for each set of data.

  b. What are your estimates for the numbers of deaths per 100,000 from each type of cancer in the year 2000?

  c. Based on the graph, about how many women actually died in 1990 from stomach and lung cancer if there are approximately 100 million adult women in the United States?

  d. Can you give any reasons why the trend for stomach cancer (like that for most other cancers) is decreasing exponentially whereas the trend for lung cancer is rising exponentially?

5. The following table gives the number, in thousands, of passenger cars produced in the United States over the course of several years. Find the best linear and exponential fits to this data and use them to predict the number of cars that were produced in 1995. What do these trends tell you about the future of the U.S. car industry?

Year	1985	1986	1987	1988	1989	1990
Number of cars	8185	7829	7099	7113	6823	6077

6. The following table gives the number of violent crimes per 100,000 Americans over the last three decades.

Year	1960	1965	1970	1975	1980	1985	1990
Number of crimes	175	200	360	490	580	550	750

    a. Find the best linear and exponential fits to this set of data.
    b. Use them to predict the number of violent crimes, per 100,000 people, that occurred in 1995.
    c. With both models, predict when the number of violent crimes will reach 1000 per 100,000 people.
    d. What is the doubling time for the exponential model?

7. The following table gives the number of deaths from AIDS in the United States for men and for women from 1986 through 1992.

Year	1986	1987	1988	1989	1990	1991	1992
AIDS deaths among men	12,099	19,256	27,435	29,942	36,770	38,013	39,160
AIDS deaths among women	1,048	1,832	3,284	3,653	4,883	5,688	6,312

    a. Find the best exponential function and the best power function that fit the data for AIDS-related deaths among men as a function of time. Which is a better fit to the data?
    b. Find the best exponential function and the best power function that fit the data for AIDS-related deaths among women as a function of time. Which is a better fit to the data?
    c. Now draw a scatterplot for the number of AIDS-related deaths among women W as a function of the number of AIDS-related deaths among men M in each year. From the scatterplot, does there appear to be any pattern? If so, does the pattern appear to be linear or nonlinear? Determine the best fit to these data pairs from among the families of linear, exponential, power, and logarithmic functions.

8. In Example 4 of Section 2.3, we examined the growth in circulation for *USA Today* and found that its overall growth rate was 4.4% from 1985 to 1993. The following table gives the actual circulation in millions for each of the nine years. Find the best exponential fit to this data and then compare its growth rate with the growth rate that was based on just 1985 and 1993. Since both rates are considerably larger than the growth rate for the U.S. population (0.7%), what do you think must happen in the foreseeable future to the circulation of *USA Today*?

Years since 1985	0	1	2	3	4	5	6	7	8
Circulation	1.418	1.459	1.586	1.656	1.755	1.843	1.867	1.957	2.001

9. The following table shows the annual world oil production in millions of barrels.

Year	1880	1890	1900	1910	1920	1930	1935	1940	1945	1950	1955	1960	1962	1964
Oil	30	77	149	328	689	1,412	1,655	2,150	2,595	3,803	5,626	7,674	8,882	10,310

1966	1968	1970	1972	1974	1976	1978	1980	1982	1984	1986	1988	1990
12,016	14,104	16,690	18,584	20,389	20,188	21,922	21,722	19,411	19,837	20,246	21,338	22,415

Let t be the time in years since 1880.

a. Plot the points and observe the pattern of the data. Is it clear that there are actually two different patterns? When does the pattern change? Can you explain why there might have been a change in the pattern at that point in time?

b. Find the best exponential fit to the data corresponding to the first pattern. What is the growth rate in oil production over this time period? If that pattern continued through the end of the century, estimate the world oil production in the year 2000.

c. Find the best fit among the different families of functions we have studied so far to account for the pattern during the second time period. Using that function, what is your best prediction for world oil production in the year 2000?

10. Suppose that a function f is increasing and concave up, and that $f(60) = 250$, $f(70) = 300$. Which of the following values are possible and which are impossible? Explain.

a. $f(65) = 270$ b. $f(65) = 275$ c. $f(65) = 280$
d. $f(100) = 400$ e. $f(100) = 450$ f. $f(100) = 500$
g. $f(40) = 100$ h. $f(40) = 150$ i. $f(40) = 200$

11. Suppose that a function f is decreasing and concave up, and that $f(10) = 80$, $f(12) = 70$. Which of the following values are possible and which are impossible? Explain.

a. $f(11) = 78$ b. $f(11) = 75$ c. $f(11) = 72$
d. $f(15) = 50$ e. $f(15) = 55$ f. $f(15) = 60$
g. $f(5) = 100$ h. $f(5) = 105$ i. $f(5) = 110$

12. Several sets of data have been linearized to find the best exponential fit. The calculator gives the following equations for the lines that best fit the transformed data. Undo the transformations to get the best-fit exponential functions:

a. $Y = 0.3522 + 1.0843X$ b. $Y = -1.3015 + 0.7840X$
c. $Y = 0.8525 - 1.2733X$

13. Several sets of data have been linearized to find the best power fit. The following equations are for the lines that best fit the transformed data. What are the best-fit power functions?

a. $Y = 0.3522 + 1.0843X$ b. $Y = -1.3015 + 0.7840X$
c. $Y = 0.8525 - 1.2733X$

14. In Example 1 on Kepler's Third Law, we found the best power function fit by writing

$$\log D = \log (t^{0.667}) + 0.2596.$$

Show that you get the same result using properties of logarithms when you write this function as

$$\log D - \log (t^{0.667}) = 0.2596.$$

15. Assume that each of the planets from Mercury to Neptune revolves about the sun in a roughly circular orbit.

 a. Extend the first table given in Example 1 to include the speed of each planet in its orbit.
 b. Find the best fit to this set of data on the speed of a planet as a function of the length of its year from among linear, exponential, and power functions.
 c. Explain how the formula you found in part (b) can be directly determined algebraically from Kepler's Third Law.

3.4 *What the Residuals Tell Us About the Fit*

Throughout this chapter, we have been concerned with finding the function that best fits a set of data values. If the data points fall into a linear pattern, then we use the regression line. If the data points fall into a non-linear but recognizable pattern, then we can transform the data values to linearize them, find the regression line for this transformed set of points, and finally, undo the transformation using the appropriate inverse function to get the corresponding function that best fits the data.

The problem with the curve-fitting procedures is assessing *how well* a particular function "fits" the set of data. When the original data or the transformed data are approximately linear, the correlation coefficient is an effective measure of the goodness of the fit. However, we have seen that over a limited interval, different functions may be relatively close to one another but eventually will diverge if the interval is extended. Thus we often face situations in which a variety of different functions—for example, linear, exponential, and power functions—are all reasonably good fits to a set of data; that is, their correlation coefficients all can be quite high. In this case, which function gives the best fit?

It turns out that the residuals associated with the fit provide the additional information we need to determine how good the fit is. Recall that the residuals represent the vertical distances between the actual or transformed data values and the predicted values based on the corresponding regression line equation. When we find a linear fit to the original data, then the residuals are the differences between the actual y-values and the predicted values based on the regression line, as shown in Figure 3.26(a). When we transform the data and obtain the regression line for the transformed data, then the residuals are the differences between the *transformed* data values (Y) and the predicted values based on the regression line for the transformed data ($mX + b$). The differences between the original y-values and the predicted values based on the *detransformed* function— say an exponential function—are not called residuals; we call these differences *errors*, as shown in Figure 3.26(b).

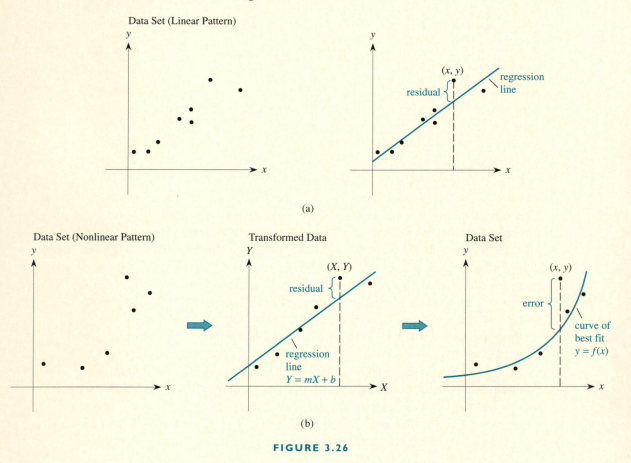

FIGURE 3.26

If the actual data value is y and the predicted value given by the regression line is y_p, then the corresponding residual is $y - y_p$. If the actual data point lies above the regression line, then the residual is positive; if the data point lies below the regression line, the residual is negative (see Figure 3.27).

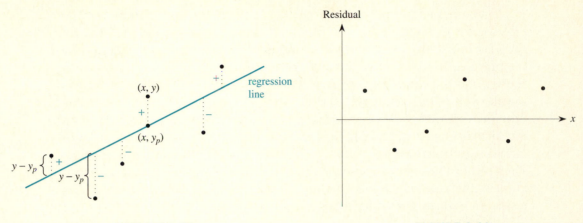

FIGURE 3.27   FIGURE 3.28

   If the fit is good, then we would expect roughly half of the residuals to be positive and roughly half to be negative because the regression line should pass, more or less, midway between the data points. We plot the residuals shown in Figure 3.27 on a different set of axes called a **residual plot**, as shown in Figure 3.28. Points on the horizontal or base line represent data points with zero residual—that is, data points that fall exactly *on* the regression line. Whenever a data point lies above the regression line, the residual is positive and the associated point in the residual plot is above the horizontal axis by that amount. Similarly, whenever a data point lies below the regression line, the residual is negative and the associated point in the residual plot is below the horizontal axis by that amount. (Note that the horizontal axis shown in the residual plot has no direct relationship to the regression line.)

   For now, let's look at the case where the regression line is the best fit to a set of data. Figure 3.29 shows both a scatterplot and the associated residual plot for a set of data in which a linear function is a good fit to the data. Notice that the regression line essentially passes through the middle of the cluster of data points, so that roughly half the points lie above the regression line and roughly half lie below it. Also, notice that in the residual plot, roughly half the residuals are positive (corresponding to data points that lie above the regression line) and roughly half are negative (corresponding to data points below the regression line). Moreover, notice that the residuals seem to be *scattered randomly* both above and below the horizontal axis; there does not appear to be any pattern to their locations.

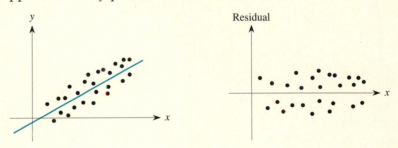

FIGURE 3.29

In comparison, consider the scatterplot and the associated residual plot shown in Figure 3.30. The data fall into a linear pattern, but the regression line does not appear to be a particularly good fit; it is distorted by the presence of two points that are far from the line. How does this poor fit show up in the residual plot? Again notice that roughly half the residuals are above the horizontal axis and roughly half are below it. However this time the fact that most of the residuals on the left are negative (reflecting the fact that most of the data points on the left fall below the regression line) is significant. Similarly, most of the residuals on the right are positive. Thus rather than being scattered randomly above and below the horizontal axis, the residuals have a pattern. The existence of this pattern in the residuals indicates that the fit is not a good one.

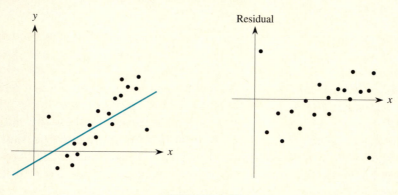

FIGURE 3.30

Consider now the scatterplot of a different set of data and its associated residual plot shown in Figure 3.31. The data values appear to fall into an exponential pattern. In the scatterplot, we have drawn the regression line. Notice that the data points on the left lie above the line, the points in the middle lie below the line, and the points on the right lie above the line. This behavior is reinforced by the residual plot in which the points fall into a U-shaped pattern with the middle residual points falling below the horizontal axis. As was the case in Figure 3.30, the existence of this pattern indicates that the linear fit is not a good one.

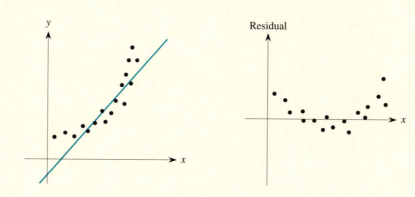

FIGURE 3.31

Figure 3.32 shows the best exponential fit to the original data and the residual plot associated with the transformed data. It is visually obvious that the exponential curve is a good fit to the data. Furthermore, notice that the residuals are small and there is no discernible pattern to them. They are scattered about the horizontal axis in an apparently random pattern.

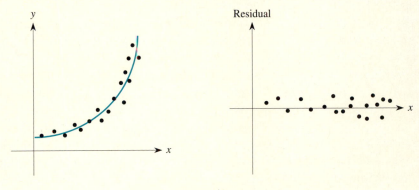

FIGURE 3.32

In general, given a set of data, it may not be evident that a particular curve is a good fit to the data just by looking at the scatterplot. We have seen previously that the scales involved can distort the image when either or both variables extend over a very large interval of values, so we are unable to recognize a good or a bad fit. However, we can overcome this difficulty by examining and interpreting the residual plot associated with the regression line for the transformed data:

- If a particular fit is good, then the residuals should display no pattern.
- If the residuals do display a pattern (for example, rising and falling, U-shaped, or periodic), then the fit is likely not a good one and you should look further for a better fit among other families of functions.

In addition,

- If the residuals are extremely large numerically, then you should have suspicions about the fit, even if the fit on the scatterplot looks good to the naked eye. (Recall the linear fit in the planetary data example in Section 3.3.)

Furthermore, if only one or two residuals are extremely large compared to all the others, that might indicate an error in keying in the data value, an error in the actual measurement, or a highly unrepresentative value, called an **outlier.** Check your values to be sure that there is no typing error. If there is none, then you might want to experiment by removing the outlier from the data set and recomputing the regression equation to see if there is a dramatic improvement in the fit. However, any final report you write on your results should indicate the apparently spurious outlier(s) even if they are not used in the calculations.

Once you have convinced yourself that you have found an appropriate fit for the data, then the linear correlation coefficient r provides a measure of the degree of agreement. However, keep in mind that we can calculate a correlation coefficient for *any* set of data, either a set of original values or a transformed set of values. It is possible that several different fits for the same set of data all have a significant level of correlation. (Think about the graphs of functions: Over a relatively small interval, a line, an exponential function, and a power function may appear almost identical, but when you extend them sufficiently, they clearly move away from one another.) Thus a good strategy is first to find the *best* fit and only then consider the associated correlation coefficient. You can, and probably should, try different fits for a given set of data, one after the other, until you come up with the best curve among the families of functions we have considered in this chapter (linear, exponential, power, and logarithmic functions). Of course, it is certainly possible that a set of data will not fall into any of these patterns, but a great many reasonable data sets tend to. In later chapters, we will extend these ideas to cases in which the possible patterns include quadratic and higher degree polynomials, periodic functions, or other patterns that arise frequently. It is also possible that two or more different patterns apply to different portions of a set of data—for instance, the pattern could start with linear growth and later could appear to have exponential growth.

EXAMPLE

Table 3.5 shows the growth of the federal debt in trillions of dollars from 1980 to 1993. Determine an appropriate curve that best fits this data. If the pattern continues unchecked, what is your best estimate for the size of the debt in the year 2000?

Year	1980	1981	1982	1983	1984	1985	1986	1987	1988	1989	1990	1991	1992	1993
Debt	0.91	0.99	1.14	1.37	1.56	1.82	2.12	2.35	2.60	2.87	3.21	3.60	4.00	4.35

TABLE 3.5: *Federal Debt (in trillions)*

Solution If we examine the scatterplot for this data in Figure 3.33(a) along with the superimposed regression line, we see that a linear function is a reasonably good fit. In addition, the correlation coefficient for this linear fit is $r = 0.991$, which indicates a high level of positive correlation. However, the scatterplot for the original data shows a pattern that is concave up. Also, the residual plot in Figure 3.33(b) displays a pattern, which suggests that we try an exponential fit. (You might want to look at the successive ratios of the debt figures, which are nearly constant, to see that an exponential function is a good idea to try.)

To construct the exponential function that best fits this data, we first transform the data values by taking the logarithm of the debt figures. The corresponding scatterplot for the transformed data is shown in Figure 3.34 with the associated regression line superimposed. We therefore see that the regression line is a good fit to the transformed data, so the corresponding exponential function should be a good fit to the original data.

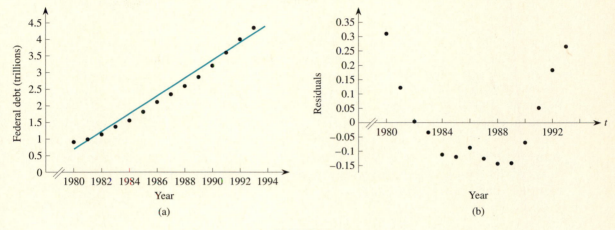

FIGURE 3.33

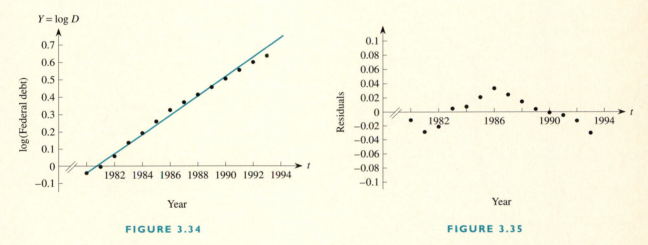

FIGURE 3.34

FIGURE 3.35

Furthermore, the corresponding residuals are displayed in Figure 3.35. Notice that the residuals are all relatively small with approximately half of the points above and half below the zero-line, which suggests that the exponential fit is reasonably good. However, there does appear to be an inverted, slightly U-shaped pattern to the residuals, so the exponential function may not fully explain the growth behavior. The corresponding correlation coefficient $r = 0.996$ (which is higher than the correlation coefficient for the linear fit) indicates a very high degree of positive linear correlation between the transformed data and the year. The resulting regression equation, based on measuring years from 1900, is

$$Y = 0.0537X - 4.323,$$

which is equivalent to

$$\log(\text{Debt}) = 0.0537t - 4.323.$$

When we undo the transformation, we find that the equation for the best exponential fit is

$$
\begin{aligned}
\text{Debt} &= 10^{0.0537t - 4.323} \\
&= (10^{-4.323})(10^{0.0537t}) \\
&= 0.0000475\,(10^{0.0537})^t \\
&= 0.0000475\,(1.132)^t.
\end{aligned}
$$

(See Figure 3.36.) The growth factor 1.132 tells us that the national debt is growing at an annual rate of over 13% based on this model. If this growth pattern persists, what will be the value of the national debt in the year 2000? In that year, when $t = 100$, the debt will be about 11.5 trillion dollars! This amount is equivalent to approximately $46,000 per person in the country. Fortunately, the national debt appears to be growing somewhat less rapidly since 1989, which is relatively good news for the future.

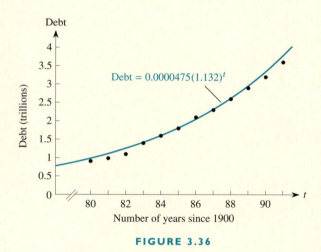

FIGURE 3.36

E X E R C I S E S

1. The best-fit line is constructed for each of four sets of nonlinear data, whose patterns can be roughly described as follows:

   a. increasing and concave up
   b. increasing and concave down
   c. decreasing and concave up
   d. decreasing and concave down

   Match each description with one of the two possible residual plots shown on the next page. Explain your answer in each case.

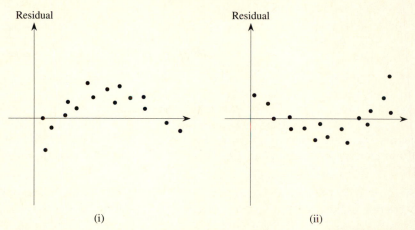

(i)   (ii)

2.   Three different types of functions are fitted to a set of data based on the following three associated residual plots. Decide which function is the best fit to the data.

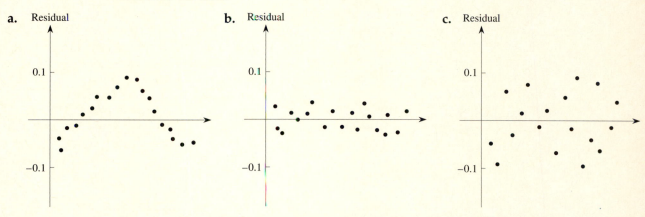

a.   b.   c.

3.   Continue the analysis of the data on the national debt given in the Example in this section to see if a power function fit is better or worse than the exponential fit. If this pattern continues, what do you predict the national debt will be in the year 2000? What does this debt amount to per person?

4.   Working with the results of Exercise 1 in Section 3.3, analyze the associated residual plot to decide which of the three fits (linear, exponential, and power function) is the best fit to the data given for the number of hours that each Apollo mission spent on the moon as a function of the number of months since the first mission.

5.   Working with the results of Exercise 2 in Section 3.3, analyze the associated residual plot to determine which of the three fits (linear, exponential, and power function) is the best fit to the set of data for the number of high school graduates as a function of the year.

6.   The following table shows the growth of the federal debt in billions of dollars from 1940 to 1990. Determine an appropriate curve that best fits this data.

Year	1940	1950	1960	1970	1980	1990
Debt	51	257	291	381	909	3207

7. The following table shows the growth of the U.S. population in millions from 1940 to 1990. Determine an appropriate curve that best fits this data.

Year	1940	1950	1960	1970	1980	1990
Population	131.7	150.7	179.3	203.3	226.5	248.7

8. Use the data given in Exercises 6 and 7 to construct a set of values representing the average amount of the national debt per person in the United States and determine an appropriate function that best fits this data. If these trends continue, what is your best estimate for your share of the debt in 2000? in 2005?

9. According to *Motor Trend* magazine, the following data are the times (in seconds) it takes a Trans Am to accelerate from zero to the indicated speed.

Speed (mph)	30	40	50	60	70
Time (seconds)	3.00	4.29	5.52	7.38	9.81

Determine the best fit (linear, exponential, or power function) for the acceleration time as a function of the final speed. Estimate how long it will take a Trans Am to accelerate to 45 mph; to 80 mph; to 90 mph. Which estimated time is most likely to be accurate? (You might want to compare this data to the corresponding information on the Mercedes in Exercise 3 of Section 3.2.)

10. Suppose that a function f is decreasing and concave down and that $f(10) = 80, f(12) = 70$. Which of the following values are possible and which are impossible? Explain. (Compare your answers to the results of Exercise 5 in Section 3.3.)

a. $f(11) = 78$	**b.** $f(11) = 75$	**c.** $f(11) = 72$
d. $f(15) = 50$	**e.** $f(15) = 55$	**f.** $f(15) = 60$
g. $f(5) = 100$	**h.** $f(5) = 105$	**i.** $f(5) = 110$

11. Craig is cruising down the interstate at a steady 75 miles per hour on his way to enjoy spring break in Fort Lauderdale. He passes a police radar site, and the officer takes off after him. Unfortunately, Craig doesn't notice the police car in hot pursuit. At what instant is Craig farthest ahead of the police car?

12. Joe is driving down a highway at 50 miles per hour (about 73 feet per second) when his girlfriend Beth passes him at 60 mph (88 feet per second), but doesn't see him. Joe speeds up steadily until he catches up to Beth's car. The accompanying diagram shows their respective speeds as a function of time. Use the graph to answer the following questions:

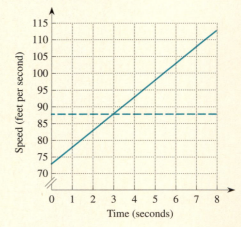

a. At what time is Joe farthest behind Beth?
b. How fast is he going when he does catch up to her?
c. How long does it take him to overtake her?
d. How far has he gone while pursuing her along the highway?

3.5 *Analyzing Data*

Up to this point, we have made the data analysis process more complicated than it need be in practice. We asked you to do many calculations by hand that can be and usually are done by using technology. Our intent has been more than just fitting the best curve (from a relatively simple family of functions) to a set of data. It has also been to develop a variety of skills with manipulating certain functions, as well as to reinforce the behavior patterns of various families of functions. In practice, all the transformations and associated work we have demonstrated so far can be done by a computer or sophisticated calculator.

In general, for a given situation, you will be faced with a set of data that must be analyzed and fitted with the best possible curve. You enter the original data into a computer program or a calculator and decide on the type of fit you want to try (say linear, exponential, power, or some other function). The machine will do the work and present you with the answer. You must then decide if this is the best fit possible; if it is not, you should try a different fit and continue this process until you are satisfied with the result. Some professional statistical software packages provide the ability to fit over a hundred different families of functions to a given set of data, but that is well beyond our needs here.

In this section, we describe how this process can be done using either a graphing calculator or a computer. We then consider some examples of sets of data from various applications to illustrate the ideas.

Typically, you begin by entering the original set of data and having your calculator or computer program display the scatterplot, the associated regression line, and the correlation coefficient. You also could see the residual plot. However, you also have the choice of a variety of options for fitting other families of functions to the data, including exponential functions, power functions, and logarithmic functions (among others). The calculator will perform the appropriate transformation and display the equation (or possibly just the coefficients) of the detransformed curve, along with the correlation coefficient. Thus, if you want a power function fit to a set of data using one widely used calculator, it will provide you with the coefficients, say

$$a = 5.234$$
$$b = 1.482$$
$$r = 0.9163675,$$

where the resulting power function is

$$f(x) = a \cdot x^b = 5.234x^{1.482}.$$

Just as you can either capture a linear regression equation as a function or enter the equation directly, you likewise can capture the nonlinear regression equation as a function and have its graph superimposed over the associated scatterplot so that you can see how well the curve fits the data.

By going through the different families of functions that are available, you can decide on the best fit among them by looking at the value for the correlation coefficient. Simply take the fit that corresponds to a value of r closest to $+1$ for positive correlation or -1 for negative correlation. Also, examine the residual plot to see if the fit explains most of the variation in the data or if there is some pattern that suggests a different function would be a better fit.

E X A M P L E   1

The following table lists the number (in thousands) of master's degrees awarded to women in the United States in the indicated years over the last few decades. Determine the best fit to these data points. Predict how many such degrees were received by women in the year 1995.

Year	1970	1975	1980	1985	1990
Number (thousands)	83	131	147	143	163

Solution   Rather than using the entire year for the independent variable, we use years since 1900 so that $t = 70, 75, \ldots$. By looking at the data, we see that there is an overall trend of increase in the number of master's degrees M awarded to women. We obtain the following values for the correlation coefficient for each fit and the associated best-fit function:

Linear fit: $r = 0.8943$ Equation: $M = 3.44t - 141.8$ $(t = 70, 75, \ldots)$

Exponential fit: $r = 0.8641$ Equation: $M = 13.04\,(1.029)^t$ $(t = 70, 75, \ldots)$

Power fit: $r = 0.8789$ Equation: $M = 0.0049t^{2.327}$ $(t = 70, 75, \ldots)$

Based on these three sets of results, we decide that the linear fit is the best because the corresponding correlation coefficient is the closest to 1. However, if we examine the scatterplot for this data with the best-fit linear function superimposed, as shown in Figure 3.37, we see that the line is not a particularly good fit. The growth pattern is clearly concave down, which suggests that a logarithmic function might be a better fit. The logarithmic function with base e that best fits this data is

$$M = 277.53 \cdot \ln t - 1081.66.$$

This function is equivalent to

$$M = 639.04 \cdot \log t - 1081.66,$$

using logs to the base 10. The corresponding correlation coefficient is $r = 0.9065$, which indicates a fairly high degree of positive correlation. We therefore conclude that over the period 1970–1990, the logarithmic fit is best.

The corresponding logarithmic curve is shown superimposed over the scatterplot in Figure 3.38. Observe that the curve looks almost linear with just the slightest hint of being concave down. Recall that any smooth curve looks linear if you zoom in on its graph sufficiently or focus on a small enough portion.

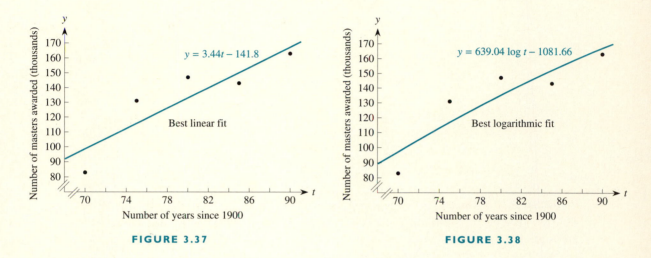

FIGURE 3.37   FIGURE 3.38

To predict the number of master's degrees that were earned by women in the year 1995, let's consider the value based on each fit when $t = 95$:

Linear prediction: 185,000   Exponential prediction: 197,123

Power function prediction: 196,046   Logarithmic prediction: 182,184

We likely would conclude that the prediction based on the logarithmic growth model is probably the most accurate. However, because the logarithmic function is almost linear over the relatively small interval considered here, the prediction based on it is quite close to that based on the linear fit.

E X A M P L E   2

The following data show the U.S. unemployment rate on a month-by-month basis from November 1991 (month 1) through November 1992 (month 13) during a recession in the United States. Find the best fit to this data using linear, exponential, power, and logarithmic functions.

Month	x	Rate	Month	x	Rate
November 1991	1	6.9	June 1992	8	7.8
December 1991	2	7.1	July 1992	9	7.7
January 1992	3	7.1	August 1992	10	7.6
February 1992	4	7.3	September 1992	11	7.5
March 1992	5	7.3	October 1992	12	7.4
April 1992	6	7.2	November 1992	13	7.2
May 1992	7	7.5			

Solution   We begin with the scatterplot for this data, as shown in Figure 3.39.

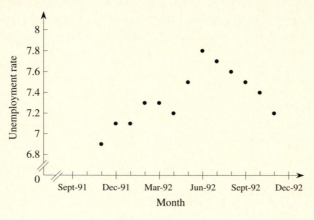

FIGURE 3.39

It indicates that the pattern for the data does not look particularly linear, exponential, power, or logarithmic. Nevertheless, let's see how good these four fits are and which is the best.

We find that the corresponding best fits are as follows.

Linear fit:	$r = 0.592$	$y = 0.0396x + 7.077$
Exponential fit:	$r = 0.598$	$y = 7.075(1.005)^x$
Power fit:	$r = 0.728$	$y = 6.933\, x^{0.034}$
Logarithmic fit:	$r = 0.719$	$y = 6.930 + 0.563 \log x$
		$(= 6.930 + 0.244 \ln x)$

When we compare the results, we see that the power function fit has the largest correlation coefficient, so it may be the best. In most of the cases we have considered so far, the values for the correlation coefficient were extremely close to 1, usually of the order of 0.9 or 0.95. In this example, the correlation coefficients range from a minimum of 0.592 to a maximum of 0.728. When we compare each of these values to the critical value of r, 0.553, corresponding to a sample of size $n = 13$ given in Table 3.2 of Section 3.2, we see that each function has a significant level of correlation. Thus we conclude that the data can be reasonably well fitted with any of the four functions, although the power function is the best fit based on only the correlation coefficient. A more detailed study also should include examining the residuals.

From an examination of the scatterplot in Example 2, it is evident that none of the four functions is an exceptionally good fit in the sense that the overall pattern starts to increase and then decrease. Clearly, a much better fit might well involve a function that first increases and then decreases, such as a parabola, which you have seen in algebra as the graph of a quadratic equation. (In later chapters we will consider how to fit a quadratic function to a set of data such as this one.) Alternatively, from the scatterplot, we might conclude that there are two separate linear patterns: a roughly increasing one from $x = 0$ to $x = 8$ and then a decreasing one from $x = 8$ to $x = 13$. We ask you to explore this possibility in Exercise 8 of this section.

E X A M P L E   3

An object is dropped from the top of the 1377-foot-high World Trade Center in New York City. The following set of measurements record how far the object has fallen after the given number of seconds.

Time (seconds)	1	2	3	4	5	6	7	8
Distance fallen (feet)	16	64	144	256	401	574	786	1022

Find a function that gives the distance the object has fallen after any time t.

Solution   If we enter this set of data into the calculator and look for the best fit among our usual families of functions (linear, exponential, and power), we find that the power function is the best choice because it has the largest correlation coefficient $r = 0.999999$ when compared to $r = 0.976$ for the linear function and $r = 0.959$ for the exponential function. The best power function fit is

$$D = 16.004t^{1.99977}.$$

In fact, it can be shown that the actual formula for the distance fallen, based on an application of Newton's Laws of Motion, is precisely

$$D = 16t^2.$$

The coefficient 16 is measured in feet per second per second, or ft/sec^2. Because this coefficient actually is one-half of a quantity known as the *acceleration due to gravity* (denoted by $g = 32$ ft/sec^2), we can write the formula as

$$D = \tfrac{1}{2}gt^2.$$

Usually, it is far more important to know how high the falling object in Example 3 is above the ground rather than how far it has fallen. If the object is dropped from a height of 1377 feet and the distance it falls in t seconds is $16t^2$, then subtracting the distance fallen from the initial height of 1377 feet gives

$$H = 1377 - 16t^2$$

as the object's height above the ground at any time t. We now can easily answer the question of how long it takes for the object to hit the ground.

We set $H = 0$, and solve the resulting equation.

$$16t^2 = 1377$$

$$t^2 = \frac{1377}{16} = 86.0625.$$

$$t = \sqrt{86.0625} \approx 9.3 \text{ seconds}$$

Thus, it takes about 9.3 seconds until impact.

In general, if an object is dropped from any initial height y_0, its height y above ground level at any time t is given by

$$y = y_0 - 16t^2.$$

EXERCISES

1. According to the U.S. Department of Education, the following data are the numbers (in thousands) of college degrees awarded during the indicated year from 1900 to 1990.

Year	1900	1910	1920	1930	1940	1950	1960	1970	1980	1990
College graduates	30	54	73	123	223	432	530	878	935	1017

Determine the best fit (linear, exponential, power, or logarithmic function) for the number of college graduates as a function of the year. Use each function to predict the number of college graduates in the year 2000. Which of the four predictions seems the most reasonable? the least reasonable? Use each of the four trends to predict the year in which there will be 2 million college graduates. What is the doubling time for the exponential model?

2. Compare the results in Exercise 1 to those results in Exercise 2 of Section 3.3 that concern the numbers of high school graduates. In particular, for the two exponential growth models, which model is growing faster? If the trends for both data sets continue unchanged, when will the number of college degrees awarded surpass the number of high school degrees awarded? Discuss the reasonableness of this scenario.

3. Use the set of data from Exercise 2 in Section 3.3 on high school degrees awarded and the data from Exercise 1 above on college degrees awarded. Determine the best relationship (linear, exponential, power, or logarithmic function) between the number of college degrees awarded and the number of high school degrees awarded the same year. What is the significance of the positive slope for the linear fit? What is the significance of the fact that the growth factor in the exponential fit is greater than 1?

4. Marc has noticed that the radio frequency numbers on the AM dial of his stereo do not seem to lie in a linear pattern. He measures in centimeters the distances from the extreme left end of the dial to each of the numbers printed and gets the readings shown in the accompanying figure.

a. Determine from among exponential, logarithmic, and power functions the one that best represents how the distance is related to the station numbers shown.
b. Using your best-fit function, estimate the distance from the left end of the dial to radio station 880; to radio station 1270.
c. What station would be 6 cm from the left end of the dial?
d. Check to see if the same function fits the comparable set of readings from your own radio.
e. Check to see what function best fits the readings on the FM band on your own radio.

5. The following table shows the median family income in the United States from 1970 to 1992.

Year	1970	1975	1980	1985	1988	1990	1992
Median family income	$9,867	$13,719	$21,023	$27,735	$32,191	$35,353	$36,812

a. Determine the best relationship (linear, exponential, power, or logarithmic) between median family income and the year.
b. Use your result in part (a) to predict the median family income in 1995 and in 2000.
c. Using each model, determine when median family income will reach $50,000.
d. What is the doubling time for the exponential model?
e. Consult the current edition of *Statistical Abstracts of the United States* to find the correct value for the median family income in 1995. How close is your prediction in part (a) to the actual value?

6. Suppose the pattern you found in Exercise 5 for the growth in median family income continues without change. In addition, suppose that inflation "remains under control" for the foreseeable future and is limited to about 3% per year. Write a short interpretation of what these two trends, if they continue without change, mean in terms of the standard of living in 20 years.

7. The following table shows the Dow-Jones average for 30 industrial stocks at the start of each year since 1980.

Year	1980	1981	1982	1983	1984	1985	1986	1987	1988	1989
Dow	839	964	875	1047	1259	1212	1547	1896	1939	2169

1990	1991	1992	1993	1994	1995	1996
2753	2634	3169	3301	3758	3834	5177

   a. Find the function that best fits this data.
   b. What is your prediction, based on this function, for the current value of the Dow? Check a newspaper or listen to the business news on the radio or television to find out how close the prediction is to the actual value.
   c. The Dow closed above 4000 for the first time on February 23, 1995. Which of the models comes closest to predicting that date?
   d. The Dow closed above 5000 for the first time on November 21, 1995. Which of the models comes closest to predicting that date?
   e. Based on your best-fitting function, when do you predict the Dow will first reach 6000?

8. Use the data on the U.S. unemployment rate in Example 2 to fit a pair of linear functions. (Be careful how you use the data point $(8, 7.8)$. Should you use it as part of both data subsets or should it be used in only one subset? If one, which one makes the most sense? You will get different results depending on how you make this assignment.) Where do the two lines intersect? What is your prediction for the unemployment rate in January 1993? the unemployment rate in January 1994? Is this reasonable? How do the values of the two correlation coefficients you get for both linear functions compare to the values of r in Example 2?

9. The height of an object falling from an initial height of y_0 is given by the formula

$$y = y_0 - 16t^2,$$

   which is based on the British system of units with feet and seconds. What is the equivalent formula based on the metric system of units with meters and seconds? (*Hint*: 1 foot = 0.3048 meters.)

10. Galileo conducted his famous experiment in which he dropped objects from the top of the 179-foot-high Leaning Tower of Pisa in about 1590. His goal was to obtain experimental data in order to show that all bodies fall with equal velocities. How long did it take for the objects that he dropped from the tower to hit the ground?

11. The Eiffel Tower is 300 meters tall. How long would it take an object dropped from its top to hit the ground?

12. An experiment is conducted in which a ball is dropped from an initial height of 9 feet and its subsequent height above floor level as a function of time is recorded and displayed. (See the accompanying figure.) When the curve is traced out, the measurements shown on the graph indicate the times when the ball hits the floor, the times when the ball reaches its maximum heights, and the values of these maximum heights.

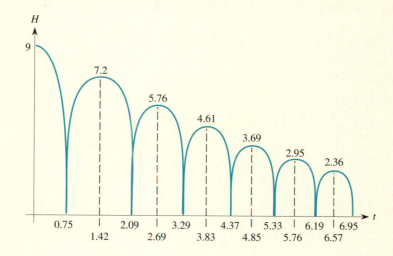

a. Notice that the times when the ball hits the floor appear to follow a linear pattern. Find the best linear fit to these times as a function of the number of the bounce; that is, bounce number $n = 1$ occurs at time $t = 0.75$, etc.

b. Notice that the times when the ball reaches its maximum heights also appear to follow a linear pattern. Find the best linear fit to these times as a function of n.

c. Notice that the maximum heights do not follow a linear pattern. Find the best nonlinear function that fits these data values, as a function of n, from among the families of functions you have studied in this chapter. For the exponential fit, what is the significance of the base you obtain?

d. Find the best fit for the maximum heights H as a function of time t.

e. Use the results you obtained in parts (a)–(d) to predict the corresponding values for the times and height on the next bounce of the ball.

13. Draw the graph of a function f that is decreasing and concave up. Mark three points on the curve: P near the left, Q near the center, and R near the right. These points determine three line segments PQ, QR, and PR.

a. List the three line segments in the order of increasing slopes.

b. List the three segments in the order of increasing steepness.

14. Repeat Exercise 13 if the function is decreasing and concave down.

3.6 *Analyzing the Challenger Data: A Case Study*

At the beginning of this chapter, we presented the data on O-ring problems that were eventually used to identify the O-rings as the likely cause
of the *Challenger* disaster. We now use this set of data as a case study to illustrate the process of data analysis in more detail than we have done to
this point.

Recall that the data used involved the number N of O-ring problems or
"incidents" as a function of launch temperature T. The set of values are
shown in the following table.

T	53	57	58	63	66	67	67	67	68	69	70	70	70	70
N	3	1	1	1	0	0	0	0	0	0	1	1	0	0

72	73	75	75	76	76	78	79	80	81
0	0	2	0	0	0	0	0	0	0

Figure 3.40 shows the scatterplot for this data along with a curve superimposed over the data points to indicate the nature of the relationship—it
appears to be a decaying exponential.

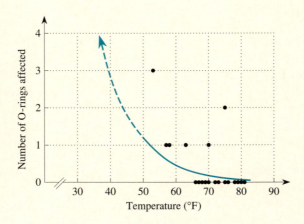

FIGURE 3.40

However, this curve is only an artist's rendition of the apparent relationship. We now seek to determine the equation that best fits the data.

To treat this problem systematically, we begin with the given data and
construct the regression line and the corresponding residual plot for it, as
shown in Figures 3.41 and 3.42. Notice that, to the eye, the regression line
does not appear to be that great a fit. Furthermore, the residual plot
demonstrates a pattern in the residuals, which suggests that a linear fit is
not adequate to describe the data accurately. Also, notice that the sizes of
the residuals are rather large, which also indicate a poor fit. However, the

correlation coefficient for the linear fit is $r = -0.567$, and according to Table 3.2 in Section 3.2, this value does indicate a significant level of negative linear correlation.

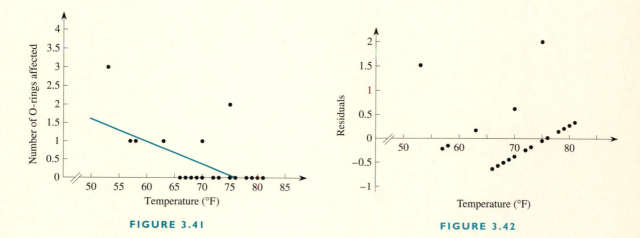

FIGURE 3.41   FIGURE 3.42

Let's try to improve on this analysis by trying an exponential fit because the data points appear to fall into a decaying exponential pattern as the temperature increases. To do so, we must transform the data in order to consider the logarithm of the number of incidents, log N versus the temperature T. However, notice that the values for N include $N = 0$. We cannot take the logarithm of 0—it is not defined! Thus we need to proceed here with care.

One way to circumvent the problem is to shift the data values up to avoid the zeros. The simplest approach is to increase each value of N by 1 so that we compare $N + 1$ to T. We construct the best fit to the resulting set of data and thus obtain the regression equation relating $N + 1$ to T. We then must shift back down to obtain an expression for N in terms of T. The data values we work with are given in the following table.

T	N	$N + 1$	$\log (N + 1)$	T	N	$N + 1$	$\log (N + 1)$
53	3	4	0.602	70	0	1	0
57	1	2	0.301	70	0	1	0
58	1	2	0.301	72	0	1	0
63	1	2	0.301	73	0	1	0
66	0	1	0	75	2	3	0.477
67	0	1	0	75	0	1	0
67	0	1	0	76	0	1	0
67	0	1	0	76	0	1	0
68	0	1	0	78	0	1	0
69	0	1	0	79	0	1	0
70	1	2	0.301	80	0	1	0
70	1	2	0.301	81	0	1	0

The scatterplot and the associated regression line for the exponential fit for the *shifted data values* are shown in Figure 3.43. The corresponding residual plot in Figure 3.44 indicates that the fit is better than the linear fit because the residuals are smaller even though there still is a pattern to the residuals.

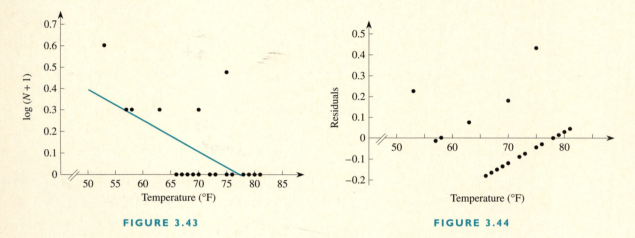

FIGURE 3.43   **FIGURE 3.44**

The regression equation for the transformed data set is

$$Y = -0.01457\,X + 1.13.$$

In terms of the variables we are using, this equation becomes

$$\log (N + 1) = -0.01457\,T + 1.13.$$

We undo the logarithm by taking exponentials of both sides of the equation.

$$N + 1 = 10^{-0.01457\,T + 1.13}$$
$$= (10^{-0.01457})^{T} \cdot 10^{1.13}$$
$$= 13.49\,(0.967)^{T}.$$

Finally, we solve for N to get the best exponential fit of N as a function of T.

$$N = 13.49(0.967)^{T} - 1,$$

which is shown superimposed over the original scatterplot in Figure 3.45.

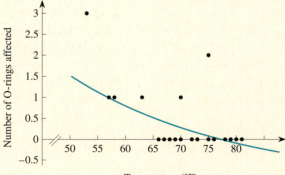

FIGURE 3.45

The graph certainly suggests that the likelihood of trouble with the O-rings will increase dramatically with falling temperature. We also know that there is a danger in extrapolating far beyond the range of data values. But the overall trend is so dramatic and the potential loss in terms of both human life and hardware is so extreme that there should not have been a launch if the data were analyzed in this way.

EXERCISES

1. Instead of adding 1 to each value of N, as was done with the *Challenger* data in this section, suppose you add some other quantity, say 2, to each value. How do the results compare to those obtained?

2. A cup of hot coffee at 200°F is left on the table in a 70°F room to cool. The temperature readings on the coffee at different times as it cools down to 70°F are given as follows:

Time t	0	5	10	15	20
Temperature T	200	163	139	118	108

   Find the exponential function that best fits this data.

3. While watching his VCR, Ken has noticed that the counter seems to move much more quickly near the beginning of the tape than toward the end of the tape, so he knows that the readings are not linear. To find the actual pattern, he records the counter reading every 15 minutes and obtains the following set of data relating the counter reading to the elapsed time in hours:

Time	0	0.25	0.50	0.75	1.0	1.25	1.5	1.75	2.0	2.25	2.5
Reading	0	445	817	1162	1448	1732	2005	2260	2503	2721	2942

   a. Find from among exponential, power, and logarithmic functions the function that best fits this data giving the VCR counter reading in terms of the elapsed time.
   b. Using that function, what would you predict the reading to be after three hours?
   c. Suppose that the label on a VCR tape indicates that a certain program Ken recorded runs from 1600 through 3400 on the counter. How long will that program run?
   d. Suppose that the VCR tape is a six-hour tape. Programs already recorded end at a counter reading of 4200. How much time is left on the tape for the next recording?

CHAPTER SUMMARY

In this chapter, you have learned the following:

- How to find the regression, or least squares, line that is the best linear fit to a set of data.
- How to interpret the correlation coefficient as a measure of how good the linear fit is.
- How to use the regression equation for making predictions.
- How to transform a set of data if you believe the underlying pattern is an exponential function, a power function, or a logarithmic function.
- How to undo the transformation to produce the best-fit exponential, power, or logarithmic function.
- How to interpret the residuals as an alternative way to assess how well a linear function fits the original data or the transformed data.
- How to use the best-fitting nonlinear function for making predictions.

REVIEW EXERCISES

1. The table below shows the budget and the attendance (both in millions) at 15 zoological parks in the United States. Find the best-fit function from among linear, exponential, power, and logarithmic functions for the attendance as a function of the budget. How good is the fit?

Budget	10.0	3.4	27.0	6.2	9.7	7.0	4.8	18.0	6.5	13.0	9.0	15.7	7.0	3.2	14.7
Attendance	1.0	0.5	2.0	0.6	1.3	1.0	1.1	4.0	0.6	3.0	0.5	1.3	1.0	0.5	2.7

2. Social Security Administration figures show the contribution and benefit base (in thousands of dollars) for Old Age and Survivors Disability Insurance (OASDI) from 1983 to 1992. Draw a graph of the benefit base as a function of the year since 1983. Determine the function that fits this data best from among linear, exponential, power, and logarithmic functions.

Year	83	84	85	86	87	88	89	90	91	92
OASDI Base	35.7	37.8	39.6	42.0	43.8	45.0	48.0	51.3	53.4	55.5

3. The following table gives the relationship between the average longevity (in years) and the gestation period (in days) for a sample of animals. The data indicate that the animals' average longevity can be predicted reasonably well as a function of the gestation period.

Gestation (days)	365	122	63	61	52	238	330	100	19	350	105
Longevity (years)	12	5	12	12	7	25	20	15	3	12	16

a. Find from among linear, exponential, power, and logarithmic functions the one that best predicts the longevity as a function of the gestation period.

b. Use your function grapher to graph the best-fit function.

4. Records of drug offenses that occurred from 1960 to 1989 have been gathered by the U.S. Department of Justice.

a. For the data given below, find from among linear, exponential, power, and logarithmic functions the function $D(t)$ that best models the number of jail sentences (in thousands) served in state prisons for drug offenses as a function of the year.

Year	1960	1964	1970	1974	1978	1981	1982	1983
Number of jail sentences	3.148	3.079	6.596	10.709	9.481	11.487	13.336	14.120

1984	1985	1986	1987	1988	1989
18.529	24.173	33.140	46.028	61.573	87.859

b. Use the best-fitting function found in part (a) to predict the number of jail sentences served in state prisons for drug offenses for the year 1996.

c. Check the current edition of the *World Almanac* to see if the trend still holds.

5. The U.S. Postal Service charges 32 cents for first-class postage on the first ounce of mail and 23 cents for every ounce thereafter. What linear function, based on weights 1, 2, . . . , 10 ounces, best models this situation? Explain why this function does not give the exact charges for an 8.5-ounce letter.

6. Consumer credit data from 1970 to 1991 show that the amount of outstanding consumer credit (in billions of dollars) in the United States at the end of each year is as follows:

Year	1970	1975	1980	1985	1987	1988	1989	1990	1991
Credit	133.8	207.5	355.4	601.6	692.0	742.1	791.8	809.3	796.7

a. Find the linear function that best fits the data.

b. Determine the year in which the outstanding consumer credit first will exceed 900 billion dollars.

c. Find more recent data in the current *World Almanac* to see if the trend you predicted continues to hold.

7. The number of airline flights generally has risen over the last 20 years. Flights (in millions) are given for the years 1977–1994 as follows:

Year	1977	1978	1979	1980	1981	1982	1983	1984	1985	1986
Number of flights	4.9	5.0	5.4	5.4	5.2	5.0	5.0	5.4	5.8	6.4

	1987	1988	1989	1990	1991	1992	1993	1994
	6.6	6.7	6.6	6.9	6.8	7.1	7.2	7.5

   a. Find the exponential function that best fits these data.
   b. How well does the exponential function fit the data?
   c. Predict when the number of airline flights will first exceed 9 million per year.

8. We find that a certain situation is modeled well by the function $y = f(x) = 3.77 + 2.5 \log x$.

   a. Assuming that the domain of the function makes sense only for values of x between 10 and 10,000, what is the corresponding range of the function?
   b. What is the value of x for which the function is equal to 7?
   c. Write an expression for the value of x as a function of y.

9. From a randomly chosen sample of 16 American cities we recorded the data for the height (in feet) and the number of stories of the tallest buildings in those cities.

Height	589	1023	880	529	800	745	689	875	1454	1127	912
Number of stories	44	55	63	40	60	52	52	60	110	100	41

	574	720	620	1368	1046
	49	71	45	110	77

   a. Find from among linear, exponential, power, and logarithmic functions the best-fit function relating a building's height to its number of stories.
   b. What is the significance of the slope of the linear function?
   c. On the average, how many feet are allocated to each story?

10. Draw a scatterplot for each of the functions f, g, and h in the following table. For each set of data, guess whether the pattern of data is linear, exponential, or logarithmic.

x	1	2	3	4	5	6
$f(x)$	6	4	2	0	−2	−4
$g(x)$	5.4	4.86	4.374	3.937	3.543	3.189
$h(x)$	−2	−0.194	0.863	1.612	2.194	2.669

4

EXTENDED FAMILIES OF FUNCTIONS

4.1 *Polynomial Functions*

Samantha has been keeping track of the price of the stock of HyperTech Corporation ever since her grandmother gave her several shares as a gift. She has plotted the stock values, as shown in Figure 4.1, and seeks to construct a mathematical model that represents the price of the stock. Unfortunately, none of the functions with which she is familiar—linear functions, exponential functions, power functions, or log functions—are reasonable candidates because none have this kind of behavior pattern. To better capture the trend in the stock prices, she needs a function that changes both its direction and its concavity (see Figure 4.2).

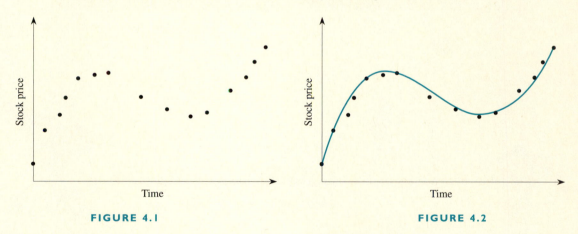

FIGURE 4.1   FIGURE 4.2

Notice that this graph increases, then decreases, and finally increases again. So, the graph has two turning points, one at the local maximum point and the other at the local minimum point. Also, the curve initially is concave down and then is concave up, so the graph has one point of inflection, where the concavity changes.

In this section, we introduce a new family of functions that possess this type of more intricate behavior. It is known as the family of *polynomial functions*. A **polynomial function,** or **polynomial,** is any sum of power functions with nonnegative integer powers. For instance,

$$y = 3x - 5, \qquad y = 6x^2 + x - 7, \qquad y = 4x^3 + 5x^2 - 7x + 12,$$
$$\text{and} \qquad y = 10x^8 - 7x^5 + 3$$

are all polynomials. The **degree** of a polynomial is the highest power of the variable present. So, the four polynomials displayed are of degrees 1, 2, 3, and 8, respectively. The constant multiples in each expression are called the **coefficients** of the polynomial. Another way to describe a polynomial is to say that it is a *linear combination* of power functions since, as we noted, it is made up of a sum of power functions. From this point of view, power functions are the basic building blocks we use to construct any polynomial.

If a polynomial has degree 1, then it is a linear function, $y = ax + b$, and its graph is a straight line with slope $m = a$.

Quadratic Polynomials

If the degree of a polynomial is 2, then it is a **quadratic function** of the form

$$y = ax^2 + bx + c,$$

where a, b, and c are constants and $a \neq 0$. The graph of any quadratic function is a curve called a **parabola.** We see such curves all around us—in the path of a fly ball in baseball, in the shape of the main support cable in a suspension bridge such as the Golden Gate Bridge or the George Washington Bridge, or the cross sections of a TV satellite dish. (See Figure 4.3.)

FIGURE 4.3

Now let's analyze the behavior of quadratic functions. As we have seen with power functions in Section 2.4, the sign of the coefficient of the highest power term, known as the *leading coefficient*, determines the overall behavior of the parabola. When the leading coefficient is positive, the parabola opens upward, so it is concave up. When the leading coefficient is negative, the parabola opens downward, so it is concave down. Whichever way the parabola opens, as x increases indefinitely in either direction, the parabola either increases toward infinity or decreases toward negative infinity. (See Figure 4.4.)

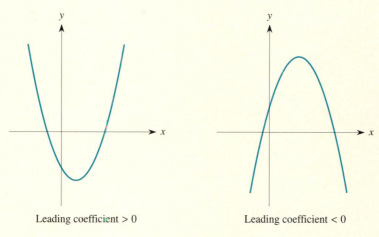

Leading coefficient > 0 Leading coefficient < 0

FIGURE 4.4

By its very nature, every parabola has one turning point—its **vertex.** If the parabola opens upward, then the turning point corresponds to the minimum value of the function. If the parabola opens downward, then the turning point corresponds to the maximum value of the function. In addition, the parabola is always symmetric about the vertical line through its turning point. (See Appendix C for a discussion of symmetry.)

Next let's examine the effects of the other two terms. In Figure 4.5 (on the next page), we show the graphs associated with the quadratic functions $y = x^2$, $y = x^2 + 6$, $y = x^2 - 5x + 6$, and $y = x^2 + 5x + 6$. While the leading term determines the basic behavior of the quadratic function (which makes it a parabola), the other terms affect the location of its graph. The constant term (in this case, 6) acts to produce a vertical shift: It raises the parabola $y = x^2$ up 6 units (or lowers the parabola if the constant term is negative). Use your function grapher to experiment with this effect on the graph of the parabola by changing the constant term. For instance, how do the graphs of $y = x^2 + 5x + 7$ and $y = x^2 + 5x - 2$ compare to the graph of $y = x^2 + 5x + 6$? Be sure you look at enough graphs to convince yourself of the effect of the constant term.

The linear term $-5x$ or $+5x$ serves to shift the parabola right or left, respectively. Again, use your function grapher to see the effects of using a variety of different values for this coefficient. For instance, how does the graph of $y = x^2 - 4x + 6$ or $y = x^2 + 2x + 6$ compare to $y = x^2 - 5x + 6$? You will note, though, that the effect of changing the coefficient of the x-term does not merely shift the graph to the left or the right (which is the major effect); it also introduces some vertical movement.

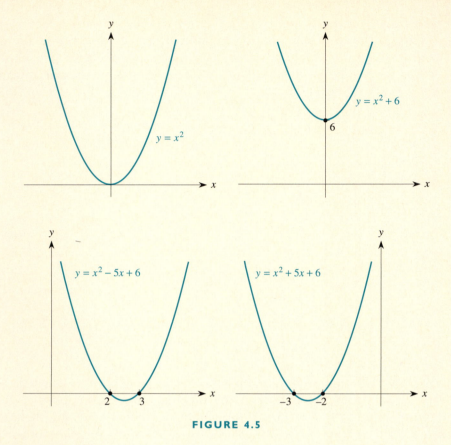

FIGURE 4.5

Another key piece of information about a quadratic function is that it always has two **zeros.** These are the values of the variable x that make the value of the function zero. Equivalently, if we set the expression for the quadratic function equal to zero, then we have a **quadratic equation** and this equation has two **roots.** Note that a *function has zeros* and an *equation has roots* and that there is a direct correspondence between them. The zeros of a function f occur at precisely the same points as the roots of the equation $f(x) = 0$.

The two roots for any quadratic equation

$$ax^2 + bx + c = 0$$

always can be found from the *quadratic formula.*

The Quadratic Formula

$$x = \frac{-b \pm \sqrt{b^2 - 4ac}}{2a}$$

If the coefficients are properly interconnected, then it may be possible to find the roots of the quadratic equation by algebraic factoring.

The two roots of a quadratic equation could be real numbers or a pair of complex numbers of the form $x = \alpha + \beta i$ and $x = \alpha - \beta i$, where $i = \sqrt{-1}$ (α and β are the Greek letters alpha and beta, respectively). Just as the point where a line crosses the x-axis gives the root of a linear equation, the points where a parabola crosses the x-axis give the real roots of a quadratic equation. For example, the graph of $y = x^2 - 5x + 6$ is shown in Figure 4.6; notice that the graph crosses the x-axis twice: once when $x = 2$ and again when $x = 3$. The associated quadratic equation

$$x^2 - 5x + 6 = 0$$

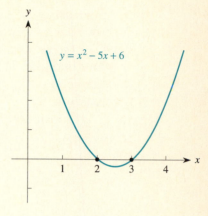

FIGURE 4.6

has real roots $x = 2$ and $x = 3$. You can find them by factoring the quadratic as

$$(x - 2)(x - 3) = 0,$$

and then using the fact that when the product of two factors is zero, one or the other must be zero. Thus $(x - 2) = 0$ or $(x - 3) = 0$, leading to the roots $x = 2$ and $x = 3$.

Thus for any quadratic function:

> The real roots of a quadratic equation correspond graphically to the points where the associated parabola crosses the x-axis.

Since 2 and 3 are the roots of $x^2 - 5x + 6 = 0$, we can factor the polynomial as

$$x^2 - 5x + 6 = (x - 2)(x - 3).$$

In fact, for any quadratic function:

> The real roots of a quadratic equation correspond algebraically to the linear factors of the quadratic polynomial.

If we know that a parabola crosses the x-axis at a point $x = r$, then $x = r$ is a zero of the associated quadratic function and $x - r$ is a factor of the quadratic expression. Since you can locate the real roots of any quadratic very accurately with your graphing calculator or with the quadratic formula, you can always find the linear factors.

Depending on the orientation of the parabola (opening up or down) and the position of the turning point, a parabola may not touch the x-axis at all. This was the case with the graph of $y = x^2 + 6$ in Figure 4.5. For such a parabola, the corresponding quadratic equation still has two roots, but they are complex roots. If a quadratic equation has complex roots, they must occur in conjugate pairs of the form $\alpha \pm \beta i$. This follows directly from the quadratic formula for the case where the term inside the radical, $b^2 - 4ac$, is negative. The expression $b^2 - 4ac$ is called the **discriminant** of the quadratic. When the discriminant is positive, the two roots are real numbers; when the discriminant is negative, the two roots are complex numbers.

For instance, the discriminant for $y = x^2 + 6$ is $0^2 - 4(1)(6) = -24 < 0$, so the two roots are complex. The quadratic formula tells us that the roots are

$$x = \frac{-0 \pm \sqrt{0^2 - 4(1)(6)}}{2(1)}$$

$$= \frac{\pm\sqrt{-24}}{2}$$

$$= \pm\frac{2\sqrt{-6}}{2} = \pm\sqrt{6}\,i.$$

A third possibility is that the parabola could be tangent to the x-axis; that is, it can touch the axis and bounce back without ever crossing the axis. As an example, consider the quadratic function $y = x^2 - 4x + 4$. What are the roots of the corresponding quadratic equation? Use your function grapher to examine the corresponding graph. What is the significance of the roots in this case? What is the value of the discriminant in this case?

Cubic Polynomials

If the degree of a polynomial is three, it is called a **cubic function** and its graph is called a **cubic.** In general, a cubic function has the form

$$y = ax^3 + bx^2 + cx + d,$$

where a, b, c, and d are constants and $a \neq 0$. For example, the graph of the cubic $y = x^3 + 3x^2 - 8x - 4$ is shown in Figure 4.7. This graph is typical of a cubic function. Notice first that the cubic goes toward positive infinity in one direction and drops toward negative infinity in the other. Also, notice that there are two turning points; there is one point of inflection; the curve is concave down on one side of the point of inflection and concave up on the other; and the particular curve shown crosses the x-axis at three points so it has three real zeros.

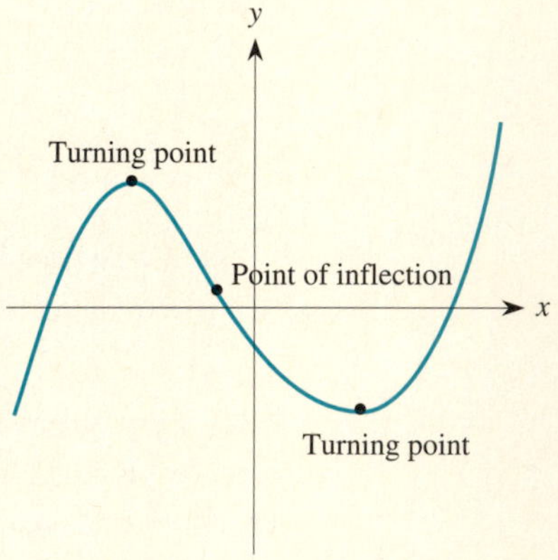

In general, a cubic function has three zeros, and a cubic equation

FIGURE 4.7

$$ax^3 + bx^2 + cx + d = 0$$

has three roots. The roots can all be real roots or can consist of a single real root and a pair of complex conjugate roots. Further, each of the real roots corresponds to a linear factor of the cubic expression.

> The real roots of a cubic equation correspond graphically to the points where the associated cubic curve crosses the x-axis.
>
> The real roots of a cubic equation correspond algebraically to the linear factors of the cubic polynomial.
>
> Any complex conjugate roots correspond to a quadratic factor of the cubic polynomial.

If a cubic has three real roots, then its curve crosses the x-axis at the corresponding three points. If it has only one real root, then the curve crosses the x-axis only once, as shown in Figure 4.8.

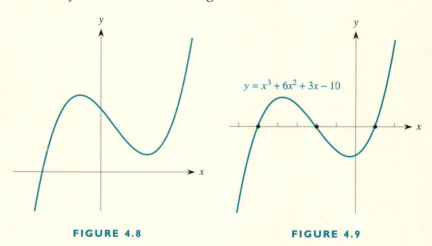

FIGURE 4.8   FIGURE 4.9

For example, the cubic

$$f(x) = (x - 1)(x + 2)(x + 5) = x^3 + 6x^2 + 3x - 10$$

has three real zeros at $x = 1, -2,$ and -5 corresponding to each of the three factors. Consequently, its graph (Figure 4.9) crosses the x-axis at $x = 1, -2,$ and -5. Further, the leading term, x^3, being the highest power present, eventually dominates the other terms as x increases. Since the leading coefficient, 1, is positive, the cubic curve must increase toward $+\infty$ as x approaches ∞, and the curve must decrease toward $-\infty$ as x approaches $-\infty$. Verify this graphically using your function grapher and numerically by substituting some large positive and negative values for x.

Although there is a formula for calculating the roots of a cubic equation, it is considerably more complicated than the quadratic formula and is seldom used. If the cubic polynomial happens to factor simply, you can find the roots directly since each factor corresponds to a root. However, this is not very likely to happen. Usually, the simplest way to find the real roots of a cubic equation is to approximate them using your function grapher—just keep zooming in on the points where the curve crosses the x-axis until you have found the roots to whatever degree of accuracy you desire. For the cubic $y = x^3 + 3x^2 - 8x - 4$, a calculator investigation indicates that the roots are located at approximately $x = -4.561577$, -0.4384546, and 1.9999994. This last x-value suggests that the third root is precisely $x = 2$. How might you check if this is indeed the case? If it is true, how can you use the fact that 2 is a root to simplify the original equation?

Finally, as with a quadratic function, the leading coefficient in a cubic determines the overall behavior pattern of the function. If the leading coefficient is positive, then the cubic curve increases as x increases (except possibly for a relatively small dip between the two turning points). If the leading coefficient is negative, the cubic curve decreases as x increases (except for a possible rise between the two turning points), as seen in Figure 4.10.

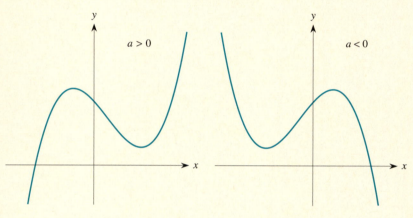

FIGURE 4.10

Polynomials of Degree n

The ideas discussed for polynomials of degree 2 (quadratics) and degree 3 (cubics) can be extended to polynomials of any degree n. In particular, we have the following facts.

- A polynomial of degree n has precisely n zeros.
- The corresponding polynomial equation of degree n has precisely n roots. They may be real or complex.
- The complex roots occur in pairs of complex conjugates, $\alpha \pm \beta i$, where $i = \sqrt{-1}$.
- The real roots correspond to the points where the curve crosses the x-axis.
- You can always find the real roots graphically by using your function grapher to zoom in on the points where the curve crosses the x-axis.

- The real roots correspond to linear factors of the polynomial expression.
- A polynomial of degree n has at most $n - 1$ turning points.
- A polynomial of degree n has at most $n - 2$ points of inflection.
- Every polynomial approaches $\pm\infty$ as x approaches ∞ and as x approaches $-\infty$. The particular orientation depends on the sign of the leading coefficient.

For example, suppose a polynomial has roots at $x = -4, -1, 1, 3$, and 6. Its degree is at least 5; it might be higher if there are complex roots or repeated roots. The five corresponding linear factors are $(x - (-4)) = (x + 4)$, $(x + 1)$, $(x - 1)$, $(x - 3)$, and $(x - 6)$. Suppose that these are the only roots. Then one *possible* formula for this polynomial is

$$P(x) = (x + 4)(x + 1)(x - 1)(x - 3)(x - 6),$$

although any constant multiple, A, of this expression would be an alternate formula. You can determine the value of the multiple A if you know the vertical intercept of the polynomial, or any other point on the curve. If the multiple A is positive, then the graph of the polynomial has the behavior shown in Figure 4.11. Notice that this fifth degree polynomial has four turning points. Also, observe the *end behavior*: $P(x) \to \infty$ as $x \to \infty$ and $P(x) \to -\infty$ as $x \to -\infty$. Alternatively, if the constant multiple A is negative, then this end behavior is reversed; the graph drops toward $-\infty$ as $x \to \infty$ and rises toward $+\infty$ as $x \to -\infty$. Do you see why this is the case?

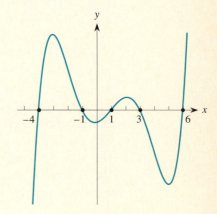

FIGURE 4.11

What if a polynomial has a double or repeated factor such as the quadratic $y = x^2$? For instance,

$$P(x) = (x + 1)(x - 2)(x - 4)^2 = 0$$

has roots at $x = -1, 2$, and 4, but $x = 4$ is a *double root*. If you examine its graph, as shown in Figure 4.12, you will observe that the curve comes down to touch the x-axis at $x = 4$ where it flattens out and then rises again. If you zoom in on the curve about this point, you will see that the x-axis is tangent to the graph at $x = 4$, just as the x-axis is tangent to the parabola $y = x^2$ at the origin. You should examine the graph of a polynomial that has a triple factor to see what happens near that triple root. Try to predict what will happen based on your knowledge of the behavior of $y = x^3$.

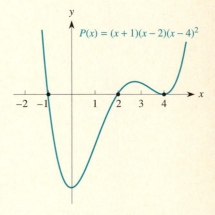

FIGURE 4.12

While we expect that you will use your function grapher to produce the graph of a polynomial, you should be very careful to interpret what the calculator or computer draws. Usually, the important characteristics of any function, and a polynomial in particular, are:

- The end behavior (Is it increasing or decreasing as $x \to \infty$ or $x \to -\infty$?)
- The intervals where the function is increasing or decreasing
- The locations of the turning points
- The intervals where the function is concave up or concave down
- The locations of the points of inflection
- The locations of the real zeros

The end behavior of any polynomial depends on the leading term since, as x increases in either the positive or negative direction, that term will eventually dominate all other terms. For instance, consider the polynomial $P(x) = 2x^4 - 6x^3 + 7x - 10$. When x is very large, we expect that the term $2x^4$ should overwhelm all other terms. We show the graphs of the functions $P(x)$ and $Q(x) = 2x^4$ for x between -3 and 4 and y between -25 and 100 in Figure 4.13, where we see that the two curves look quite different. We see a slightly larger view in Figure 4.14, where x is between -6 and 6 and y is between -50 and 950. Here the two curves look more similar than in the previous view. In the much larger view in Figure 4.15, where x is between -25 and 25 and y is between 0 and 500,000, we no longer can see much difference between the two curves. The term $2x^4$ dominates the behavior of the polynomial; the effect of the rest of the terms seems negligible. In general, for any polynomial, when x is large enough, the curve is indistinguishable from the curve corresponding to just the leading term. Therefore, in the large, the behavior of any polynomial is identical to that of the power function consisting of the leading term.

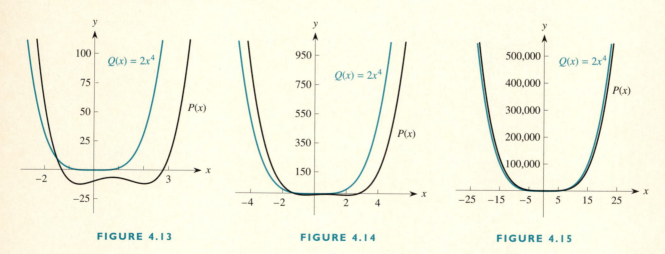

FIGURE 4.13   FIGURE 4.14   FIGURE 4.15

You can see the end behavior easily with your function grapher if you use a reasonably large viewing window. However, the location of the turning points and the zeros is a *local* aspect of the graph and can be easily missed if the viewing window is too large. On the other hand, if you use too small a viewing window, you will certainly lose the overall growth pattern of the polynomial. For instance, by focusing too closely on one particular turning point or root, you may lose sight of all the others. It is rare that a single view suffices to show all the important details of a function. Therefore, as a matter of routine, you should use the information found in several different views on your calculator or computer to sketch a rough hand-drawn picture of the function, called the *complete graph*, which highlights the key information, even if it is intentionally *not* drawn to scale.

EXERCISES

1. The overall trend in the growth of the GDP (gross domestic product, formerly called the gross national product) has been upward except for a small dip. Sketch a graph representing the value of the GDP as a function of time. What type of function might model it? What can you conclude about any of the coefficients?

2. The overall pattern in the growth of the Dow-Jones average over the last 10 years has been one of increase except for two relatively sharp, but short-term, drops. Sketch a graph representing the value of the Dow-Jones average as a function of time. What type of function might model it? What can you conclude about any of the coefficients?

3. Each of the following graphs represents a polynomial. For each one:
    a. What is the minimum possible degree of the polynomial? Why?
    b. Is the leading coefficient of the polynomial positive or negative? Why?

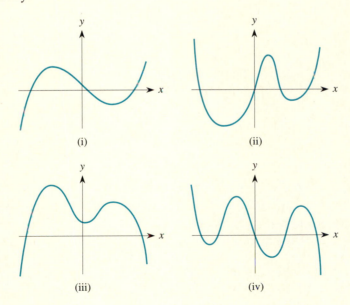

(i)   (ii)

(iii)   (iv)

4. Match each of the polynomial expressions with its graph. Use your knowledge about roots; do not use your function grapher.

   a. $f(x) = (x - 1)(x - 3)(x + 3)$

   b. $f(x) = (x + 1)(x + 2)(2 - x)$

   c. $f(x) = (x - 1)(x^2 + 4)$

   d. $f(x) = (x - 1)(x + 1)(x - 3)(x + 3)$

   e. $f(x) = 3x^3 - x^4$

   f. $f(x) = (x - 2)(x - 4)(x + 3)^2$

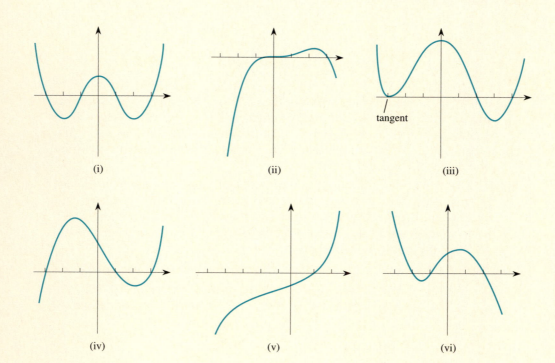

      (i)   (ii)   (iii)

      (iv)   (v)   (vi)

Based on your knowledge about roots and factors, sketch the graph of each of the polynomial functions in Exercises 5–8. Do not use your function grapher.

5. $f(x) = (x + 2)(x - 1)(x - 3)$

6. $f(x) = 5(x^2 - 4)(x^2 - 25)$

7. $f(x) = -5(x^2 - 4)(x^2 - 25)$

8. $f(x) = 5(x - 4)^2(x^2 - 25)$

9. The polynomial $P(x) = 2x^6 + 5x^5 - 8x^4 - 21x^3 - 12x^2 + 22x + 12$ can be factored as $P(x) = (x - 2)(x - 1)(2x + 1)(x + 3)(x^2 + 2x + 2)$.

   a. What is the degree of the polynomial?

   b. What are the real roots? The complex roots?

   c. What happens as $x \to +\infty$? As $x \to -\infty$?

   d. What is the maximum number of turning points you expect? Why? What is the maximum number of points of inflection? Why?

10. Determine cubic polynomials to represent the following graphs:

a.

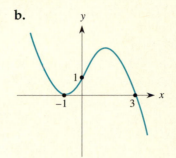

b.

11. Each of the following graphs represents a function. For each one,

    i. Read off approximate intervals on which the function is increasing and on which it is decreasing.

   ii. Estimate intervals on which the function is concave up; concave down.

  iii. Find a possible formula for the function.

a.

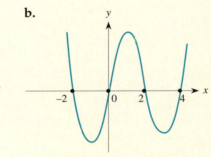

b.

12. Determine which of the graphs shown below suggest the end behavior for each of the following polynomials:

  a. $y = 5x^5 - 8x^4 + 2x^2 + 3x - 4$ **b.** $y = -4x^6 + 3x^4 + 7x^3 - 8x^2 - 4x$

  c. $y = 3x^8 + 4x^5 + 6x^3 - 5x^2 + 6$ **d.** $y = -x^7 - 4x^6 + 3x^4 - 6x^3 + 7x - 9$

  e. $y = -4x^9 + 6x^6 - 5x^3 + 35$ **f.** $y = 100 - x^4$

  g. $y = (9 - 6x^2)^3$ **h.** $y = (9 - 6x^3)^3$

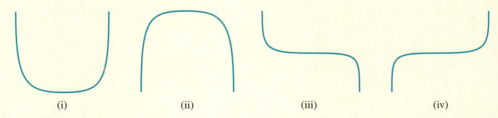

       (i)   (ii)   (iii)   (iv)

13. **a.** The graph of the polynomial $P(x) = 2x^4 - 6x^3 + 7x - 10$ in Figure 4.13 suggests that there are three turning points. Use your function grapher to locate them by zooming in on the graph.

    b.   Estimate all intervals over which $P(x)$ is increasing or decreasing.

    c.   Estimate the locations of all points of inflection.

    d.   Estimate all intervals over which $P(x)$ is concave up or concave down.

    e.   Estimate all real roots.

14. Describe the end behavior of each of the functions in (a)–(e) below. Specifically for the graph of each function f:

    i. As $x \to +\infty$, does $f(x) \to +\infty$ or $-\infty$? Why?

    ii. As $x \to -\infty$, does $f(x) \to +\infty$ or $-\infty$? Why?

    a. $f(x) = -3x^3 + 70x^2 - 20$ **b.** $f(x) = 20x^4 + 3x^3 + x^2 + 1000$

    c. $f(x) = -3x^4 + 20x^3 - 5x^2 + x - 20$ **d.** $f(x) = x^4 + x^5$

    e. $f(x) = 4x^4 + 5x^5 - 6x^6$.

15. Find the equation of a quadratic that has a real root at $x = 2$ and a turning point at $(1, 5)$.

16. A cubic polynomial P has turning points at $(1, 4)$ and $(5, 12)$.

    a. What is the behavior of $P(x)$ as x approaches ∞?

    b. Where is the point of inflection? (*Hint:* Cubics are symmetric about their point of inflection.)

17. Suppose that a quadratic has roots at $x = 6$ and $x = -2$.

    a. Write a possible formula for the quadratic function.

    b. Use the fact that a quadratic is symmetric about the vertical line through its turning point to determine the location of the turning point of this quadratic function.

    c. Suppose that the quadratic has a maximum value of 20. What must its equation be?

    d. Suppose that the quadratic has a minimum value of -20 instead. What must its equation be?

18. An apple is tossed from ground level straight up into the air at time $t = 0$ with velocity 64 feet per second. Its height at time t is $f(t) = -16t^2 + 64t$. Find the time when it hits the ground and the instant that it reaches its highest point. What is the maximum height?

19. The height s of an object above the ground at time t is given by

$$s = v_0 t - \frac{1}{2} g t^2,$$

where v_0 represents the initial velocity and g is a constant called the *acceleration due to gravity*.

    a. At what height does the object start?

    b. How long is the object in the air before it hits the ground?

    c. When will the object reach its maximum height?

    d. What is that maximum height?

20. **a.** Sketch a graph of today's air temperature from midnight to midnight.

    b. When is it a minimum? A maximum?

    c. When do you think it has a point of inflection?

    d. What type of polynomial might be a good fit to the curve you drew?

e. Can you think of a function that is a better fit if you expand the domain to include the temperatures for yesterday and tomorrow as well?

21. Factor each of the following polynomial expressions completely:

   a. $x^2 + 7x + 12$ **b.** $x^2 - 4x - 5$

   c. $x^3 - 36x$ **d.** $x^3 - 4x^2 + 3x$

   e. $x^3 + 10x^2 + 25x$ **f.** $x^3 + x^2 - 20x$

22. The *average rate of change* of a function f over an interval $x = a$ to $x = b$ is defined to be the slope of the line segment connecting the endpoints of the curve on that interval:

$$\frac{\Delta y}{\Delta x} = \frac{f(b) - f(a)}{b - a}$$

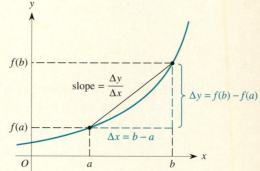

(See the accompanying figure.) The following table gives some values for the function $f(x) = x^3 - 4x$.

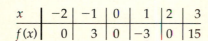

x	-2	-1	0	1	2	3
$f(x)$	0	3	0	-3	0	15

   a. Find the average rate of change of f from $x = -2$ to $x = 3$.

   b. Calculate the average rate of change of f between each successive pair of points in the table; that is, between $x = -2$ and $x = -1$, between $x = -1$ and $x = 0$, and so on. What is the average value of all these slopes?

   c. Extend the table to include the point where $x = 4$ and repeat parts (a) and (b). Does the same result hold?

   d. Extend the table further to include $x = -3$. Show that the same conclusion holds.

   e. Do you think the same conclusion holds for any function and any set of points? Try to state this result as a potential theorem.

23. Prove the result you conjectured in part (e) of Exercise 22. Let f be defined on an interval from a to b. The average rate of change in f is $[f(b) - f(a)]/(b - a)$. Let $x_0 = a, x_1, x_2, \ldots, x_n = b$ be any set of uniformly spaced points so that $\Delta x = (b - a)/n$.

24. Find all polynomials, p, of degree ≤ 2 that satisfy each set of conditions.

   a. $p(0) = p(1) = p(2) = 1$

   b. $p(0) = p(1) = 1$ and $p(2) = 2$

   c. $p(0) = p(1) = 1$

   d. $p(0) = p(1)$

(*Hint:* Think about the graphs.)

4.2 *Fitting Polynomials to Data*

The ideas we developed in Chapter 3 on fitting various families of functions (linear, exponential, power, and logarithmic) to a set of data can be extended to allow us to fit a polynomial function to data values as well. While we do not have the mathematical ideas in place yet to discuss this fully (these ideas will be introduced in Chapter 6), we will be content for now to simply use some of the capabilities built into most sophisticated calculators. We illustrate these notions with the following example on the spread of AIDS.

E X A M P L E   1

The table below shows the accumulated total number of reported cases of AIDS in the United States since 1983, based on Center for Disease Control statistics.

Year	1983	1984	1985	1986	1987	1988	1989
Number of AIDS Cases	4589	10,750	22,399	41,256	69,592	104,644	146,574

1990	1991	1992	1993	1994
193,878	251,638	326,648	399,613	457,280

At first thought, we might expect that the growth in the total number of reported cases of AIDS is exponential and so we would attempt to fit an exponential function to this set of data. The resulting best-fit exponential function, found with a calculator, is

$$f(t) = 9420.9(1.492)^t,$$

where t is measured in years since 1983. The corresponding correlation coefficient $r = 0.9686$ is quite close to 1, and so suggests that this function is a very good fit. We superimpose this exponential function over the data points in Figure 4.16 and see that the curve does not seem to be as good a match to the pattern of the data as we expected. Also, a residual plot would show a pattern, further verifying that the exponential function is not an ideal fit.

Alternatively, suppose we use the capability of the calculator to fit a polynomial to this set of data. Most calculators allow us to fit polynomials of degree 2, 3, or 4 to such a set of data, and it is simple enough to experiment with different degrees. When we do so, we find that a cubic polynomial is an excellent fit to this set of data and, in particular, the calculator gives the best cubic function as

$$y = -68.8t^3 + 4781.7t^2 - 2666.2t + 7094.5,$$

where t is again the number of years since 1983. We show this polynomial superimposed over the AIDS data points in Figure 4.17. Observe that the fit seems exceptionally good, certainly far better than the exponential fit shown in Figure 4.16.

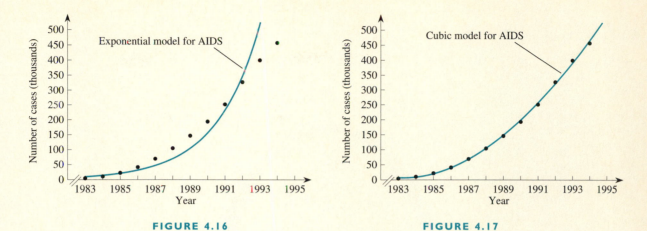

FIGURE 4.16   FIGURE 4.17

This graph strongly suggests that the number of cases in the spread of AIDS follows a cubic pattern. (When scientists discovered this several years ago, they were extremely excited because polynomial growth is so much slower than exponential growth, which is what they too had expected.) The approach used to determine this best-fit cubic polynomial is different from the types of transformations we used in Chapter 3, so the correlation coefficient does not apply directly. However, a comparable measure of the goodness of fit leads to a value of 0.9998, testifying to how well the cubic fits the data.

Notice that the portion of the cubic curve that fits the data points is concave up. We know from the formula for the cubic that the leading coefficient is negative, and so the cubic will eventually approach $-\infty$. Thus if the cubic model continues to hold in the future (and that is a very big IF), then the curve has yet to reach its point of inflection. Once it does and if the cubic model continues to apply, then the growth in AIDS may slow down. However, recall how dangerous it can be to extrapolate with a mathematical model. The model only describes the situation based on these data points; it is not a guarantee of the actual process, especially for extrapolating into the future.

Let's look at another example of this notion of fitting polynomials to data. Figure 4.18 shows a drawing of the famous Gateway Arch in St. Louis. The shape of it suggests a portion of a downward opening parabola. Let's see if we can determine a specific function that best models the arch.

FIGURE 4.18

E X A M P L E   2

Determine a polynomial function that best fits the Gateway Arch.

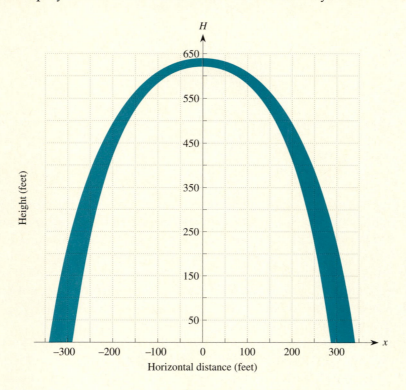

FIGURE 4.19

Solution   To find an appropriate function, we need some measurements on the arch. Overall, the arch stands 630 feet tall, and the distance between its two legs also is 630 feet. We superimpose a grid on the arch in Figure 4.19 and choose our coordinate system so that the vertical axis passes through the center of the arch. We now construct the following table of estimates of the height H corresponding to various horizontal distances x. We make our estimates from the middle of the arch; slightly different results might occur if we use values from the inner edge or the outer edge. We ask you to investigate these possibilities in the exercises at the end of the section.

x	−325	−300	−250	−200	−150	−100	0	100	150	200	250	300	325
H	0	100	330	500	570	610	630	610	570	500	330	100	0

The first thing we notice from both the figure and the table is that the measurements are symmetric about the vertical axis $x = 0$. As a result, we would expect that the best-fit parabola should have no x term. When we enter the data into the quadratic regression routine of a calculator, we find the quadratic function that best fits the data is

$$H = -0.0064x^2 + 0x + 699.01.$$

We show the result of plotting this function over the data points in Figure 4.20 and conclude that it is a reasonably good fit, though certainly not a great one. Among other things, the curve rises much too high above the central data point. Also, the pattern of data points flattens out far more than the parabola does near the center.

What about using a higher degree polynomial? From the basic shape of the arch, we know that a cubic would not be appropriate: It does not have the correct behavior. How about a *quartic polynomial*? We know that the graph of $y = x^4$ is much flatter than the graph of $y = x^2$ near the origin, so a quartic may be a better type of function to fit to the data on the arch. When we do this, the calculator responds with the equation

$$H = (-3.27 \times 10^{-8})x^4 + 0x^3 - 0.00282x^2 + 0x + 644.25.$$

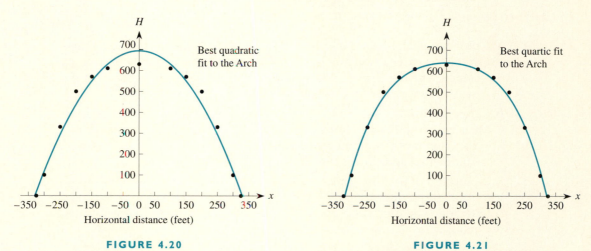

FIGURE 4.20   FIGURE 4.21

We show this function superimposed over the data points in Figure 4.21 and conclude that, visually, it appears to be an exceptionally good fit to the shape of the arch. (Actually, the true shape of the arch is a curve known as a hyperbolic cosine that you may encounter in calculus.)

EXERCISES

1. We saw in the text that the cubic

    $$y = -68.8t^3 + 4781.7t^2 - 2666.2t + 7094.5$$

    is an excellent fit to the total number of reported cases of AIDS in the United States from 1983 to 1994.
    a. Based on this model, what is the prediction for the total number of cases up through 1995?
    b. Check a recent copy of the *Statistical Abstracts of the United States* or an almanac to see how accurate the prediction in part (a) is.

    c. Assuming that the pattern continues, how many total cases would you expect by the year 2000?

    d. When would you expect a total of 1,000,000 cases of AIDS?

2. Find the equations of the best quadratic and quartic functions to fit measurements taken at the outer edge of the Gateway Arch instead of at the middle.

3. Repeat Exercise 2 with measurements taken at the inner edge of the arch instead of at the middle.

4. The following table gives the number of AIDS-related deaths in the United States for men and for women from 1986 through 1992.

Year	1986	1987	1988	1989	1990	1991	1992
AIDS deaths among men	12,099	19,256	27,435	29,942	36,770	38,013	39,160
AIDS deaths among women	1048	1832	3284	3653	4883	5688	6312

    a. Determine the best cubic fit to the number of AIDS-related deaths among men as a function of time. By a visual inspection, how does it compare to the best fit you obtained to this data from among exponential and power functions in Exercise 7 of Section 3.3?

    b. Determine the best cubic fit to the number of AIDS-related deaths among women as a function of time. By a visual inspection, how does it compare to the best fit you obtained to this data from among exponential and power functions in Exercise 7 of Section 3.3?

5. In Example 2 of Section 3.5, we observed that the scatterplot (Figure 3.39) for the data on the unemployment rate in the United States over the course of a year suggests a parabolic pattern. Use your calculator to find the quadratic function that best fits the data.

6. The figure below shows a grid superimposed on the image of the McDonald's arches.

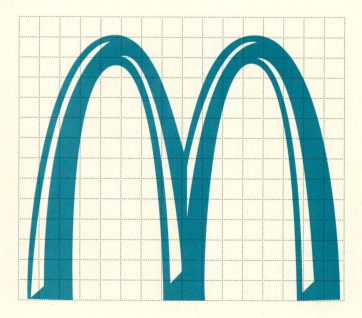

a. Decide on a scale you can use to estimate measurements on the arches. (*Hint:* Think about where you want to set up your coordinate axes.)

b. Use your estimated measurements to determine the equation of a polynomial that best fits one of the arches. (*Hint:* Think again about where you want to set up your coordinate axes.)

c. Can you use the formula you obtained for one of the arches to construct a formula for the other arch? Explain.

4.3 *The Roots of Polynomial Equations: Real or Complex?*

The Roots of Quadratics

In Section 4.1, we saw that, given any quadratic equation

$$ax^2 + bx + c = 0,$$

we can always find its roots using the quadratic formula

$$x = \frac{-b \pm \sqrt{b^2 - 4ac}}{2a}.$$

Further, the roots may be two distinct real numbers, a repeated real root, or a pair of complex conjugate numbers of the form $\alpha \pm \beta i$, where $i = \sqrt{-1}$. Geometrically, the real roots correspond to points where the graph of the quadratic function—a parabola—crosses the x-axis. If there is a double root, then the x-axis is tangent to the parabola: The curve touches the axis and bounces back. When the parabola does not touch the x-axis at all, the roots are complex (see Figure 4.22).

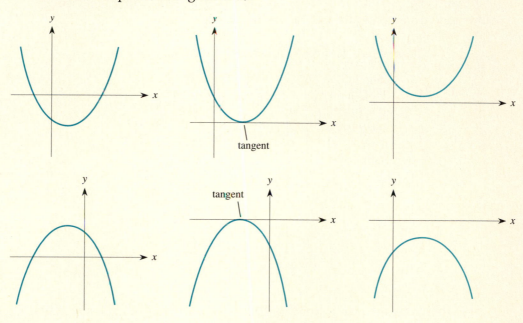

FIGURE 4.22

Most students tend to believe that complex roots occur very rarely. In this section, we investigate how frequently complex roots actually arise. To do so, we consider many different quadratic equations and see what percentage of them do indeed have complex roots. As you have seen, a quadratic equation has complex roots when the discriminant, $b^2 - 4ac$, is negative. In such a case, the quadratic formula requires taking the square root of the negative discriminant and hence produces two complex numbers. For instance, for the quadratic equation $x^2 - 2x + 2 = 0$, the discriminant is -4, and so the quadratic formula gives the roots as

$$x = \frac{2 \pm \sqrt{4 - 8}}{2} = \frac{2 \pm \sqrt{-4}}{2} = \frac{2 \pm 2i}{2} = 1 \pm i.$$

The complex roots are $x = 1 + i$ and $x = 1 - i$. Thus we can use the sign of the discriminant as our criterion for deciding whether any particular quadratic has complex roots.

To come to any meaningful conclusions about the percentage of quadratics with complex roots, we must examine very large numbers of quadratics. This requires using a computer rather than hand computation. Let's begin with the simplest case, when the quadratic has integer coefficients. Even then, there are an infinite number of possible quadratics, so the best we can do is examine a finite selection of them. Suppose that we wish to examine all possible quadratics where the coefficients a, b, and c are integers from 0 to 5, say. We can write this in *interval notation* as $[0, 5]$. We then use a computer program that considers all possible integer values for a, b, and c in the desired interval, say $[0, 5]$, and keeps track, using the discriminant criterion, of how many of the quadratics have complex roots. The results of such an investigation involving all possible integer coefficients in various intervals are shown in Table 4.1.

Interval for a, b, and c	Percentage of Complex Roots
all in $[0, 5]$	70
all in $[0, 10]$	73
all in $[0, 20]$	74
all in $[0, 50]$	74
all in $[-3, 3]$	37.4
all in $[-5, 5]$	37.5
all in $[-10, 10]$	37.8
all in $[-20, 20]$	37.7
all in $[-50, 50]$	37.5
$[0, 5]$, $[0, 5]$, $[-5, 0]$	0
$[0, 5]$, $[-5, 0]$, $[0, 5]$	70

TABLE 4.1

We therefore see that, rather than being a rarity, complex roots actually occur with rather surprising frequency. In fact, if we restrict our attention to cases where the coefficients are all positive integers, it appears that almost three-quarters of all such quadratics have complex roots. Even when we allow negative values, it appears that typically about 40% have complex roots.

There is one exception in the table. If the constant coefficient c is negative while a and b are both positive, it appears from the table that the quadratic always has two real roots. Can you explain why? Look at the discriminant. (Notice that we have checked only specific integer values for a and b between 0 and 5 and c between -5 and 0, so we cannot generalize to what may happen over all similar ranges of values.)

We should not restrict this investigation to cases where the coefficients are just integers. Rather, we should also see what happens when the quadratic has non-integer coefficients, either rational numbers or irrational numbers. In such cases, we cannot simply check all possible cases because there are infinitely many possibilities. Instead, we use a random selection process that generates large numbers of quadratics with randomly selected (non-integer) coefficients in desired ranges, tests each for the nature of its roots, and keeps count of how many of the roots are complex. This is just a modification of the Monte Carlo method we applied in Section 1.6 to estimate the value for π. Instead of randomly generating the coordinates for a point in a square, as we did there, we now randomly generate three numbers to represent coefficients of a quadratic equation.

Some typical results for different ranges of values for non-integer coefficients are shown in Table 4.2. Each is based on the outcomes for 1000 randomly generated quadratics. Very comparable outcomes would result from other runs involving 1000 sets of coefficients in the same ranges of values.

Interval for a, b, and c	Percentage of Complex Roots
all in $[0, 5]$	75.9
all in $[0, 10]$	74.4
all in $[0, 20]$	73.8
all in $[0, 50]$	74.6
all in $[-1, 1]$	38.2
all in $[-2, 2]$	36.2
all in $[-3, 3]$	36.8
all in $[-5, 5]$	37.3
all in $[-10, 10]$	36.7
all in $[-20, 20]$	37.2
all in $[-50, 50]$	37.4
$[0, 5]$, $[0, 5]$, $[-5, 0]$	0
$[0, 5]$, $[-5, 0]$, $[0, 5]$	73.9

TABLE 4.2

Again, we see that complex roots occur with surprising frequency. In fact, if all the coefficients are positive, then the quadratic polynomials are very likely to have a pair of complex roots. Even when there are negative coefficients, a significant percentage of the quadratics (at least when the coefficients are fairly small) have complex roots. Again, the one exception seems to be when the constant coefficient c is negative and a is positive. Can you think of any reason why this is the case?

We suggest that you conduct your own investigations of these ideas if an appropriate program is available or if you care to write a fairly short program for your calculator. Think about the following questions.

- With integer coefficients, what happens as the size of the interval increases? Does the frequency of complex roots stay roughly the same or does it increase or decrease significantly?
- What happens if you use different ranges of values for each coefficient?

We caution you not to be too generous in your choices when you begin; such systematic processes tend to take very long. For example, if you want to check all quadratics where a, b, and c are integers between 0 and 10, say, you are actually requesting that the computer or calculator investigate 1210 different equations. (There are 10 possible values for a since the equation would not be quadratic if a were zero. There are 11 possible values for b and 11 for c. This leads to $10 \times 11 \times 11 = 1210$ different cases.) If you ask for 0 to 100 on each of the coefficients, the computer or calculator will investigate 100×101^2 different quadratics, and it may take all night to complete this study of more than one million cases.

Alternatively, suppose you consider the frequency of complex roots when the coefficients are non-integer. What happens to the frequency as the range of values for a, b, and c increases? What happens if they are all restricted to $[-1, 1]$ or even $[-0.1, 0.1]$? What happens if you use other ranges of values on each coefficient? For example, suppose you consider the case where all three coefficients are in the interval $[-10, 10]$. A typical random quadratic might be

$$5.2744806x^2 + 2.8515691x - 7.0337904,$$

whose discriminant is positive and hence both its roots are real. Now consider the quadratic

$$0.52744806x^2 + 0.28515691x - 0.70337904$$

that might have arisen had we restricted our attention to all coefficients in the interval $[-1, 1]$. Its discriminant is also positive (in fact, it is precisely $1/100$ of the discriminant of the previous quadratic), so it also has a pair of real roots. Is it therefore surprising that in Table 4.2, the outcomes for the percentages of complex roots in cases where the coefficients are taken in intervals such as $[0, 5]$, $[0, 10]$, $[0, 20]$, and $[0, 50]$ are all very close?

The Roots of a Cubic

We now consider an arbitrary cubic equation

$$ax^3 + bx^2 + cx + d = 0,$$

where $a, b, c,$ and d are any four real numbers and $a \neq 0$. Just as a quadratic equation has two roots, a cubic equation has three. They can be either real or complex roots. Furthermore, any complex roots must occur as a pair of complex conjugates, $\alpha + \beta i$ and $\alpha - \beta i$. Thus, given any cubic equation, the three roots may be either three real numbers, or a single real number and a pair of complex conjugate numbers.

Moreover, we know geometrically that the real roots correspond to points where the graph of the cubic crosses the x-axis. If there are three distinct real roots, then the cubic curve crosses the x-axis in three places. (See Figures 4.23a and 4.23b.) If there is a double real root and a separate real root, then the x-axis is tangent to the cubic at the point corresponding to the double root and the curve crosses the x-axis at the point corresponding to the other real root. (See Figures 4.23c and 4.23d.) If there is a triple real root (as with $y = x^3$), the cubic flattens out as it crosses the x-axis at the single point. (See Figures 4.23e and 4.23f.) Finally, if there is a single real root and a pair of complex conjugate roots, the cubic crosses the x-axis once. (See Figures 4.23g and 4.23h.) Thus a cubic can have either three real roots or one real root.

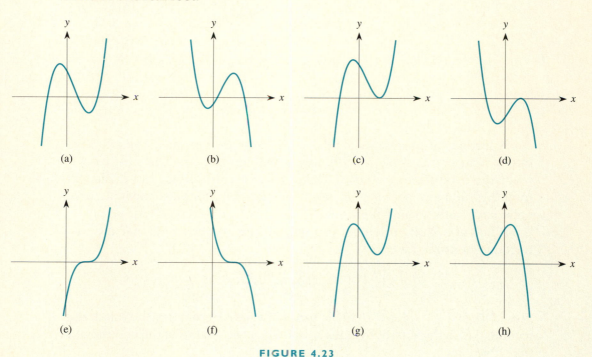

(a)   (b)   (c)   (d)

(e)   (f)   (g)   (h)

FIGURE 4.23

We saw that quadratic equations are likely to have complex roots. A natural question to ask is: How likely is it for a cubic equation to have complex roots? To answer this question on an experimental basis, we again use a computer program to investigate many different cubics. First, though, we must devise a method that allows us to test whether a given cubic has complex roots. The sign of the discriminant in the quadratic formula provides such a test for quadratic equations, but we do not have anything similar available here for cubic equations.

Suppose that a cubic has three real roots. As we saw previously, this means that the curve crosses the x-axis at three places (or at two places if there is a double real root or at one place if there is a triple real root). Consider the cubics shown in Figure 4.24. They all have the same shape; the only difference is the height of the turning points. The cubic on the left has its first turning point above the x-axis and its second below; therefore it has three real roots. The second cubic has both turning points above the x-axis and so must have a pair of complex roots. Likewise, the third cubic has both of its turning points below the x-axis, and so it also must have a pair of complex roots.

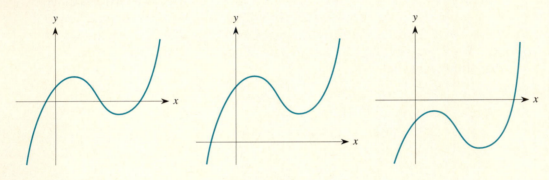

FIGURE 4.24

A further case occurs when the x-axis is tangent to the curve at one of the turning points; such a cubic has a double real root. The final case is when the two turning points coincide; this case corresponds to a triple real root. We therefore see that, in order to have two complex roots, a cubic must have both turning points above the x-axis or both below it.

When you study calculus, you will be able to determine that the two turning points are located at

$$x = \frac{-b \pm \sqrt{b^2 - 3ac}}{3a},$$

provided that $b^2 - 3ac \geq 0$. (This formula clearly resembles the quadratic formula.) Call these two x-values x_1 and x_2. Since we know the equation of the cubic curve,

$$y = f(x) = ax^3 + bx^2 + cx + d,$$

complete graph. Realize that it may be quite difficult to get all of the important details on the behavior of this function from a single view in your function grapher; try it and see what kinds of information may be wiped out because of the scale you use for the domain and range.

b. Notice that $R(x)$ has a zero at $x = 2$ and vertical asymptotes at $x = \pm 1$. Use your calculator to check numerically for values of x on either side of $x = 1$ and on either side of $x = -1$. Further, the numerator is dominated by x while the denominator is dominated by x^2, so for large values of x, the rational function behaves like $\frac{x^2}{x} = \frac{1}{x}$.

Therefore, for large positive values of x, the function is positive and decays toward the x-axis as a horizontal asymptote. Similarly, for large negative values of x, the function is negative and rises toward the x-axis as a horizontal asymptote. A complete graph is shown in Figure 4.33. As before, though, we urge you to examine the behavior carefully with your function grapher to see how viewing the overall characteristics depends on the windows you use. Also, examine the graphs of the quotient function and the limiting function, $y = \frac{1}{x}$ in this case, in the same large viewing window. What do you observe?

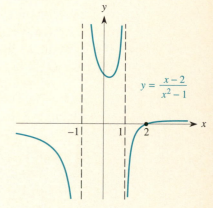

$y = \dfrac{x-2}{x^2-1}$

FIGURE 4.33

A Function of a Function

There is yet another way in which we can construct functions out of simpler functions. Consider the function $f(x) = \sqrt{x^3 + 1}$. Let's see what this means. Suppose we take $x = 1$, so that $f(1) = \sqrt{1 + 1} = \sqrt{2}$. If we try $x = 2$, then $f(2) = \sqrt{8 + 1} = \sqrt{9} = 3$. To evaluate this function in each case, we actually had to perform two successive steps: First, for the given value of x, we evaluated the expression $x^3 + 1$ and second, we took the square root of the result. The reason for this is that we were really working with two functions successively: first the function $x^3 + 1$ and then the function $\sqrt{u}$, where $u = x^3 + 1$. The function f is actually obtained as a *function of a function*, and we call f a **composite function.**

Let's set up the mathematical framework for this concept. Suppose we let $y = F(u)$, where $u = G(x)$. Our example, for instance, involves $y = F(u) = \sqrt{u}$ where in turn $u = G(x) = x^3 + 1$. Consequently,

$$y = F(u) = F(G(x)) = F(x^3 + 1) = \sqrt{x^3 + 1}.$$

Our original function f is really the result of applying the functions G and F successively. This composite function is sometimes written as $F \circ G$ and read "F of G".

With this idea, we can construct many kinds of functions using our basic functions as building blocks. For instance, consider 10^{3x} in which the linear function $3x$ is used as the exponent for the exponential function with base 10. Similarly, in $\log(x^2 - 5x + 2)$, the quadratic function $x^2 - 5x + 2$ is used as the argument of the log function.

Pictorially, you can think of a composite function as shown in Figure 4.34. We start with a value of x, which is carried into a value u by the first, or inner, function G, which in turn is carried to a value y by the second, or outer, function F. In order for this to make sense mathematically, note that the domain of the outer function F must include the range of the inner function G.

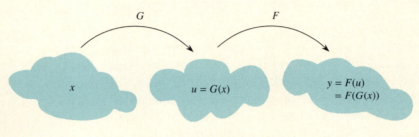

FIGURE 4.34

Is $F \circ G$ the same as $G \circ F$? That is, is the order important when we form the *composition* of two functions? Again consider $f(x) = \sqrt{x^3 + 1} = F(G(x)) = F \circ G(x)$, where

$$u = G(x) = x^3 + 1 \qquad \text{and} \qquad y = F(u) = \sqrt{u}.$$

If we interchange the order, we get

$$G \circ F(x) = G(F(x)) = G(\sqrt{x}) = (\sqrt{x})^3 + 1 \neq F \circ G(x) = \sqrt{x^3 + 1}.$$

In general, except in very rare cases,

$$G(F(x)) \neq F(G(x)).$$

However, if F and G are inverse functions of each other, the equality does hold.

Shifting Functions

There are several other ways in which we can build new functions out of old ones. Consider $y = x^2$, as well as the related functions $y = x^2 + 1$, $y = x^2 + 3$, $y = x^2 - 2$, and $y = x^2 - 5$, all of which are drawn in Figure 4.35. What is the effect of the constant in each case? Clearly, the constant shifts the basic parabola $y = x^2$ up or down by that amount.

We can get a different feel for what is happening if we rewrite each of these expressions by moving the constant term to the left side as follows: $y - 1 = x^2$, $y - 3 = x^2$, $y + 2 = x^2$, and $y + 5 = x^2$. Thus we see that it is really the value of y that is changing because of the additive constant. No wonder the height of the graph $y = x^2$ is shifted either up or down.

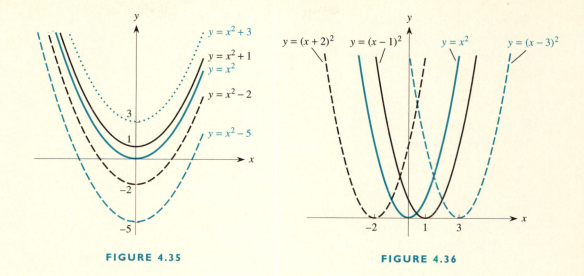

FIGURE 4.35

FIGURE 4.36

In general, the following two principles hold for any function of x:

> **Vertical Shift**
>
> Replacing y with $y - b$ shifts the graph *up* by the amount b.
> Replacing y with $y + b$ shifts the graph *down* by the amount b.

What is the effect of adding a constant to x or subtracting a constant from x? That is, what happens if, in $y = x^2$, we replace x by $(x - 1)$ or by $(x - 3)$ or by $(x + 2)$? The results are shown in Figure 4.36. We see that instead of a vertical shift, this change causes a horizontal shift. For instance, $y = (x - 1)^2$ has a double zero at $x = 1$, so the graph of $y = x^2$ is shifted to the right by 1 unit; similarly, $y = (x + 2)^2$ has a double zero at $x = -2$, so the graph of $y = x^2$ is shifted to the left by 2 units.

In general, the following two principles hold for any function of x:

> **Horizonital Shift**
>
> Replacing x with $x - a$ shifts the graph to the *right* by the amount a.
> Replacing x with $x + a$ shifts the graph to the *left* by the amount a.

Thus, for instance, the graph of $y = 10^{x-2}$ has the identical shape as the graph of $y = 10^x$, but is shifted to the *right* by 2 units. Similarly, the graph of $y = \sqrt{x + 3}$ has the same shape as the graph of $y = \sqrt{x}$, but is shifted to the *left* by 3 units. Check these and other graphs on your function grapher.

When we combine a horizontal shift (x is replaced by $x - a$) with a vertical shift (y is replaced by $y - b$), we effectively have a diagonal shift.

For example, the graph of $y = (x - 4)^2 + 7$ is a parabola whose vertex is at $(4, 7)$; it is obtained by shifting the graph of $y = x^2$ to the right 4 units and up 7 units. (See Figure 4.37.) Similarly,

$$x^2 + y^2 = r^2$$

is the equation of a circle with radius r centered at the origin (see Appendix C), while

$$(x - 5)^2 + (y - 3)^2 = r^2$$

is the equation of a circle with radius r centered at the point $P(5, 3)$. The new circle is produced from the original circle by a combination of a horizontal and a vertical shift. (See Figure 4.38.)

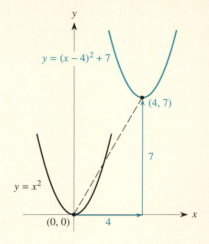

FIGURE 4.37

A Constant Multiple of a Function

We can also create a new function from a given function by multiplying by a constant. For example, consider $y = 2^{-x}$ and $y = 5 \cdot 2^{-x}$. Both are exponential decay functions, as shown in Figure 4.39. The base function $y = 2^{-x}$ passes through the point $(0, 1)$ while the transformed function $y = 5 \cdot 2^{-x}$ passes through the point $(0, 5)$, so you might be tempted to think of the second as resulting from a vertical shift of the first. However, think about what each looks like for large values of x; both curves have the x-axis as a horizontal asymptote. Therefore there must be a relationship between them that is different from merely a vertical shift. In particular, the height for every point on the curve $y = 5 \cdot 2^{-x}$ is five times the height of the corresponding point on the curve $y = 2^{-x}$. The effect of the constant multiple 5 is to increase the height all along the curve by a factor of 5. If we multiply the original function by 20, then the curve will be *stretched* to a new curve that is everywhere 20 times as tall.

On the other hand, if we multiply by $\frac{1}{4}$, then the curve will shrink to a new curve that is everywhere one-quarter the height. Finally, if we multiply the function by a negative constant, such as -3, then the curve is stretched by a factor of 3, but is also flipped upside down. Verify this on your function grapher.

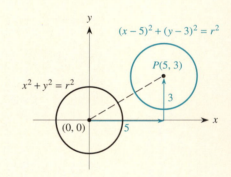

FIGURE 4.38

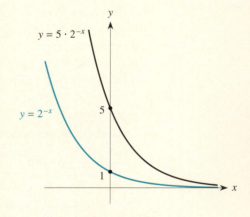

FIGURE 4.39

Multiplying a function by a constant changes the height of its graph by that multiple, but it does not change the general shape.

If the multiple is greater than 1, then the height is increased.
If the multiple is a fraction between 0 and 1, then the height is decreased.
If the multiple is negative, then the curve is flipped over.

4

A chicken is taken from the freezer at 0°F and directly put into an oven at a constant temperature of 350°F. Construct a function to model the temperature of the chicken as it cooks in the oven.

Solution The temperature of the chicken rises rapidly at first and then increases ever more slowly the closer the chicken's temperature is to the oven temperature of 350°. Eventually, the temperature of the chicken would level off at the temperature setting for the oven. The temperature T plotted against time t, in hours, looks like the graph in Figure 4.40. (This is actually an oversimplification because the temperature rise will temporarily stop at the freezing point of 32° while the ice melts. Also, the chicken should be removed from the oven when its temperature reaches about 180°, or it will begin to burn.)

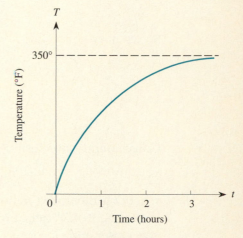

FIGURE 4.40

The horizontal line representing the oven temperature of 350° is a horizontal asymptote because the curve gets closer and closer to it as time increases, but never quite reaches it. The rate at which the temperature of the chicken increases slows down as it approaches 350°, so the curve is concave down.

Suppose we now want to model this process by creating a formula giving the temperature T as a function of the time t. At a very simplistic level, we will find a mathematical model by inspecting the graph of the process and deciding what kind of function has the right shape. In later chapters, you will see how to construct such a function directly.

If you look at the graph in Figure 4.40, you may be reminded of an exponential decay function turned upside down so that it rises toward the oven temperature 350° instead of dropping asymptotically toward the horizontal axis. That is, we can form such a function from a pure exponential function by using a negative coefficient (to turn the curve upside down) and a vertical sh... curve approaches 350 instead of 0. With this reasoning, we might suspe... formula for T would look like

$$T = 350 - B \cdot a^t,$$

where t is in hours and $a < 1$. As t increases, the term a^t approaches 0, and... the entire expression $350 - B \cdot a^t$ approaches 350. What might be possible va... ues for B and a? We know that at time $t = 0$, the chicken's temperature is $T = 0$... when it comes out of the freezer, so that

$$T(0) = 350 - B \cdot a^0 = 350 - B = 0.$$

Thus $B = 350$ and the formula becomes

$$T = 350 - 350 \cdot a^t.$$

Furthermore, suppose that we test the temperature of the chicken after half an hour and find that $T\left(\frac{1}{2}\right) = 110°$. Then, using this value, we find that

$$T\left(\frac{1}{2}\right) = 350 - 350 \cdot a^{1/2} = 110.$$

So we have

$$350a^{1/2} = 350 - 110 = 240$$

$$a^{1/2} = \frac{240}{350} = 0.686,$$

$$a = 0.47.$$

Consequently, our formula for the temperature becomes

$$T = 350 - 350(0.47)^t = 350 \cdot [1 - (0.47)^t],$$

where t is measured in hours. This function is an upside-down exponential: As t increases, $(0.47)^t$ gets progressively smaller, and so $1 - (0.47)^t$ increases and gets closer and closer to 1. That is, $1 - (0.47)^t \to 1$ as $t \to \infty$. As a consequence,

$$T = 350 \cdot [1 - (0.47)^t] \to 350 \quad \text{as } t \to \infty,$$

confirming that the graph has a horizontal asymptote at $T = 350°$.

Verify this behavior by using your function grapher. Look at the overall shape, then zoom in to verify the height of the asymptote. Estimate by eye from the graph when T reaches 180°, when it reaches 250°, and when it reaches 300°, 340°, and 349°.

In general, consider the function $y = f(t) = L + Ba^t$, where $a < 1$. We know that as t increases, a^t decays toward zero, so that the function approaches a limiting value of L. The question is: How does it approach L?

Whenever $B < 0$, the values of the function are less than L. As the term Ba^t decreases, the amount subtracted from L decreases, so the values of the function increase toward L in a concave down manner, as happened with the temperature of the chicken in the oven. See Figure 4.41. On the other hand, whenever $B > 0$, the values of the function are greater than L and so decrease toward it in a concave up manner, as also shown in Figure 4.41.

As you will see, this type of function arises as the mathematical model for quite a number of different situations and both of these cases will occur frequently in the sequel.

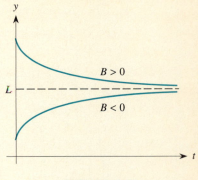

FIGURE 4.41

EXERCISES

1. If $f(x) = 3x - 4$ and $g(x) = \frac{1}{x}$, find:

    a. $f(5) + g(5)$ **b.** $f(5) - g(5)$

    c. $f(5) \cdot g(5)$ **d.** $\dfrac{f(5)}{g(5)}$

    e. $f(g(5))$ **f.** $g(f(5))$

    g. $f(f(5))$ **h.** $g(g(5))$

    i. $f(x) + g(x)$ **j.** $f(x) - g(x)$

    k. $f(x) \cdot g(x)$ **l.** $\dfrac{f(x)}{g(x)}$

    m. $f(g(x))$ **n.** $g(f(x))$

    o. $f(f(x))$ **p.** $g(g(x))$

2. Repeat Exercise 1 if $f(x) = x^2 + 4$ and $g(x) = \sqrt{x}$.

3. Repeat Exercise 1 if $f(x) = 10^x$ and $g(x) = \log x$.

4. Two functions f and g are defined in the following table. Use the given values to complete the table. If any of the entries are not defined, write "undefined."

x	$f(x)$	$g(x)$	$f(x) - g(x)$	$f(x) \cdot g(x)$	$f(g(x))$	$g(f(x))$	$f(x+1)$	$g(2x)$	$f^{-1}(x)$
0	1	0							
1	2	3							
2	3	1							
3	0	2							

5. Match each of the following functions with its graph:

a. $y = \dfrac{x^2 - 1}{x^2 - x - 6}$

b. $y = \dfrac{x^2 + 1}{x^2 - x - 6}$

c. $y = \dfrac{9 - x^2}{x^2 - 4}$

d. $y = \dfrac{x^2 - x - 6}{x^2 - 1}$

e. $y = \dfrac{x^3 - x}{x^2 - 4}$

f. $y = \dfrac{(x - 1)(x - 4)}{x^2 - 4}$

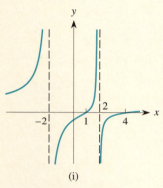

(i)

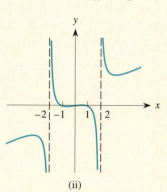

(ii)

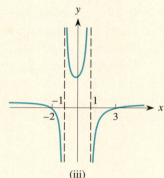

(iii)

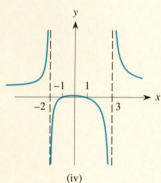

(iv)

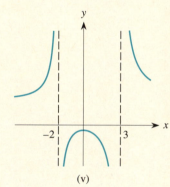

(v)

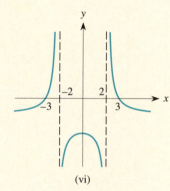

(vi)

6. For the two functions f and g that are defined by the graphs shown, find the following:

a. $f(g(1))$ **b.** $g(f(1))$ **c.** $f(g(-1))$ **d.** $g(f(-1))$

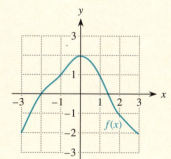

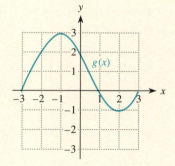

7. For the functions f and g in Exercise 6, sketch the graph of

a. $f(g(x))$ **b.** $g(f(x))$ **c.** $2g(x)$ **d.** $f(x) + 1$

e. $f(x + 1)$ **f.** $f(x - 1)$ **g.** $g(2x)$ **h.** $g\left(\dfrac{1}{2}x\right)$

For Exercises 8 through 11, determine functions F and G such that $h(x) = F(G(x))$.
There are different correct answers to this; however, do not use $F(x) = x$ or $G(x) = x$.

8. $h(x) = x^4 + 5$ 9. $h(x) = (x + 5)^4$

10. $h(x) = \log(x + 3)$ 11. $h(x) = 3 + \log x$

12. Consider the function $y = f(x) = x^2$.

    a. Write an equation for the function that you get when you stretch the graph of f by a factor of 2 and then shift it up 3 units. Call this new function F and sketch its graph.

    b. What is the equation you get if you reverse the order of the two operations in part (a)? Call this new function G and sketch it.

    c. What is $F - G$?

13. **a.** Translate the line $y = mx$ into a line with slope m that passes through the point $P(5, 12)$.

    b. Repeat part (a) if the new line passes through the point $P(x_0, y_0)$. What do you call this new equation?

14. **a.** Translate the parabola $y = x^2$ into a parabola with vertex at $P(5, 12)$.

    b. Repeat part (a) if the new parabola has its vertex at the point $P(x_0, y_0)$.

15. For the function f shown, sketch the graph of

    a. $y = -f(x)$
    b. $y = 2f(x)$
    c. $y = f(x) - 1$
    d. $y = f(x - 1)$
    e. $y = f(x + 1)$
    f. $y = f(x) + 1$

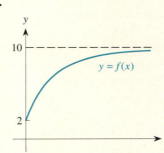

16. For each of the functions f shown below, sketch the graph of:

    (i) $y = -f(x)$ **(ii)** $y = 2f(x)$ **(iii)** $y = -2f(x)$ **(iv)** $y = f(x + 2)$
    (v) $y = f(x) + 2$ **(vi)** $y = f(x) - 2$ **(vii)** $y = f(x - 2)$

a.

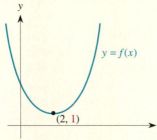

b.

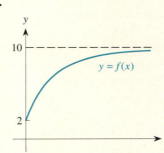

c.

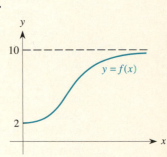

17. If $f(x) = x^2 - 3x + 4$ and h is a constant, find:

    a. $f(x) + h$ **b.** $f(x + h)$ **c.** $f(x + h) - f(x)$ **d.** $\dfrac{f(x + h) - f(x)}{h}$

    e. What is the value of the expression in part (d) if $x = 2$ and if $h = 0.1$? If $h = 0.01$? If $h = 0.0001$?

18. **a.** An unbaked apple pie is taken from the counter in a kitchen where the temperature is 70°F and placed in an oven. Suppose that, after 60 minutes, the temperature of the pie is 180°F. Sketch a graph of the temperature of the pie as a function of time.
    b. The pie is removed from the oven and placed back on the counter. Suppose it takes another 60 minutes for its temperature to come back down to 70°F. Sketch a graph of the temperature of the pie as a function of time.
    c. When the first pie is removed from the oven, a second, unbaked pie is put into the oven to bake. Sketch a graph of the *sum* of the temperatures of the two pies as a function of time over the 60-minute time frame.
    d. Find a formula that models the temperature of the pie, while it cools, as a function of time.

19. A Thanksgiving turkey is taken from the refrigerator at a temperature of 40°F and placed into a hot oven at 350°F to cook. After one hour, the internal temperature of the bird is 124°F. Write a possible formula for the temperature of the turkey as a function of time measured in minutes.

20. According to Einstein's Theory of Relativity, the mass M of an object increases as its speed increases according to the formula

$$M = f(v) = \frac{M_0}{\sqrt{1 - \frac{v^2}{c^2}}} = M_0\left(1 - \frac{v^2}{c^2}\right)^{-1/2},$$

where M_0 is the mass of the object when it is at rest ($v = 0$) and c is the speed of light (about 186,282 miles per second). Suppose an object has a rest mass of $M_0 = 1$ unit.

   a. Construct a table of values for the mass of the object for each of the following speeds expressed as a fraction of the speed of light: $v = 0$, $0.5c$, $0.9c$, $0.95c$, $0.99c$, $0.999c$.
   b. Sketch a graph showing the behavior of the mass of an object as its speed approaches the speed of light.
   c. What is the mathematical significance of the speed of light? What is the physical significance of the speed of light in the context of the speeds of moving objects?

21. Some physicists hypothesize the existence of particles called *tachyons* that exist only at speeds greater than that of light. The slower that a tachyon moves, the greater its mass and the speed of light is a lower limit on the possible speed of a tachyon. Sketch a graph of the mass as a function of speed for all possible values of $v \geq 0$. Indicate which region corresponds to normal particles and which to tachyons.

22. According to Newton's Laws of Motion, the speed of an object can only be changed by applying a force. Also, the greater the mass of an object, the more force is needed to accelerate it to a given velocity in a fixed amount of time. Suppose an object is to be accelerated from speed 0 to almost the speed of light.

   a. Sketch the graph of the force needed to accelerate it as a function of the velocity v. Pay careful attention to concavity.

b. Sketch the graph of the velocity as a function of the force needed, paying attention to concavity.

23. In an attempt to claim responsibility for winning the war against the growing national balance of trade deficit, the president drew a graph similar to the one shown to illustrate the trend in the *annual* deficit.

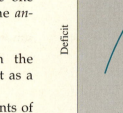

Annual deficit

a. Based on this graph, sketch the graph of the *total* national debt as a function of time.

b. Does your graph have any points of inflection? If so, what do they represent?

c. Do you agree or disagree with the president's assertion that we have won the war? Explain.

24. Use your function grapher to graph the functions $f(x) = x^n (0.5)^x$, for $n = 1, 2, 3, 4, 5$, and estimate the location of the turning point for each curve for $x > 0$. Then perform a linear regression analysis on the values of these turning points, as functions of n. Is the linear fit appropriate? What does it predict if $n = 1.5$? Is it accurate compared to the actual graph?

25. Use your function grapher to graph the functions $f(x) = x^2 a^x$, for $a = 0.3$, 0.4, 0.5, 0.6, and 0.7. Estimate the location of the turning point for each curve by zooming in on it. Then determine the function from among the usual families of functions—linear, exponential, and power—that best fits this data as a function of the base a.

26. Describe how you might use the results of Exercises 24 and 25 to find a function of the form $f(x) = x^p a^x$ that matches the function (shown in Figure 4.30 in the text) on the level of Lyme disease antibody in the bloodstream.

27. For any two linear functions $f(x) = ax + b$ and $g(x) = cx + d$, is $f \circ g = g \circ f$?

28. Find conditions on the coefficients a, b, and c in $P(x) = ax^2 + bx + c$ if P is to satisfy each equation for all values of x:

    a. $P(x) = P(-x)$ **b.** $P(x) = -P(x)$ **c.** $P(2x) = 2P(x)$

29. **a.** Graph the two functions $y = \sqrt{x^2 + 25}$ and $y = \sqrt{x^2} + \sqrt{25} = x + 5$. Are they the same?

    b. Repeat part (a) with $y = \sqrt{x^2 + 4}$ and $y = x + 2$. Are they the same?

    c. Can you find any value for a for which $\sqrt{x^2 + a^2} = x + a$?

30. **a.** Graph the two functions $y = \frac{1}{x + 4}$ and $y = \frac{1}{x} + \frac{1}{4}$. Are they the same?

    b. Repeat part (a) with $y = \frac{1}{x - 5}$ and $y = \frac{1}{x} - \frac{1}{5}$. Are they the same?

    c. Can you find any value for a so that $\frac{1}{x + a} = \frac{1}{x} + \frac{1}{a}$?

4.5 *Finding Roots of Equations*

Throughout this chapter, we have repeatedly encountered one of the most important problems that arises in mathematics—finding the roots of an equation. The problem might be to find the zeros of a polynomial or some other type of function; it might be to find the point of intersection of two (or more) curves. When such problems arose previously, we usually suggested that you try to locate each root on a trial-and-error basis numerically or by using your function grapher to zoom in on the root. Such relatively inefficient approaches do not make sense for a problem that arises frequently, so mathematicians have developed many more effective methods for finding roots of equations.

We know that the quadratic formula gives the roots of any quadratic equation. Unfortunately, it applies only to quadratic equations or those that can be reduced or interpreted as quadratic. For centuries, mathematicians searched for more sophisticated formulas for the zeros of higher degree polynomials, and eventually formulas giving the roots of any cubic or quartic equation were developed. However, early in the nineteenth century, the French mathematician Evariste Galois proved that no such formula *could* exist for the general polynomial of degree higher than four. Polynomial equations are among the simplest that arise; there are no general formulas for the roots of nonpolynomial equations. Thus the search for "closed-form solutions" is usually hopeless and we cannot expect to obtain an exact answer for the roots of an arbitrary equation.

On the other hand, in the process of performing the trial-and-error methods suggested above, you should have noticed that such methods can be continued indefinitely, and that each successive repetition typically improves the level of accuracy. Thus, for all practical purposes, we can *approximate* the desired roots numerically to *any desired degree of accuracy* by applying trial-and-error methods repeatedly. Therefore, if we can get *any* number of correct decimal places for an answer, there is no practical difference between an exact result and an approximate result. (In this regard, mathematicians have calculated an approximation to the irrational number π that is correct to several billion decimal places! For all practical purposes, there is no difference between the exact number π and this, or a much poorer, approximation.)

We now develop a very effective numerical method, known as the *bisection method*, for calculating the roots of an arbitrary equation $f(x) = 0$, provided that the function f is *continuous*. A function f is continuous if its graph can be drawn smoothly without lifting the pen from the paper; it has no jumps or holes. If a function has such interruptions, as shown in Figure 4.42, then we say that it is *discontinuous* at the point

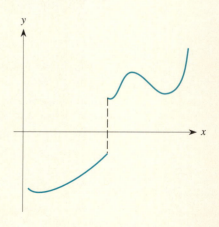

FIGURE 4.42

where the interruption occurs and the bisection method will not apply.

Fortunately, most functions that arise in practice are continuous. Only relatively unusual functions, such as $y = \frac{1}{x}$, are discontinuous, and typically the discontinuity occurs at only one point (or, at worst, at a few points). For $y = \frac{1}{x}$, the only discontinuity occurs at $x = 0$ where the function is undefined. Examine its graph on your function grapher. Similarly, any rational function having a vertical asymptote is discontinuous at the point where the function is undefined. In the following development, we restrict our attention to functions that are continuous.

The bisection method is based on the following observation for any smooth curve: If the function f has a zero at the point $x = R$, then the graph of the function typically crosses the x-axis at that point. If it does, the curve must be above the x-axis, so $f(x) > 0$, on one side of $x = R$ and below the x-axis, so that $f(x) < 0$, on the other side of R (see Figure 4.43). The only exception to this is the case where the x-axis is tangent to the curve at the root (see Figure 4.44); we will see that the bisection method does not apply in this case.

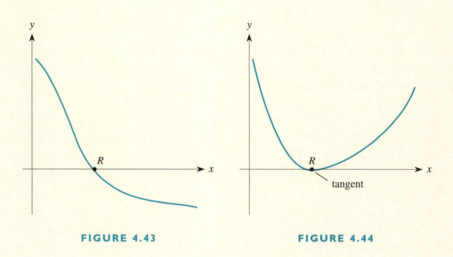

FIGURE 4.43   **FIGURE 4.44**

Suppose we start with an interval $a \le x \le b$, or equivalently $[a, b]$, such that $f(a)$ and $f(b)$ have opposite signs (see Figure 4.45, on the next page). Since f is continuous, there must be at least one point R within this interval where the graph crosses the x-axis. The bisection method is based on the idea that if we find the midpoint of the interval $[a, b]$, call it M_1, then the root R must occur either in the left half-interval $[a, M_1]$ or in the right half-interval $[M_1, b]$. In Figure 4.45, the root is in the left half-interval $[a, M_1]$. Having narrowed the interval containing the root by half, we then bisect the new half-interval at its midpoint M_2, say, and so generate an interval that contains the root R but is one-quarter the size of the original interval. If this procedure is continued indefinitely (see Figure 4.46, on the next page), it produces a sequence of ever smaller subintervals, each containing the desired root. Eventually, the endpoints of the interval will agree to any desired degree of accuracy. Since the root we seek is between the endpoints, its decimal representation must have the same number of significant digits as the endpoints.

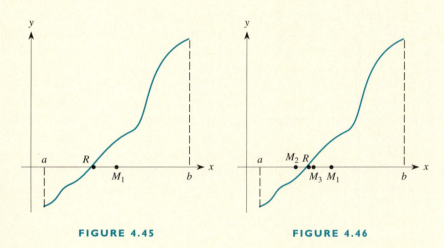

FIGURE 4.45   FIGURE 4.46

We summarize the above ideas in the following statement, which guarantees that the bisection method always produces an approximate root when carried far enough:

> If f is continuous on an interval $[a, b]$ and $f(a)$ and $f(b)$ have opposite signs, then the sequence determined by the bisection method always converges to a zero of f inside the interval.

The key to determining which half of the interval to use to continue the bisection process is based on the fact that the function changes sign on either side of the root. Thus, in Figure 4.46, we see that $f(a)$ is negative and $f(M_1)$ is positive, so the left half-interval $[a, M_1]$ should be used. Then, since $f(M_2)$ is negative and $f(M_1)$ is positive, the root must occur in the subinterval $[M_2, M_1]$. Continuing the process, since $f(M_2)$ and $f(M_3)$ have opposite signs, we focus on the subinterval $[M_2, M_3]$, and so on.

E X A M P L E   I

Find a root of $x^{5/3} - x^{2/3} - 1 = 0$.

Solution To apply the bisection method, we first must locate an initial interval where the function $f(x) = x^{5/3} - x^{2/3} - 1$ is continuous and changes sign. Note that $f(0) = -1$, $f(1) = -1$, and $f(2) = 0.59$. Thus, the function changes sign between $x = 1$ and $x = 2$ and so the initial interval we take is $[a, b] = [1, 2]$. (We could use $[0, 2]$ since the function also changes sign on this interval, but the larger the interval, the longer it will take to calculate the desired root.) See Figure 4.47 for a graph of the function. The following table provides the corresponding interval endpoints. We also include a column showing, in order, the sign of the function f at the left endpoint, the midpoint, and the right endpoint of each interval. Notice that these signs tell us that for the first row, f is negative at the left endpoint ($x = 1$), f is negative at the midpoint ($x = 1.5$), and f is positive at the right endpoint ($x = 2$). Therefore, the root must be between 1.5 and 2 and these values of x form the interval for the second row.

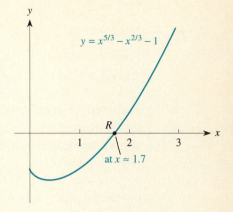

$y = x^{5/3} - x^{2/3} - 1$

R

at $x \approx 1.7$

FIGURE 4.47

	Left	Right	Signs
$n = 0$	1	2	$- - +$
$n = 1$	1.5	2	$- + +$
$n = 2$	1.5	1.75	$- - +$
$n = 3$	1.625	1.75	$- - +$
$n = 4$	1.6875	1.75	$- + +$
$n = 5$	1.6875	1.71875	$- + +$
$n = 6$	1.6875	1.703125	$- - +$
$n = 7$	1.695313	1.703125	$- - +$
$\vdots$			
$n = 10$	1.701172	1.702148	$- + +$
$n = 15$	1.701599	1.701630	$- + +$
$n = 20$	1.701607	1.701608	$- + +$

Based on these values, we conclude that the desired root lies between $x = 1.701607$ and $x = 1.701609$. Thus, correct to five decimal places, the root is $R = 1.70161$.

Clearly, if greater accuracy were needed, the process could be continued. However, note that the above level of accuracy required 20 *iterations*, or repetitions, to give five decimal accuracy, which is a fairly slow rate of *convergence*. There are other methods available that provide comparable levels of accuracy in fewer iterations, which means that the rate of convergence is faster. Of course, there is usually a trade-off: Faster rates of convergence usually entail more sophisticated approaches.

Notice that the bisection method essentially produces two sequences of numbers, both of which converge to the desired root. One sequence converges from above and the other from below, so the root R is always squeezed in between the corresponding terms of the two sequences for all values of n.

E X A M P L E   2

Find the value of x for which the function $f(x) = x \log x$ reaches a height of 2.

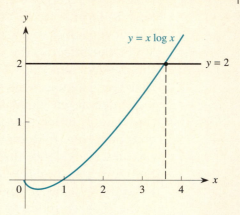

Solution   Geometrically this is asking for the point where the curves $y = x \log x$ and $y = 2$ intersect, as shown in Figure 4.48. This happens when

$$y = x \log x = 2,$$

or equivalently, when

$$x \log x - 2 = 0.$$

To find this point, we consider the function $g(x) = x \log x - 2$ and determine the value of x when

FIGURE 4.48

$$g(x) = x \log x - 2 = 0.$$

To use the bisection method, we observe from Figure 4.48 that the intersection occurs between 3 and 4 and, in particular, $g(3) = -0.5686$ and $g(4) = 0.4082$. So, there must be a root in the interval $[3, 4]$, as shown in Figure 4.49. The bisection method then produces the following values.

n	Left	Right
0	3	4
1	3.5	4
2	3.5	3.75
3	3.5	3.625
4	3.5625	3.625
5	3.59375	3.625
$\vdots$		
10	3.59668	3.59766
15	3.59726	3.59729
20	3.597284	3.597285

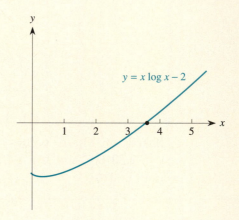

FIGURE 4.49

Therefore the desired root is between $x = 3.597284$ and $x = 3.597285$, so it is approximately $R = 3.59728$. To verify the accuracy of this result, notice that $g(3.59728) = -0.00000497$, which is quite close to zero.

There are several possible complications in using the bisection method. If a continuous function f has several zeros in an interval $[a, b]$, the method will certainly converge to one of them provided the function changes sign across the entire interval. But it is not at all evident in advance which root will be found. So it makes sense to start with an initial interval as small as possible that brackets a root. You can usually find such a starting interval by looking at the graph of the function. However, if the roots are very close together, as in Figure 4.50, then it is clearly possible that you might not notice that there are three of them if your viewing window is too large. Therefore, you cannot be certain that you have found all the roots of a function simply by a visual examination of its graph. This information only comes from a thorough understanding of the behavior of the function. Methods for analyzing the behavior of a function in considerably more detail form an important part of calculus.

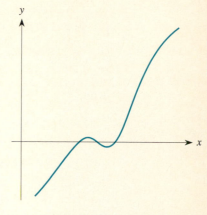

FIGURE 4.50

Another complication, as we mentioned before, occurs when the x-axis is tangent to the graph at the root. Since the function doesn't change sign about this point, the bisection method does not apply.

The Error in the Approximation

In applying any numerical method such as the bisection method to determine a root, it is important to realize that the best we can usually achieve is an approximation of the exact root. At each iteration of the method, we obtain a better estimate of the root. Thus it becomes desirable that we be able to estimate how accurate the approximation is at each stage so that we know when to stop the process.

With the bisection method, suppose we know that there is a root in some interval $[a, b]$, where a and b are successive integers, say 2 and 3. If we select the midpoint M_1 of this interval, then it is obvious that the root R is no further than half the length of the interval from M_1. That is, the error in the approximation E_1 (the difference between the correct value for R and our approximation) is at most $\frac{1}{2}$, so that $E_1 \le \frac{1}{2}$. When we bisect the second time to obtain a new midpoint, M_2, in either $[a, M_1] = [2, 2.5]$ or $[M_1, b] = [2.5, 3]$, then the length of each subinterval is half the length of the last interval or $\frac{1}{4}$. Therefore, the corresponding error, E_2, after two iterations is

$$E_2 \le \frac{1}{4}.$$

Moreover, since the length of each interval is always half the length of the preceding interval, the error with using the bisection method decreases by an additional factor of one-half from one approximation to the next. Thus after n iterations, the approximation for the root R will have a maximum possible error of

$$E_n \le \left(\frac{1}{2}\right)^n = \frac{1}{2^n}.$$

For instance, if the original interval is $[a, b] = [2, 3]$, then the maximum possible error after $n = 6$ iterations of the bisection method is

$$E_6 \le \frac{1}{2^6} = \frac{1}{64} = 0.0156;$$

after $n = 10$ iterations, the maximum possible error is

$$E_{10} \le \frac{1}{2^{10}} = 0.000976,$$

which means that the approximation would be accurate to about three decimal places. If the initial interval is smaller, then the maximum error is likewise much smaller and we gain the desired level of accuracy with fewer iterations.

We can turn this error analysis around to estimate the minimum number of iterations needed to obtain a desired level of accuracy in the approximation. Suppose we want four decimal places based on an initial interval $[a, b] = [2, 3]$. This means that the approximation must be within 0.00005 of the correct value R. (We do not use 0.0001, as you might think at first; we have to anticipate rounding in the fifth decimal place to obtain four correct decimals.) Therefore we must find the minimum number of iterations n that guarantees that

$$E_n \le \frac{1}{2^n} \le 0.00005$$

or equivalently,

$$\frac{1}{0.00005} = 20,000 \le 2^n.$$

The first value of n for which this holds is $n = 15$ since $2^{15} = 32,768$ while $2^{14} = 16,384$. Therefore we are guaranteed that 15 iterations will give us the desired level of accuracy. We may well get this level of accuracy in fewer than 15 iterations, so this is the worst case scenario.

EXERCISES

In Exercises 1–5, use the bisection method to find the indicated points to two decimal place accuracy.

1. a. The real root of $x^3 - 3x + 1 = 0$ between 0 and 1.
    b. The real root of $x^3 - 3x + 1 = 0$ between 1 and 2.
    c. The third real root of $x^3 - 3x + 1 = 0$.
2. The roots of $x^2 + \log x - 2^x = 0$.

3. The points of intersection of $f(x) = 2^x$ and $g(x) = x^3$.

4. The points of intersection of $f(x) = \dfrac{1}{2^x}$ and $g(x) = x^2$.

5. The point of intersection of $f(x) = 10^{-x}$ and $g(x) = \log x$.

6. Use the results of Exercise 1 to find all intervals where $x^3 - 3x + 1 > 0$.

7. Find all intervals where $x^4 - 2x^3 - 3x^2 + 5x + 1 < 0$.

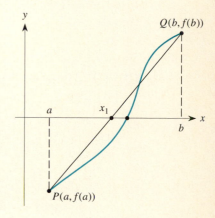

8. The bisection method is a fairly inefficient approach for locating a root because it makes no use of intelligent strategy. The *secant method* is a more efficient approach. Suppose that a function f changes sign between $x = a$ and $x = b$. The two points $P(a, f(a))$ and $Q(b, f(b))$ lie on the curve, as shown in the accompanying figure, and also determine the line segment PQ shown. The secant method is based on finding the point x_1 where this line crosses the x-axis and using it, instead of the midpoint of the interval, as the basis for a smaller interval that contains the desired root.

   a. Write an equation for the line segment PQ.
   b. Determine an expression for the value of x where the line segment PQ crosses the x-axis.
   c. Use the accompanying figure to draw the next secant line and indicate the point x_2 where it crosses the x-axis.

9. Perform the first three iterations of the secant method to locate the root requested in Exercise 1(a). Compare the accuracy you achieve after these three iterations to that you obtained in Exercise 1(a) after three iterations. Which method leads to a smaller interval containing the desired root in three iterations?

10. Consider the unit circle $x^2 + y^2 = 1$ and the point $P(0, -4)$. Most of the lines drawn through P either cross the circle at two points or miss the circle entirely. There are only two lines through P that touch the circle at exactly one point where the lines are tangent to the circle.

   a. Use your function grapher to estimate, by trial and error, the slope of one of those lines, correct to two decimal places. (*Hint:* Use the equation for the lower half of the circle.)
   b. Estimate the coordinates of the point of tangency.
   c. What is the slope of the line from the center of the circle to the point of tangency? How does it compare to the slope of the line you found in (a)?
   d. What is the second point where a line through P will be tangent to the circle? What is the slope of that line?
   e. Use geometry to find the point of tangency in closed form.

11. The value of a diamond depends on a variety of factors, although the most important is its weight, as measured in carats. (Other factors are its color, its clarity, and how it is cut.) The accompanying figure, which shows the value of a diamond as a function of its weight, is an example of a *piecewise continuous* function. Write a short paragraph to interpret the information contained in the graph.

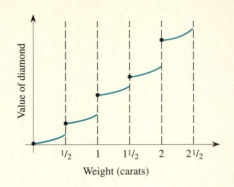

4.6 *Finding Polynomial Patterns*

In Section 2.2, we developed a criterion for determining whether a set of points (x_i, y_i) follows a linear pattern when the x-values are uniformly spaced:

> A set of points lie on a line if the differences between successive y-values are all equal when the x-values are uniformly spaced. The slope of that line
>
> $$m = \frac{\Delta y}{\Delta x}$$
>
> is the constant difference between successive y-values divided by the uniform spacing between successive x-values.

If the differences between the y-values are approximately, but not exactly, the same, then our techniques for fitting a regression line to the data become the method of choice.

We now consider the related problem of determining whether a set of points follows a quadratic, a cubic, or a higher degree polynomial pattern. Suppose we take the set of points $(0, 1)$, $(1, 2)$, $(2, 5)$, $(3, 10)$, $(4, 17)$, and $(5, 26)$ which actually lie on the parabola $y = x^2 + 1$. We construct the following table:

x	y	Δy
0	1	1
1	2	3
2	5	5
3	10	7
4	17	9
5	26	

Obviously, the Δy values are not constant. In fact, they clearly follow a linear pattern since the differences between successive Δy values are all constant. Thus, we extend the previous table to look at the differences of the differences $\Delta(\Delta y)$, or the *second differences* $\Delta^2 y$, of the y-values:

$$
\begin{array}{c|c}
x & y \\
\hline
0 & 1 \\
1 & 2 \\
2 & 5 \\
3 & 10 \\
4 & 17 \\
5 & 26
\end{array}
\qquad
\begin{array}{c}
\Delta y \\
1 \\
3 \\
5 \\
7 \\
9
\end{array}
\qquad
\begin{array}{c}
\Delta^2 y \\
2 \\
2 \\
2 \\
2
\end{array}
$$

We see that all the second differences are constant.

In general, we have the following criterion when the x-values are uniformly spaced:

A set of points (x_i, y_i) lie on a quadratic $y = ax^2 + bx + c$ if the second differences of the y-values are all constant when the x-values are uniformly spaced.

In the Exercises, we ask you to explore the significance of this constant second difference.

EXAMPLE I

Show that the points $(0, 2)$, $(1, 0)$, $(2, 4)$, $(3, 14)$, $(4, 30)$, and $(5, 52)$ lie on a parabola. Then find the equation of the parabola.

Solution   We construct a table of differences as follows:

$$
\begin{array}{c|c}
x & y \\
\hline
0 & 2 \\
1 & 0 \\
2 & 4 \\
3 & 14 \\
4 & 30 \\
5 & 52
\end{array}
\qquad
\begin{array}{c}
\Delta y \\
-2 \\
4 \\
10 \\
16 \\
22
\end{array}
\qquad
\begin{array}{c}
\Delta^2 y \\
6 \\
6 \\
6 \\
6
\end{array}
$$

Since the difference of the differences is constant, the points follow a quadratic pattern of the form

$$y = ax^2 + bx + c,$$

where the coefficients a, b, and c must be determined. Substituting the coordinates from the first point $x = 0$ and $y = 2$, we find that

$$2 = a(0) + b(0) + c,$$

so $c = 2$. Using the second point $(1, 0)$ and the fact that $c = 2$, we get

$$y = 0 = a(1^2) + b(1) + c = a + b + 2,$$

and so

$$a + b = -2. \tag{1}$$

Using the third point $(2, 4)$ along with $c = 2$, we get

$$y = 4 = a(2^2) + b(2) + c$$
$$= 4a + 2b + 2.$$

So

$$4a + 2b = 2$$

or equivalently,

$$2a + b = 1. \tag{2}$$

We now solve these two equations in two unknowns for a and b. Subtracting Equation (1) from Equation (2) yields

$$a = 3.$$

Substituting this into Equation (1) gives

$$b = -5,$$

and so the desired quadratic is

$$y = 3x^2 - 5x + 2.$$

We can extend the above ideas to develop similar criteria for deciding when a set of points (x_i, y_i) follow a polynomial pattern of degree n for any n. In particular:

A set of points (x_i, y_i) lie on a cubic $y = ax^3 + bx^2 + cx + d$ if the third differences (differences of the differences of the differences) of the y-values are all constant when the x-values are uniformly spaced.

Sums of Integers

We will use the above ideas on differences and polynomial patterns to develop a number of formulas involving sums of numbers that arise frequently in mathematics. Among these are the sum of the first n integers

$$1 + 2 + 3 + 4 + 5 + \cdots + n$$

and the sum of the squares of the first n integers

$$1^2 + 2^2 + 3^2 + 4^2 + \cdots + n^2.$$

Let's begin with the first expression for the sum of the integers. Suppose we use S_n to denote the sum of the first n integers, so that

$$S_n = 1 + 2 + 3 + \cdots + n.$$

For instance, $S = 1 + 2 + 3 + 4 = 10$. We seek to find an expression for S_n for any value of n. We will do this in two ways. The first is a particularly simple way that involves a lovely little trick. If

$$S_n = 1 + 2 + 3 + \cdots + (n - 2) + (n - 1) + n,$$

then we can also write this in the reverse order as

$$S_n = n + (n - 1) + (n - 2) + \cdots + 3 + 2 + 1.$$

We now add these two equations together term by term in the following way:

$$S_n + S_n = [1 + n] + [2 + (n - 1)] + [3 + (n - 2)] + \cdots$$
$$+ [(n - 1) + 2] + [n + 1]$$
$$= (n + 1) + (n + 1) + (n + 1) + \cdots + (n + 1) + (n + 1).$$

Since there are n of these terms on the right side,

$$2S_n = n \cdot (n + 1).$$

We therefore obtain

$$S_n = \frac{n(n + 1)}{2}.$$

> The sum of the first n integers is:
>
> $$1 + 2 + 3 + \cdots + n = \frac{n(n + 1)}{2}.$$

For instance, with $n = 100$,

$$1 + 2 + 3 + \cdots + 100 = \frac{100(101)}{2} = 5050.$$

We can write the above formula using *summation notation* where the Greek letter Σ (sigma) is used to indicate that a set of terms are being added together. We then have

$$\sum_{k=1}^{n} k = 1 + 2 + 3 + \cdots + n = \frac{n(n + 1)}{2}.$$

Alternatively, we can derive this result using the ideas on fitting functions to data. Although the derivation is somewhat longer, the advantage of doing it is that it provides a technique that can be applied to more sophisticated cases later. The sum of the first integer is $S = 1$; the sum of the first two integers is $S = 1 + 2 = 3$; the sum of the first three integers is $S = 1 + 2 + 3 = 6$ and so forth. In particular, $S = 10$, $S = 15$, and $S = 21$. If we form a table of differences with these entries, we find that

$$
\begin{array}{c|c}
n & S \\
\hline
1 & 1 \\
2 & 3 \\
3 & 6 \\
4 & 10 \\
5 & 15 \\
6 & 21 \\
\end{array}
\quad
\begin{array}{c}
\Delta S \\
2 \\
3 \\
4 \\
5 \\
6 \\
\end{array}
\quad
\begin{array}{c}
\Delta^2 S \\
1 \\
1 \\
1 \\
1 \\
\end{array}
$$

Since the second differences $\Delta^2 S$ are constant, the desired pattern is a quadratic function of n, $S_n = an^2 + bn + c$, where a, b, and c are constants that we must now determine. Using the point $n = 1$, $S_1 = 1$, we find

$$S_1 = 1 = a(1^2) + b(1) + c,$$

and so

$$a + b + c = 1.$$

When $n = 2$, we have $S_2 = 3$, so that

$$S_2 = 3 = a(2^2) + b(2) + c,$$

and hence

$$4a + 2b + c = 3.$$

Similarly, when $n = 3$ and $S_3 = 6$, we have

$$S_3 = 6 = a(3^2) + b(3) + c.$$

So

$$9a + 3b + c = 6.$$

We therefore have a system of three linear equations in three unknowns:

$$a + b + c = 1 \tag{3}$$
$$4a + 2b + c = 3 \tag{4}$$
$$9a + 3b + c = 6. \tag{5}$$

We solve this system using the elimination method from algebra. We first subtract Equation (4) from Equation (5) to get

$$5a + b = 3 \tag{6}$$

and then subtract Equation (3) from Equation (4) to get

$$3a + b = 2. \tag{7}$$

This gives a system of two linear equations in the two unknowns a and b; c has been eliminated. We now subtract Equation (7) from Equation (6) to find

$$2a = 1, \quad \text{or} \quad a = \frac{1}{2}.$$

Substituting this into Equation (7) gives

$$3\left(\frac{1}{2}\right) + b = 2, \quad \text{or} \quad b = \frac{1}{2}.$$

Substituting both of these values into Equation (3), we find

$$\frac{1}{2} + \frac{1}{2} + c = 1, \quad \text{or} \quad c = 0.$$

Thus the resulting quadratic function for the sum of the first n integers is

$$S_n = \frac{1}{2}n + \frac{1}{2}n = \frac{1}{2}n(n + 1),$$

the same result we obtained above.

Sums of Squares of Integers

We now turn to the problem of finding a formula for the sum of the first n squares:

$$S_n = 1^2 + 2^2 + 3^2 + 4^2 + \cdots + n^2.$$

We have

$$S_1 = 1, \quad S_2 = 1^2 + 2^2 = 5, \quad S_3 = 1^2 + 2^2 + 3^2 = 14,$$
$$S_4 = 30, \quad S_5 = 55, \quad S_6 = 91,$$

and so on. To make things a little simpler, we will also use the sum of the squares of the first zero terms, $S_0 = 0^2 = 0$. Arranging these values in a table, we find that

$$
\begin{array}{c|c}
n & S_n \\
\hline
0 & 0 \\
1 & 1 \\
2 & 5 \\
3 & 14 \\
4 & 30 \\
5 & 55 \\
6 & 91 \\
\end{array}
\qquad
\begin{array}{c}
\Delta S_n \\
1 \\
4 \\
9 \\
16 \\
25 \\
36 \\
\end{array}
\qquad
\begin{array}{c}
\Delta^2 S_n \\
3 \\
5 \\
7 \\
9 \\
11 \\
\end{array}
\qquad
\begin{array}{c}
\Delta^3 S_n \\
2 \\
2 \\
2 \\
2 \\
\end{array}
$$

Since the third differences $\Delta^3 S_n$ are constant, these data values must follow a cubic pattern; that is, the formula for the sum of the squares of the first n integers is a cubic function,

$$S_n = an^3 + bn^2 + cn + d.$$

There are four unknowns here, so that we will need four equations to determine them. Thus,

$$\text{when } n = 0 \text{ and } S_0 = 0: \qquad 0 = d$$

$$\text{when } n = 1 \text{ and } S_1 = 1: \qquad 1 = a + b + c + d.$$

Since $d = 0$, we have

$$a + b + c = 1 \tag{8}$$

$$\text{when } n = 2 \text{ and } S_2 = 5: \qquad 8a + 4b + 2c = 5 \tag{9}$$

$$\text{when } n = 3 \text{ and } S_3 = 14: \qquad 27a + 9b + 3c = 14. \tag{10}$$

We eliminate c first. We subtract two times Equation (8) from Equation (9) to get

$$6a + 2b = 3. \tag{11}$$

Similarly, we subtract three times Equation (8) from Equation (10) to get

$$24a + 6b = 11. \tag{12}$$

Now we subtract three times Equation (11) from Equation (12) to find that

$$6a = 2, \qquad \text{or} \qquad a = \frac{1}{3}.$$

Substituting into Equation (11), we find

$$6\left(\frac{1}{3}\right) + 2b = 2 + 2b = 3, \qquad \text{or} \qquad b = \frac{1}{2}.$$

Consequently, from Equation (8),

$$\frac{1}{3} + \frac{1}{2} + c = \frac{5}{6} + c = 1,$$

and so

$$c = \frac{1}{6}.$$

Therefore the sum of the squares of the first n integers is

$$S_n = \left(\frac{1}{3}\right)n^3 + \left(\frac{1}{2}\right)n^2 + \left(\frac{1}{6}\right)n$$

$$= \left(\frac{1}{6}\right)n[2n^2 + 3n + 1]$$

$$= \left(\frac{1}{6}\right)n(n + 1)(2n + 1)$$

when we factor the quadratic expression in the brackets.

> The sum of the squares of the first n integers is:
>
> $$\sum_{k=1}^{n} k^2 = 1^2 + 2^2 + 3^2 + \cdots + n^2 = \frac{n(n + 1)(2n + 1)}{6}.$$

For example, if $n = 100$, then

$$1^2 + 2^2 + \cdots + 100^2 = \frac{100(101)(201)}{6} = 338{,}350.$$

Note that while this formula is true for *all* values of n, we have only established it for $n = 0, 1, \ldots, 6$ using this approach. We return to it in Section 6.2 after we have developed some powerful techniques that allow us to prove such results for all values of n.

Incidentally, you could obtain this formula by applying the ideas from Section 4.2 on fitting polynomials to data with a graphing calculator. In this case, there should be a perfect fit of a cubic polynomial to the "data values" because the pattern is cubic. However, the calculator routines give decimal values instead of fractions: $a = 0.33333333$ (or $1/3$), $b = 0.5$, $c = 0.166666667$ (or $1/6$) and $d = -1E-11 = -1 \times 10^{-11} = -0.00000000001$, which is not quite 0. Realize, though, that it is important for you to be able to perform the algebraic methods as well as simply using the calculator for curve fitting.

E X A M P L E   2

When cannonballs are stacked in a pyramidal pile, they are orga-
nized as follows: There is a single ball at the top of the pile; there
are four balls in the second layer from the top arranged in a square
to support the single ball on top; there are nine balls in the third
layer arranged in a square of size three by three; and so forth.
Suppose there is a pile of cannonballs that is 10 layers high. How
many cannonballs are in the pile?

Outline of third layer
on top of fourth layer

Solution   The number of cannonballs is equal to

$$1^2 + 2^2 + 3^2 + \cdots + 10^2.$$

We can evaluate this total using the above formula for the sum of the squares of
the first n integers with $n = 10$. Thus,

$$1^2 + 2^2 + 3^2 + \cdots + 10^2 = \frac{10(10 + 1)[2(10) + 1]}{6}$$

$$= \frac{10(11)(21)}{6}$$

$$= 385 \text{ cannonballs.}$$

The following example illustrates some additional applications of
these ideas to find the total for a quantity when the individual amounts
are known. It uses these basic properties of sums of numbers:

$$\sum_{k=1}^{n} (a_k + b_k) = \sum_{k=1}^{n} a_k + \sum_{k=1}^{n} b_k$$

$$\sum_{k=1}^{n} (m \cdot a_k) = m \cdot \sum_{k=1}^{n} a_k, \text{ for any constant } m.$$

You will be asked to prove these results, which depend only on the fact
that Σa_k is a shorthand notation for a sum of numbers, in the Exercises.

E X A M P L E   3

A study of the financial records of a company finds that its monthly revenues,
in thousands of dollars, are modeled by the equation $R(x) = 0.001x^2 + 0.02x + 32$,
where x is the number of months and $x \geq 1$. Find the total revenue for this company over its first 10 years of operation.

Solution   The 10-year period is equivalent to 120 months. We need to add the
revenues $R(1)$ in month 1, $R(2)$ in month 2, $R(3)$ in month 3, . . . , $R(120)$ in

month 120. Therefore we seek

$$R = R(1) + R(2) + \cdots + R(120) = \sum_{k=1}^{120}(0.001k^2 + 0.02k + 32),$$

where the variable k takes on all values between 1 and 120. Using the properties of sums that we listed above, we can simplify this as

$$\sum_{k=1}^{120}0.001k^2 + \sum_{k=1}^{120}0.02k + \sum_{k=1}^{120}32,$$

which is equivalent to

$$0.001\sum_{k=1}^{120}k^2 + 0.02\sum_{k=1}^{120}k + 32\sum_{k=1}^{120}1.$$

Using our formulas with $n = 120$ for the sum of a set of integers Σk and the sum of the squares of those integers Σk^2, this becomes

$$0.001\left[\frac{(120)(120+1)(2\cdot 120+1)}{6}\right] + 0.02\left[\frac{(120)(120+1)}{2}\right] + 32\cdot 120$$

since the last term represents 32 times the sum of 120 ones. Therefore the total revenue for this company over the 10-year period is

$$R = 0.001(583{,}220) + 0.02(7260) + 3840 = 4568.42 \text{ thousand dollars,}$$

or about 4.568 million dollars.

Interpolating Polynomials

The above ideas were all based on using a large number of points and discovering, via differences, that the points all lie on a polynomial of a certain degree. However, problems often arise when we have only a relatively small number of points. For instance, what happens if we have three points, say $A(0,5)$, $B(1,8)$, and $C(2,6)$? If we construct a difference table for these points, we get the table shown above.

x	y	Δy	$\Delta^2 y$
0	5		
		3	
1	8		-5
		-2	
2	6		

Clearly, there is not enough information about $\Delta^2 y$ to conclude that all second differences are constant. Furthermore, if the x-values were not uniformly spaced, we would have even more of a problem.

Let's take a different approach. We know that two points determine a line. To determine the equation of that line, $y = ax + b$, we need to find values for the two unknown coefficients a and b. For instance, if the two points are $P(1,4)$ and $Q(2,1)$, then we can find the equation of the line joining them. Since the slope

$$m = \frac{1 - 4}{2 - 1} = -3$$

and $(1, 4)$ is a point on the line, we obtain the equation

$$y - 4 = -3(x - 1) \qquad \text{or} \qquad y = -3x + 7.$$

Alternatively, we could proceed as follows. We know that the equation of the line will be of the form $y = ax + b$. Since the line passes through the two points $P(1, 4)$ and $Q(2, 1)$, we use the coordinates of the points to construct two equations in the two unknowns:

$$\text{at } P(1, 4): \qquad 4 = a(1) + b$$
$$\text{at } Q(2, 1): \qquad 1 = a(2) + b.$$

When we solve these two linear equations simultaneously, we find that

$$a = -3 \qquad \text{and} \qquad b = 7,$$

which leads to $y = -3x + 7$, the same equation for the line we found above using the point-slope form.

Suppose now that we have three points $P(1, 4)$, $Q(2, 1)$, and $R(5, 10)$ and would like to construct a polynomial that passes through them. This can be done in an unlimited number of ways using polynomials of different degrees. However, we would like to do this using the polynomial of *lowest* degree. It can be shown that, just as two points uniquely determine a line, three *noncollinear* points (points that do not lie on a line) uniquely determine a parabola of the form $y = ax^2 + bx + c$. That is, given any three non-collinear points with different x-coordinates, there is a unique quadratic polynomial whose graph passes through the points (see Figure 4.51). Similarly, four points uniquely determine a cubic curve (provided the points do not lie on a line or on a parabola). In general, any set of $n + 1$ points with different x-coordinates uniquely determine a polynomial of degree n (or lower if they fall on a curve of lower degree). The corresponding polynomial is known as an *interpolating polynomial* because, once its equation is found, we can use it to estimate the value at intermediate points. We illustrate how to find the equation of an interpolating polynomial based on three points in Example 4.

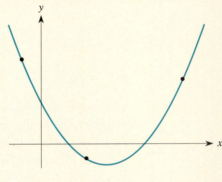

FIGURE 4.51

E X A M P L E   4

Given the three points $P(1, 4)$, $Q(2, 1)$, and $R(5, 10)$, construct an interpolating polynomial of degree 2 and estimate the value of the underlying function when $x = 0$ and when $x = 3$.

Solution Since there are three points, we expect the resulting polynomial to be a quadratic and so we assume that it has the form

$$y = ax^2 + bx + c,$$

where a, b, and c are three coefficients that must be determined. We now use the fact that the resulting parabola must pass through the three given points. Therefore:

At $P(1, 4)$:
$$a(1^2) + b(1) + c = 4$$
$$a + b + c = 4$$

At $Q(2, 1)$:
$$a(2^2) + b(2) + c = 1$$
$$4a + 2b + c = 1$$

At $R(5, 10)$:
$$a(5^2) + b(5) + c = 10$$
$$25a + 5b + c = 10$$

We thus have a system of three simultaneous linear equations in the three unknowns a, b, and c:

$$a + b + c = 4 \qquad (13)$$

$$4a + 2b + c = 1 \qquad (14)$$

$$25a + 5b + c = 10 \qquad (15)$$

If we subtract Equation (13) from Equation (14), we eliminate the variable c to get

$$3a + b = -3. \qquad (16)$$

Similarly, if we subtract Equation (13) from Equation (15), we get

$$24a + 4b = 6. \qquad (17)$$

We now have a system of two linear equations in two unknowns. If we multiply Equation (16) by 4, we get

$$12a + 4b = -12.$$

We subtract this from Equation (17) to eliminate b and so obtain

$$12a = 18$$

$$a = \frac{3}{2}.$$

Substituting into Equation (16), we find that

$$b = -3 - 3\left(\frac{3}{2}\right) = -\frac{15}{2}.$$

Finally, we substitute a and b into Equation (13) to get

$$c = 4 - \frac{3}{2} - \left(-\frac{15}{2}\right) = 10.$$

Therefore, the interpolating polynomial is

$$y = \frac{3}{2}x^2 - \frac{15}{2}x + 10$$

$$= \frac{1}{2}(3x^2 - 15x + 20).$$

Now, if $x = 0$, we find that $y = 10$ is the predicted value of the underlying function. Similarly, if $x = 3$, we find the predicted value is $y = 1$ (see Figure 4.52).

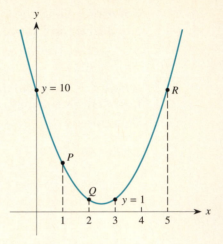

FIGURE 4.52

We note that interpolation polynomials are precisely the techniques used in many calculators to compute the values of *transcendental functions* such as sin x, cos x, tan x, log x, and e^x. Basically, the idea is to incorporate specific, well-chosen values for each of these functions and to construct an appropriate interpolating polynomial (often of degree 6), which is then used to provide very accurate approximations (good to eight or ten decimal places) to the desired function.

However, there are some major disadvantages to using an interpolating polynomial, particularly if you have a large number of data points. We show the graph of the polynomial that passes through five given points in Figure 4.53. Notice that to pass through the points precisely, the curve has some huge swings in order to "come around" as it traverses the polynomial path from one point to the next. In fact, just looking at the five points might suggest a roughly linear trend that could be picked up using our regression or least squares methods from Chapter 3. So, merely passing through a set of points may miss a much simpler trend in a set of data.

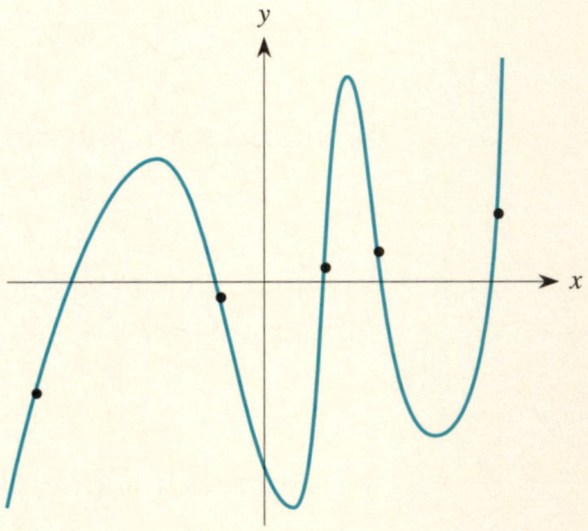

FIGURE 4.53

Finally, let's put all of the ideas from this section into perspective. Suppose you have a dozen measurements or points relating to some quantity. Theoretically, you know that there is a unique polynomial of degree 11 or less whose graph passes through these 12 points and you can construct this interpolating polynomial. But, in the process, the resulting curve will likely miss any simpler trend, if there is one, in the data points. So interpolation methods are likely to be your last choice in this case. In fact, your first move should be to look at the scatterplot of the data to see, visually, if there does appear to be a fairly simple pattern. You can then try different fits to the data using some appropriate families of functions as we did in Chapter 3. If the data points are uniformly spaced, you might also want to construct a table of differences to detect if there is a polynomial pattern, particularly if you suspect that there might be some such relationship, as we found in our study of sums of integers and squares of integers.

EXERCISES

1. Determine which of the following sets of values come from a quadratic function and which come from a cubic function. For those that come from a quadratic, determine the equation of the quadratic.

x	$f(x)$	$g(x)$	$h(x)$	$k(x)$
0	0	1	1	3
1	−2	6	0	1
2	2	13	5	3
3	12	22	22	9
4	28	33	57	19
5	50	46	116	33

2. The following set of measurements were taken on a quantity which follows a cubic pattern. However, one of the values was recorded in error. Find the incorrect entry and correct it. (*Hint:* It is not necessary to actually determine the formula for the cubic.)

x	0	1	2	3	4	5	6	7
y	40	34	24	22	40	90	184	344

3. Consider the array of numbers known as Pascal's Triangle where each row begins and ends with 1 and each intermediate entry is simply the sum of the two numbers diagonally above it.

$$
\begin{array}{ccccccc}
 & & & 1 & 1 & & & \\
 & & 1 & 2 & 1 & & \\
 & & 1 & 3 & 3 & 1 & \\
 & 1 & 4 & 6 & 4 & 1 & \\
 1 & 5 & 10 & 10 & 5 & 1 & \\
1 & 6 & 15 & 20 & 15 & 6 & 1
\end{array}
$$

The rows are numbered $n = 1, 2, \ldots$. The second diagonal consists of the entries $1, 2, 3, 4, 5, 6, \ldots$

a. Find a formula for the terms in the third diagonal: $1, 3, 6, 10, 15, \ldots$ in terms of the row number n.

    b. Find a formula for the terms in the fourth diagonal 1, 4, 10, 20, . . . in terms of the row number n.

4. Construct the quadratic polynomial that passes through the points $P(0, 1)$, $Q(1, 4)$, and $R(2, 9)$. Use it to estimate the value of the underlying function when $x = 0.5$ and when $x = 3$.

5. The main support cable of a suspension bridge is a parabola. For the Golden Gate bridge, suppose the cable's lowest point is 15 feet above the roadway. Use the dimensions shown in the figure to find an equation of the cable for the Golden Gate bridge.

6. Find the sum of the first 25 integers; the sum of the first 100 integers; the sum of the first 1000 integers.

7. Find the sum of the squares of the first 25 integers; the sum of the squares of the first 50 integers.

8. Suppose the produce manager in a supermarket receives a delivery of 1000 large grapefruit which are to be displayed in a pyramid with a square base. How many layers are needed?

9. **a.** Find the sum of the integers from 83 through 225, inclusive.
    b. Find the sum of the squares of these integers.

10. The annual rainfall R, in inches, in a particular region in year t since the start of this century can be modeled by the formula $R(t) = -0.02t^2 + 1.8t + 42$. Find the total rainfall from 1900 (when $t = 0$) through 1990 in that region.

11. Cannonballs are sometimes stacked in rectangular piles. The accompanying figure shows the fourth layer of a stack of n rectangular layers.

    a. Suppose such a stack ends with a single row of two balls at the top. Devise a formula in terms of summation formulas for the number of balls in a stack that is n layers high.

    b. Use the properties of summations to expand the formula you found in part (a).

Fourth layer

    c. Suppose a stack of cannonballs ends with a single row of three balls as the top layer. Devise a formula for the number of balls in a stack that is n layers high.

    d. Use the summation formulas from parts (a) and (c) to predict what the result would be if the top layer consists of a single row of four balls.

12. **a.** Consider the function $y = ax^2$. Construct a table of values of the function for $x = -2, -1, 0, 1, 2, 3$ and extend it to a table of differences until you can construct a formula for $\Delta^2 y$ for this function.

    b. Repeat part (a) for the function $y = ax^3$ to devise a formula for $\Delta^3 y$.

    c. Repeat part (a) for the function $y = ax^4$ to devise a formula for $\Delta^4 y$.

    d. Based on your results in parts (a) through (c), can you predict a formula for $\Delta^5 y$ when $y = ax^5$?

13. Given a sequence of numbers $\{y_0, y_1, y_2, \ldots, y_n, y_{n+1}, y_{n+2}, \ldots\}$, show that:

    a. $\Delta^2 y_0 = y_2 - 2y_1 + y_0$ **b.** $\Delta^2 y_n = y_{n+2} - 2y_{n+1} + y_n$, for any n.

14. Suppose that a set of data values (x_0, y_0), (x_1, y_1), (x_2, y_2), (x_3, y_3), ... have uniformly spaced x-values ($= \Delta x$) and have constant second differences, $\Delta^2 y = k$, so that the points follow a quadratic pattern $y = ax^2 + bx + c$. Use the result of Exercise 12 to show that the leading coefficient is

$$a = \frac{1}{2} \frac{\Delta^2 y}{(\Delta x)^2}.$$

(*Hint:* Write $x_1 = x_0 + \Delta x$ and $x_2 = x_0 + 2\Delta x$ and use the first three points to construct a system of linear equations in a, b, and c.)

15. Determine a formula for the sum of the cubes of the first n integers:

$$1^3 + 2^3 + 3^3 + \cdots + n^3.$$

16. Find the sum of the cubes of the first 25 integers.

17. By writing out

$$\sum_{k=1}^{n} a_k = a_1 + a_2 + \cdots + a_n \qquad \sum_{k=1}^{n} b_k = b_1 + b_2 + \cdots + b_n$$

show that

$$\sum_{k=1}^{n} (a_k + b_k) = \sum_{k=1}^{n} a_k + \sum_{k=1}^{n} b_k$$

and

$$\sum_{k=1}^{n} (m \cdot a_k) = m \cdot \sum_{k=1}^{n} a_k \text{ for any constant } m.$$

18. The points $P(1, 3)$, $Q(2, 5)$, and $R(3, 7)$ are collinear. What happens when you attempt to construct the interpolating polynomial of degree 2 that passes through the three points?

19. Construct the interpolating polynomial of degree 2 that passes through the following points:

x	0	0.3	0.6
y	1	0.955	0.825

These are the values of the cosine function when $x = 0, 0.3$, and 0.6 (in radians). Use the interpolating polynomial to approximate the value of $\cos x$ for $x = 0.10$, $x = 0.20$, $x = 0.25$, $x = 0.40$, and $x = 0.75$. Compare each approximation with the correct value found with your calculator set in radian mode. Which ones would you consider accurate approximations? Which are not?

CHAPTER SUMMARY

In this chapter, you have learned the following:

- How quadratic, cubic, quartic, and higher degree polynomials behave.
- How the real roots of a polynomial equation relate to the linear factors.
- What the real roots of a polynomial equation mean in terms of the graph.
- How the end behavior of a polynomial depends on its leading term.
- How to find the real roots of a polynomial graphically, numerically, and in the case of quadratic equations, algebraically.
- How to fit polynomial functions to sets of data.
- The relative frequency with which complex roots occur.
- What it means to add, subtract, or multiply functions.
- The behavior of rational functions.
- What it means to have a function of a function.
- The effects of shifting and stretching on a function.
- How to find the roots of equations.
- How to interpret the higher order differences of a set of numbers to determine polynomial patterns in sets of data.
- How to find the sum of the first n integers and the sum of the squares of the first n integers.
- The fact that $n + 1$ points typically determine a unique polynomial of degree n.

REVIEW EXERCISES

Sketch the graphs of the following functions without using your function grapher:

1. $f(x) = (x + 3)(x - 2)(x - 4)$ **2.** $g(x) = (2 - x)(x + 3)(x + 1)$

3. $F(x) = (x + 2)(x - 3)(x - 4)(x - 1)$ **4.** $G(x) = (x + 3)(x - 2)(x - 4)^2$

Factor each of the following polynomials to determine their roots algebraically.

5. $P(x) = x^2 + x - 6$ **6.** $Q(x) = 2x^2 + 9x - 5$

7. $R(x) = x^3 - 3x^2 + 2x$

8. Use the quadratic formula to verify your answers to Exercises 5 through 7.

9. Use the bisection method to find the real root, correct to two decimal places, of $x^2 - 12x + 7 = 0$ between $x = 9$ and $x = 12$.

10. Estimate the location of the turning points of the graph of the function $y = x^3 + 4x^2 - 5$.

11. Determine the graphs of the functions f and g and use them to draw the graph of $f + g$.

    a. $f(x) = x^2 - 5$, $g(x) = 3x + 2$

    b. $f(x) = -x^3 + 4$, $g(x) = x^2$

12. Analyze the behavior of each of the following rational functions including identifying all zeros, vertical asymptotes, and end behavior as x approaches ∞ and $-\infty$. Estimate all turning points graphically.

a. $R(x) = \dfrac{x^2 - 4}{x^2 + 9}$ **b.** $Q(x) = \dfrac{x^2 - 4}{x^2 - 9}$

c. $S(x) = \dfrac{x^2 + 4}{x^2 + 9}$ **d.** $T(x) = \dfrac{x^2 + 4}{x^2 - 9}$

13. For each of the following functions, sketch the graph of

   (i) $-f(x)$ **(ii)** $3f(x)$ **(iii)** $f(x) - 4$

   (iv) $f(x - 3)$ **(v)** $f(x + 3)$ **(vi)** $-f(x - 4)$

a. 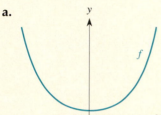 **b.** 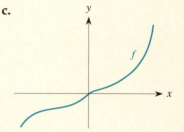 **c.**

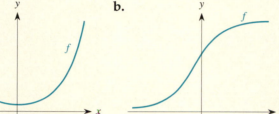

14. Suppose that $f(x) = 2x^2 + 1$ and $g(x) = \dfrac{x - 1}{x + 2}$. Find the following:

a. $f(3) + g(3)$ **b.** $f(f(3))$ **c.** $g(f(3))$ **d.** $g(g(3))$

e. $g(3)f(3)$ **f.** $\dfrac{f(3)}{g(3)}$ **g.** $f(g(x))$ **h.** $f(f(x))$

i. $g(f(x))$ **j.** $g(g(x))$ **k.** $g(x)f(x)$ **l.** $\dfrac{f(x)}{g(x)}$

15. Suppose that $f(0) = 2$, $f(1) = 2$, $f(2) = 3$, $f(3) = 0$ and that $g(0) = 1$, $g(1) = 0$, $g(2) = 2$, $g(3) = 3$. Find the following quantities for $x = 0, 1, 2$, and 3.

a. $f(g(x))$ **b.** $g(f(x))$ **c.** $f(x) + g(x)$ **d.** $\dfrac{f(x)}{g(x)}$

16. Answer the questions in Exercise 15 (a)–(d) for the functions f and g in the graphs below for $x = 1, 2, 3$, and 4.

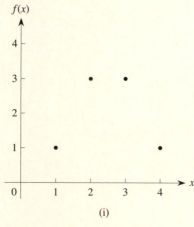

(i)

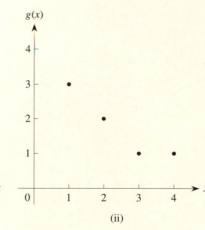

(ii)

17. The return in dollars on an investment seems to be well approximated by the function $F(t) = 2t^2 + t + 4.2$, while the return on another investment is modeled by $G(t) = 7.8t + 3.5$. Determine for which values of $t > 0$ the second investment is better than the first.

18. Evaluate the sum

$$3 + 6 + 9 + 12 + 15 + \cdots + 300.$$

19. The values of y shown for selected values of x indicate that y is a polynomial function of x. Find that function.

x	0	1	2	3	4	5	6
y	5	6	13	26	45	70	101

20. The values of y shown for selected values of x indicate that y is a polynomial function of x. Find that function.

x	0	1	2	3	4	5	6
y	-4	-5	-4	5	28	71	140

5

SEQUENCES AND THEIR APPLICATIONS

5.1 *Introduction to Sequences*

Consider the following table of values first studied in Chapter 3 for the United States population (in millions) from 1780 through 1900:

Decade	0	1	2	3	4	5	6	7	8	9	10	11	12
Year	1780	1790	1800	1810	1820	1830	1840	1850	1860	1870	1880	1890	1900
Population	2.8	3.9	5.3	7.2	9.6	12.9	17.1	23.2	31.4	39.8	50.2	62.9	76.0

These successive population values form a *sequence* of numbers that we write as

$$P = \{2.8, 3.9, 5.3, 7.2, 9.6, 12.9, 17.1, 23.2, 31.4, 39.8, 50.2, 62.9, 76.0\}.$$

Clearly, we could continue this sequence to reflect the population values each decade from 1910 through 1990 and, in the future, this sequence could be continued still further.

The notion of a sequence is one of the most important and useful ideas in mathematics. As you will see, it provides us with a variety of extremely powerful tools for applying some very simple mathematics to an incredibly wide array of problems that arise in many different fields.

Informally, a **sequence** is any ordered set of real numbers. That is, there is a first number a_1, a second a_2, a third a_3, and so forth. These numbers are called the **terms** or **elements** of the sequence. For instance, with the sequence of U.S. population values, the first term is 2.8 million (for 1780), the second term is 3.9 million, and so on. As other examples of sequences, we have

$$A = \{2, 4, 6, 8, 10, \ldots\}$$
$$B = \{2, 4, 8, 16, 32, \ldots\}$$
$$C = \left\{\frac{1}{2}, \frac{2}{3}, \frac{3}{4}, \frac{4}{5}, \ldots\right\}.$$

In general, a sequence can be written as

$$X = \{x_1, x_2, x_3, \ldots, x_n, \ldots\},$$

where x_n represents the *general* or *nth term* of the sequence for any positive integer n. For example, in sequence A above, the 12th term is $a_{12} = 24$; the 60th term is $a_{60} = 120$; and the nth term, for any positive integer n, is $a_n = 2n$. We often write a sequence $X = \{x_1, x_2, x_3, \ldots, x_n, \ldots\}$ in the shorthand notation $X = \{x_n\}$. Thus, we can write

$$A = \{a_n\} = \{2n\}, \qquad \text{or simply } a_n = 2n,$$

for the above sequence A.

Likewise, it is simple to determine the pattern that allows us to express all the terms in the sequences B and C as well. We have for each value of $n = 1, 2, 3, \ldots,$

$$B = \{b_n\} = \{2^n\} \quad \text{and} \quad C = \{c_n\} = \left\{\frac{n}{n+1}\right\}.$$

However, consider the sequence

$$D = \{3, 3.1, 3.14, 3.141, 3.1415, \ldots\}.$$

Although there is no obvious general term for the elements of this sequence, we observe that each successive term is a better and better approximation to the value of $\pi = 3.14159\ldots$.

More formally, we can think of a sequence as being a function defined on a set of integers such as $n = 1, 2, 3, \ldots$. The corresponding range is some set of real numbers determined by the rule for the sequence. In other words, for each positive integer $n = 1, 2, 3, \ldots$, there corresponds a real number x_n as the nth term in the sequence. You can see this in the general terms found for the above examples. For instance, in sequence A, the first term is $a_1 = 2$, the second term is $a_2 = 4$, and so forth.

Furthermore, this interpretation allows us to consider a sequence defined on the nonnegative integers. This lets us start with a 0th term, x_0, for a sequence $X = \{x_0, x_1, x_2, \ldots\}$. This case will arise often when we want to think of x_0 as an initial or starting value for a sequence. It is occasionally necessary to consider sequences defined on sets other than the positive or nonnegative integers. For instance, the sequence whose general term is $x_n = \frac{n}{n-2}$ would be defined for the positive integers $n = 3, 4, 5, \ldots$ in order to avoid the specific value of $n = 2$ where the sequence is not defined.

While we write a sequence as $X = \{x_n\}$ or simply as x_n, you should keep in mind that a sequence is actually a function of n, which could be written as $X = f(n)$.

In the above discussion, we have tacitly assumed that all sequences continue indefinitely; that is, they are **infinite sequences.** This is not necessarily the case. Frequently, we encounter sequences that may have only a finite number of terms and so are called **finite sequences.** For instance, consider the sequence $\{4, 14, 34, 42, 59, 125, \ldots, 205\}$, which is a finite sequence; it represents the stops on one of the subway lines in New York City. In our work here, all sequences will be infinite unless specifically mentioned. Of course, in general, it is only possible to consider a finite number of terms in any sequence since we can handle only a finite number of terms in finite time with finite accuracy. Thus the terms in the sequence

$$D = \{3, 3.1, 3.14, 3.141, 3.1415, \ldots\}$$

converge to π in the sense that each successive term is closer to π than the preceding one. As we have seen, computers have generated several billion digits of π, and theoretically could continue to calculate many billions more. Despite this, the best that can ever be achieved in practice is a finite number of terms; we can never calculate the value of π precisely. Consequently, when working with any sequence in practice, we primarily will be concerned with the first M terms of an infinite sequence, for some positive integer M.

Beyond that, we often try to anticipate what the effects would be if we continued the process to consider the entire infinite sequence. That is, we want to know the general behavior of the entire sequence. If the successive terms of a sequence approach a single fixed, finite value as n approaches ∞, we say that the sequence **converges** or is **convergent.** The value that the terms approach is known as the **limit** of the sequence. For example, the sequence

$$C = \{c_n\} = \left\{\frac{n}{n+1}\right\}$$

converges because the successive terms converge to the value 1 as n becomes large; just substitute larger and larger values for n, say $n = 100$, 1000, 10,000, 100,000, $\ldots$, and see what happens to the quotient. Similarly, the sequence D above converges since the successive terms approach π. On the other hand, if the terms of a sequence do not converge to a single finite limiting value, we say that the sequence **diverges** or is **divergent.** The two sequences

$$A = \{a_n\} = \{2n\} \quad \text{and} \quad B = \{b_n\} = \{2^n\}$$

both diverge because the successive terms approach ∞ as n approaches ∞.

The above examples of convergent sequences were relatively simple since we could recognize the limit of each — the terms in sequence C converge to 1; the terms in D converge to π. Often, though, we encounter cases where the terms of a sequence converge to a number that is not immediately recognizable. Consider the sequence E whose general term is

$$e_n = \left(1 + \frac{1}{n}\right)^n$$

for $n = 1, 2, \ldots$. The first ten terms are

2, 2.25, 2.3704, 2.4414, 2.4883, 2.5216, 2.5465, 2.5658, 2.5812, 2.5937;

they are increasing steadily, but "slowing down." A graph of the points, as shown in Figure 5.1, suggests a concave down pattern. Are the terms of this sequence converging to some finite value? Let's look further. We eventually get e_{100} = 2.7048, e_{500} = 2.7156, e_{1000} = 2.7169, $e_{10,000}$ = 2.7181, $e_{100,000}$ = 2.718268, and $e_{1,000,000}$ = 2.718280. The actual limit is the number 2.718281828459 . . . , denoted universally by the symbol e. (The irrational number e is used as the base for an exponential function, as discussed in Section 2.5).

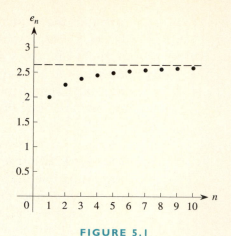

FIGURE 5.1

Finally, consider the following problem that can be traced back to ancient India. A wizard once performed a special service for the maharajah and was asked what he wanted as a reward. The wizard pointed to a chessboard and asked the maharajah to place one grain of wheat on the first square, two on the second, four on the third, and so forth until the 64th square was covered. How many grains of wheat would there be on the 64th square?

Let's consider this problem from the point of view of sequences, where a_n represents the number of grains of wheat on the nth square. Thus we see that

$$a_1 = 1$$
$$a_2 = 2$$
$$a_3 = 4 \qquad (= 2^2)$$
$$a_4 = 8 \qquad (= 2^3)$$
$$a_5 = 16 \qquad (= 2^4)$$
$$a_6 = 32 \qquad (= 2^5)$$
$$a_7 = 64 \qquad (= 2^6),$$

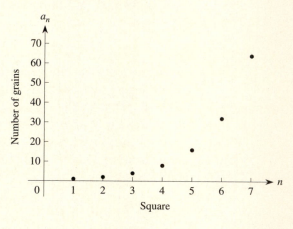

FIGURE 5.2

and so on. In general, each term in this sequence is twice the preceding term. We can visualize the pattern of terms in the sequence by plotting them as shown in Figure 5.2.

If we tried to predict the general term of this sequence for any square on the chessboard numbered between 1 and 64 by simply continuing to place grains of wheat on the squares, we would have trouble. On the other hand, if we consider the pattern given by the terms, then it is clear that the number of grains of wheat on the nth square is $a_n = 2^{n-1}$. For instance, on square number 25, there would be 2^{24} grains of wheat. On the last square, the wizard would find 2^{63} grains of wheat. If we evaluate 2^{63} using a calculator, the result is approximately 9.223372×10^{18}, or 9,223,372,000,000,000,000. (There are 18 digits, including zeros, after the 9.) Thus there are over nine million trillion grains of wheat on the last square!

A related question that may have occurred to many of you is: What is the total number of grains of wheat that the wizard receives if the chessboard is completely covered? It turns out that there is a very simple way to calculate this sum, which we will study in a later section. For now, though, we state that the sum of the terms for this finite sequence is given by

$$1 + 2 + 2^2 + 2^3 + 2^4 + \cdots + 2^{63} = 2^{64} - 1,$$

which is approximately 1.844674×10^{19}. (We suspect that all the wizard received was an all-expenses paid vacation in the palace dungeon when the maharajah finally realized what the request involved.)

EXERCISES

Write the first six terms of each sequence whose general term is given:

1. $x_n = 4n$

2. $x_n = 3n + 5$

3. $x_n = \frac{1}{2}n$

4. $x_n = n^2 + 5$

5. $x_n = n^3 - 10$

6. $x_n = \frac{n^2 + 1}{n^2 + 2}$

7. $a_n = \frac{2^n}{3^n}$

8. $a_n = \frac{n^2}{2^n}$

9. $y_n = \frac{1}{n}, \ n \geq 1$

10. $y_n = \frac{\log n}{n}, \ n \geq 1$

11. $p_n = 1 - (0.2)^n$

12. $p_n = 1 + (0.2)^n$

13–24. Decide which sequences given in Exercises 1–12 seem to converge and which clearly diverge. Give reasons for your decisions. For those that you are not sure about, what could you do to come to a decision?

25–36. Plot the points that you calculated for each of the sequences in Exercises 1–12. Decide which graphs appear to be strictly increasing or strictly decreasing. Decide which graphs are concave up or concave down.

In Exercises 37–42, predict the next two terms in each of the sequences. Then determine an expression for the general term for each sequence. In each case, decide whether the sequence converges or diverges.

37. $\{3, 5, 7, 9, 11, \ldots\}$

38. $\{2, 5, 8, 11, 14, 17, \ldots\}$

39. $\{192, 96, 48, 24, 12, \ldots\}$

40. $\left\{\frac{2}{5}, \frac{4}{25}, \frac{8}{125}, \ldots\right\}$

41. $\left\{\frac{1}{3}, \frac{2}{4}, \frac{3}{5}, \frac{4}{6}, \ldots\right\}$

42. $\{2, 5, 10, 17, 26, 37, 50, 65, 82, \ldots\}$

43. Consider the sequence $e_n = \left(1 + \frac{1}{n}\right)^n$ again. Use your calculator to evaluate e_{1000}, $e_{1,000,000}$, $e_{10,000,000}$, and $e_{100,000,000}$. Keep track of the results. What do you observe about the terms of this sequence? What is your best estimate for the limiting value? Continue the process of taking larger and larger values for n, say up to $n = 10^{15}$. You will find, depending on your calculator, that the terms eventually jump apparently to 1 instead of continuing as you would expect; this is due to round-off errors in the calculator. By trial and error, can you find the value of n where this occurs on your calculator?

44. Consider the sequence $f_n = \left(1 - \frac{1}{n}\right)^n$. What is the limiting value for the sequence as $n \to \infty$? How is this limiting value related to the one in Exercise 43? (*Hint:* There is a simple arithmetic relationship.)

45. Repeat Exercise 44 using the sequence $g_n = \left(1 + \frac{2}{n}\right)^n$. How is the limiting value related to the one in Exercise 43? Based on this result, conjecture what you think the limiting value will be for $\left(1 + \frac{5}{n}\right)^n$ as $n \to \infty$.

46. What is the limit of the sequence $h_n = (1 + n)^{1/n}$ as $n \to \infty$?

47. Suppose the successive terms of a sequence are increasing and the graph drawn through the corresponding points is concave up for all n. Can the sequence converge to a limit? Explain your answer.

5.2 *Applications of Sequences*

Sequences give us a very powerful mathematical tool for modeling a variety of important problems in modern life. In particular, we now consider one special type of sequence that is simply an extension of the problem of the number of wheat grains on a chessboard.

The population of the world was 5.7 billion people in 1995 and was growing at a rate of 1.5% per year, so the growth factor is $k = 1.015$. We assume that this trend continues into the future. Let $P_0 = 5.7$, the initial population. We therefore find in 1996, after one year,

$$P_1 = (1.015)P_0,$$

and after a second year,

$$P_2 = (1.015)P_1,$$

and beyond that,

$$P_3 = (1.015)P_2$$
$$P_4 = (1.015)P_3.$$

In general, if this trend continues, we have after any number of years,

$$P_{n+1} = (1.015)P_n.$$

The world population sequence $P = \{P_n\}$ thus has the property that each term is a constant multiple, in this case 1.015, of the preceding term.

Suppose the sequence $X = \{x_n\}$ has the property that each term is some constant multiple k of the preceding term. That is, $x_2 = kx_1$, $x_3 = kx_2$, and in general,

$$x_{n+1} = kx_n$$

for any value of n. Such a sequence is known as a **geometric sequence** or an **exponential sequence.** Further, suppose that the initial term in this sequence is x_0 corresponding to $n = 0$. Therefore, it follows that

$$x_1 = kx_0$$
$$x_2 = kx_1 = k(kx_0) = k^2x_0$$
$$x_3 = kx_2 = k(k^2x_0) = k^3x_0$$
$$x_4 = kx_3 = k(k^3x_0) = k^4x_0,$$

and so forth. This pattern leads to an obvious formula for the general term of the sequence,

$$x_n = k^nx_0$$

for any $n = 0, 1, \ldots$. For instance, in the case of the world population illustration above, a formula for the population as a function of time n is

$$P_n = P_0(1.015)^n = 5.7(1.015)^n,$$

an exponential function, which accounts for the name exponential sequence.

Recall that whenever the constant multiple k is greater than 1, the values for x_n get successively larger. On the other hand, if k is between 0 and 1, then succeeding values of x_n become successively smaller. Such situations are called, respectively, *growth* and *decay* processes and arise in many different situations. We saw some illustrations of this in Chapter 2.

The equation $x_{n+1} = kx_n$, which relates x_{n+1} to x_n, is an example of a **difference equation;** it is sometimes known as a **recursion equation** or a **discrete dynamical system.** We will encounter many situations involving difference equations throughout this book.

It is important to realize that there are different ways by which the terms of a sequence $\{x_n\}$ can be specified. First, we can have an expression for the general term of the sequence, say $x_n = n2^n$. From this, we can generate any desired term in the sequence. For instance, $x_{10} = 10 \cdot 2^{10} = 10 \times 1024 = 10{,}240$. Second, we can have a difference equation that relates successive terms in the sequence. From the difference equation we also can generate any desired term in the sequence once a starting value is specified. For instance, suppose that $x_{n+1} = 2x_n - 5$ and $x_0 = 9$. Then $x_1 = 2(9) - 5 = 13; x_2 = 2(13) - 5 = 21$; and so forth.

Admittedly there is more work involved in having to calculate each term in the sequence successively starting from the initial term x_0 in order to get to some particular term well along in the sequence. However, as we will see in later sections, many situations can best be modeled by expressing the quantity involved as a difference equation. Once we know the appropriate difference equation and one particular value in the sequence, we can then determine all values for the sequence. Notice that we can do this without ever having a formula for the general term of the sequence.

Before going on, be sure you understand the terminology we are using. An equation such as $x_{n+1} = 1.43x_n$ is a difference equation. Its **solution** is the sequence whose general term is $x_n = (1.43)^n x_0$. Similarly, if the difference equation is $y_{n+1} = 0.36y_n$, then its solution is $y_n = (0.36)^n y_0$.

> The difference equation for exponential growth or decay
>
> $$x_{n+1} = kx_n$$
>
> has as its solution the sequence
>
> $$x_n = k^n x_0,$$
>
> where x_0 is the starting value for the sequence.

We now consider several growth and decay examples that use the above ideas.

E X A M P L E   1

Suppose you deposit $1000 in a bank account paying 5% interest, compounded annually. Write a difference equation to represent the balance in your account after any number of years and find an expression from the difference equation for the balance at any time.

Solution   From our study of exponential growth in Chapter 2, we know that the balance in the account after any number of years is given by $b(t) = 1000(1.05)^t$. We now look at this situation from the point of view of difference equations.

After one year, the original $1000 principal (b_0) has earned 5% of $1000, or $0.05(1000) = \$50$ in interest, so there is a new balance of

$$b_1 = 1000 + 0.05(1000) = (1 + 0.05)1000 = (1.05)1000 = 1050.$$

Thus after one year,

$$b_1 = 1.05 b_0.$$

By the end of the second year, the balance grows to

$$b_2 = b_1 + 5\% \text{ of } b_1 = 1.05 b_1.$$

Similarly, by the end of the third year, the balance is

$$b_3 = b_2 + 5\% \text{ of } b_2 = 1.05 b_2.$$

In general, by the end of the $(n + 1)$st year, for any n, the balance in your account is

$$b_{n+1} = b_n + 5\% \text{ of } b_n = 1.05 b_n,$$

and this is the desired difference equation relating the balance in any given year to the balance in the preceding year.

Now let's find an expression for the balance b_n after any number of years. We have, from the difference equation,

$$b_1 = 1.05b_0.$$

After the second year, the balance is

$$b_2 = 1.05b_1 = 1.05(1.05b_0) = (1.05)^2 b_0.$$

Similarly, after the third year, the balance is

$$b_3 = 1.05b_2 = 1.05[(1.05)^2 b_0] = (1.05)^3 b_0.$$

In general, after n years,

$$b_n = (1.05)^n b_0,$$

which is the solution to the difference equation. Note that we could have obtained this solution directly from the difference equation

$$b_{n+1} = (1.05)b_n,$$

which is the difference equation for a geometric sequence with growth constant $k = 1.05$. Since the initial deposit $b_0 = 1000$, the balance after n years is once again

$$b_n = 1000(1.05)^n.$$

If all we were going to consider is difference equations as simple as the one in this example, there would be no point in introducing them at all. However, we can easily extend this difference equation to model related, but more complicated, scenarios. For instance, suppose you set up a savings plan in which you deposit \$1000 $(= b_0)$ in an account paying 5% per year and then deposit an additional \$400 into that account every succeeding year. Then, instead of the simple difference equation

$$b_{n+1} = 1.05b_n$$

we had previously, we would consider the related, though more complicated, difference equation

$$b_{n+1} = 1.05b_n + 400.$$

We will discuss using difference equations to model more complicated scenarios in the next section. For now, we consider only simple difference equations to familiarize you with the ideas, terminology, and notation.

EXAMPLE 2

During one of New York City's recent financial crises, someone discovered a million dollar loan the city made to the U. S. Government in 1812. At first it appeared that the loan had not been repaid. Using a 6% annual compound interest rate, what would this amount have become by 1996?

Solution   The 6% growth rate corresponds to a growth factor of $k = 1.06$, so after n years, the amount would be

$$b_n = (1.06)^n b_0 = (1.06)^n \$1,000,000.$$

For the year 1996, $n = 1996 - 1812 = 184$, and so the resulting balance would be

$$b_{184} = (1.06)^{184} \times \$1,000,000$$
$$= 45{,}318.88 \times 1{,}000{,}000$$
$$= 45{,}318{,}880{,}000 \text{ dollars,}$$

which would easily have solved the municipal money problem for many years to come (see Figure 5.3). Unfortunately for New York City, the loan was later found to have been repaid, with interest, in 1815.

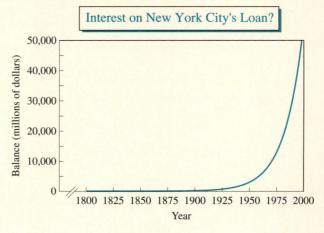

Interest on New York City's Loan?

FIGURE 5.3

Think About This

Assuming the interest rate of Example 2, how much interest did New York City receive in 1815 from this loan?

EXAMPLE 3

The world's population is currently increasing at a rate of about 1.5% per year, and so the population is modeled by the difference equation

$$P_{n+1} = 1.015 P_n.$$

Assuming the world population to be 5.7 billion in 1995, find (a) the population in 15 years; and (b) how long it takes the population to double.

Solution   a.   If we start with an initial population of $P_0 = 5.7$ billion, the solution to the difference equation is

$$P_n = (1.015)^n P_0,$$

which represents the population after n years. So after 15 years,

$$P_{15} = (1.015)^{15} \times 5.7 = 7.126 \text{ billion.}$$

b.   The population will have doubled at the time when

$$P_n = (1.015)^n P_0 = 2 P_0.$$

We first divide both sides of this equation by P_0 and then take logs of both sides to get

$$(1.015)^n = 2$$
$$n \cdot \log (1.015) = \log 2$$

Consequently,
$$n = \frac{\log 2}{\log (1.015)} = 46.555.$$

The world population will double in about every $46\frac{1}{2}$ years at this rate of growth.

E X A M P L E   4

In an effort to reduce the breeding rate of a strain of pesticide-resistant mosquitoes in the southeastern United States, a group of scientists released large numbers of sterilized male mosquitoes each season to mate with the fertile females who would consequently produce no offspring. Suppose the result of this was to lower the mosquito population by 2% per month. (a) What percentage of the population would remain after one year? (b) How long would it take to lower the population by half? (c) In how many years would the population be down to 10% of the original level?

Solution   a.   Let the initial population be P_0 and the population after n months be P_n. We then find

$$P_{n+1} = P_n - 2\% \text{ of } P_n$$
$$= P_n - 0.02P_n = (0.98)P_n.$$

The solution to this difference equation is given by the exponential sequence

$$P_n = (0.98)^n P_0,$$

whose graph is shown with a few dots connected in Figure 5.4. After one year, when $n = 12$ months, we have

$$P_{12} = (0.98)^{12} P_0 = 0.7847P_0,$$

so about 78.5% of the population remains.

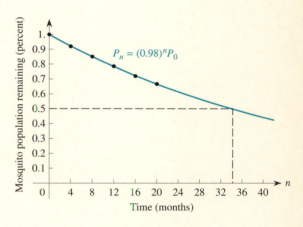

FIGURE 5.4

b.   To find how long it would take for the mosquito population to be reduced to half the original size, we set up the equation

$$P_n = (0.98)^n P_0 = \frac{1}{2} \text{ of } P_0.$$

Equivalently, we need to solve the equation $(0.98)^n = 0.5$ for n by taking logs of both sides:

$$\log (0.98)^n = n \log (0.98) = \log 0.5,$$

and so

$$n = \frac{\log 0.5}{\log (0.98)} = 34.31,$$

or about 34 months.

c. To reduce the population to 10% of the original level, we need

$$(0.98)^n P_0 = \frac{1}{10}P_0 = (0.1)P_0.$$

Consequently, when we divide both sides of the equation by P_0 and then take logs of both sides, we get

$$n \log (0.98) = \log (0.1)$$

$$n = \frac{\log (0.1)}{\log (0.98)} = 113.97 \text{ months} \approx 9.5 \text{ years}.$$

In these examples, we expressed a quantity (such as a population) as a sequence x_n that is defined only discretely for $n = 0, 1, 2, \ldots$. Think of it as $x_n = f(n)$. However, to answer questions about when the quantity being studied reaches a certain level, we have tacitly "connected the dots" with a smooth curve to create a continuous function $y = f(t)$ that is defined not just for the nonnegative integers, but for all the numbers in between as well.

EXERCISES

1. Consider the sequence:

   $$\{ \ \ 2, \ \ 5, \ \ 11, \ \ 23, \ \ 47, \ \ \underline{\quad}, \ \ \underline{\quad}, \ \ \ldots, \ \ \underline{\quad}, \ \ \underline{\quad}, \ \ \ldots \ \ \}.$$
   $$\ \ \ x_0 \ x_1 \ \ x_2 \ \ x_3 \ \ x_4 \ \ \ \ x_5 \ \ \ \ \ x_6 \ \ \ \ \ \ \ \ \ \ \ x_{10} \ \ \ \ x_{11}$$

   a. Determine the terms x_5 and x_6.
   b. If you are told that $x_{10} = 3071$, what is x_{11}?
   c. Suppose you know the value of x_n, for any value of n. Write an equation that you would use to calculate the following term x_{n+1}. (This equation is the difference equation.)
   d. Suppose you are told that the solution to the difference equation for this sequence is $x_n = 3 \cdot 2^n - 1$ for each n. Verify that it gives the correct terms in the sequence when $n = 1, 2, 3, 5, 10,$ and 11.

Find the solutions to the following difference equations for any n and indicate which solutions are convergent and which are divergent sequences.

2. $x_{n+1} = 4x_n, \quad x_0 = 1$ 3. $x_{n+1} = 4x_n, \quad x_0 = 10$

4. $x_{n+1} = 1.5x_n, \quad x_0 = 20$ 5. $z_{n+1} = \frac{1}{2}z_n, \quad z_0 = 64$

6. $y_{n+1} = \frac{3}{4}y_n, \quad y_0 = 64$ 7. $b_{n+1} = 0.62b_n, \quad b_0 = 10$

8. $w_{n+1} = w_n, \quad w_0 = 5$ 9. $q_{n+1} = q_n, \quad q_0 = 6420$

For each of the following difference equations with the given initial condition, decide whether the solution is strictly increasing, strictly decreasing, or neither; and whether its graph is strictly concave up, strictly concave down, or neither.

10. $x_{n+1} = \frac{1}{2}x_n + 100, \quad x_0 = 100$

11. $x_{n+1} = \frac{1}{2}x_n + 100, \quad x_0 = 250$

12. $x_{n+1} = x_n + (-2)^n, \quad x_0 = 10$

13. Find the balance in a bank account in which $100 is deposited at 6% annual interest compounded for 10 years.

14. Find the balance after one year if $100 is deposited at an annual rate of 5% compounded quarterly instead of yearly. What is the balance after 10 years?

15. Assume that the world population increases at 2% per year. Find the percentage increase in population in 25 years and the time needed for the population to double. Do the same for a 3% growth rate.

16. Suppose world food production increases at a rate of 1% per year. What percentage increase will occur in 10 years?

17. Suppose a bacterial culture increases at the rate of 50% per hour. If there are initially 1,000,000 bugs, how many are there after 4 hours? After a full day? How long does it take for the population to increase a hundredfold?

18. According to an article in the *New York Times* on May 27, 1990, a wealthy Pennsylvania merchant named Jacob DeHaven loaned $450,000 to the Continental Congress in 1776 to rescue the troops at Valley Forge. The descendants of Mr. DeHaven took the U.S. government to court to sue for what they believed they were owed. Since the interest rate in effect in 1776 was 6% per year, how much did the family stand to collect in 1991, assuming interest is compounded annually?

19. The lily pads in a pond grow in such a way as to double the area of the pond covered daily.

    a. If the lily pads exactly cover the entire pond on the 25th day, how much of the pond do they cover on the 24th day?

    b. Write a difference equation to model the fraction of the pond covered on any given day and find the solution to this difference equation.

    c. If the area of the pond is 40,000 square feet, find the area covered by the lily pads on the initial day.

    d. What area of the pond is covered by the lilies at the end of one week?

20. Suppose the dose of a drug is 640 milligrams. If the effectiveness decreases at the rate of 25% of the drug per hour, how long would it take to bring down the effective drug level to under 100 mg? To under 5% of the original level?

21. Consider the difference equation $x_{n+1} = \sqrt{x_n^2 + 1}$, with $x_0 = 1$. Find enough terms of the solution sequence so that you are able to write a formula for x_n for any n. How would this expression change if $x_0 = 10$?

22. Use your function grapher to investigate the family of functions $y = x^{1/n}$, $n = 2, 3, 4, \ldots$, for $x > 0$. In words, describe the behavior of the members of the sequence of functions $\{x^{1/n}\} = \{x^{1/2}, x^{1/3}, x^{1/4}, \ldots\}$ as n increases for x near 0, for x moderately large, and as $x \to \infty$. What fundamental law of algebra does this suggest as $n \to \infty$?

23. Pick any two positive integers, say 2 and 5. Add 1 to the second number and divide the result by the first to produce a third number, $(5 + 1)/2 = 3$. Repeat this procedure by adding 1 to the third number and dividing by the second. Keep repeating the process until you notice something interesting. Now pick two other positive integers and perform the same process. What happens? How long does it take to happen?

24. Suppose a and b are any two positive integers. Show algebraically that the process described in Exercise 23, when applied to a and b, always produces a sequence of terms that repeats in blocks of five entries.

25. Write the process described in Exercise 23 as a difference equation in terms of x_n, x_{n+1} and x_{n+2}.

5.3 *Mathematical Models and Difference Equations*

As we have discussed before, a *mathematical model* is a mathematical representation for an object or a process. For example, when a quarterback throws a long pass down the field, the path or trajectory of the football can be represented by a simple mathematical formula involving a quadratic function relating the height y to the horizontal distance x. The formula is a mathematical model for the path of the ball. More generally, a mathematical model for a process is an expression or an equation that represents that process.

Typically, we start with a process that we wish to analyze mathematically. We try to determine some of the underlying principles or assumptions on which the process is based and then attempt to find a relatively simple mathematical relationship that reflects those principles or assumptions. The result may be one or more simple algebraic equations such as the formula for the path of the football; it may be a difference equation such as $P_{n+1} = 1.015 P_n$ that models the growth of the world population, as we saw in Section 5.2; or it may be some other type of equation involving calculus. In this section, we see how some of our previous ideas about sequences can be brought to bear in a study of population growth.

One of the earliest known attempts to develop a mathematical model for a biological process is the work of the Italian mathematician Fibonacci, who lived around 1200 A.D. Fibonacci became interested in the breeding patterns of rabbits and developed a simple model for predicting the local rabbit population. Fibonacci adopted the following assumptions for his model:

1. Newborn rabbits mature in one month.
2. Rabbits have litters monthly once they are mature.
3. Each litter consists of precisely one male and one female.

The first two assumptions are fairly accurate, although you might argue about the third on a variety of grounds. First, rabbit litters tend to be considerably larger than two. However, there is a certain mortality rate for newborn rabbits (they can't run fast enough to escape from the predators in the neighborhood) that lowers the number per litter that survive to maturity. Second, it is unreasonable to expect that precisely one male and one female will survive. However, if we consider a large population of rabbits, the numbers of males and females for the entire population balance out and average about a 50-50 split per litter.

Now, let's apply Fibonacci's assumptions and see what happens. Suppose we start with one pair of newborn rabbits—one male, the other female—on January 1. By February 1, we still have the same pair, but now they have matured and will begin to do what rabbits do best: produce new rabbits. Thus on March 1, we have two pairs of rabbits, the original pair and their first set of offspring. By April 1, the original pair has produced another litter while the previous litter has matured and is ready to enter the family business. At this stage, things start getting complicated, so we use the diagram in Figure 5.5 to keep track of the rabbits. Let the symbol ⟨⟩ denote an immature pair and ⟨⟩ represent a breeding pair.

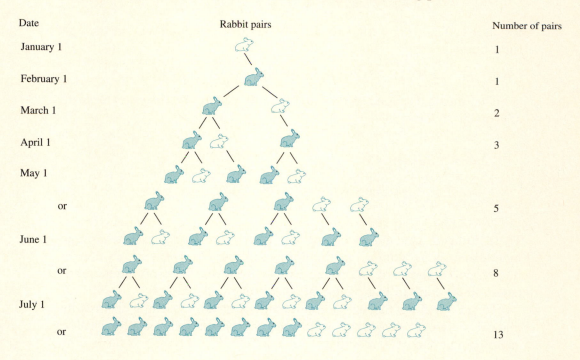

Date	Rabbit pairs	Number of pairs
January 1		1
February 1		1
March 1		2
April 1		3
May 1		
or		5
June 1		
or		8
July 1		
or		13

FIGURE 5.5

If you examine the rabbit population on July 1, you might notice several things. First, the number of mature pairs is 8, which is the total number of pairs the previous month. Also, the number of immature pairs, 5, is equal to the population two months earlier on May 1. Therefore the rabbit population on July 1 is equal to the population on June 1 (mature) plus the population on May 1 (immature). This pattern is not coincidental, and it persists indefinitely. Let's see why. The population in the current month, P_n, consists of breeding pairs and newborn pairs. Every pair alive the previous month, P_{n-1}, is still alive in the current month and in fact is a breeding pair. Every pair alive two months ago, P_{n-2}, was a breeding pair in the previous month, and so produced a new pair in the current month. That is, the number of newborn pairs in the current month equals the number of pairs two months ago. Therefore, for any $n \geq 2$,

$$P_n = P_{n-1} + P_{n-2}.$$

This equation is a difference equation relating the successive population values. Actually, it is desirable to write this equation in a somewhat different form. Instead of considering the current population P_n in terms of the preceding two months' populations, we look at the population two months ahead, P_{n+2}, which is determined by the current population P_n and next month's population, P_{n+1}. As a result, we can rewrite the difference equation in the equivalent form

$$P_{n+2} = P_n + P_{n+1}$$

for all $n \geq 0$ starting with $P_0 = P_1 = 1$. Note that with either form we get the values:

$$1, 1, 2, 3, 5, 8, 13, 21, 34, 55, 89, 144, 233, 377, 610, 987, \ldots.$$

This particular sequence of numbers is called the *Fibonacci sequence* after its discoverer.

> **Fibonacci Sequence**
>
> $$\{1, 1, 2, 3, 5, 8, 13, 21, 34, 55, 89, 144, \ldots\}$$

These numbers arise in a surprising variety of ways—in nature (the arrangement of petals on sunflowers, the number of rings in seashells), economics, human psychology, art, and so on. However, we will keep our attention focused on the constantly growing rabbit population of old Italy.

To get a feel for the size of the rabbit population based on this mathematical model, let's look at a table and graph (Figure 5.6) of these values covering a period of 30 months.

Month	Number of Pairs	Month	Number of Pairs
0	1	16	1,597
1	1	17	2,584
2	2	18	4,181
3	3	19	6,765
4	5	20	10,946
5	8	21	17,711
6	13	22	28,657
7	21	23	46,368
8	34	24	75,025
9	55	25	121,393
10	89	26	196,418
11	144	27	317,811
12	233	28	514,229
13	377	29	832,040
14	610	30	1,346,269
15	987		

Fibonaccis's rabbit model

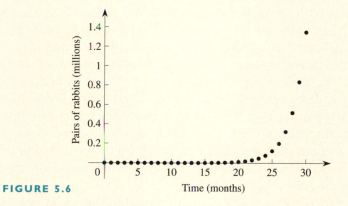

FIGURE 5.6

Both the table and Figure 5.6 show that the population eventually sky-rockets and we soon face a population explosion among the rabbits. Comparing entries in the table we notice that after the first few entries, the ratio of successive terms is approximately 1.618. For example, $233/144 = 1.618$ and $1597/987 = 1.618$. This suggests that, eventually, the growth pattern is roughly exponential with a growth rate of $0.618 = 61.8\%$ per month. These numbers are, if anything, on the conservative side since each litter of rabbits will likely contain more than two. On the other hand, deaths among the rabbits have been ignored. Consequently, the Fibonacci model for the rabbit population may not be a particularly good match to the actual population.

Let's look at a different model for the population. Suppose there are R rabbits present at a given time. The larger this number is, the more new rabbits will be born. Thus the *increase* or *change* in the rabbit population depends on the number of rabbits, R, present. That is, the increase in the number of rabbits is proportional to R. Thus

$$\text{Change in } R = a \times R,$$

where a is a constant of proportionality.

Also, the change in R is just the new value for R minus the old one. That is,

$$R_{new} - R_{old} = aR_{old}$$
$$R_{new} = R_{old} + aR_{old}$$
$$= (1 + a)R_{old}.$$

We can rewrite this relationship in the more familiar form for a difference equation

$$R_{n+1} = (1 + a)R_n.$$

Recognize that this result is a difference equation for exponential growth that we considered in the last section where the coefficient $(1 + a)$ is the growth factor and is equivalent to the constant k.

For populations in general, the growth rate a depends on the species. For humans, the growth rate might be 0.015, so the growth factor could be $k = 1 + a = 1.015$ on an annual basis. We have included a table in Appendix D that provides some fascinating information on human population growth. The table gives information on 1995 population figures for virtually every country and region in the world including birth and death rates, overall population growth rates (the constant a), the doubling time for each population, and projections for the populations in the years 2010 and 2025. We suggest that you examine this table to get a better idea of some of the implications of human population growth based on the exponential growth model.

We now apply the exponential growth model to the rabbit population. We saw with the Fibonacci model that the ratio of successive terms became roughly 1.618 after the first few entries and so the terms grow roughly exponentially thereafter. Let's therefore use this value 1.618 in the exponential model

$$R_{n+1} = 1.618R_n$$

and see how the results compare to those with Fibonacci's model. Suppose we again start with the same number of pairs, namely 1. The results are shown in the following table; after the first few entries, we round to the nearest whole number of pairs of rabbits.

Month	Number of Pairs	Month	Number of Pairs
0	1	19	9,345
1	1.6	20	15,121
2	2.6	21	24,465
3	4.3	22	39,585
4	6.9	23	64,048
5	11	24	103,630
6	18	25	167,673
7	29	26	271,295
8	47	27	438,955
9	76	28	710,229
10	123	29	1,149,151
11	199	30	1,859,326
12	322	⋮	
13	521	36	33,360,045
14	843	⋮	
15	1364	48	10,739,123,314
16	2206	⋮	
17	3570	60	3.45709279E+12
18	5776		$\approx$ 3.5 trillion!!!

Comparing this table of values with the previous table of values for the rabbit population using Fibonacci's assumptions, we see that this population is growing faster than the Fibonacci model. Fibonacci's model leads to roughly exponential growth with a growth factor of approximately 1.618 after about the first 10 terms, so it lags behind the pure exponential growth model at that stage. Since the Fibonacci model starts out more slowly and since both grow thereafter at the same rate of about 61.8% per month, the Fibonacci model will never catch up to the exponential model. We have thus developed two completely different mathematical models, the Fibonacci model and the exponential growth model, to describe the growth of the rabbit population. Each model is based on an entirely different set of assumptions and so gives different results.

Using either of these models, it is clear that Italy would have had a major overpopulation problem with rabbits back in Fibonacci's time, let alone by now. Since this has not happened, it suggests that there is something wrong either with the mathematical model or the assumptions on which it is based. Actually, both mathematical models are fairly accurate, at least up to a point. As long as the rabbit population remains relatively small, these two models give numbers that can reasonably estimate the population. However, neither one should be carried too far since it is literally impossible for any process to continue expanding exponentially forever. Instead, it is necessary to take into account other factors that will act to curb the growing population. For example, as the number of rabbits increases, so too will the number of foxes and other predators that live off of them. In turn, this will serve to hold the rabbit population in check. Also, when the rabbit population grows too large, they quickly consume most of the available food supply and there will be inadequate food to sustain such a large population. The result is starvation until the population decreases to a more sustainable size. We will discuss the mathematical details of this type of scenario in Section 5.5.

E X A M P L E I

In 1994, the population of the United States was 260.8 million with an annual growth rate of 0.7%. At the same time, the population of Mexico was 91.8 million with an annual growth rate of 2.2%. If the growth rates remain constant, when will the population of Mexico overtake that of the United States?

Solution   Using the exponential growth model, the projected population of the United States is given by

$$P_n = 260.8(1.007)^n,$$

and that for Mexico is given by

$$Q_n = 91.8(1.022)^n.$$

We seek the value of n when the two populations will be equal, so

$$260.8(1.007)^n = 91.8(1.022)^n,$$

as illustrated in Figure 5.7.

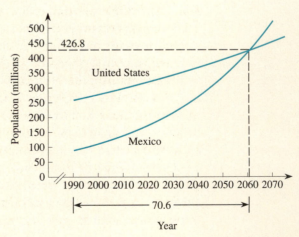

FIGURE 5.7

If we first divide both sides of the equation by 91.8 and then divide both sides by $(1.007)^n$, we obtain

$$2.841 = \frac{(1.022)^n}{(1.007)^n} = (1.01489)^n.$$

We now take logs of both sides to find

$$\log(2.841) = \log(1.01489)^n$$
$$= n \log(1.01489).$$

So

$$n = \frac{\log(2.841)}{\log(1.01489)}$$
$$= 70.645 \text{ years.}$$

Therefore we can expect Mexico's population to overtake that of the United States in about 70.6 years from 1994, or in 2064, assuming both growth rates persist. At that time, both countries would have populations of

$$260.8(1.007)^{70.6} = 91.8(1.022)^{70.6} = 426.8 \text{ million.}$$

You can check this result graphically using your graphing calculator.

We suggest that you experiment with the exponential growth model using either your graphing calculator or an appropriate computer program. Select values for the growth rate a and initial population P_0. Then vary the length of time over which you want to study the population. We suggest that you begin with a time period of relatively small size and then extend it to get a better feel for the shape and the size of the population over different time periods. Also, see what happens if you change the value for the growth rate a or the initial population P_0, or the length of time over which you consider the solution. In particular, observe the effects over time of making even small changes in the growth rate; eventually, even a seemingly minimal change in a will produce very different population numbers.

In addition to the population data given in Appendix D, current values can be found in any information almanac, as well as comparable figures for states and major cities. We encourage you to experiment with these values to see what the population will become in future years if current growth rates continue.

To summarize these notions, recall that a *difference equation* is an equation relating successive terms, x_n, x_{n+1}, and possibly x_{n+2}, of a sequence. The *solution* of a difference equation is a *sequence* (a function of n) that satisfies this difference equation. While the solution can always be expressed in terms of the specific numbers $\{x_0, x_1, x_2, \ldots\}$ in the sequence, it is desirable, whenever possible, to express the solution as a formula for x_n in terms of n. Such a formula is called a **closed form expression** for the general term x_n. Thus, a sequence x_n can be defined in two ways:

1. In closed form with a formula for x_n in terms of n
2. By a difference equation that relates successive terms of the sequence

For instance, the exponential growth or decay model is represented by the difference equation:

$$x_{n+1} = kx_n, \text{ with initial value } x_0.$$

Its solution can be given in closed form by the formula

$$x_n = k^n x_0$$

for all $n = 0, 1, 2, \ldots$ or by the individual terms in the sequence

$$x_n = \{x_0, kx_0, k^2 x_0, k^3 x_0, \ldots\}.$$

Now, suppose we have a sequence

$$\{1, 3, 5, 7, 9, \ldots, x_n, \ldots\}$$

that clearly is not exponential. We notice that each term in the sequence is 2 more than the preceding term and so, for all n,

$$x_{n+1} = x_n + 2, \quad x_0 = 1$$

Thus we have a difference equation that represents the relationship between every pair of successive terms in the sequence.

It turns out that the solution to this difference equation is given by

$$x_n = 2n + 1$$

for all $n \geq 0$. Try it out—substitute $n = 0, 1$, and 2 into the solution and see that you get the first three values for the original sequence. Then, using the expression for $x_n = 2n + 1$, show that x_{n+1} is indeed equal to $x_n + 2$ for any value of n.

- A difference equation is a relationship between successive terms of a sequence.
- The solution to a difference equation is the sequence that satisfies the difference equation; preferably the solution is written as an expression for the general term of the sequence.

A difference equation that relates one term of a sequence x_{n+1} to the preceding term is called a **first order difference equation.** These are the types of difference equations we will study in this and the next chapter. A difference equation such as Fibonacci's that relates one term x_{n+2} to the preceding two terms x_n and x_{n+1} is called a **second order difference equation.** We illustrate additional first order difference equations in the following examples.

EXAMPLE 2

Tara has just bought a treadmill and sets up a daily exercise program. She will use the treadmill for 5 minutes the first day and increase her use by $\frac{1}{4}$ of a minute each successive day. Find a formula for the length of time she uses the treadmill on the nth day for any n.

Solution We could answer this directly by recognizing that the given information represents a linear function with slope $m = \frac{1}{4}$ and constant term 5, so that $T = \left(\frac{1}{4}\right)n + 5$. However, we will view this as
a simple first order difference equation for illustrative purposes.

Let T_n be the number of minutes that Tara uses the treadmill on the nth day and T_{n+1} be the time she uses it on the following day. The corresponding difference equation is

$$T_{n+1} = T_n + \frac{1}{4}.$$

Since she uses the treadmill for 5 minutes initially, we have $T_0 = 5$ and the difference equation tells us that

$$T_1 = T_0 + \frac{1}{4},$$

$$T_2 = T_1 + \frac{1}{4} = T_0 + 2 \cdot \frac{1}{4},$$

$$T_3 = T_2 + \frac{1}{4} = T_0 + 3 \cdot \frac{1}{4},$$

and in general, since $T_0 = 5$, the solution is the sequence

$$T_n = T_0 + n \cdot \frac{1}{4} = \frac{1}{4}n + 5,$$

as before.

In the next example, we model the level of contaminants in a lake under a variety of circumstances to illustrate how various difference equations can arise. We simply quote the formula for the solution in each case (the solutions can be found using methods we will develop later in this chapter and in the following chapter). We then see how the solution can be used to determine the behavior pattern for the level of contamination over time.

EXAMPLE 3(a)

Initially, 500 pounds of a contaminant are dumped into a lake. Each year, 10% of the contaminant is washed away. Find an expression for the amount of contaminant present after any number of years.

Solution   Let C_n represent the amount of contaminant present after n years; we are told that $C_0 = 500$. Since 10% of the contaminant present is washed out of the lake over the course of any year, 90% of the amount present at the start of a year will still be present a year later. Therefore the situation is modeled by the difference equation

$$C_{n+1} = (0.9)C_n, \quad C_0 = 500.$$

Since this is the difference equation for an exponential sequence, its solution is

$$C_n = 500(0.9)^n.$$

This is an exponential decay function and thus tells us that the level of contaminant in the lake will slowly decay toward zero over time.

E X A M P L E   3 (b)

Initially, there are 500 pounds of the contaminant in the lake and 10% of it is washed out each year. However, a manufacturing plant annually dumps 100 pounds of the contaminant into a river that feeds the lake. Find the level of contaminant present after any number of years.

Solution   The amount of contaminant present in the lake is reduced by 10% over the course of a year, so that 90% of the amount present each year remains into the following year. However, this amount is then increased each year by an additional 100 pounds. This situation is modeled by the difference equation

$$C_{n+1} = 0.9C_n + 100, \quad C_0 = 500$$

whose solution, using the formal methods in Chapter 6, turns out to be

$$C_n = 1000 - 500(0.9)^n.$$

As n increases, the exponential term dies out; but, since it is subtracted from 1000, the solution rises toward 1000 as a horizontal asymptote. The lake eventually will contain 1000 pounds of the contaminant. See Figure 5.8.

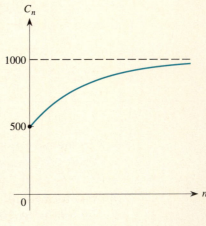

FIGURE 5.8

E X A M P L E   3 (c)

The situation is the same as in Example 3(b), but now the plant increases its production yearly and so increases the amount of the contaminant it dumps into the river by 25 pounds each year starting with the initial level of 100 pounds.

Solution The 25-pound per year increase in the amount of contaminant dumped into the river means that 100 pounds are dumped the initial year, 125 the following year, 150 the year after that, and so on following this pattern of linear growth. Thus during the $(n + 1)$st year, the company will dump $100 + 25n$ pounds of the contaminant into the river and therefore eventually into the lake. As before, 90% of the amount present in the lake at the start of a year remains at the end of that year. It is then augmented by the new amount dumped into the river. The difference equation that models this situation is

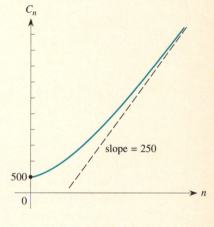

$$C_{n+1} = 0.9C_n + 100 + 25n, \quad C_0 = 500$$

whose solution, using the methods of Chapter 6, turns out to be

$$C_n = 2000(0.9)^n - 1500 + 250n.$$

While the exponential term slowly dies out, the linear term continues to increase as time passes. As a consequence, over the long term, the level of contaminant increases almost linearly with a slope of 250 pounds per year. See Figure 5.9.

FIGURE 5.9

E X A M P L E 3 (d)

The situation is the same as in part (b), but the plant now increases the amount of the contaminant dumped into the river by 20% per year starting with the initial level of 100 pounds.

Solution The yearly amount of contaminant dumped into the river is now given by the exponential function $100(1.20)^n$, so that the difference equation modeling this situation is

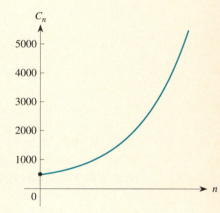

$$C_{n+1} = 0.9C_n + 100(1.20)^n, \quad C_0 = 500$$

whose solution turns out to be

$$C_n = 166.7(0.9)^n + 333.3(1.2)^n.$$

The first term is an exponentially decreasing function which eventually dies out while the second is an exponentially increasing one. Early on, the decay term has some contribution, but because its coefficient (166.7) is smaller than the growth term's coefficient (333.3), the contribution is minimal and rather quickly diminishes. The overall behavior pattern is one of roughly exponential growth in the amount of contaminant (as seen in Figure 5.10). The eventual growth factor is about 1.2 (verify this by calculating a pair of successive terms in the solution for moderately large values of n, say $n = 20$ and $n = 21$), so the annual growth rate is about 20%.

FIGURE 5.10

E X A M P L E   3 (e)

The situation is the same as in part (b) with the plant initially dumping 100 pounds of the contaminant, but EPA regulations requiring that the company reduce the level of dumping by 25% per year.

Solution   The amount dumped into the river is now represented by the exponential decay function $100(0.75)^n$, so that the corresponding difference equation is

$$C_{n+1} = 0.9C_n + 100(0.75)^n, \quad C_0 = 500$$

whose solution turns out to be

$$C_n = 1166.7(0.9)^n - 666.7(0.75)^n.$$

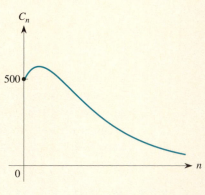

Both terms are exponentially decaying to zero and so the amount of contaminant will eventually die out but rather slowly. However, because the coefficient of $(0.9)^n$ is so large, the values for C_n increase early on before long term decay sets in, as shown in Figure 5.11. Substitute different values of n into the formula for the solution to check that $C_1 = 550$, $C_2 = 570$, $C_3 = 569.3$, $C_4 = 554.5$, $C_5 = 530.7$, $C_{10} = 369.3$, $C_{20} = 139.7$, and $C_{30} = 49.3$.

FIGURE 5.11

E X E R C I S E S

1. Repeat the calculations associated with Fibonacci's rabbit model over the course of the first year assuming an initial population of 10 pairs of newborn rabbits. How do your values relate to the ones shown in the text?

2. Suppose that rabbits of a particular breed take two months to mature, but that Fibonacci's other assumptions still hold. Calculate the rabbit population over the course of the first year, assuming an initial population of one newborn pair.

3. In 1994, the population of India was 911.6 million with an annual growth rate of 1.9%. Predict the population for the year 2000. How long is the doubling time? When will the population of India reach 5 billion if current trends continue?

4. In 1994, the population of Kenya was 27.0 million with an annual growth rate of 3.3%. What will the population be in 2000? When will the population of Kenya reach 100 million if current trends continue?

5. In 1994, the population of Israel was 5.4 million with an annual growth rate of 1.5%. At the same time, the population of Jordan was 4.2 million with an annual growth rate of 3.3%. In what year will the population of Jordan overtake that of Israel if current trends continue?

6. [1]Suppose that the population of a certain species of fish in a lake grows exponentially with a growth rate a. Write a difference equation to model each of the following situations:

   a. The fish population grows exponentially.
   b. Each year, fishermen catch and remove 100 fish from the lake.
   c. Each year, fishermen catch and remove 40% of the fish from the lake.
   d. The number of fish taken from the lake each year is proportional to the square root of the number of fish in the lake.
   e. Each year, fishermen remove 40% of the fish, and the state's wildlife department restocks the lake with 500 fish.

7. Jack and Jill are planning a retirement account that will grow in value over time. Write a difference equation for the balance b_n in this account for each of the following scenarios they are contemplating:

   a. They will deposit $2000 into an account guaranteed to pay 6% interest per year.
   b. They will deposit $2000 initially into an account that pays 6% per year and then deposit an additional $1000 each and every year.
   c. They will deposit $2000 initially into the account and then each year increase their yearly contribution by $1000.
   d. They will deposit $2000 initially into the account and then each year increase their yearly contribution by 10%.

8. Juanita has $80,000 in a retirement fund that pays 6% interest per year.

   a. If she plans to withdraw $10,000 per year to live on, write a difference equation for the balance in her account.
   b. How long will it take for the balance in Juanita's account to be depleted if she withdraws $10,000 every year?
   c. Suppose she plans to withdraw 20% of the account balance every year. Write a difference equation for the balance in the account.
   d. How long will it take for the balance to be depleted with this withdrawal plan?
   e. Is there a fixed amount that Juanita can withdraw from the account every year without reducing the balance? If so, find it.

9. Marine biologists estimate that there were about 8000 bowhead whales in the waters near Alaska in 1992 and that their numbers grow naturally at an annual rate of 3%. Alaskan Eskimos are allowed to catch about 50 whales per year.

   a. Write a difference equation giving the population of the whales from one year to the next.
   b. Calculate the projected whale population each year until the year 2000.

---

[1] Adapted from an example in *Teaching Differential Equations with a Dynamical Systems Viewpoint* by Paul Blanchard, *College Mathematics Journal*, vol 25, 1994, 385–393.

c. Determine the largest number of whales that the Eskimos could catch each year without the whale population going into decline.

d. Suppose you represent the Eskimos in petitioning the government to increase the annual whale harvest. What arguments would you use before the panel making the decision in order to justify increasing the annual harvest?

e. Suppose you represent a conservation group that is opposed to increasing the annual whale harvest. What arguments would you use in requesting a denial of the petition?

10. A company expects that the productivity of new employees should increase each day as they gain experience. When a new person starts "cold," the company expects he or she can produce P_0 items per hour. The following day, hourly production should increase by 1 to $P_0 + 1$ items per hour; the day after that, production should increase by 2 items per hour; then by 3 and so on.

a. Write a difference equation to model this situation.

b. Find an expression for the solution to this difference equation for any number of days n.

c. How sensible does this expectation for continued improvement seem to be?

11. Psychologists have found that, when a person learns a new body of knowledge, the amount of new knowledge that he or she gains in a given time period is proportional to the amount that the person does not know. That is, it is easier to improve when you know a little than it is to improve when you know a lot. Suppose that a student, while preparing for the SAT vocabulary test, is trying to learn 400 new words from a set of flash cards.

a. Write a difference equation for this learning model based on the number of words W_n the student knows out of the 400 total on the nth pass through the deck.

b. Is the constant of proportionality in the difference equation positive or negative? Is it less than 1 or greater than 1?

c. Explain why it is reasonable to expect the student to learn more words during the first few passes through the deck of cards than through later passes through the deck.

d. Based on this learning model, sketch a possible graph of W, the number of new words that the student knows, as a function of n, the number of passes through the deck.

12. Tribbles are adorable, furry little creatures. The only trouble with tribbles is that they breed like tribbles. Specifically, suppose that a tribble matures in three days and then reproduces asexually daily by splitting off a new tribble on the fourth day and every day thereafter.

a. Construct a difference equation for the tribble population if we start with one newborn tribble by expressing T_{n+3} in terms of T_n, T_{n+1}, and T_{n+2}. (*Hint:* Draw a diagram. Let Δ = newborn tribble, $\square$ = day-old immature tribble, and $\bigcirc$ = mature tribble, and keep track of the number of each over the first 10 days.)

    b. Use the difference equation to calculate the tribble population over the course of the first 15 days based on an initial newborn tribble the first day.

    c. Examine the ratio of successive terms to determine if the tribble population appears to be growing exponentially. If it does, what is the exponential growth rate?

    d. Using the result of part (c), what do you estimate the tribble population to be after a full year?

    e. Assuming that the human population of Earth is currently 5.6 billion and growing exponentially at an annual rate of 1.7%, estimate how long it will take until every person has a tribble of his or her own.

13. The population of Retland was 40 million in 1960 and has been growing at an ever increasing rate. Sketch a graph of the population, $P(t)$, as a function of time. Suppose the population was 50 million in 1970 and 64 million in 1980. Use the graph to determine which of the following are possible values for the population and which are impossible:

    a. $P(1975) = 57$ **b.** $P(1990) = 75$ **c.** $P(1990) = 85.$

14. Consider the difference equation $x_{n+1} = x_n + kn$, for any constant multiple k. Show that the solution will always be a quadratic function of n. (*Hint:* Use a result from Section 4.6.)

15. Consider the difference equation $x_{n+1} = x_n + kn^2$, for any constant multiple k. Show that the solution will always be a cubic function of n.

5.4 *Eliminating Drugs from the Body*

Jason is suffering from a severe headache and takes two aspirin tablets to reduce the pain. We know that this may give him about four hours of relief. Eventually, though, the effect of the aspirin will wear off, as it is eliminated from the body. If Jason is lucky, the headache will be gone by then; otherwise, he may have to take another dose.

In this section, we will examine the mathematics involved in modeling how effective the aspirin is over time or equivalently, how the body eliminates a drug or other medication. Let's assume that a person takes a certain dosage of a medication and that it has been completely absorbed into the blood. The kidneys remove the medicine from the bloodstream by filtering out foreign chemicals. It is reasonable to assume that during any fixed time period, say a four-hour period, the kidneys remove a fixed percentage of the medicine from the blood.

Let's suppose that the original dosage of some liquid medicine is 16 mL (milliliters). Let's also suppose the kidneys eliminate one-fourth of this particular medication during any four-hour period. Therefore, during the first four-hour period, the kidneys will eliminate one quarter of the drug, or 4 mL, leaving 12 mL in the blood. What happens during the next four-hour period? The kidneys remove another quarter of the drug. But, the amount of drug remaining after the *first* four hours is 12 mL; therefore, during the *second* four hours, the kidneys eliminate 25% of the 12 mL, or 3 mL. Consequently, after eight hours, the amount of the drug remaining in the body is 9 mL. Then, during the *third* four-hour period, the kidneys eliminate 25% of the remaining 9 mL, or 2.25 mL, leaving 6.75 mL in the body after a total of 12 hours, and so forth.

Notice that, although 4 mL of the drug are removed during the first four-hour period, less is removed each subsequent four-hour period. Consequently, we cannot conclude that all of the drug will be eliminated after four four-hour periods. In fact, only about two-thirds of the original dosage will be removed.

Let's model this situation with a difference equation. Again suppose that the original dosage is $D_0 = 16$ mL and that D_n is the amount of the drug in the body after n four-hour time periods. Thus $D_1 = 12$ mL after one time period, $D_2 = 9$ mL after two time periods, and so forth. In general, the amount of the drug left in the body after n time periods is D_n and during the following time period, 25% of D_n is eliminated to produce D_{n+1}. That is,

$$D_{n+1} = D_n - 25\% \text{ of } D_n$$
$$= D_n - 0.25D_n$$
$$= 0.75D_n.$$

This difference equation allows us to determine the amount of the drug left in the body after any number of four-hour time periods.

Using the above equation for the terms in the sequence representing the remaining drug level, we find that

$$D_1 = 0.75D_0$$
$$D_2 = 0.75D_1 = 0.75(0.75D_0) = (0.75)^2 D_0$$
$$D_3 = 0.75D_2 = 0.75[(0.75)^2 D_0] = (0.75)^3 D_0$$
$$D_4 = 0.75D_3 = (0.75)^4 D_0$$

and in general,

$$D_n = (0.75)^n D_0.$$

In the preceding illustration, $D_0 = 16$ mL, so the solution to this particular problem is

$$D_n = (0.75)^n 16$$

after n four-hour time periods. We show the pattern of successive values of this sequence of drug levels, D_n, in Figure 5.12.

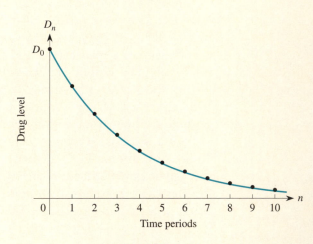

FIGURE 5.12

Notice that the pattern here is one of exponential decay with a decay factor of 0.75 and a decay rate of

$$a = 1 - 0.75 = 0.25.$$

Suppose now that a drug test can detect as little as 1 mL of this drug in the bloodstream. How long after the initial dose of 16 mL will the drug test still be effective? That is, we want to find the value of n for which

$$D_n = (0.75)^n \, 16 = 1,$$

and so

$$(0.75)^n = \frac{1}{16} = 0.0625.$$

To solve for n, we take logarithms of both sides and find

$$n \log (0.75) = \log (0.0625)$$

$$n = \frac{\log (0.0625)}{\log (0.75)} = 9.64.$$

That is, the drug test is effective for $n = 9.64$ four-hour time periods, or about $38\frac{1}{2}$ hours.

EXAMPLE

Suppose we use a more sensitive test for the presence of this drug in the blood, one that can detect as little as 0.0001 mL. How long is the test effective if the original dosage was 16 mL?

Solution Proceeding as we did before, we want to determine the value of n for which

$$D_n = (0.75)^n \, 16 = 0.0001.$$

So

$$(0.75)^n = \frac{0.0001}{16} = 0.00000625.$$

To solve for n, we take logarithms of both sides and find

$$n \log (0.75) = \log (0.00000625)$$

$$n = \frac{\log (0.00000625)}{\log (0.75)} = 41.65$$

four-hour time periods or a total of 166.6 hours, which is about a full week.

Repeated Drug Dosages

A more interesting situation involving the elimination of drugs from the body is based on the fact that people often use a drug on a maintenance basis—they take a fixed dosage of the medication every time period, not just a single dose. It might be a repeated dosage of a cold medication every four hours or a daily dosage of a high blood pressure medication every 24 hours.

Suppose, then, that a person takes a 16 mL dose of the same liquid cold medication every four hours. After the first four-hour time period, 25% of the drug, or 4 mL, is eliminated. This leaves 12 mL and the next dose adds 16 mL to that. Therefore, after the first four hours, the drug level is $D_1 = 12 + 16 = 28$ mL. That's almost double the original dose! What can we say about the level of the drug in the system?

During the second four-hour period, the kidneys eliminate 25% of the drug present, or 7 mL, leaving 21 mL, and the person takes the next dose of 16 mL. Thus after two four-hour periods, the level of the drug in the system is $D_2 = 21 + 16 = 37$ mL. Does this suggest that the level of the drug in the bloodstream keeps rising indefinitely? Is that reasonable?

Before overreacting, let's see what happens if the level of drug in the system does get very high. For argument's sake, suppose that the level reaches 1000 mL after some length of time. During the following four-hour period, the kidneys would remove 25% of the drug, or 250 mL, leaving 750 mL. The person would take the next dose of 16 mL, so that the total amount of the drug would be 766 mL, considerably down from the 1000 mL we hypothesized. Consequently, it seems that if the level rises too high, there are some counteracting effects that tend to reduce the level. Thus the calculations suggest there is no way that the process we have described can lead to an infinite drug level in the body.

Let's model this process with a difference equation. The initial dosage is $D_0 = 16$ mL. During the first four-hour time period, 25% of this amount is removed from the blood and the person takes the next dose of 16 mL. Thus

$$D_1 = D_0 - 0.25D_0 + 16 = 0.75D_0 + 16.$$

Similarly, during the second four-hour period, 25% of the drug is eliminated and the person takes another 16 mL dose, so that

$$D_2 = D_1 - 0.25D_1 + 16 = 0.75D_1 + 16.$$

After the third period,

$$D_3 = D_2 - 0.25D_2 + 16 = 0.75D_2 + 16.$$

In general, after $n + 1$ time periods,

$$D_{n+1} = D_n - 0.25D_n + 16 = 0.75D_n + 16,$$

and this is the difference equation for this drug concentration model.

Notice that this is not a difference equation for an exponential growth or decay model, so we do not have a familiar formula for its solution. We will develop a fairly easy technique for obtaining the solution in Chapter 6 and a more complicated approach in Section 5.7, but for now we will consider the behavior of the terms of the sequence. We start with an initial level of $D_0 = 16$ mL. Using the difference equation, we then have

$D_0 = 16$ mL

$D_1 = 0.75D_0 + 16 = 28$ mL

$D_2 = 0.75D_1 + 16 = 37$ mL

$D_3 = 0.75D_2 + 16 = 43.75$ mL

$D_4 = 0.75D_3 + 16 = 48.81$ mL

$D_5 = 0.75D_4 + 16 = 52.61$ mL

$D_6 = 0.75D_5 + 16 = 55.46$ mL

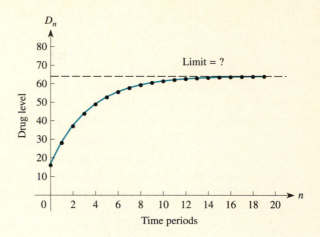

FIGURE 5.13

Notice that each successive term, while larger than the preceding value, has grown by somewhat less than the term before. That is, the change from D_0 to D_1 is a growth of 12 mL; the change from D_1 to D_2 is only 9 mL; the change from D_2 to D_3 is 6.75 mL, and so forth. We plot these points in Figure 5.13. Notice that the curve drawn through the points is concave down and that the successive values seem to be leveling off. That is, the drug level continues to rise, but at a less steep rate (the curve is concave down). The overall pattern is characteristic of an inverted exponential decay process. Rather than gradually dying out toward the horizontal axis as an asymptote, this process gradually rises toward a horizontal asymptote, the limiting amount of drug in the body. If you continue the above process numerically, you will see that the limiting amount L appears to be 64 mL.

In terms of the original problem, this limit L represents the maximum level that the drug will achieve in the body. This is known as the *maintenance level* for that drug. Further, once that level has been reached, the level of drug in the body will return to this maintenance level every four hours after each repeated dose is taken.

We can determine this limiting value precisely using the following argument. Suppose that for some value of n, D_n reaches the limit L so that all successive drug levels are the same. Thus for n large enough, we assume that $D_{n+1} = L$ and that $D_n = L$. Substituting these into the difference equation for the drug model,

$$D_{n+1} = 0.75D_n + 16,$$

we get

$$L = 0.75L + 16$$

$$0.25L = 16,$$

and so the limiting value L is

$$L = \frac{16}{0.25} = 64 \text{ mL}.$$

Incidentally, the curve shown in Figure 5.13 is incorrect; it is based simply on connecting the points to demonstrate the overall pattern for the points. However, it completely ignores what happens during the four-hour intervals that we are considering. Can you sketch a more detailed version that accurately reflects what happens?

In practice, researchers determine that a specific level L of a drug is most appropriate, considering factors of both safety and effectiveness. An initial dose of 16 mL of the medicine means that for some period of time, the amount of drug in the bloodstream is below the optimal level. Because of this, doctors often prescribe an initial dose that is above the normal dose so that the drug level approaches the maintenance level L more rapidly. For example, an initial dose of 48 mL followed by periodic doses of 16 mL every four hours will achieve the desired level very quickly. However, there may be a safety factor involved in taking such a large dosage of the drug, especially as the first dose.

We developed all these ideas in the context of a single drug whose level in the bloodstream decreases at a particular rate. In actuality, different drugs are "washed out" of the blood at different rates. For example, aspirin is removed quite rapidly so that its level of effectiveness is reduced by about 50% every 29 minutes. That is, its half-life is 29 minutes. In the four-hour periods we have considered, over 99% of the aspirin in the blood would be removed by the kidneys.

In the above discussion, we worked with the terms in the sequence $D_n = \{16, 28, 37, 43.75, \ldots\}$, which is the solution of the difference equation

$$D_{n+1} = 0.75D_n + 16,$$

but we have not had a formula for this solution as a function of n. In the next chapter, we will develop a simple technique that allows us to find the solutions to such difference equations. For now, we will simply construct such a solution using our knowledge of the behavior of functions. As we have noticed, the solution to this particular difference equation converges to a limiting value of 64 mL as a horizontal asymptote. The shape of the graph in Figure 5.13 suggests that it might be an upside-down decaying exponential that approaches the limiting value of 64 as n approaches infinity. Therefore, we might expect that a formula for this function would be

$$D_n = 64 - Ba^n,$$

for some constant base $a < 1$ and some constant multiple B. We can determine values for a and B by using two points on the curve. We first use the fact that $D_0 = 16$. Substituting $n = 0$ into the above expression for D_n, we find

$$D_0 = 64 - Ba^0 = 64 - B.$$

Consequently,

$$64 - B = 16$$
$$B = 48.$$

Therefore the formula for D_n becomes

$$D_n = 64 - 48a^n.$$

Next, we use the fact that $D_1 = 28$. Substituting $n = 1$ into the above expression for D_n, we have

$$D_1 = 64 - 48a^1 = 64 - 48a.$$

So

$$64 - 48a = 28,$$

from which we find that

$$a = 0.75,$$

the portion of the medication that remains in the bloodstream after each four-hour period. Thus, our supposed solution for this difference equation is

$$D_n = 64 - 48(0.75)^n.$$

Let's see if this checks out with the values we calculated before. Substituting $n = 2, 3,$ and 4 into the above expression for D_n gives

$$D_2 = 64 - 48(0.75)^2 = 37$$
$$D_3 = 64 - 48(0.75)^3 = 43.75$$
$$D_4 = 64 - 48(0.75)^4 = 48.81$$

and these all match the correct values. Thus it seems that we have the correct formula for D_n; however, in order to be certain, we will come back to this problem later when we have developed a more formal technique for solving such difference equations.

Constructing the Solution in General

Can we extend the above approach to solve the comparable difference equation for any medication with any given periodic dosage? Suppose we have a medication for which the kidneys remove a fixed percentage every time period, leaving a fraction a in the bloodstream. Also, suppose the initial and repeated dosage is an amount B. The corresponding difference equation is

$$D_{n+1} = aD_n + B.$$

For any value of a between 0 and 1 and any positive value for B, the successive terms in the sequence for D_n will have behavior comparable to that shown in Figure 5.9: The solution is an increasing function that is concave down and approaches a horizontal asymptote. If your calculator displays graphs from difference equations, select some typical values for a and B and check out the behavior of the solution.

Since the level of the medication in the blood rises toward the maintenance level L, we can solve for L by realizing that, should this level actually be achieved, then the value of both D_n and D_{n+1} would be L, so that

$$L = aL + B.$$

From this, we find that the maintenance level is

$$L = \frac{B}{1-a}.$$

Furthermore, since the shape of the curve is apparently an inverted decaying exponential, we assume that an expression for the solution is of the form

$$D_n = L - Cb^n$$

for all $n \geq 0$, where b and C are constants. When we substitute our expression for $L = \frac{B}{1-a}$, we get

$$D_n = \frac{B}{1-a} - Cb^n.$$

In particular, when $n = 0$, we have

$$D_0 = \frac{B}{1-a} - Cb^0 = B,$$

and so

$$\frac{B}{1-a} - C = B.$$

Consequently,

$$C = \frac{B}{1-a} - B.$$

If we put the terms on the right side over a common denominator of $(1-a)$ and then factor out the common factor $B/(1-a)$, we find that

$$C = \frac{B}{1-a} - \frac{B(1-a)}{1-a}$$

$$= \frac{B}{1-a}[1 - (1-a)] = \frac{aB}{1-a}.$$

Thus

$$D_n = \frac{B}{1-a} - \frac{aB}{1-a}b^n$$

$$= \frac{B}{1-a}(1 - ab^n).$$

When $n = 1$, this becomes

$$D_1 = \frac{B}{1-a}(1 - ab^1) = \frac{B}{1-a}(1 - ab).$$

Alternatively, directly from the difference equation, we have

$$D_1 = aD_0 + B = aB + B = B(a+1)$$

because $D_0 = B$. If we equate these two expressions for D_1, we have

$$\frac{B}{1-a}(1 - ab) = B(a+1).$$

When we divide out the common factor B (which is not zero), we get

$$\frac{1 - ab}{1 - a} = a + 1.$$

When we multiply both sides by $1 - a$, we get

$$1 - ab = (a + 1)(1 - a) = 1 - a^2,$$

and therefore

$$ab = a^2$$
$$b = a.$$

The formula for the solution is thus

$$D_n = \frac{B}{1 - a}(1 - a \cdot a^n) = B\frac{(1 - a^{n+1})}{1 - a}.$$

To verify that this is indeed a formula for the solution for every value of n, we must show that it satisfies the difference equation. To do so, we substitute into the difference equation both this expression for D_n and the corresponding expression for D_{n+1} when n is replaced by $n + 1$:

$$D_{n+1} = B\frac{(1 - a^{(n+1)+1})}{1 - a} = B\frac{(1 - a^{n+2})}{1 - a}.$$

Rather than using the difference equation in the given form,

$$D_{n+1} = aD_n + B,$$

it is simpler to work with a slight modification,

$$D_{n+1} - aD_n = B.$$

When we substitute the expressions for D_n and D_{n+1} into the left-hand side of this difference equation, we get

$$D_{n+1} - aD_n = B\frac{(1 - a^{n+2})}{1 - a} - aB\frac{(1 - a^{n+1})}{1 - a}$$

$$= B\frac{(1 - a^{n+2} - a + a \cdot a^{n+1})}{1 - a}$$

$$= B\frac{(1 - a^{n+2} - a + a^{n+2})}{1 - a}$$

$$= B\frac{(1 - a)}{1 - a} = B,$$

and we have shown that

$$D_{n+1} - aD_n = B.$$

Thus, this expression for D_n is truly a formula for the solution.
    We summarize these results on the next page.

Level of Medication in the Bloodstream:

Assumptions

- The kidneys remove a fixed proportion, $1 - a$, of a medication from the bloodstream every time period;
- The initial and repeated dosage of this medication every time period is B.

Mathematical Model

- The difference equation is:

$$D_{n+1} = aD_n + B, \quad 0 < a < 1$$

- The solution is:

$$D_n = B\frac{(1 - a^{n+1})}{1 - a}$$

- The maintenance level for the medication is:

$$L = \frac{B}{(1 - a)}$$

Realize that this mathematical model was developed by simplifying the situation considerably. First, when a person takes a medication, it requires a certain amount of time for it to be completely absorbed into the blood, as well as to reach the intended part of the body. Second, the rate at which a particular drug is washed out of the blood depends on many factors, including a person's weight, metabolism, and the state of the kidneys and liver, since they are involved in the elimination process. There are many other factors that come into play and you definitely should not make any medical judgments about the presence of a drug based on the simplified results of this section.

Just as the difference equation $x_{n+1} = ax_n$ can be used to model many phenomena other than population growth, so the difference equation

$$x_{n+1} = ax_n + B,$$

where B is a constant, arises as the model for many phenomena other than drug concentration levels with repeated dosages. For the drug model, we solved the difference equation

$$D_{n+1} = aD_n + B$$

to get the solution

$$D_n = B\frac{(1 - a^{n+1})}{1 - a}.$$

In the same way, the difference equation

$$x_{n+1} = ax_n + B,$$

whatever the situation being modeled, has the solution

$$x_n = B\frac{(1 - a^{n+1})}{1 - a}.$$

We note that the difference equation $x_{n+1} = ax_n + B$ typically models a real situation whenever $0 < a < 1$. In such a case, $a^{n+1} \to 0$ as $n \to \infty$, and so the solution

$$x_n = B\frac{(1 - a^{n+1})}{1 - a} = \frac{B}{1 - a}(1 - a^{n+1})$$

approaches a limiting value

$$L = \frac{B}{1 - a}$$

as a horizontal asymptote as $n \to \infty$. On the other hand, if $a > 1$, then the term a^{n+1} increases as n increases and so the solution grows toward infinity.

We can get a better feel for the behavior of this solution in general by rewriting it slightly. We first write the solution as

$$x_n = L(1 - a^{n+1}) = L - aL(a^n).$$

If we set $n = 0$, we have the initial value

$$x_0 = L(1 - a) = L - aL,$$

so

$$aL = L - x_0.$$

Consequently,

$$x_n = L - (L - x_0)a^n = L + (x_0 - L)a^n.$$

Notice that whenever a is between 0 and 1, a^n approaches zero as n increases, so that the last term dies out and the solution approaches the limiting value of L. See Figure 5.14.

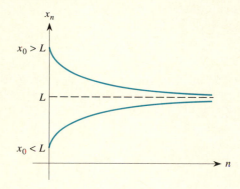

FIGURE 5.14

Moreover, if $x_0 > L$, then the solution x_n is greater than L and so the solution decreases toward L. If $x_0 < L$, then the coefficient of a^n, $(x_0 - L)$, is negative and so $x_n < L$, and the solution rises toward L as n increases.

We summarize these ideas as follows.

The complete solution to the difference equation

$$x_{n+1} = ax_n + B,$$

where a and B are constants is

$$x_n = \frac{B}{1-a}(1 - a^{n+1}).$$

If $0 < a < 1$, the solution can be written as

$$x_n = L + (x_0 - L)a^n,$$

where

$$L = \frac{B}{(1-a)}$$

is the limiting value for the solution as $n \to \infty$.

Realize that this solution applies only for difference equations of this particular form—it will not apply to a difference equation such as

$$x_{n+1} = 1.05x_n + 3n$$

because $3n$ is not a constant.

Finally, notice the form of the solution

$$x_n = L + (x_0 - L)a^n$$

to the general difference equation

$$x_{n+1} = ax_n + B.$$

It consists of two parts: One part is a multiple of a^n and the other is a constant term L. You can think of them as arising, respectively, from the simplified difference equation

$$x_{n+1} = ax_n$$

and from the constant term B in the actual difference equation. This interpretation forms the basis of the more sophisticated solution methods we will develop in Chapter 6.

EXERCISES

1. Suppose that the kidneys remove 30% of a drug from the bloodstream every four hours. If a person takes a single dose of 16 mL, find the

amount of the drug in the body after 12 hours. After 24 hours. How long does it take for the level to drop below 1 mL? Below 0.01 mL?

2. Suppose that the person in Exercise 1 takes repeated doses of 16 mL of the same drug every four hours. What is the drug level after 12 hours? After 24 hours? What is the limiting value for the dosage?

3. In the Example in the text, suppose you assume that D_n, for some unspecified value of n, is considerably larger than the limiting value, say $L = 100$. What does the mathematics predict the successive values will be? Can you explain why this happens? Does it correspond to what would happen in the body after an overdose of a drug?

4. Suppose that the kidneys remove 25% of a drug in the bloodstream every time period and the initial dose is 48 mL followed by 16 mL dosages every time period thereafter. How long will it take until the drug level in the bloodstream exceeds 60 mL?

5. Figure 5.9 ignores the removal by the kidneys of the drug during each four-hour time period. Sketch a more accurate continuous graph of drug level versus time.

6. The following graph shows the level of a medication in the bloodstream just after each repeated dose of 16 mL is taken. Sketch the graph of the drug level of the same medication just before the next dose is taken. How do the two graphs compare? Use the two graphs, drawn on the same set of axes, to construct a graph showing the actual level of the medication in the blood at all times, not only at the times just before or just after the medication is taken.

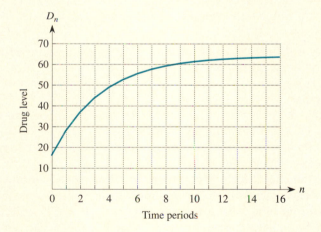

7. The drug dosage for a certain drug is 10 mg per day. If the kidneys remove 60% of the drug every 24 hours, find the maintenance level for the medication.

8. Suppose the daily dosage of the drug in Exercise 7 is halved to 5 mg per day. Is the maintenance level also halved?

9. Suppose the person in Exercises 7 and 8 decides to take 10 mg every second day instead of 5 mg each day. Does she achieve the same maintenance level for the medication? Can you explain why or why not?

10. Two 5-grain aspirin tablets (whose half-life is 29 minutes) contain 650 mg of the drug. How much is left in the bloodstream after two hours? How long does it take until the level of effectiveness is equivalent to 10 mg of aspirin? If an individual takes two tablets every four hours, what is the maintenance level for the aspirin?

11. The maintenance level for a certain drug is 400 mg and 80% of it is washed out of the bloodstream every 24 hours. What daily dosage is necessary to achieve the maintenance level?

12. The half-life of a medication is 16 hours. What daily dosage is necessary to achieve a maintenance level of 250 mg?

13. The maintenance level for a certain drug is 600 mg. A patient starts with an initial dose of 100 mg and repeats it daily. Sketch the graph of the level of the drug in the bloodstream as a function of time. Suppose that $D_5 = 400$ and $D_{10} = 520$. Use the graph to determine which of the following are possible values for the drug level and which are impossible:

   a. $D_7 = 400$ b. $D_7 = 460$ c. $D_{12} = 540$ d. $D_{12} = 560$

14. The daily dosage for a certain medication is 200 mL and the maintenance level is 500 mL. A person taking this medication reaches a level of 450 mL in 10 days. Let r_1 represent the average daily rate of increase of the drug level over the full 10-day period, let r_2 be the average daily rate of increase over the first five-day period, and let r_3 be the average daily rate of increase over the last five days. Without calculating their values, list these three rates in increasing order. (See Exercise 24 of Section 4.1.)

15. Your car has a 14 gallon gas tank that you fill as soon as the level drops to half-full. Also, every time you fill up, you add one quart $\left(\frac{1}{4}\text{ gallon}\right)$ of an additive that mixes thoroughly with the gas and is then used up along with the gas.

   a. Write a difference equation that models the amount of the additive A_n in the tank from one fill-up to the next.
   b. Use the difference equation to calculate the amounts of additive in the tank over the first 10 fill-ups.
   c. Sketch the graph of A_n as a function of n based on the values from part (b). What does the behavior suggest?
   d. Find the limiting value for the amount of the additive in the tank as n increases indefinitely.
   e. Find the closed form solution of the difference equation.
   f. How would the difference equation and the limiting value change if you fill up when the tank is 40% full instead of 50% full?
   g. How would the limiting value change if your gas tank holds 16 gallons instead of 14 and you fill up when the tank is half full?

5.5 *The Logistic or Inhibited Growth Model*

In Sections 5.2 and 5.3, we considered the exponential growth model based on the assumption that the change in the population is

$$\Delta P_n = P_{n+1} - P_n = aP_n.$$

This is equivalent to the difference equation

$$P_{n+1} = (1 + a)P_n = kP_n,$$

where $a > 0$ is the growth rate and $k = 1 + a$ is the growth factor. This model is very effective in predicting the growth of a rabbit or human (or other) population over the short run. However, if such a growth process were to continue indefinitely, you would expect that other factors must come into play to slow down, or *inhibit*, the rate of growth. We now modify the exponential growth model to reflect such a situation. In particular, we seek a process in which population growth starts out in an exponential mode, but eventually slows down and levels off as the population reaches the maximum size that can be sustained by the environment. This type of behavior is illustrated in Figure 5.15.

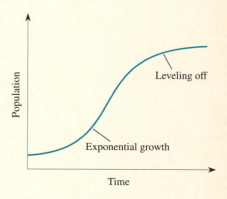

FIGURE 5.15

It is possible to treat this process mathematically by a relatively simple extension of the work we've done on exponential growth. We now introduce an extra term in the equation

$$\text{Change in } P = aP$$

to account for this leveling-off effect, one that serves to decrease the rate of growth in the population. Biologists and other scientists who have studied such processes have observed that this can be modeled by subtracting a term that is proportional to the square of the population. Thus we obtain the expression for the change in P,

$$\Delta P_n = P_{n+1} - P_n = aP_n - bP_n^{\,2},$$

where b is some new positive constant. Alternatively, we can rewrite this difference equation as

$$P_{n+1} = (1 + a)P_n - bP_n^{\,2}.$$

In either form, the resulting process is known as the **logistic growth model** or the **inhibited growth model** and applies to many different species as well as to other types of situations. (Note that if $b = 0$, the logistic model reduces to the exponential growth model.)

To study this model, we must get some feel for appropriate values for the **inhibiting constant** b. You may want to experiment using your graphing calculator (if it allows for difference equations) or appropriate computer software. For example, suppose the exponential growth rate is $a = 1$

for a species such as rabbits that breeds rapidly and the inhibiting constant is $b = 0.04$. Using these values, the resulting graph has the desired shape shown in Figure 5.15. It is called a **logistic curve**. However, if you try $b = 3$ instead, say, then the result will be quite different. (We discuss the resulting type of chaotic behavior in Section 6.11.) In fact, after a little experimentation, you will realize that the model is an effective description of logistic population growth only when b is much smaller than a.

Let's consider the case with $a = 1$, $b = 0.0004$ and an initial population of one pair of rabbits. Using $P_0 = 1$, we get

$$P_1 = (1 + 1)P_0 - 0.0004\,P_0^2 = 2 - 0.0004 \approx 2,$$

so

$$P_2 = 2P_1 - 0.0004\,P_1^2 \approx 2(2) - 0.0004(2)^2 \approx 4,$$

and so forth. When we continue this process, we obtain the following table of results and the accompanying graph in Figure 5.16.

Month	Number of Pairs	Month	Number of Pairs
0	1	9	463
1	2	10	840
2	4	11	1398
3	8	12	2014
4	16	13	2406
5	32	14	2496
6	63	15	2500
7	125	16	2500
8	243		

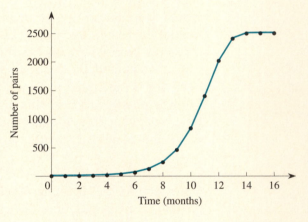

FIGURE 5.16

Notice that the population increases to a maximum of 2500 and seems to remain at that level thereafter. Notice also that the population grows rapidly to about half this maximum during the first 10 or 11 months, and then grows more slowly thereafter until it reaches the 2500 level. This is typical of the logistic model in that there is an initial spurt in the population followed by a slower rate of growth and an eventual leveling off to a constant fixed population known as the **maximum sustainable population** or the **limit to growth**.

The logistic curve has the property that it begins at time $n = 0$ as if it were an exponential growth curve. At first it is concave up, but eventually the rate of growth diminishes, and the curve becomes concave down. Eventually it approaches a horizontal asymptote (the limit to growth or the maximum sustainable population) as n approaches infinity. Thus the population being modeled eventually levels off and approaches a constant value.

Since the logistic curve changes from concave up to concave down, it has a point of inflection. We have seen that one characteristic of such a point is that the function is growing most rapidly or decreasing most rapidly there. Equivalently, the quantity being modeled is likewise growing most rapidly or decreasing most rapidly at a point of inflection. In this instance, the population is increasing most rapidly at this point. To determine roughly where the point of inflection occurs, we can examine the values in the table accompanying Figure 5.16 and observe that the greatest increase in the rabbit population occurs during the eleventh month when the population jumps from 1398 to 2014.

For comparison, let's consider what happens when we change the value for the inhibiting constant b. Suppose $b = 0.000032$. The corresponding results over the first 20 months are shown in the following table:

Month	Number of Pairs	Month	Number of Pairs
0	1	11	1,982.4
1	2	12	3,839.0
2	4	13	7,206.3
3	8	14	12,750
4	16	15	20,299
5	32	16	27,413
6	63.9	17	30,779
7	127.7	18	31,243
8	255.0	19	31,250
9	507.8	20	31,250
10	1,007.4		

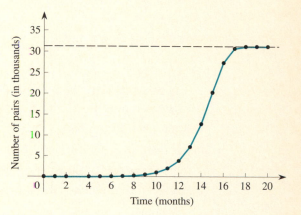

FIGURE 5.17

The resulting curve (Figure 5.17) has the same logistic shape as that in Figure 5.16. In fact, the first few population values are the same as in the previous table because the initial growth rate a is the same and the inhibiting term has minimal impact while the population is small. Moreover, because b is now smaller than before, the population grows at a faster rate for a longer time and so the population here has a higher maximum sustainable level (31,250) than the preceding example. Further, the point of inflection now occurs during the 14th month when the population grows from 12,750 to 20,299.

To understand the behavior of the solution of the logistic difference equation, suppose we choose the constant b much smaller (say by a factor of $1/1000$) than a. As long as the population P_n remains relatively small, the second term, bP_n^2, in the logistic equation

$$\Delta P_n = aP_n - bP_n^2$$

is negligible compared to the first term, aP_n. Therefore the equation is essentially equivalent to the difference equation

$$\Delta P_n = aP_n$$

for exponential growth. However, as P_n grows larger, the term $-bP_n^2$ has an ever greater impact, and its effect is to reduce the value for ΔP_n. So the change in P_n begins to decrease. Of course, the fact that the change decreases does not necessarily mean that P_n itself decreases; it simply does not grow as fast. Further, as we have mentioned, the values for P_n eventually approach a horizontal limit and there is no further growth in the population.

To achieve zero population growth, there should be no change in the value for P_n, so that the difference between successive terms, $P_{n+1} - P_n$, should be zero. Therefore we set the right-hand side of the logistic equation $\Delta P_n = aP_n - bP_n^2$ to zero and obtain

$$aP_n = bP_n^2.$$

Since $P_n \neq 0$, we have

$$P_n = \frac{a}{b} = L.$$

It is this ratio that represents the maximum possible population and is the value, or height, of the horizontal asymptote. In the above two illustrations, we first used $a = 1$ and $b = 0.0004$, so that $L = \frac{a}{b} = 2500$. Then we used $a = 1$ and $b = 0.000032$, so that $L = \frac{a}{b} = 31,250$. Thus under the logistic model for population growth, the population increases from its initial size P_0 until it reaches the limiting value $\frac{a}{b}$ and then stabilizes there forever.

There is a minor catch to this. What if the original population P_0 is larger than the maximum sustainable population? For instance, in the first example above, suppose that the initial rabbit population is 3000 pairs instead of 1 pair. If the phrase "maximum sustainable population" means what it says, then we would expect that there are too many rabbits present and that their population would have to decline due to starvation, predators, and other inhibiting factors. The graph for this situation is shown in Figure 5.18. You may want to explore this situation using a graphing calculator or computer program.

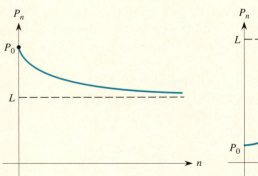
FIGURE 5.18

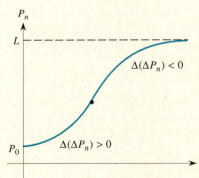
FIGURE 5.19

Can we locate the point of inflection where the logistic curve changes concavity? Since this represents the point where the population is growing most rapidly, we want to find where the change in the population, $\Delta P_n = P_{n+1} - P_n$, is greatest (see Figure 5.19). To the left of this point, the logistic curve is concave up, so not only is the population increasing, but it is increasing at an ever faster rate. To the right of this point, the logistic curve is concave down, so that the population is increasing at an ever slower rate. At the point of inflection, the population is growing most rapidly. Biologically, the most vigorous growth in a population occurs just before decline begins to set in.

We can locate the point of inflection in the following way. Consider the logistic difference equation,

$$\Delta P_n = a P_n - b P_n^2.$$

Although we typically think of P_n as a function of time n, we can alternatively think of this equation as expressing the change in the population, ΔP_n, as a function of the population size, P_n. This is a quadratic function whose graph is shown in Figure 5.20. Notice that the corresponding parabola opens downward, since the coefficient of the second degree term, P_n^2, is negative. Furthermore, the quadratic

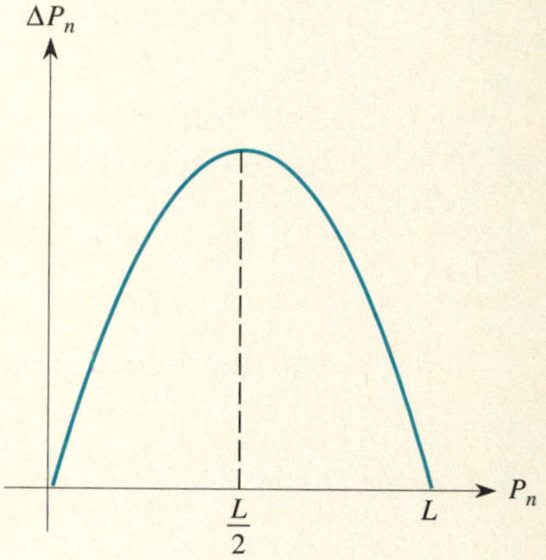

$$a P_n - b P_n^2 = P_n(a - b P_n),$$

FIGURE 5.20

has two real roots, one when $P_n = 0$ and the other when $P_n = L = \frac{a}{b}$, which is the limit to growth. At both of these extremes, the population is not growing and ΔP_n must be zero. Moreover, because a parabola is symmetric, it follows that the maximum value for ΔP_n must occur at the midpoint of this interval, namely when P_n is half of L or $\frac{1}{2}\left(\frac{a}{b}\right)$. This maximum value for ΔP_n corresponds to the point of inflection. In the previous example on rabbits with $L = 31{,}250$, the point of inflection must occur at $\frac{1}{2}L = 15{,}625$. We previously observed that the point of inflection occurred during the fourteenth month when the population was growing most rapidly, jumping from 12,750 to 20,299.

The Logistic Growth Model

The **logistic difference equation** is:

$$\Delta P_n = a P_n - b P_n^2 \qquad \text{or} \qquad P_{n+1} = (1 + a) P_n - b P_n^2,$$

where a is the initial growth rate and b is the inhibiting constant.

The maximum sustainable population is $L = \frac{a}{b}$.

The point of inflection occurs when $P_n = \frac{1}{2}L = \frac{1}{2}\left(\frac{a}{b}\right)$.

We have only looked at one specific application of the logistic model concerned with population growth for a single species. The same mathematical model applies to any other species—only the values of the constants a and b change.

E X A M P L E   I

A bacterial culture grows according to the logistic model with $a = 0.4$ and $b = 0.00008$. If there are initially 500 bacteria in the culture, find the number present for $n = 1, 2, \ldots, 6$ and the limiting population for the culture.

Solution   Since we are told that the bacterial culture satisfies the logistic model, we know that

$$P_{n+1} = (1 + a)P_n - bP_n{}^2$$
$$= 1.4P_n - 0.00008P_n{}^2$$

starting with $P_0 = 500$. Therefore, we find successively that

$$P_1 = 1.4(500) - 0.00008(500)^2 = 700 - 20 = 680,$$
$$P_2 = 1.4(680) - 0.00008(680)^2 = 952 - 37 = 915,$$
$$P_3 = 1.4(915) - 0.00008(915)^2 = 1214,$$
$$P_4 = 1.4(1214) - 0.00008(1214)^2 = 1582,$$
$$P_5 = 1.4(1582) - 0.00008(1582)^2 = 2015,$$
$$P_6 = 1.4(2015) - 0.00008(2015)^2 = 2496,$$

where we have rounded each successive entry to the nearest whole number since we are dealing with the number of bacteria in the culture. Further, the maximum sustainable population of the environment is given by the ratio

$$\frac{a}{b} = \frac{0.4}{0.00008} = 5000.$$

Notice that the first six terms of the sequence for the population grew to almost one-half of the maximum value. This is very close to the point of inflection where the population is growing most rapidly. However, it will take considerably more than another six terms to get as close to the limiting value. For example, $P_{12} = 4680$ and $P_{18} = 4983$.

It is this type of mathematical approach that is the foundation for most of the projections on limits to growth for the world population. Moreover, since the growth factor $1 + a$ incorporates the exponential growth rate a, you might also want to interpret some of these ideas in the context of the values given in the population table in Appendix D. In particular, you might explore the results of using some of the growth rates shown for different countries and assume different values for the inhibiting constant b to see the effects on the limits to growth. Remember, though, that b should be much smaller than a. We will discuss the question of how one estimates values for a and b based on actual population data in Section 6.8.

Let's now examine the possible behavior of the solution to the logistic difference equation in greater detail. Suppose that the maximum sustainable population for an environment is $L = 1000$. The point of inflection then occurs at the time when $P_n = 500$. If the initial population P_0 is less than 500, then the logistic curve will begin growing concave up; when it passes a height of 500, the concavity changes and it continues to grow, but is now concave down as it approaches its horizontal asymptote at the level of 1000 (see Figure 5.21).

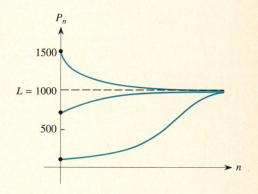

FIGURE 5.21

If the initial population P_0 is between 500 and 1000, then it starts out above the point of inflection. As a result, the logistic curve grows toward the limit to growth, 1000, only in a concave down fashion. Finally, suppose the initial population is greater than the maximum sustainable level of 1000. This can happen if there is a sudden influx of immigrants, an unexpected baby boom, or a change in conditions such as a drought or famine. The resulting logistic curve starts above the limiting value of 1000 and decreases toward 1000 in a concave up manner. Although we typically think of a logistic curve as the S-shaped curve corresponding to initial population values below the inflection point, the other cases we just described are also called logistic curves because they arise as solutions to the logistic difference equation.

The same mathematical ideas used in the logistic model for population growth can be applied in other situations, as Example 2 illustrates.

EXAMPLE 2

Suppose there are 4000 people in a university dormitory complex when one student starts a rumor circulating at 9 A.M. Saturday morning. If each person who hears the rumor passes it on to two other people every hour, how long will it take until everyone has heard it?

Solution At first thought, this problem may seem equivalent to the exponential growth problems we considered earlier based on the difference equation

$$P_{n+1} - P_n = aP_n \qquad \text{or} \qquad P_{n+1} = (1 + a)P_n,$$

which has the solution

$$P_n = (1 + a)^n P_0.$$

If we follow this line of reasoning, we would set P_n equal to 4000 and solve for n. However, this is based on the idea of exponential growth in the number of people who hear the rumor. In practice, this might happen for a while, but soon it will be difficult to find people who have not already heard the rumor. In fact, eventually the people who are passing the rumor on will likely be repeating it to two people who have already heard it and who are themselves trying to pass it on.

We therefore have to change the underlying mathematical formulation to introduce an inhibiting term. We do this in the following way. The number of new people who hear the rumor every hour (that is, the change in the number who have heard it) depends not just on how many people have already heard it, P_n, but also on the number of people left of the original 4000 who have not heard it, $(4000 - P_n)$. That is, ΔP_n, the change in P_n, is proportional to both P_n and $(4000 - P_n)$. This leads to the difference equation

$$P_{n+1} - P_n = mP_n(4000 - P_n) = 4000mP_n - mP_n^2,$$

where m is the constant of proportionality. Adding P_n to both sides of the equation, we get

$$P_{n+1} = (1 + 4000m)P_n - mP_n^2 = (1 + a)P_n - bP_n^2,$$

where we have written $a = 4000m$ and $b = m$ to emphasize that this is a logistic model. Initially, each person tells two new people, so $a = 2$ and $1 + a = 3$. Since the limiting value is $L = 4000$, we find that

$$b = \frac{a}{L} = \frac{2}{4000} = 0.0005.$$

The table below shows the number of people (rounded to the nearest whole person) who have heard the rumor after each hour, based on the logistic model

$$P_{n+1} = 3P_n - 0.0005P_n^2.$$

Time	Logistic	Exponential
0	1	1
1	3	3
2	9	9
3	27	27
4	80	81
5	238	243
6	686	729
7	1823	2,187
8	3807	6,561
9	4174	19,683
10	3811	59,049

The third column gives the corresponding totals using the exponential model $P_{n+1} = 3P_n$ (with $b = 0$) for comparison, although this model is clearly not appropriate. Thus we conclude that the 4000 students will have heard the rumor in about 9 hours.

It is interesting to examine the preceding table to compare the results with the two models. For the first three hours, the two models give identical results. Thereafter, however, the inhibiting term in the logistic equation starts having an effect to limit the growth of the "population." As a result, the prediction from the logistic model begins to lag behind the exponential growth pattern and then falls further and further behind (see Figure 5.22).

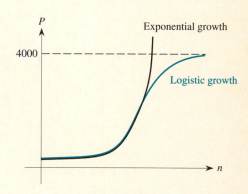

FIGURE 5.22

Notice from the last two terms in the table that the logistic solution eventually oscillates slightly above and below the maximum population of 4000. If we calculate additional terms of the logistic solution, we would observe that they gradually converge to 4000. It turns out that the value we picked for $a = 2$, so that $1 + a = 3$, is a critical one. For any value of a smaller than 2, the process definitely converges from below to the appropriate horizontal asymptote. Whenever a is 2 or larger, the logistic solution has a slight oscillatory character that often is a better match to actual growth processes. We will not go into this aspect of the theory here. You may want to experiment with this yourself using your calculator or an appropriate computer program.

One problem with the logistic model is that no one has ever been able to discover a closed form formula for P_n. However, the fact that no explicit solution is known is not a major handicap. Actually, other than the aesthetic pleasure of having the solution expressed as a formula, the closed form solution would likely not contribute much to this model. The formula would be used primarily to calculate the values for the population at each time period. However, the logistic difference equation itself allows us to calculate the values for the solution recursively using the previous value at each time period.

Moreover, not being tied to a requirement for explicit solutions allows us to consider a much broader spectrum of models. For example, we might want to consider the effects of working with a difference equation of the form

$$\Delta P_n = aP_n - bP_n{}^r,$$

where r is any exponent, not necessarily 2. Such models have applications in certain types of growth processes, although we will not go into them here.

As we have suggested, the logistic growth model applies to a variety of situations. For example, there is a process known as the spread of technological innovations. We can see this in action with various consumer products. For example, when the first vacuum cleaners were introduced, they were purchased by relatively few people. Based on word of mouth and advertising, more and more people bought them and the number in use grew rapidly. Eventually, the market for vacuum cleaners became saturated as virtually every household had one, so that the growth flattened out and new sales are essentially on a replacement basis. The same logistic pattern applies to items such as televisions or stereos. Newer products such as home computers, VCRs, CD players, and cellular phones are now at earlier stages of the logistic growth process.

WalMart reportedly uses this application of logistic growth to maintain its profit levels. It keeps track of the sales of any line of merchandise, say a particular color and style of jeans. Once sales pass the point of inflection on the logistic curve, WalMart discontinues orders for that style and orders a different style instead. In this way, WalMart only stocks and sells the first line of jeans while they are "hot." Consequently few, if any, of these jeans need to be sold on a clearance basis where profits are reduced.

EXAMPLE 3

The first diesel locomotive was put into service in the United States in 1925. In the early years, the use of diesels among the 25 major railroads then in existence grew at about 25% per year. It took about 30 years until all railroads were using diesels. Construct a logistic model for the spread of diesel locomotives.

Solution   Since the early growth rate was 25%, we assume that $a = 0.25$, so that $1 + a = 1.25$. Further, since the limiting value must be $L = 25$ railroads,

$$L = 25 = \frac{a}{b},$$

and so

$$b = \frac{a}{25} = \frac{0.25}{25} = 0.01.$$

Thus, our logistic model is given by the difference equation

$$P_{n+1} = 1.25P_n - 0.01P_n^2.$$

We start with $P_0 = 1$ for the one railroad using a diesel locomotive in 1925 when $n = 0$. We then obtain the set of values shown at the right.

Year	Number of Diesels	
1	1.24	$(= 1.25 \times 1 - 0.01 \times 1^2)$
2	1.53	$(= 1.25 \times 1.24 - 0.01 \times 1.24^2)$
3	1.89	$(= 1.25 \times 1.53 - 0.01 \times 1.53^2)$
4	2.33	
5	2.86	
10	7.29	
⋮	⋮	
15	14.60	
⋮	⋮	
20	21.01	
⋮	⋮	
25	23.88	
⋮	⋮	
30	24.79	

Thus we see that the model agrees with the historical fact that it took about 30 years after 1925 (or until about 1955) for all 25 of the major railroads to have introduced diesels. Incidentally, the other values from the model match up quite well with the actual spread in the use of diesels.

Probably the hardest part of this type of analysis is determining the appropriate values for the growth rate a and the inhibiting constant b. We will address this question in the next chapter when we develop methods of data analysis for estimating the logistic parameters a and b from a set of data.

Most of the cases we have considered with the logistic model are actually somewhat simplistic. For instance, if the values for a and b are taken larger than the ones we have used, then it is possible for the solution to overshoot the limiting level L for some value of n. However, once there, the succeeding term must decrease down toward (or actually even past) the level L. Thus it is possible to obtain a solution that oscillates above and below the level L while converging to it. In many ways, this is a more realistic scenario than the ideal one we portrayed above where the solution strictly increases toward L. A real population is likely to grow beyond its limit to growth, then decrease to below the maximum sustainable level, and then oscillate in this manner thereafter.

EXERCISES

Calculate the first 10 values predicted for the logistic solution subject to each set of values and plot the points. What is the limiting value for the population? How close do you come to it during the first five time periods?

1. $a = 0.02, \quad b = 0.0005, \quad P_0 = 10$
2. $a = 0.02, \quad b = 0.0005, \quad P_0 = 3$
3. $a = 0.02, \quad b = 0.001, \quad P_0 = 5$
4. $a = 0.02, \quad b = 0.002, \quad P_0 = 5$

In Exercises 5–7, calculate the first 10 terms of the Fibonacci-like sequences based on $P_{n+2} = P_{n+1} + P_n$ using the following sets of starting values. Plot the points in each case. Does there seem to be a limiting value or do the values appear to increase indefinitely? Does the growth appear to be approximately exponential?

5. $P_0 = 1, \quad P_1 = 2$ 6. $P_0 = 1, \quad P_1 = 0$ 7. $P_0 = 1, \quad P_1 = 5$

8. For each of the Fibonacci-like sequences in Exercises 5 through 7, calculate the ratios of successive terms $\frac{P_{n+1}}{P_n}$. What do you observe?

9. A population grows according to the logistic growth model $\Delta P_n = 0.05P_n - 0.00002P_n^2$. Sketch the behavior of the population if (a) $P_0 = 500$; (b) $P_0 = 1500$; and (c) $P_0 = 3500$.

10. Consider again the rumor spreading through the dorms in Example 2. Suppose that each person repeats it to three new people each hour instead of to two people. How long will it take until all 4000 students have heard it?

11. Suppose that a certain population grows according to the logistic model from an initial size of 100 to a final size of 1000.

   a. Sketch the graph of the population as a function of time.

   Use the concavity of your graph from part (a) to answer the following questions.

   b. Suppose the population is 250 after 10 time periods and 300 after 12 time periods. Use this information to estimate the size of the population after 11 time periods. Is the actual value higher or lower than your estimate? How do you know?

   c. Suppose the population is 900 after 30 time periods and 910 after 31 time periods. Use this information to estimate the population after 35 time periods. Is the actual value higher or lower than your estimate? How do you know?

12. Suppose that the deer population in a wildlife refuge follows a logistic growth pattern with an annual growth rate of 40% and an inhibiting rate of 0.02%. Write difference equations to model each of the following situations:

   a. The deer population simply grows according to the logistic model.

   b. The population grows according to the logistic model, but hunters are allowed to eliminate 120 deer from the region each year.

   c. The logistic model applies and hunters eliminate 30% of the deer in the region each year.

   d. The logistic model applies and the number of deer that hunters eliminate in the region each year is proportional to the square root of the number of deer living there.

   e. The logistic model applies, hunters eliminate 40% of the deer, and the state's wildlife department moves 75 new deer into the area each year.

13. The growth of a population follows a logistic trend. You observe that the limiting value seems to be about 16,000 and the first few values are 100, 105, 110.2, 115.17, and 121.5. Use this information to estimate both coefficients in the logistic equation and see how close the terms you calculate come to matching these observed values.

14. The values for a and b are typically obtained as estimates based on a set of observations. As such, they are likely to be somewhat inaccurate. It is important to know how sensitive the results of the logistic model are to slight changes (or errors) in either a or b. Suppose we estimate $a = 0.05$ and $b = 0.00002$ so that $L = \frac{a}{b} = 2500$. What would be the effect on L if a were actually 10% larger than 0.05 (0.055 instead of 0.05) while b remains fixed? 20% larger? 30% larger? 10% smaller? 20% smaller? How does the value of L depend on the estimate for a if b remains fixed?

15. Repeat Exercise 14 by considering the effect on L of changes in b if a is fixed. In particular, what would be the effect on L if b were actually 10% larger than 0.00002? 20% larger? 30% larger? 10% smaller? 20% smaller? How does the value of L depend on the estimate for b if a is fixed?

16. Based on your results from Exercises 14 and 15, does the logistic model seem more sensitive to errors or changes in the estimates for a or for b? In estimating values for these two parameters based on a set of data, which do you expect to be found more accurately?

17. The population of a region can be modeled by the logistic growth model with coefficients a and b. A major drought hits the region.

   a. Which coefficient, a or b, is more likely to change because of the drought? Does it become larger or smaller?
   b. Depending on how close the population was to the maximum sustainable population level before the drought, describe the effects of the drought on the long-term behavior of the population.

18. Consider the growth model based on the difference equation

$$\Delta P_n = a P_n - b P_n^3,$$

   where b is smaller than a.

   a. Sketch the graph of ΔP_n as a function of P_n.
   b. Determine the maximum sustainable population L in terms of a and b for a species modeled by this difference equation.
   c. Use your graph from part (a) to determine the sign of ΔP_n if P_n is between 0 and L. What does this tell you about the behavior of P_n?
   d. Use your graph from part (a) to determine the sign of ΔP_n if P_n is greater than L. What does this tell you about the behavior of P_n?
   e. Use your graph from part (a) to show that the solution to this difference equation must have a point of inflection if P_0 is small enough.
   f. Use the fact that the turning points of the general cubic curve $y = Ax^3 + Bx^2 + Cx + D$ occur at

$$x = \frac{-B \pm \sqrt{B^2 - 3AC}}{3A}$$

   to determine the location of the point of inflection for the solution to this difference equation.
   g. You know that the point of inflection for the logistic model occurs at precisely half the height of the limiting value. For the model in this exercise, does this point of inflection occur at a comparable, a higher, or a lower level? What does this tell you about the kind of behavior for a population that can be well-modeled by this difference equation?

19. Repeat parts (a)–(e) of Exercise 18 for the difference equation model

$$\Delta P_n = a P_n - b P_n^4.$$

5.6 *Radioactive Decay*

One of the characteristics of any radioactive substance, such as radium or uranium, is that it transforms or decays over time into some other element,
often lead. This decay is accompanied by the release
of energy. It is this energy that we detect and call radiation. More specifically, the rate at which an element
decays is distinctive for that element. That is, during
any fixed length of time, the same percentage of the
mass of any given radioactive element will decay. For
example, suppose a radioactive substance loses 20%
of its mass in a year. Then, in the course of *any* one-
year period, this substance will lose 20% of whatever
mass it had at the start of that period. A graph of such
behavior is shown in Figure 5.23.

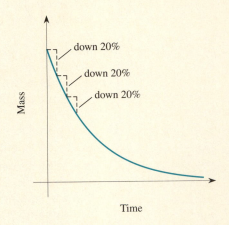

FIGURE 5.23

As a particular case, consider radium. Over the course of a century,
approximately 4.3% of any radium present will decay into lead, leaving
95.7% of the radium at the end of 100 years. Thus if someone had put aside
100 grams of radium in the year 1900, we would expect to find only 95.7
grams by the year 2000.

Suppose there was an initial amount of radium, R_0, at some particular
time in the past. After one century, the amount of radium left was

$$R_1 = R_0 - 4.3\% \text{ of } R_0$$
$$= 0.957R_0.$$

By the end of a second century, the amount of radium left was

$$R_2 = R_1 - 4.3\% \text{ of } R_1$$
$$= 0.957R_1 = (0.957)^2 R_0,$$

and by the end of a third century,

$$R_3 = R_2 - 4.3\% \text{ of } R_2$$
$$= 0.957R_2 = (0.957)^3 R_0.$$

In general, the amount of radium present after $n + 1$ centuries is modeled
by the difference equation

$$R_{n+1} = 0.957R_n,$$

whose closed form solution is

$$R_n = (0.957)^n R_0$$

for any n. This formula tells us the amount of radium remaining after n centuries from an initial amount R_0. For instance, suppose $R_0 = 100$ grams. We
display the values for the amount of radium left after n centuries on the
next page.

n	R_n
0	100
1	95.7
2	91.59
3	87.65
4	83.88
5	80.27
⋮	⋮
10	64.44
⋮	⋮
15	51.72
⋮	⋮
20	41.52
⋮	⋮
25	33.33

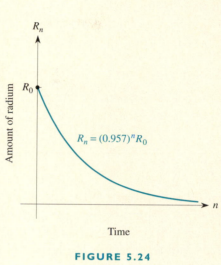

FIGURE 5.24

Notice that during the first five centuries, the amount of radium decreases by almost 20 grams; during the next five centuries, it drops by about 16 grams; during the following five centuries, it drops by only about 13 grams, and so on. In Figure 5.24, we show a graph of the amount of radium as a function of time. Since the amount of radium is given by the exponential decay function $R_n = (0.957)^n R_0$, the values begin by decreasing relatively rapidly, but the decrease slows down. The values eventually approach the time-axis as a horizontal asymptote.

When we considered exponential decay phenomena, one of the questions that we addressed was that of half-life: How long does it take for the quantity to diminish by half? Let's find the half-life for radium based on the solution

$$R_n = (0.957)^n R_0$$

of the difference equation. We wish to determine the value of n for which

$$R_n = (0.957)^n R_0 = \frac{1}{2} R_0.$$

We first divide both sides of this equation by R_0 and then take logs to obtain

$$n \log (0.957) = \log \frac{1}{2},$$

from which we find that

$$n = \frac{\log \frac{1}{2}}{\log (0.957)} \approx 15.77 \text{ centuries.}$$

That is, the half-life for radium is approximately 1577 years. (The actual value for its half-life is closer to 1590 years; our calculations were based on the fact that *approximately* 4.3% decays into lead each century. Notice how the results of rounding can be magnified over a long time period.)

Think About This

What is the actual percentage of radium that decays into lead each year based on the half-life of 1590 years?

EXAMPLE 1

Some scientists put the age of the moon at 3.5 billion years based on radioisotope dating of rock samples brought back by the Apollo astronauts. If the half-life of uranium 238 (U^{238}) is 4.5 billion years, what percentage of the original amount of U^{238} on the moon is still present?

Solution   Since the half-life is 4.5 billion years and the moon is only 3.5 billion years old, more than half of the original amount of the uranium must still be present. Draw a rough sketch to see if you can estimate the amount more accurately.

The amount of U^{238} present can be modeled by the difference equation

$$R_{n+1} = (1 - a)R_n,$$

where a is the decay rate. The solution to this difference equation for exponential decay is

$$R_n = (1 - a)^n R_0.$$

Moreover, since the half-life of U^{238} is 4.5 billion years, we use the fact that

$$(1 - a)^{4.5} R_0 = \frac{1}{2} R_0$$

or, when we divide through by R_0,

$$(1 - a)^{4.5} = (1 - a)^{9/2} = \frac{1}{2}.$$

To solve for the unknown base, $1 - a$, we raise both sides to the $\frac{2}{9}$ power to get

$$1 - a = \left(\frac{1}{2}\right)^{2/9} = 0.857.$$

Consequently, the amount of U^{238} present after n billion years is

$$R_n = (0.857)^n R_0.$$

After the moon's 3.5 billion years, we find the amount of U^{238} remaining is

$$(0.857)^{3.5} R_0 = 0.583 R_0,$$

and so about 58% of the original U^{238} would still be present.

Notice that we did not need to find the value of the decay rate a. Rather, we found the decay factor $1 - a = 0.857$ and this was enough to answer the question posed.

One of the most useful applications of radioactive decay is the carbon-dating process to establish the age of fossils. It is based on the fact that carbon 14 decays with a half-life of 5730 years into carbon 12.

E X A M P L E 2

Crater Lake in Oregon was formed as the result of a volcanic eruption. A charcoal sample from a tree that burned during the eruption contains about 46% of the C^{14} found in live trees. What was the approximate date for the formation of Crater Lake?

Solution   Since the half-life of C^{14} is 5730 years and slightly more than 50% of the radioactive carbon has disintegrated, we should expect that the time involved is somewhat more than 5730 years; we might guess something on the order of 6000 years. Now let's find out more precisely. The difference equation for radioactive decay is

$$R_{n+1} = (1 - a)R_n,$$

whose solution is

$$R_n = (1 - a)^n R_0$$

for some decay factor $(1 - a)$. Since the half-life of C^{14} is 5730 years, we have

$$R_{5730} = (1 - a)^{5730} R_0 = \frac{1}{2}R_0,$$

and so

$$(1 - a)^{5730} = \frac{1}{2}$$

$$1 - a = \left(\frac{1}{2}\right)^{1/5730} = 0.99988.$$

Therefore,

$$R_n = (0.99988)^n R_0.$$

We now have to find how long it takes for the original amount of C^{14} to decay to the point where only 46% of it is present. Thus we want to find n when

$$R_n = (0.99988)^n R_0 = 0.46 R_0.$$

If we divide through by R_0 and then take logs of both sides, we find

$$n \log (0.99988) = \log (0.46)$$

$$n = 6471.$$

We therefore conclude that Crater Lake was formed almost 6500 years ago, or around the year 4500 B.C.

(*Note:* The results in this example are extremely sensitive to rounding. If you were to perform the calculations using either fewer or more decimal places, you would likely get considerably different answers. For example, until about ten years ago, scientists believed that the half-life of carbon 14 was 5570 years, not 5730. We suggest that you repeat this example using the older value of 5570 to see the effect. In fact, repeat the example using $\left(\frac{1}{2}\right)^{1/5730} = 0.999879$, which is slightly more accurate than the value 0.99988 that we used, and see how different a value you get for the age of Crater Lake.)

Our discussions on radioactive decay are all based on what happens "in the large." Thus over the course of about 1590 years, 50% of any mass of radium will have decayed. However, the process takes place one atom at a time. That is, at any given instant, there is a certain chance that any given atom of radium will decay. It may happen, or it may not. There is no way of predicting whether an individual atom will actually decay at any given time or, for that matter, during any reasonable time interval. The actual process is a totally random one. It is only when there are a large number of such atoms present, and some of them decay while others do not, that we can "see" the pattern of exponential decay take place.

A random process such as radioactive decay is ideal for experimenting with a computer program that allows you to select the half-life for a radioactive substance and "watch" the atoms decay. Figure 5.25 shows the results of using such a program.

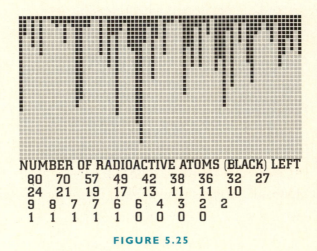

NUMBER OF RADIOACTIVE ATOMS (BLACK) LEFT

80	70	57	49	42	38	36	32	27
24	21	19	17	13	11	11	10	
9	8	7	7	6	6	4	3	2	2
1	1	1	1	1	0	0	0	0

FIGURE 5.25

The program starts with 80 radioactive "atoms" in the row across the top and simulates the random process of radioactive decay based on the half-life selected. That is, for each interval of time, the computer plays a game of chance to decide whether each remaining radioactive atom (darker box) decays into a nonradioactive atom (lighter box). Each atom has the same chance of decaying in a given period of time; some will decay and others will not. If a particular atom's "turn" has come up, randomly, it decays. Each row in Figure 5.25 shows the set of outcomes in one time period. The "surviving" atoms in each row are counted, and the counts are listed at the bottom. Notice the pattern in the list. It suggests that the number of radioactive atoms remaining decays exponentially. You can also see this by plotting the number of radioactive atoms remaining after each time period as a function of time.

When you run such a program with the same half-life repeatedly, you will see that different combinations of atoms decay, randomly, at each stage. While we cannot predict what happens with any individual atom, we can observe and even predict the overall pattern. In the large, the overall pattern for many different atoms is the same — it is one of exponential decay.

EXERCISES

1. One of the major concerns about above-ground nuclear testing is that it produces strontium 90 (Sr^{90}), a radioactive element whose half-life is 29 years and which has worked its way into our food chain. That is, Sr^{90} from fallout is deposited on grass, eaten by cows, carried into their milk, and eventually finds its way onto the kitchen table. Suppose that, as a result of a single nuclear explosion, the amount of Sr^{90} in a particular valley has exceeded health limits by a factor of 10. How long will it take until the Sr^{90} decays to the safety level?

2. The famous Cro-Magnon cave paintings are found in the Lascaux Cave in France. If the level of carbon 14 radioactivity in charcoal in the cave is down to approximately 14% of that of living wood, estimate the date at which the paintings were made.

3. The level of carbon 14 in a charred roof beam found in a 1950 excavation of an ancient Babylonian city is about 61% of the level in living wood. Estimate when the fire occurred.

4. The well-preserved body of an ice-age man was found in melting snow in the Northern Italian Alps in 1991. An examination of a tissue sample from the body indicated that 47% of the carbon 14 present in the body at the time of death had decayed. When did this man die?

5. Several groups of scientists were allowed to test the Shroud of Turin, the supposed burial cloth of Jesus, in 1991. They found that the cloth contained 91% of the amount of carbon 14 contained in newly-made cloth of the same material. Based on this information, how old is the Shroud of Turin?

6. Suppose that the total amount of a certain radioactive isotope now present on Earth is 100 pounds. If the half-life of the isotope is 25 million years and the age of Earth is 4.6 billion years, estimate the amount of the substance that was present originally.

7. Suppose that the total amount of radium now present on Earth is 200 pounds. Estimate the amount of radium that was present 50 million years ago during the age of the dinosaurs.

8. The age of the universe is believed to be about 12 billion years. What percentage of the original amount of U^{238} is still present?

9. Suppose that a scientist has some initial amount R_0 of a radioactive substance whose half-life is measured on a scale of days.

   a. Sketch the graph of the amount of this substance present as a function of time.

   Use the concavity of your graph from part (a) to answer the following questions.

   b. Suppose that you measure the amount of the substance after 10 days and find that there are 800 grams left and that, after 11 days, there are 750 grams left. Use this information to estimate the number of grams that will remain after 20 days. Is the actual value higher or lower than your estimate? How do you know?

   c. Suppose that you are told that the amount of the substance present after 30 days is 400 grams. Use this information and the amount left after 10 days to estimate the amount that was present after 20 days. Is the actual value higher or lower than your estimate? How do you know?

   d. How might you use the results from (b) and (c) to come up with a better estimate of the amount of radioactive material present after 20 days?

10. A certain radioactive isotope has a half-life of 20 days. Suppose 800 mg are present initially and consider a 60-day time period. Let r_1 represent the average daily rate of decrease of the isotope over the full 60-day period, let r_2 be the average daily rate of decrease over the first 30-day period, and let r_3 be the average daily rate of decrease over the last 30 days. List these three rates in increasing order without calculating their values.

11. The simulation used to produce Figure 5.25 is based on a half-life of 5 years.

   a. Find the formula for the amount of this substance present at any time based on 80 initial units of it.

   b. Find the exponential function that best fits the data shown in Figure 5.25. How close does it come to matching the predicted formula you found in part (a)?

5.7 *Geometric Sequences and Their Sums*

Consider the difference equation for exponential growth

$$x_{n+1} = rx_n$$

whose solution is given by

$$x_n = Cr^n,$$

where C is any arbitrary constant. Using the notation for sequences, we can write this solution as

$$\{C, Cr, Cr^2, Cr^3, Cr^4, \ldots, Cr^n, \ldots\},$$

where the initial term corresponding to $n = 0$ is $x_0 = Cr^0 = C$. For simplicity, suppose that $C = 1$, so that this geometric sequence reduces to

$$\{1, r, r^2, r^3, r^4, \ldots, r^n, \ldots\}.$$

For instance, if $r = 5$, we have the geometric sequence

$$\{1, 5, 5^2, 5^3, 5^4, \ldots, 5^n, \ldots\} = \{1, 5, 25, 125, 625, \ldots, 5^n, \ldots\}.$$

From the difference equation $x_{n+1} = rx_n$, we see that each term in any geometric sequence is a *constant multiple* of the preceding term or equivalently, there is a common ratio r between successive terms. Alternatively, x_n is an exponential function of n and we know that the ratio of successive values is a constant, namely the growth or decay factor, r. In the above illustration, each term is five times the preceding one, so that the common ratio r of each pair of successive terms is 5.

We can deduce a considerable amount of information regarding the behavior of the terms in a geometric sequence $\{1, r, r^2, r^3, r^4, \ldots, r^n, \ldots\}$ just from the value of the common ratio r. We summarize these facts as follows.

1. If r is larger than 1, then the terms are successively larger and approach infinity. (We say they *increase monotonically*.)
2. If $r = 1$, then the terms are all equal to 1 for all n.
3. If r is between 0 and 1, that is, if $0 < r < 1$, then the terms are successively smaller and approach zero. (We say they *decrease monotonically*.)
4. If $r = 0$, then all terms after the initial term, 1, are zero.
5. If r is between -1 and 0, that is, $-1 < r < 0$, then the terms oscillate between positive and negative, each is numerically smaller than the preceding term, and they approach 0.
6. If $r = -1$, then the terms oscillate between 1 and -1.
7. If r is less than -1, then the terms oscillate between positive and negative and each term is numerically larger than the preceding term.

The following seven sequences illustrate these properties.

1. $r = 5$: $\{1, 5, 5^2, 5^3, 5^4, \ldots, 5^n, \ldots\} \to \infty$

2. $r = 1$: $\{1, 1, 1, 1, 1, \ldots\} \to 1$

3. $r = \frac{1}{2}$: $\left\{1, \frac{1}{2}, \left(\frac{1}{2}\right)^2, \left(\frac{1}{2}\right)^3, \left(\frac{1}{2}\right)^4, \ldots, \left(\frac{1}{2}\right)^n, \ldots\right\}$

    $$= \{1, 0.5, 0.25, 0.125, 0.0625, \ldots\} \to 0$$

4. $r = 0$: $\{1, 0, 0, 0, 0, \ldots\} \to 0$

5. $r = -\frac{1}{4}$: $\left\{1, -\frac{1}{4}, \left(-\frac{1}{4}\right)^2, \left(-\frac{1}{4}\right)^3, \left(-\frac{1}{4}\right)^4, \ldots, \left(-\frac{1}{4}\right)^n, \ldots\right\}$

    $$= \{1, -0.25, 0.0625, -0.015625, 0.00390625, \ldots\} \to 0$$

6. $r = -1$: $\{1, -1, (-1)^2, (-1)^3, (-1)^4, \ldots, (-1)^n, \ldots\}$
    $= \{1, -1, 1, -1, 1, -1, \ldots\}$ (does not converge)

7. $r = -2$: $\{1, -2, (-2)^2, (-2)^3, (-2)^4, \ldots, (-2)^n, \ldots\}$
    $= \{1, -2, 4, -8, 16, -32, 64, -128, \ldots\} \to \pm\infty$
    (does not converge)

We have seen that geometric sequences arise in a great variety of applications. An important related question frequently occurs: What is the sum of the terms $1, r, r^2, \ldots, r^n$ for some value of n? That is, what is

$$1 + r + r^2 + r^3 + r^4 + \cdots + r^n$$

for any given value of n? We saw one instance of this in Section 5.1 where we considered the number of grains of wheat that could be piled onto a chessboard according to the pattern: one grain on the first square, two on the second, four on the third, and so forth up to 2^{63} on the 64th square. This sum is

$$1 + 2 + 2^2 + 2^3 + 2^4 + \cdots + 2^{63}.$$

To answer the general question about the sum of the terms in any geometric sequence, we introduce the following notation:

$$S_0 = 1$$
$$S_1 = 1 + r$$
$$S_2 = 1 + r + r^2$$

and, in general, for the terms $1, r, r^2, \ldots, r^n$,

$$S_n = 1 + r + r^2 + \cdots + r^n.$$

We now multiply this expression for S_n by the common ratio r to obtain

$$r \cdot S_n = r(1 + r + r^2 + \cdots + r^n)$$
$$= r + r^2 + r^3 + \cdots + r^n + r^{n+1}.$$

When we subtract the expression for $r \cdot S_n$ from the expression for S_n, the intermediate terms $r, r^2, r^3, \ldots, r^n$ all cancel out and so we are left with

$$S_n - r \cdot S_n = 1 - r^{n+1}.$$

That is,

$$(1 - r)S_n = 1 - r^{n+1},$$

from which, if $r \neq 1$, it immediately follows that

$$S_n = \frac{1 - r^{n+1}}{1 - r}.$$

Sum of the terms $1, r, r^2, \ldots, r^n$ of a finite geometric sequence

$$1 + r + r^2 + r^3 + \cdots + r^n = \frac{1 - r^{n+1}}{1 - r}$$

provided that $r \neq 1$.

To illustrate this result, consider the example with the number of grains of wheat on a chessboard from Section 5.1. The common ratio is $r = 2$ and, since there are 64 squares on the board numbered from 0 to $n = 63$, we need the sum of the terms $1, 2, 2^2, \ldots, 2^{63}$:

$$S_{63} = 1 + 2 + 2^2 + \cdots + 2^{63}$$
$$= \frac{1 - 2^{64}}{1 - 2}$$
$$= 2^{64} - 1 = 1.844674 \times 10^{19},$$

which is more than 18 quintillion.

When we listed the seven possible cases for a geometric sequence, we pointed out that in two of the cases, $0 < r < 1$ and $-1 < r < 0$, the *terms of the sequence* approach zero. Also, when $r = 0$, all terms after the initial 1 are already 0. Let's see what this means in terms of the *sum of the terms* of a geometric sequence. Using the formula for the sum of the terms from 1 to r^n,

$$S_n = \frac{1 - r^{n+1}}{1 - r},$$

we see that whenever $-1 < r < 1$, the term r^{n+1} becomes smaller and smaller, approaching 0 as n increases. This enables us to give meaning to the sum of all the terms of an infinite geometric sequence. We say

$$1 + r + r^2 + r^3 + \cdots + r^n + \cdots = \frac{1 - 0}{1 - r} = \frac{1}{1 - r}$$

provided that $-1 < r < 1$ or equivalently, $|r| < 1$. However, if r is greater than or equal to 1 or if r is less than or equal to -1, the values r^{n+1} do *not* approach zero as n approaches infinity, and no value for the sum of the entire sequence is suggested. Thus the only case in which the sum of an entire geometric sequence has a finite value is the one with the common ratio r strictly between -1 and 1.

> **Sum of the terms of an infinite geometric sequence**
> $$1 + r + r^2 + r^3 + \cdots + r^n + \cdots = \frac{1}{1 - r}$$
> provided that $-1 < r < 1$.

For instance, the sum of the terms in the entire geometric sequence

$$1 + \frac{1}{2} + \left(\frac{1}{2}\right)^2 + \left(\frac{1}{2}\right)^3 + \cdots = \frac{1}{1 - \frac{1}{2}} = \frac{1}{\frac{1}{2}} = 2.$$

Add enough of the terms from this sum to convince yourself that this result is reasonable.

The above facts and results regarding geometric sequences occur very frequently throughout applications of mathematics and we will encounter such sequences repeatedly later in this book. For now, let's consider several additional situations where they arise.

E X A M P L E   I

In our discussion of the elimination of a drug from the body in Section 5.4, we assumed that the kidneys remove 25% of the drug in the bloodstream every four hours and that a person takes the same dosage $D_0 = 16$ mL every four hours. Find the level of drug in the body after 5 four-hour time periods, after 10 time periods, and after n time periods. What is the limiting value L for the drug in the body?

Solution We saw in Section 5.4 that this situation can be modeled with the difference equation

$$D_{n+1} = 0.75D_n + 16$$

and we created a formula

$$D_n = 64 - 48(0.75)^n$$

for the solution that was based on our knowledge of the behavior of exponential functions. Now let's see how this can be found directly using our knowledge of the sum of a geometric sequence. From the difference equation and the fact that $D_0 = 16$, we have

$$D_1 = 0.75D_0 + 16 = 16(1 + 0.75).$$

Similarly,

$$D_2 = 0.75D_1 + 16$$
$$= 0.75(1 + 0.75)16 + 16$$
$$= 16(1 + 0.75 + 0.75^2).$$

Furthermore,

$$D_3 = 0.75D_2 + 16$$
$$= 0.75(1 + 0.75 + 0.75^2)16 + 16$$
$$= 16(1 + 0.75 + 0.75^2 + 0.75^3).$$

In general, after n time periods,

$$D_n = 16(1 + 0.75 + 0.75^2 + 0.75^3 + \cdots + 0.75^n).$$

Clearly, this is the sum of the terms from 1 to 0.75^n in a geometric sequence with common ratio $r = 0.75$. Therefore the sum of these terms is given by

$$D_n = 16\left(\frac{1 - (0.75)^{n+1}}{1 - 0.75}\right)$$
$$= 16\left(\frac{1 - (0.75)^{n+1}}{0.25}\right)$$
$$= 64(1 - (0.75)^{n+1})$$
$$= 64 - 64(0.75)(0.75)^n$$
$$= 64 - 48(0.75)^n,$$

which is the identical formula we created in Section 5.4.

Consequently, after $n = 5$ periods, we have

$$D_5 = 64 - 48(0.75)^5 = 52.61 \text{ mL},$$

which agrees with the value we obtained in Section 5.4. Similarly,

$$D_{10} = 64 - 48(0.75)^{10} = 61.30 \text{ mL}.$$

Finally, because $r = 0.75 < 1$, the limiting value for the sum of this geometric sequence is

$$L = D_0[1 + r + r^2 + r^3 + \cdots]$$

$$= 16\left(\frac{1}{1-r}\right)$$

$$= \frac{16}{1 - 0.75}$$

$$= \frac{16}{(0.25)} = 64 \text{ mL},$$

and this is the same value we found in Section 5.4 for the drug maintenance level.

_____

E X A M P L E 2

In 1986, a total of 70,000 pages of new mathematical research was published. If the amount of research grew at the rate of 8% per year, find the total amount of new mathematics published between 1986 and 1996.

Solution Let M_n represent the number of pages of mathematics research published n years after 1986. We are told that $M_0 = 70,000$ pages in 1986 and we seek

$$M_0 + M_1 + M_2 + \cdots + M_{10} = 70,000 + 70,000(1.08) + 70,000(1.08)^2 + \cdots + 70,000(1.08)^{10}$$

$$= 70,000[1 + 1.08 + 1.08^2 + \cdots + 1.08^{10}]$$

$$= 70,000\left(\frac{1 - 1.08^{11}}{1 - 1.08}\right)$$

$$\approx 1,165,000 \text{ pages}.$$

_____

E X A M P L E 3

Suppose that a properly inflated basketball is designed to bounce back to three quarters of the height from which it is dropped. If such a ball is initially dropped from a height of 10 feet, find the total vertical distance it travels on the first 10 bounces; the first 20 bounces; the first 30 bounces. What total vertical distance does the ball cover if it keeps bouncing forever?

Solution   Figure 5.26 shows that the vertical distance the ball travels is:

10 feet until the first bounce;
plus 2 times the distance (up and then down) between the first and second bounces, or

$$2\left(\frac{3}{4} \cdot 10\right);$$

plus 2 times the distance between the second and third bounces, or

$$2\left[\frac{3}{4}\left(\frac{3}{4} \cdot 10\right)\right] = 2\left(\frac{3}{4}\right)^2 \cdot 10,$$

and so forth.

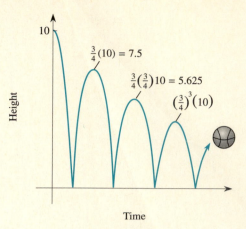

FIGURE 5.26

The total vertical distance traveled on the first n bounces is therefore

$$D_n = 10 + 2\left(\frac{3}{4}\right) \cdot 10 + 2\left(\frac{3}{4}\right)^2 \cdot 10 + 2\left(\frac{3}{4}\right)^3 \cdot 10 + \cdots + 2\left(\frac{3}{4}\right)^n \cdot 10$$

$$= 10 + 20\left(\frac{3}{4}\right) + 20\left(\frac{3}{4}\right)^2 + 20\left(\frac{3}{4}\right)^3 + \cdots + 20\left(\frac{3}{4}\right)^n$$

$$= 10 + 20\left[\left(\frac{3}{4}\right) + \left(\frac{3}{4}\right)^2 + \left(\frac{3}{4}\right)^3 + \cdots + \left(\frac{3}{4}\right)^n\right]$$

$$= 10 + 20\left(\frac{3}{4}\right)\left[1 + \left(\frac{3}{4}\right) + \left(\frac{3}{4}\right)^2 + \cdots + \left(\frac{3}{4}\right)^{n-1}\right].$$

Notice that the expression in the final brackets is the sum of a finite number of terms of a geometric sequence with common ratio $r = \frac{3}{4}$. Therefore, we find that the total vertical distance traveled by the ball during the first n bounces is

$$D_n = 10 + 15\left(\frac{1 - r^n}{1 - r}\right)$$

$$= 10 + 15\left(\frac{1 - \left(\frac{3}{4}\right)^n}{1 - \frac{3}{4}}\right)$$

$$= 10 + 15\left(\frac{1 - \left(\frac{3}{4}\right)^n}{\frac{1}{4}}\right)$$

$$= 10 + 60\left[1 - \left(\frac{3}{4}\right)^n\right].$$

Consequently, the vertical distance traveled by the ball during the first 10 bounces is

$$D_{10} = 10 + 60\left[1 - \left(\frac{3}{4}\right)^{10}\right]$$

$$= 10 + 60[1 - 0.0563]$$

$$= 10 + 56.621 = 66.621,$$

or about 66.6 feet. During the first 20 bounces, the ball travels

$$D_{20} = 10 + 60\left[1 - \left(\frac{3}{4}\right)^{20}\right] = 69.810,$$

or about 69.8 feet. And during the first 30 bounces, it travels

$$D_{30} = 10 + 60\left[1 - \left(\frac{3}{4}\right)^{30}\right] = 69.989,$$

or almost 70 feet. Thus we see that, after the first few bounces, all the subsequent bounces contribute very little additional distance, as we would expect. In fact, since the common ratio $r = \frac{3}{4}$ is between -1 and 1, we know that $\left(\frac{3}{4}\right)^n \to 0$ as n increases and so we can sum all of the terms of the geometric sequence for D_n to obtain

$$10 + 15\left[1 + \frac{3}{4} + \left(\frac{3}{4}\right)^2 + \left(\frac{3}{4}\right)^3 + \cdots + \left(\frac{3}{4}\right)^n + \cdots\right]$$

$$= 10 + 15\left(\frac{1}{1 - \frac{3}{4}}\right)$$

$$= 10 + 15\left(\frac{1}{\frac{1}{4}}\right)$$

$$= 10 + 15(4) = 70 \text{ feet.}$$

That is, theoretically, if the ball were to continue bouncing forever, the total vertical distance it would cover would be 70 feet.

EXERCISES

1. Find the sum of the terms $1 + r + r^2 + \cdots + r^n$ with $r = \frac{1}{2}$ for $n = 10$, $n = 20$, and $n = 30$.

2. Repeat Exercise 1 with $r = 0.2$.

3. Repeat Exercise 1 with $r = 0.8$.

4. Repeat Exercise 1 with $r = -0.8$.

5. Repeat Exercise 1 with $r = 1.5$.

6. Repeat Exercise 1 with $r = -2.5$.

7. Suppose there were 6000 new cases of a certain disease in 1950. If the number of new cases has diminished 20% per year ever since, what is the *total* number of people who have come down with this disease from 1950 through 1990? How many would you predict will contract this disease between 1990 and 2000?

8. Repeat Exercise 7 if the number of new cases of the disease has increased 20% per year since 1950.

9. In 1980, the United States used approximately 2.5 billion kilowatt-hours of electricity. If electric usage is growing 2% per year, find the total amount of electricity used in this country between 1980 and 1995.

10. The United States produced 195,000 metric tons of wheat in 1984. If production is growing 10% per year, find the total amount of wheat produced between 1984 and 1993.

11. The U.S. produced 70,600 metric tons of rice in 1984. If rice production is decreasing 9% per year, find the total amount of rice produced between 1984 and 1993.

12. At age 22, Ken gets his first job paying $25,000 a year. If he gets an annual increase of 6% each year, what will be his *total* earnings over his entire career when he retires at age 65?

13. Repeat the bouncing ball example from the text if the initial height of the ball is 6 feet.

14. Repeat the bouncing ball example if the initial height is 12 feet.

15. Repeat the bouncing ball example if the ball bounces back to 80% of its height. By how much does the total distance traveled by the ball change compared to a 75% bounce?

16. Repeat the bouncing ball example if the ball bounces back $\frac{2}{3}$ of its height.

17. A geometric sequence is based on the fact that the ratio of successive terms is constant: $\frac{x_{n+1}}{x_n} = r$. Suppose instead that the ratio is a linear function, say $\frac{x_{n+1}}{x_n} = rn$, for $n \geq 1$.

  a. How does the growth rate compare to that for a geometric sequence?

  b. What is the solution of $\frac{x_{n+1}}{x_n} = n$, for $n \geq 1$?

  c. What is the solution of $\frac{x_{n+1}}{x_n} = 5n$, for $n \geq 1$?

  d. What is the solution of $\frac{x_{n+1}}{x_n} = rn$, for $n \geq 1$, for any r?

  e. What is the solution of $\frac{x_{n+1}}{x_n} = rn + b$, for $n \geq 1$?

18. How would the solution of $\frac{x_{n+1}}{x_n} = n^2$ compare to the solution of $\frac{x_{n+1}}{x_n} = n$, $n \geq 1$? Can you construct this solution?

19. Consider $\frac{x_{n+1}}{x_n} = \frac{1}{n}$. How does its solution behave? How does it compare to the solution of $\frac{x_{n+1}}{x_n} = \frac{1}{r}$, for any $r > 1$? Can you construct this solution?

5.8 *Newton's Laws of Cooling and Heating*

We are all familiar with the scene in a host of crime shows where the medical examiner studies the homicide victim's body and knowledgeably announces that "Mr. Jones died at approximately 1:30 in the morning." We now consider the mathematical theory that lies behind this type of conclusion. More generally, we will investigate the rate at which any object cools off or heats up.

Suppose you heat a pizza in an oven set for 450°F and then remove it from the oven to cool in a kitchen where the temperature is a constant 70°F. We seek to determine how fast the temperature of the pizza drops until it reaches room temperature. It is clear that the temperature of the pizza drops most rapidly at first because there is a large difference between the temperature of the pizza and room temperature. As the pizza cools, the rate at which the temperature decreases slows down. That is, the temperature drops faster the hotter the pizza is; it drops most slowly the closer the pizza's temperature is to room temperature. Geometrically, we expect the graph of the temperature as a function of time to be decreasing and concave up, as shown in Figure 5.27. According to *Newton's Law of Cooling*, the change in temperature, ΔT_n, is proportional to the difference in temperature between the pizza (T_n) and the surrounding air (70°F), or $T_n - 70$. Thus,

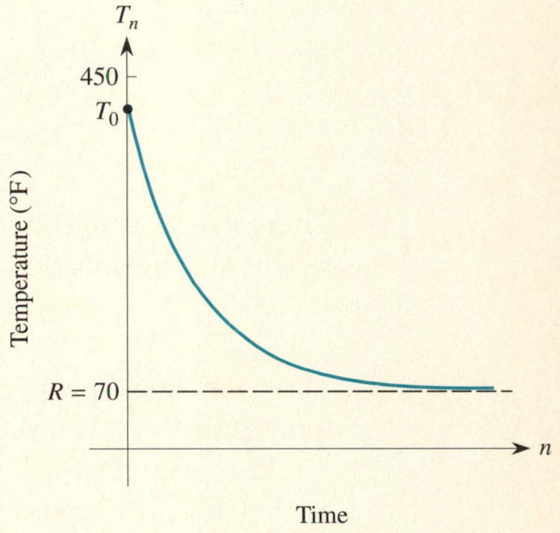

$$\Delta T_n = T_{n+1} - T_n = \alpha(T_n - 70),$$

FIGURE 5.27

where the constant of proportionality α must be negative since $T_n - 70$ is positive and the temperature change is negative. Equivalently, we obtain the difference equation

$$T_{n+1} = T_n + \alpha(T_n - 70)$$
$$= (1 + \alpha)T_n - 70\alpha.$$

More generally, if the room temperature is any constant R, then the corresponding difference equation is

$$T_{n+1} = (1 + \alpha)T_n - \alpha R. \tag{1}$$

This equation and the principle behind it are both known as *Newton's Law of Cooling*.

In the next chapter, we will develop some relatively simple methods for solving this type of difference equation. For now, we will base our work on the discussion at the end of Section 5.4. We saw there that if B is any constant and a is a constant between 0 and 1, then the solution to the difference equation

$$x_{n+1} = ax_n + B$$

is given by

$$x_n = L + (x_0 - L)a^n,$$

where

$$L = \frac{B}{1-a}$$

is the limiting value. Notice that the difference equation (1) for Newton's Law of Cooling is of the same form with

$$a = 1 + \alpha \quad \text{and} \quad B = -\alpha R.$$

Since α is a negative fraction, a is between 0 and 1. Also, we have

$$L = \frac{B}{1-a} = \frac{-\alpha R}{1-(1+\alpha)} = \frac{-\alpha R}{-\alpha} = R.$$

Therefore, if the initial temperature of the pizza (or any other cooling object) is T_0, the solution to Equation (1) is

$$T_n = R + (T_0 - R)(1 + \alpha)^n.$$

The constant of proportionality α is usually determined from an additional temperature measurement, as we will see in the examples below.

In Figure 5.28, we show the graph of this solution. Notice that it starts at the initial temperature T_0, is a decreasing function, and is concave up. The temperature decreases rapidly at first, but then levels off to approach a horizontal asymptote at the room temperature R. The overall behavior is one of exponential decay with R as the limiting value. This graph clearly matches the pattern we predicted above and that was shown in Figure 5.27.

We can also see this behavior directly from the formula for T_n. When $n = 0$, it reduces to

$$(T_0 - R)(1) + R = T_0.$$

As n increases, the term $(1 + \alpha)^n$ approaches zero, since α is a negative number between -1 and 0.

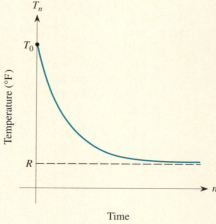

FIGURE 5.28

EXAMPLE 1

Suppose a cake is baking in a 350°F oven. It is removed when its temperature is 180°F and is left to cool in a kitchen at 70°F. After 10 minutes, the temperature of the cake is 125°F. Find the temperature after 15 minutes. How long does it take the cake to cool to 75°F?

Solution We are told that $T_0 = 180$ and $R = 70$. Therefore, the difference equation for the temperature at any time n, in minutes, is

$$T_{n+1} = (1 + \alpha)T_n - 70\alpha.$$

The corresponding solution is given by

$$T_n = (180 - 70)(1 + \alpha)^n + 70$$
$$= 110(1 + \alpha)^n + 70.$$

Further, we are told that when $n = 10$, the temperature $T_{10} = 125$, so

$$T_{10} = 110(1 + \alpha)^{10} + 70 = 125.$$

As a result,

$$(1 + \alpha)^{10} = \frac{55}{110} = 0.5.$$

If we now take the tenth root of both sides of this equation, we find that

$$1 + \alpha = 0.5^{1/10} = 0.933,$$

and so

$$\alpha = 0.933 - 1 = -0.067,$$

which is negative as we expected. Therefore the solution to the difference equation is

$$T_n = 110(1 + \alpha)^n + 70$$
$$= 110(0.933)^n + 70.$$

After $n = 15$ minutes, the temperature of the cake is

$$T_{15} = 110(0.933)^{15} + 70$$
$$= 38.87 + 70 \approx 108.9°.$$

To find the time needed for the cake to cool down to 75°F, we need to find the value of n when $T_n = 75$. That is,

$$T_n = 110(0.933)^n + 70 = 75,$$

and so

$$110(0.933)^n = 5$$

$$(0.933)^n = \frac{5}{110} = 0.045.$$

We take logarithms of both sides of this equation to obtain

$$n \log (0.933) = \log (0.045),$$

and therefore we find $n \approx 44.7$ minutes. Thus, it takes about three-quarters of an hour for the cake to cool to 75°F.

E X A M P L E   2

Mr. Jones' body was found in his kitchen at 9 A.M. by the police who noted that the body temperature was 77.3°F and that the room temperature was 70°F. An hour later, the medical examiner found the body temperature was 76.1°F. Assuming that Mr. Jones' body temperature was the normal 98.6°F at the time of death, at what time was he murdered?

Solution   Using the solution to the difference equation we obtained above, we know that the body temperature after any number of hours, n, is given by

$$T_n = (T_0 - R)(1 + \alpha)^n + R,$$

where the room temperature $R = 70°$ and $T_0 = 98.6°$. Therefore

$$T_n = (98.6 - 70)(1 + \alpha)^n + 70$$
$$= 28.6(1 + \alpha)^n + 70.$$

Since we do not know the time of death, corresponding to $n = 0$, we do not know the value of n at 9 A.M.

However, at 9 A.M., which is n hours after death, the body temperature is

$$T_n = 28.6(1 + \alpha)^n + 70 = 77.3.$$

One hour later, at 10 A.M., which is $n + 1$ hours after death, the body temperature is

$$T_{n+1} = 28.6(1 + \alpha)^{n+1} + 70 = 76.1.$$

These two conditions give

$$28.6(1 + \alpha)^n = 7.3 \tag{2}$$
$$28.6(1 + \alpha)^{n+1} = 6.1. \tag{3}$$

If we divide Equation (3) by Equation (2), we find that

$$1 + \alpha = \frac{6.1}{7.3} = 0.8356.$$

Substituting this value into Equation (2) gives

$$28.6(0.8356)^n = 7.3,$$

or equivalently,

$$(0.8356)^n = 0.2552.$$

We take logs of both sides of this equation to get

$$n \log (0.8356) = \log (0.2552)$$

$$n = \frac{\log (0.2552)}{\log (0.8356)} = 7.6$$

The body was found at 9 A.M., which is $n = 7.6$ hours after death. Thus the murder occurred 7.6 hours (which is 7 hours and 36 minutes) before 9 A.M., so we conclude that Mr. Jones was killed at approximately 1:24 A.M.

In summary,

> **Newton's Law of Cooling**
>
> *Assumptions*
>
> - The temperature, R, of the medium remains constant.
> - The change in temperature is proportional to the difference between the temperature of the object and the medium.
>
> *Mathematical Model*
>
> - Difference equation: $\Delta T_n = \alpha(T_n - R)$ or $T_{n+1} = (1 + \alpha)T_n - \alpha R$
> - Solution: $T_n = R + (T_0 - R)(1 + \alpha)^n$

E X A M P L E   3

A cup of hot coffee is left standing on a table in a room where the temperature is 70°F. The temperature of the coffee is measured every minute and the results are shown in the table below. Find a function of time that best fits these temperature readings.

t (min.)	0	1	2	3	4	5	6	7	8	9
Temp.	186	182	178	175	171	168	165	162	159	156

	10	11	12	13	14	15
	153	152	148	145	143	141

Solution   Based on our knowledge about cooling curves, we would expect the best fit to be an exponential function that decays down to 70°F. However, if we were to use our data fitting techniques from Chapter 3 to fit an exponential curve to this data, the result would be erroneous because the function we obtain would decay down to 0°F rather than 70°F. Therefore it is necessary to transform the data to reflect the fact that the temperature of the coffee only drops to 70°F. For this reason, we first subtract 70 from each temperature reading, which is equivalent to introducing a vertical shift for the temperature function, and so get the following table of values.

t (min.)	0	1	2	3	4	5	6	7	8	9
Temp.	116	112	108	105	101	98	95	92	89	86

	10	11	12	13	14	15
	83	81	78	75	73	71

We now perform a nonlinear regression analysis on this transformed set of data and so find that the exponential function that best fits the data is

$$T = 115.6(0.96771)^t,$$

where T is the temperature above 70°F. The corresponding correlation coefficient is $r = -0.99981$. The value for r is negative because the temperature readings are decreasing; the corresponding linearized data values, log T versus t, fall into a linear pattern with negative slope. Also, this value for r is extremely close to -1, indicating an extremely good fit. Finally, we undo the vertical shift by adding 70 to obtain the function

$$C = 115.6(0.96771)^t + 70$$

as the best model for the temperature of the coffee, as shown in Figure 5.29.

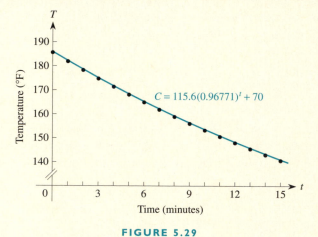

FIGURE 5.29

Newton's Law of Heating

We note that the identical mathematical methods apply if an object is being warmed rather than cooled. The corresponding principle is known as *Newton's Law of Heating* and is based on the assumption that the increase in temperature ΔT_n is proportional to the difference between the temperature, R, of the medium (the room, the freezer, the oven, etc.) and the temperature of the object, T_n. This leads to the difference equation

$$\Delta T_n = \alpha(R - T_n)$$

since for an object that is heating, T_n is always less than R. Note that the constant of proportionality α must be positive since $R - T_n$ is positive and the change in temperature of the object, ΔT_n, is also positive. We can solve this difference equation in the same manner as we solved the one for cooling above and eventually obtain

$$T_n = R + (T_0 - R)(1 - \alpha)^n.$$

We ask you to do this in one of the exercises at the end of this section.

E X A M P L E   4

A chicken is removed from the refrigerator at a temperature of 40°F and placed into an oven kept at a constant temperature of 350°F. After 10 minutes, the temperature of the chicken is 70°F. The chicken is considered cooked when its temperature reaches 180°F. How long must it remain in the oven?

Solution   Using the previously given solution of the difference equation for heating with $R = 350$ and $T_0 = 40$, we have

$$T_n = 350 - 310(1 - \alpha)^n.$$

After 10 minutes, we find that

$$T_{10} = 350 - 310(1 - \alpha)^{10} = 70,$$

so

$$(1 - \alpha)^{10} = \frac{280}{310} = 0.903.$$

Taking the tenth root of both sides, we obtain

$$1 - \alpha = 0.9898.$$

Consequently, the solution to the difference equation for the temperature of the chicken is

$$T_n = 350 - 310(0.9898)^n.$$

To find how long it takes until the temperature reaches 180°F, we solve the equation

$$T_n = 350 - 310(0.9898)^n = 180$$

and eventually find, with our usual methods, that

$$n = 58.6 \text{ minutes}.$$

That is, the chicken will be ready in about one hour.

E X E R C I S E S

1. A pot of bubbling pudding (212°F) is removed from the stove and put immediately into a refrigerator at 40°F. After 10 minutes, the temperature of the pudding is 160°F. Find the temperature after one hour. How long does it take for the temperature to drop to 75°F?

2. Sam takes a can of soda at room temperature (70°F) and puts it into a

freezer (0°F) to chill quickly. After 10 minutes, the temperature of the soda is 60°F. How long does it take until the temperature drops to 40°F?

3. A bowl of cold soup is taken out of the refrigerator (36°F) and placed into a heated oven (375°F) to warm. After 10 minutes, the temperature of the soup is 120°F. How long does it take for the soup to reach 200°F?

4. Example 2 assumed that the initial body temperature was a normal 98.6°F. Suppose that Jones had a slight fever of 100°F when he was murdered. How much of a difference does this make on the estimated time of death?

5. Example 2 also assumed that the room temperature was kept constant at 70°F. In practice, we might expect the home heating system to cycle on and off, so the actual temperature might vary between 67°F and 72°F, say. How might you take this variation into account in predicting the time of death? How much of a difference would it make?

6. Professor Smith's body was found in a large walk-in refrigerator in the laboratory at 9 A.M. by the police who noted that the body temperature was 67.3°F and that the refrigerator temperature was 40°F. An hour later, the medical examiner found the body temperature to be 63.1°. Assuming the body temperature at death was 98.6°F, at what time was Professor Smith killed?

7. Example 4 presumed that the temperature of the oven is 350°F when cooking a chicken. Suppose the temperature is set at 325°F instead. How much of a difference does this make in the time it takes to cook the chicken to 180°F?

8. A cup of boiling water (100° C) was placed in a refrigerator kept at 7°C at 8 A.M., and the following readings were obtained:

Time (min)	1	7	21	45	73	90	123	152	190
Temperature (°C)	89	71	53	36	25	20	14	11	9

Determine the best exponential fit to this data and use it to predict when the water temperature will be 5°C.

9. The following set of data is printed on a carton of milk to indicate how many days the milk will last without spoiling at different temperatures.

Temperature (°F)	32	40	45	50	60	70
Time (days)	24	11	5.5	2	1	0.5

Determine the best fit (linear, exponential, or power function) to this data. How long should milk last in a refrigerator kept at 35°F?

10. A cool potato at temperature 60°F is placed in an oven kept at a constant 350°F.

   a. Sketch the graph of the temperature of the potato as a function of time.

Use the concavity of your graph from part (a) to answer the questions in parts (b)–(d).

b. Suppose you measure the temperature of the potato after 5 minutes and find that it is 109°F and that, after 7 minutes, it is 127°F. Use this information to estimate the temperature after 10 minutes. Is the actual value higher or lower than your estimate? How do you know?

c. Suppose that you are told that the temperature of the potato after 12 minutes is 150°F. Use this information and the temperature after 5 minutes to estimate the temperature after 10 minutes. Is the actual value higher or lower than your estimate? How do you know?

d. How might you use the results from (b) and (c) to come up with a better estimate of the temperature after 10 minutes?

11. Your Thanksgiving turkey is taken from a refrigerator at 40°F and is cooked in an oven kept at a constant temperature of 350°F. The temperature T of the bird is 70°F after 30 minutes and is 96°F after 60 minutes. From your knowledge of the pattern of temperature rise, decide which of the following temperature readings are possible and which are impossible.

a. $T(45) = 80°F$ b. $T(45) = 85°F$ c. $T(75) = 105°F$ d. $T(75) = 115°F$

12. A cup of hot chocolate (temperature 180°F) is placed on a table where the air temperature is 70°F. Suppose it takes 12 minutes until the temperature of the drink is down to 100°F. Let r_1 represent the average rate of decrease in temperature per minute over the full 12-minute period, let r_2 be the average rate of decrease over the first 6 minutes, and let r_3 be the average rate of decrease over the last 6 minutes. List these three rates in increasing order.

13. Suppose that it takes t_1 minutes for a raw potato, starting at 70°F, to reach 200°F in an oven. At that time, it is removed from the oven and put back on the table where it cools down. Suppose it takes t_2 minutes to reach 70°F. Is $t_1 < t_2$, $t_1 = t_2$, or $t_1 > t_2$?

14. A potato at room temperature (T_0) is placed in an oven at temperature R. Construct the actual solution to the corresponding difference equation for heating in terms of the various parameters. What is the formula for the temperature of the potato if $T_0 = 70°F$ and $R = 350°F$?

5.9 *Sequences and Differences*

Before we continue with our study of difference equations and their applications, we will develop two additional mathematical tools: (1) some systematic methods for determining the pattern in a set of values, especially the elements of a sequence, and (2) some procedures for solving certain relatively simple classes of difference equations.

We begin with the problem of finding the general term of a sequence. Consider the sequence

$$A = \{2, 4, 6, 8, 10, 12, \ldots\}.$$

It is obvious that the term following $a_6 = 12$ should be $a_7 = 14$, since each element is two more than the preceding one. Alternatively, we can express this fact by noting that

$$a_2 - a_1 = 4 - 2 = 2$$
$$a_3 - a_2 = 6 - 4 = 2$$
$$a_4 - a_3 = 8 - 6 = 2$$

and so on. Thus if the general term is a_n, then the difference Δa_n between it and the following term, a_{n+1}, is

$$\Delta a_n = a_{n+1} - a_n = 2$$

for any value of n. That is, the difference between any two successive terms of this sequence is always 2, and we can write $\Delta a_n = 2$, for all $n \geq 1$. Moreover, since the successive differences are constant, we know that a_n is a linear function of n. In fact, the general term of this sequence is the linear function $a_n = 2n$.

Now consider the sequence

$$B = \{1, 3, 6, 10, 15, 21, \ldots\}.$$

If we look at the differences between successive terms, we find that

$$\Delta b_1 = b_2 - b_1 = 2$$
$$\Delta b_2 = b_3 - b_2 = 3$$
$$\Delta b_3 = b_4 - b_3 = 4$$
$$\Delta b_4 = b_5 - b_4 = 5$$
$$\Delta b_5 = b_6 - b_5 = 6$$

and so the pattern in the differences becomes evident. The next term, b_7, will be 7 more than b_6, so that $b_7 = b_6 + 7 = 28$; furthermore, $b_8 = b_7 + 8 = 36$, and so on. In fact, the nth term b_n of this sequence gives the sum of the first n integers. Therefore, as we saw in Section 4.6,

$$b_n = \frac{n(n + 1)}{2},$$

which is a quadratic function of n.

This approach using differences of terms does not apply for finding the general term of all sequences. For instance, it does not work with geometric sequences because the underlying relationship is an exponential function. However, for any sequence where the underlying relationship is a polynomial function, the use of differences is very effective. Let us expand on the ideas for the differences of the terms of a sequence, which we introduced in Section 4.6.

> For any sequence $X = \{x_1, x_2, \ldots\}$, the **difference operator** Δ is defined by
>
> $$\Delta x_1 = x_2 - x_1$$
> $$\Delta x_2 = x_3 - x_2$$
> $$\Delta x_3 = x_4 - x_3$$
>
> and, in general,
>
> $$\Delta x_n = x_{n+1} - x_n$$
>
> for any value of n.

Note that the result of applying the difference operator Δ to the successive terms of the sequence $X = \{x_n\}$ is to produce a new set of numbers,

$$\Delta x_1, \Delta x_2, \Delta x_3, \ldots.$$

In view of our definition of a sequence, this set of numbers can be thought of as a new sequence, the sequence of differences $\{\Delta x_1, \Delta x_2, \Delta x_3, \ldots\}$. We write

$$\Delta X = \{\Delta x_n\} = \{x_{n+1} - x_n\}.$$

Let's consider the previous example with $B = \{1, 3, 6, 10, 15, 21, \ldots\}$. If we apply the difference operator to the elements of this sequence, we obtain the sequence of differences

$$\Delta B = \{2, 3, 4, 5, 6, \ldots\}.$$

Since this is itself a sequence of numbers, we can apply the difference operator Δ to it. Thus for the first term,

$$\Delta(\Delta b_1) = 3 - 2 = 1.$$

The notation $\Delta(\Delta b_1)$ is very cumbersome and so we write

$$\Delta(\Delta b_1) = \Delta^2 b_1.$$

Similarly,

$$\Delta(\Delta b_2) = \Delta^2 b_2 = 4 - 3 = 1,$$

and so forth. We therefore see that, for any value of n

$$\Delta^2 b_n = 1,$$

or equivalently the sequence of second differences is

$$\Delta^2 B = \{\Delta^2 b_n\} = \{1, 1, 1, \ldots\}.$$

Since the second differences are all constant, we recall from Section 4.6 that the terms in the sequence B must follow a quadratic pattern in n.

In a similar way, we can define higher order differences for any sequence X. Thus the third difference is

$$\Delta^3 x_n = \Delta(\Delta^2 x_n)$$

and is the sequence formed of all third differences (the difference between each pair of successive second differences). We also know, based on our discussions in Section 4.6, that if the third differences are all constant, then the general term of the sequence will be a cubic polynomial in n.

The various differences for a given sequence can be clearly displayed using a *difference table*, as illustrated in the next example.

E X A M P L E   I

Use differences to determine the term after 56 in the sequence $X = \{1, 4, 10, 20, 35, 56, \ldots\}$.

Solution   We write the sequence and the successive differences in columns as follows:

$$
\begin{array}{llll}
x_1 = 1 & & & \\
& \Delta x_1 = 3 & & \\
x_2 = 4 & & \Delta^2 x_1 = 3 & \\
& \Delta x_2 = 6 & & \Delta^3 x_1 = 1 \\
x_3 = 10 & & \Delta^2 x_2 = 4 & \\
& \Delta x_3 = 10 & & \Delta^3 x_2 = 1 \\
x_4 = 20 & & \Delta^2 x_3 = 5 & \\
& \Delta x_4 = 15 & & \Delta^3 x_3 = 1 \\
x_5 = 35 & & \Delta^2 x_4 = 6 & \\
& \Delta x_5 = 21 & & \Delta^3 x_4 = ? \\
x_6 = 56 & & \Delta^2 x_5 = ? & \\
& \Delta x_6 = ? & & \\
x_7 = ? & & &
\end{array}
$$

Notice that all the third differences for this sequence appear to be one, so that $\Delta^3 x_4 = 1$ also. To determine the value for x_7, we work backward in the table. Since the third differences are all one, the second differences increase by 1. Therefore the next second difference in the table, $\Delta^2 x_5$, will be 7. In turn, this means that the next first difference in the table, Δx_6, will be 7 larger than $\Delta x_5 = 21$, so that $\Delta x_6 = 28$. Thus x_7 must be 28 larger than 56 or 84.

Alternatively, the fact that $\Delta^3 x_n = 1$ for all n indicates that the original sequence is a cubic function of n; we could find the equation for this cubic and use it to get the value for x_7.

Think About This

By working backward in the above table, determine x_8.

The various differences of a sequence $X = \{x_n\}$ can also be stated in terms of handy formulas. Starting with $\Delta x_n = x_{n+1} - x_n$ for any n, we have

$$\Delta^2 x_n = \Delta(\Delta x_n) = \Delta(x_{n+1} - x_n)$$
$$= \Delta(x_{n+1}) - \Delta(x_n)$$
$$= (x_{n+2} - x_{n+1}) - (x_{n+1} - x_n)$$
$$= x_{n+2} - 2x_{n+1} + x_n.$$

In a similar manner, we can show that, for any n,

$$\Delta^3 x_n = x_{n+3} - 3x_{n+2} + 3x_{n+1} - x_n.$$

Think About This

Derive the result

$$\Delta^3 x_n = x_{n+3} - 3x_{n+2} + 3x_{n+1} - x_n.$$

All higher order differences can be obtained in a similar fashion. However, the actual derivations can be avoided if we use the array of numbers known as Pascal's Triangle.

Pascal's Triangle

```
              1   1
            1   2   1
          1   3   3   1
        1   4   6   4   1
      1   5  10  10   5   1
    1   6  15  20  15   6   1
```

Notice that each row begins and ends with a 1 and every intermediate entry is simply the sum of the two numbers diagonally above it. Thus in the sixth row, the number 15 is the sum of the 5 and the 10 flanking it in the fifth row above.

Consider the various formulas we obtained above for the differences of a sequence x_n:

$$\Delta x_n = x_{n+1} - x_n$$
$$\Delta^2 x_n = x_{n+2} - 2x_{n+1} + x_n$$
$$\Delta^3 x_n = x_{n+3} - 3x_{n+2} + 3x_{n+1} - x_n$$

Note that the coefficients are simply the entries of the corresponding row of Pascal's Triangle with alternating positive and negative signs. Thus we can predict $\Delta^4 x_n$ by merely reading off the entries from the fourth row of Pascal's Triangle and alternating the signs. Doing this, we get

$$\Delta^4 x_n = x_{n+4} - 4x_{n+3} + 6x_{n+2} - 4x_{n+1} + x_n.$$

Think About This

1. Demonstrate that this formula for $\Delta^4 x_n$ is correct by applying Δ to the expression for $\Delta^3 x_n$.

2. Find formulas for $\Delta^5 x_n$ and $\Delta^6 x_n$.

Differences and Difference Equations

It is no coincidence that the terms "difference operator Δ" and "difference equation" both involve the word "difference." To see the connection, consider the difference equation for the exponential growth model

$$P_{n+1} = (1 + a)P_n.$$

We derived this equation by considering the change $P_{n+1} - P_n$ in the population from one value of n to the next:

$$\text{Change in } P_n = P_{n+1} - P_n = aP_n.$$

We now recognize that this is the same as the difference operator Δ applied to the population value P_n. That is, the exponential difference equation can also be written as

$$\Delta P_n = aP_n$$

for all n. Although it is somewhat easier to work with the equation in the previous form, it is useful to have this second form available as well. In particular, it is important to interpret the left-hand side as the change in the population when going from time n to time $n + 1$.

Similarly, for the logistic growth model, the change ΔP_n in the population, P_n, is

$$\Delta P_n = P_{n+1} - P_n = aP_n - bP_n^2.$$

So

$$P_{n+1} = P_n + aP_n - bP_n^2,$$

or equivalently

$$P_{n+1} = (1 + a)P_n - bP_n^2.$$

In Section 5.5, we found that the limiting value L for a logistic population is

$$L = \frac{a}{b}$$

and the point of inflection, where the concavity changes, occurs at

$$\frac{1}{2}L = \frac{1}{2}\left(\frac{a}{b}\right).$$

We can obtain this latter result from the difference equation using the ideas we have just developed. Recall that the point of inflection represents the point where the population is growing most rapidly so that the change in the population, ΔP_n, is greatest (see Figure 5.30). To the left of this point, the logistic curve is concave up, which means that the population is not just increasing, but it is increasing at an ever faster rate. That is, ΔP_n is also increasing and so its change, $\Delta[\Delta P_n]$, must be positive to the left of the inflection point. To the right of this point, the logistic curve is concave down, so that the population is increasing at an ever slower rate. That is, ΔP_n is decreasing and so its change, $\Delta[\Delta P_n]$, must be negative to the right of the inflection point. As a result, we see that the point of inflection corresponds to the point where the change in ΔP_n, which is $\Delta[\Delta P_n]$, changes from positive to negative. We will use this fact to locate it.

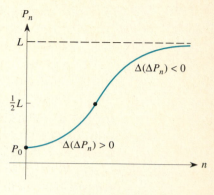

FIGURE 5.30

Let's begin by looking at the difference equation that defines the logistic model,

$$\Delta P_n = aP_n - bP_n^2.$$

Consider

$$\Delta[\Delta P_n] = \Delta[aP_n - bP_n^2].$$

Using the definition of the difference operator Δ, we find

$$\Delta[\Delta P_n] = \Delta[aP_n - bP_n^2]$$
$$= [aP_{n+1} - bP_{n+1}^2] - [aP_n - bP_n^2].$$

When we collect like terms, we obtain

$$\Delta[\Delta P_n] = a[P_{n+1} - P_n] - b[P_{n+1}^2 - P_n^2]$$
$$= a[\Delta P_n] - b[P_{n+1}^2 - P_n^2].$$

Since the term in the second brackets is a difference of squares, we can factor it to obtain

$$\Delta[\Delta P_n] = a\Delta P_n - b(P_{n+1} - P_n)(P_{n+1} + P_n)$$
$$= a\Delta P_n - b(\Delta P_n)(P_{n+1} + P_n)$$
$$= \Delta P_n[a - b(P_n + P_{n+1})].$$

This expression for $\Delta[\Delta P_n]$ is zero if either the term in the brackets is zero or if $\Delta P_n = 0$. Looking at the original difference equation

$$\Delta P_n = aP_n - bP_n^2 = P_n(a - bP_n),$$

we see that the latter occurs when either $P_n = 0$ or $P_n = \frac{a}{b}$, which is the limiting value for the population. In each case, there is no further change in the population, but neither value is a point of inflection. Therefore we consider the term in the square brackets and so find that

$$P_n + P_{n+1} = \frac{a}{b}.$$

If we divide both sides by 2, we obtain

$$\frac{1}{2}(P_n + P_{n+1}) = \frac{1}{2}\left(\frac{a}{b}\right).$$

Therefore the point of inflection occurs when the average of two successive values for the population is precisely half of the limiting value $L = \frac{a}{b}$ for the population. In order for this to happen, the earlier value for the population, P_n, must be below the midpoint level and the later value, P_{n+1}, must be above the midpoint level. At this point, the population is growing most rapidly. You can see this by looking at the tables of population values we constructed in Section 5.5. The greatest change in the population values occurs when the population crosses the midpoint value, or half of the limit to growth of the population. As we pointed out before, the most vigorous growth in a population occurs just before decline begins to set in.

EXERCISES

Apply Δ to the first six terms in each of the following sequences. See if you can find a pattern for the differences.

1. $x_n = 4n$
2. $x_n = 3n + 5$
3. $x_n = \frac{1}{2}n$
4. $x_n = n^2 + 5$
5. $x_n = 5n^2 + 3n$
6. $x_n = n^3 - 10$
7. $x_n = 3^n$
8. $x_n = \frac{1}{n}$

Determine an expression for the general term in each of the following sequences and use it to predict the next two terms. (Note: Not all can be solved using differences.)

9. $\{3, 5, 7, 9, 11, \ldots\}$
10. $\{2, 5, 8, 11, 14, 17, \ldots\}$
11. $\{192, 96, 48, 24, 12, \ldots\}$
12. $\left\{\frac{2}{5}, \frac{4}{25}, \frac{8}{125}, \ldots\right\}$
13. $\left\{\frac{1}{3}, \frac{2}{4}, \frac{3}{5}, \frac{4}{6}, \ldots\right\}$
14. $\{2, 5, 10, 17, 26, 37, 50, 65, \ldots\}$

Use a difference table to find the next term in the following sequences:

15. $\{2, 5, 11, 21, 36, 57, \ldots\}$
16. $\{1, 5, 15, 35, 70, 126, \ldots\}$

17. Suppose that we have a sequence $\{x_n\}$.
    a. If for some value of n, $\Delta x_n > 0$, how does x_{n+1} compare to x_n?
    b. If for some value of n, $\Delta x_n < 0$, how does x_{n+1} compare to x_n?
    c. Complete the following statements:
        (i) The terms of a sequence are increasing if . . . .
        (ii) The terms of a sequence are decreasing if . . . .

18. Suppose that $\Delta x_n > 0$ for $n = 0, 1, \ldots, 99$ and $\Delta x_n < 0$ for $n = 100$. What can you conclude about x_{101}?

19. Suppose that $\Delta x_n < 0$ for $n = 0, 1, \ldots$. What can you conclude about x_0?

20. Suppose that, in some sequence $\{x_0, x_1, x_2, \ldots, x_5, \ldots\}$, you know that $\Delta x_0 = 2$, $\Delta x_1 = 5$, $\Delta x_2 = 10$, $\Delta x_3 = 4$, $\Delta x_4 = -8$. If $x_0 = 20$, find $x_1, \ldots, x_5$.

21. Repeat Exercise 20 if you know that $x_0 = 100$. How does this sequence compare to the one in Exercise 20?

22. There is one and only one sequence x_n (other than constant multiples of it) that has the special property that $\Delta x_n = x_n$. Find this special sequence. Then select several other members of the same family of functions that are "close to" x_n and verify what happens when you apply Δ to them.

23. The following graphs show the first few terms of the *differences* Δx_n of an unknown sequence x_n. (The graphs do *not* show the actual x-values.) For each, indicate the value of n for which the actual sequence of x-values achieves its maximum.

a.

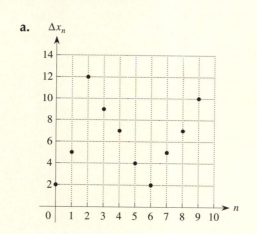

b.

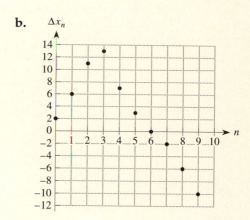

CHAPTER SUMMARY

In this chapter, you have learned the following:

- What a sequence is.
- What it means for a sequence to be convergent or divergent.
- What a difference equation is.
- What the solution of a difference equation is and how it can be given either by a closed-form formula for the nth term or generated term-by-term.
- How difference equations arise as mathematical models for population growth and other phenomena.
- How to analyze the level of a drug in the bloodstream using difference equations.
- How to find the maintenance level associated with a drug and what it means.

- How to model inhibited population growth using the logistic model.
- How to interpret the behavior of a logistic curve.
- How to model radioactive decay with a difference equation.
- How to sum the first n terms of a geometric sequence.
- How to sum all the terms of a geometric sequence provided that the common ratio is between -1 and 1.
- How to model temperature decrease using a difference equation based on Newton's Law of Cooling.
- How to model temperature increase using a difference equation based on Newton's Law of Heating.
- How to interpret the behavior of the solutions for the heating and cooling models.
- How to construct the higher order differences for a sequence and what the higher order differences mean in terms of the behavior of the terms in the sequence.

REVIEW EXERCISES

1. Write out the first five terms of the following sequences.

   a. $a_n = 6n - 1$ **b.** $t_n = \dfrac{3^n}{n}, n > 0$ **c.** $r_n = 1 - (0.3)^n$

2. Decide if the following sequences are increasing or decreasing and give a reason for your answer. If the sequence is converging, determine the limit.

   a. $a_n = 6n - 1$ **b.** $t_n = \dfrac{3^n}{n}, n > 0$ **c.** $r_n = 1 - (0.3)^n$

   d. $y_n = \dfrac{n}{\log n}$ **e.** $p_n = \dfrac{n^3 + 3}{7n^3 - 4}$ **f.** $h_n = \dfrac{7n^3 - 4}{n^3 + 3}$

3. Find a formula for the general term of each of the following sequences. Specify the domain for n.

   a. $\{1, 3, 9, 27, \ldots\}$ **b.** $\left\{\dfrac{1}{4}, \dfrac{1}{16}, \dfrac{1}{64}, \ldots\right\}$

   c. $\{1, 9, 25, 49, 81, 121, \ldots\}$ **d.** $\{40, 120, 360, 1080, \ldots\}$

4. The population of Ghana grows at a rate of about 3% per year. Determine the doubling time for the population.

5. In 1970, the population of the United States was growing at a rate of about 1.125% per year. The population of the United States in 1970 was about 203.2 million. Use this information to predict the population size in 1990. In 2000. How long would it take for the population to double?

6. The aim of a college administration is to reduce the number of students who need remedial work in English by 10% each year. At this time, there are 1600 students enrolled in remedial English classes. If this program is successful, how many students will be enrolled in remedial English in three years? How long will it take for the number of students enrolled in such classes to be reduced to one section of 15 students?

7. The level of a drug decreases at a rate of 30% of the drug per hour. Assume the initial dose is 150 mg. How long does it take to bring the drug level down to under 20 mg? How long does it take to bring the drug level down to 5% of the original level?

8. Determine the first five terms in the solution of each of the following difference equations.

   a. $x_{n+1} = x_n + 8, \ x_0 = 2$ b. $x_{n+1} = x_n - 8, \ x_0 = 12$

   c. $x_{n+1} = \frac{1}{3}x_n, \ x_0 = 5$ d. $x_{n+1} = x_n + (-3)^n, \ x_0 = 10$

9. Determine the first five elements in each sequence.

   a. $y_{n+2} = y_{n+1} + y_n, \ y_0 = 2, y_1 = 7$

   b. $y_{n+2} = y_{n+1} + y_n, \ y_0 = 3, y_1 = 7$

10. A drug is administered every six hours. The kidneys discharge 60% of the drug over that period. If the original dose is 100 mg, how much of the drug remains in the body after 8 days?

11. A drug is administered every four hours. The kidneys discharge about 70% of the drug over the four-hour period. The initial dose of the medicine is 100 mg. How much should the repeated dosage be to insure that the maintenance level of the drug is 30 mg?

12. Chlorine is added to the water supply in a reservoir at the rate of 30 units a day. It is estimated that 20% of this amount disappears each day through evaporation or filters. Is there a maintenance level of chlorine in the reservoir?

13. The difference equation $v_{n+1} = 1.30v_n - 0.00002v_n^2$ models the number of people in a town who have VCRs as a function of the year, n. Initially, 40 households had VCRs. Determine how many people will eventually have VCRs. Determine the year in which exactly half the population have VCRs.

14. A population grows according to the logistic model from an initial size of 1000 to a final size of 12,000. The annual growth rate is 20%. What is the inhibiting constant? Write the difference equation that describes this population for any year, n.

15. The size of the fish population in a stream grows in accordance with the logistic model with a growth rate of 30% and an inhibiting constant of 0.04%. Write the difference equation for each situation.

    a. The fish population grows according to the logistic model.
    b. The fish population grows according to the logistic model, but 2000 fish are caught and removed from the stream each year.
    c. The fish population grows according to the logistic model, and the county wildlife federation stocks the stream with 400 fish per year.
    d. The fish population grows according to the logistic model, but about 10% of the population are caught every year.

16. Write the difference equation for each of the following scenarios and draw a graph of the behavior of the solution. Answer the question.

a. Pancake syrup used to be 100% maple syrup. But over the years the amount of maple syrup has been reduced. Suppose that in 1960 a company began reducing the amount of maple syrup in their product by 15% per year. What is the percentage of maple syrup in its product in 1996?

b. John's investment in the stock market has been growing by 10% per year. He adds $2000 a year to his investment. If he had $50,000 invested in 1990, how much money does he have invested in 1996?

c. Advertising in the print media has been less important recently. In 1988 a company decided to decrease its budget for advertising in the print media by $20,000 per year. Also, the company increased its budget for television by 10% a year. If the budgets were each $2 million in 1988, how much is the company budgeting for print and how much for television at the end of three years? How is the total advertising budget changing?

6

MODELING WITH DIFFERENCE EQUATIONS

6.1 *Solutions of Difference Equations*

In Chapter 5, we introduced difference equations and some phenomena that can be modeled using them. In this chapter, we develop mathematical models based on difference equations for a variety of situations arising in many different fields including population growth, biology, physics, computer science, business, and finance. In each instance, the key idea will be to determine the mathematical factors that relate the change in a quantity to the quantity itself. This produces a difference equation model for the quantity that enables us to generate the solution, to analyze the behavior of the solution, and to interpret this behavior in terms of the original quantity being studied. To do so, we must develop some general ideas about difference equations and formal techniques for solving them. We first recall some of the difference equations that already have arisen:

$$x_{n+1} = (1 + a)x_n \qquad \text{exponential growth or decay models}$$
$$x_{n+1} = (1 + a)x_n - bx_n{}^2 \qquad \text{logistic growth model}$$
$$x_{n+1} = ax_n + B \qquad \text{drug level with repeated dosage model}$$
$$x_{n+1} = (1 + \alpha)x_n - \alpha R \qquad \text{Newton's law of cooling}$$
$$x_{n+2} = x_{n+1} + x_n \qquad \text{Fibonacci sequence}$$

One way to classify difference equations is in terms of their *order*. The **order** of a difference equation is the largest spacing between indices. Look at the difference equations for the exponential growth model, the logistic growth model, the drug model, and the cooling model. Each involves x_n and x_{n+1}, so the spacing between the indices n and $n + 1$ is 1. Therefore each is called **a first order difference equation.** The difference equation for the Fibonacci model is a **second order difference equation** because the largest spacing, which is between the indices n and $n + 2$, is 2. In a similar way, the difference equation

$$x_{n+4} - 5nx_{n+3} + n^7 x_{n+1} - x_n = 3^n$$

is a **fourth order difference equation** and

$$x_{n+7} - n^2 x_{n+5} + 2^n x_{n+4} = 17$$

is a **third order difference equation.**

As a general rule, the lower the order of a difference equation, the easier it is to solve the equation; that is, to find a formula for the sequence x_n that satisfies the difference equation for all n. Even among first order difference equations, some are easier to solve than others. The easiest to solve are *first order, linear difference equations*, which are so called because they involve only linear terms. They are of the form

$$x_{n+1} = a_n x_n + b_n.$$

This form is equivalent to

$$x_{n+1} - a_n x_n = b_n,$$

where a_n and b_n represent any two known sequences. For instance,

$$x_{n+1} - n^2 x_n = n \cdot 5^n$$

is a linear difference equation with $a_n = n^2$ and $b_n = n \cdot 5^n$.

A difference equation is called *linear* because of the structure of the equation. Notice the parallel between the difference equation

$$x_{n+1} = a_n x_n + b_n$$

and the equation of a line

$$Y = mX + b.$$

Notice that Y is a multiple of X plus a term that does not involve X, while x_{n+1} is a multiple of x_n plus a term that does not involve x_n. If a difference equation involves terms other than linear terms in x_n and x_{n+1}, say something like

$$x_{n+1} = 5(x_n)^2 + 3\sqrt{x_n},$$

then we call it a **nonlinear difference equation.** The logistic difference equation,

$$x_{n+1} = (1 + a)x_n - bx_n^2$$

is nonlinear because of the x_n^2 term. The idea of linearity pervades all of mathematics and will surface again in later sections.

In a particularly simple, yet important, case of first order, linear difference equations, the sequence a_n is just a constant, for example, a. We call the resulting equation,

$$x_{n+1} - ax_n = b_n,$$

a **first order, linear difference equation with constant coefficient.** For example, we might have

$$x_{n+1} - 73x_n = 5n + 2.$$

Notice that when we speak of a constant coefficient, the only coefficient we are referring to is the coefficient of x_n; the remaining term b_n can be any function of n. The difference equations for the drug and heating/cooling models are both first order, linear difference equations with constant coefficients, as is the difference equation for the exponential growth or decay models.

Finally, a further simplification involves the case where the sequence b_n on the right-hand side is identically zero; that is, $b_n = 0$ for all n. The resulting equation,

$$x_{n+1} - ax_n = 0, \qquad \text{or} \qquad x_{n+1} = ax_n,$$

is known as a **first order, linear, homogeneous difference equation with constant coefficient.** The word *homogeneous* refers to the fact that the term on the right side of the equation is zero. This is precisely the type of difference equation that models exponential growth or decay. If the right-hand side is not zero in the difference equation

$$x_{n+1} - ax_n = b_n$$

(that is, b_n is not identically zero), then the difference equation is called *nonhomogeneous.*

The other concept we will need is that of a **solution** of a difference equation, namely any sequence $\{x_n\}$ that satisfies the difference equation. For example, we previously found that the solution to the exponential growth equation

$$x_{n+1} - ax_n = 0$$

is

$$x_n = x_0 a^n$$

for all n, where x_0 represents the initial value. In particular, the solution to

$$x_{n+1} - 1.05x_n = 0, \qquad \text{or} \qquad x_{n+1} = 1.05x_n$$

is

$$x_n = x_0(1.05)^n$$

for any value of n. We know that it is the solution because, when we substitute the expression for x_n and the corresponding expression for x_{n+1} into the difference equation, we get perfect agreement. Further, this happens no matter what value we give to the lead coefficient x_0. Thus, we rewrite this solution as

$$x_n = C(1.05)^n,$$

where the constant C is known as an **arbitrary constant** because any value whatsoever will work. This solution is called the **general solution** because every solution of the difference equation can be expressed in this form. The general solution of any first order difference equation will always involve an arbitrary constant such as C. Of course, in any particular example, an initial condition (or starting value) may be specified for the solution sequence; in such a case, C is no longer arbitrary and we then have a **specific solution** of the difference equation.

E X A M P L E   I

Verify that $x_n = n^2 + C$, where C is an arbitrary constant, is the general solution of the first order, nonhomogeneous, linear difference equation with constant coefficient

$$x_{n+1} - x_n = 2n + 1.$$

Solution   If $x_n = n^2 + C$, then when we replace n with $n + 1$, we find that

$$x_{n+1} = (n + 1)^2 + C$$
$$= n^2 + 2n + 1 + C.$$

When we substitute for x_n and x_{n+1} in the original difference equation, we obtain

$$x_{n+1} - x_n = [n^2 + 2n + 1 + C] - [n^2 + C]$$
$$= 2n + 1$$

since the n^2 and C terms cancel each other out. Because $2n + 1$ is precisely the expression on the right-hand side of the original difference equation, the given sequence $x_n = n^2 + C$ satisfies the difference equation. It is the general solution because it satisfies the difference equation no matter what value C has.

The **general solution** of a first order difference equation always involves one arbitrary constant C.

The **specific solution** of a difference equation corresponds to any given initial condition x_0 that serves to determine the value of C.

Moreover, it is very important that you distinguish clearly between the *solution of an equation,* as in algebra, and the *solution of a difference equation.* The solution of an algebraic equation requires finding one or possibly several specific values of the unknown that satisfy the equation. The solution of a difference equation requires finding either an expression for a sequence or a list of the actual terms of the sequence that satisfies the difference equation. Thus, the solution of an algebraic equation consists of one or more numbers, and the solution of a difference equation consists of a sequence, which is a function.

Incidentally, notice in Example 1 that we provided you with the solution $x_n = n^2 + C$ without showing you how we found it. Since

$$x_{n+1} = x_n + 2n + 1$$

actually is a fairly simple difference equation, we can construct the solution relatively easily as follows. Suppose the initial term is $x_0 = 10$. Then

if $n = 0$: $x_1 = x_0 + 2 \cdot 0 + 1 = 10 + 0 + 1 = 11$

if $n = 1$: $x_2 = x_1 + 2 \cdot 1 + 1 = 11 + 2 + 1 = 14$

if $n = 2$: $x_3 = x_2 + 2 \cdot 2 + 1 = 14 + 4 + 1 = 19$

if $n = 3$: $x_4 = x_3 + 2 \cdot 3 + 1 = 19 + 6 + 1 = 26$

if $n = 4$: $x_5 = x_4 + 2 \cdot 4 + 1 = 26 + 8 + 1 = 35,$

and so forth. Alternatively, if the initial term is $x_0 = 20$, then

if $n = 0$: $x_1 = x_0 + 0 + 1 = 20 + 1 = 21$

if $n = 1$: $x_2 = x_1 + 2 + 1 = 21 + 3 = 24$

if $n = 2$: $x_3 = x_2 + 4 + 1 = 24 + 5 = 29,$

and so on. We organize these results in the following table, extending the entries as far as $n = 9$. We also leave room for comparable results starting with other initial values x_0. Realize that each column is a different specific solution because it corresponds to a different initial condition.

n	x_n	x_n	x_n	x_n	x_n
0	$x_0 = -10$	$x_0 = 0$	$x_0 = 10$	$x_0 = 20$	$x_0 = 30$
1			11	21	
2			14	24	
3			19	29	
4			26	36	
5			35	45	
6			46	56	
7			59	69	
8			74	84	
9			91	101	

Complete the remaining entries in the table for the other initial values using the difference equation $x_{n+1} = x_n + 2n + 1$ that defines each sequence. In the process, do you notice any relationship between the different specific solutions?

Since each of these solutions is a sequence, we can plot the resulting points in Figure 6.1 to see the behavior of the solutions. To better highlight the behavior, we have drawn a smooth curve through the points for each solution. Notice that the curves all grow in an identical manner, and thus suggest a family of functions, one for each initial value. What do you think would happen in Figure 6.1 if you used another starting value, say $x_0 = 15$ or $x_0 = 40$?

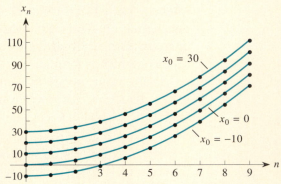

Intercepts on vertical axis are initial values x_0

FIGURE 6.1

Let's now construct a formula for the general solution to the difference equation

$$x_{n+1} = x_n + 2n + 1$$

for any arbitrary initial value x_0. We find

if $n = 0$: $x_1 = x_0 + 0 + 1 = x_0 + 1$
if $n = 1$: $x_2 = x_1 + 2 + 1 = x_1 + 3 = x_0 + 4$
if $n = 2$: $x_3 = x_2 + 4 + 1 = x_2 + 5 = x_0 + 9$
if $n = 3$: $x_4 = x_3 + 6 + 1 = x_3 + 7 = x_0 + 16$
if $n = 4$: $x_5 = x_4 + 8 + 1 = x_4 + 9 = x_0 + 25.$

The obvious pattern on the right suggests that $x_n = x_0 + n^2$ for any value of n. Because we have not specified any specific value for the initial term x_0, it is an arbitrary constant. We thus can write the general solution to the difference equation as $x_n = n^2 + C$, for any constant C. Recall that, in Example 1, we verified that this is the general solution by substituting x_n and x_{n+1} into the difference equation.

What is the significance of the arbitrary constant C in the general solution of a difference equation? In the derivation above, C came from the initial value x_0. If the initial term were $x_0 = 10$, say, then the solution would be

$$x_n = n^2 + 10,$$

which fits the set of values we calculated before. If $x_0 = 45$, then $x_n = n^2 + 45$; if $x_0 = -22$, then $x_n = n^2 - 22$. That is, every possible initial value x_0 leads to a different solution to the difference equation. We actually get a family of solutions, one sequence for each possible initial value x_0. Graphically in Figure 6.1, we can imagine a curve beginning at each value for x_0 on the vertical axis.

Whenever we have a difference equation, it is usually helpful to graph several specific solutions corresponding to different initial conditions to get a feel for the behavior of the solutions and how they depend on the initial conditions. In some cases, all of the different solutions display the same behavior pattern; in other cases, very different behavior patterns occur, depending on the choice of the initial conditions. Figure 6.2 shows the graphs of some of the solutions of the difference equation

$$x_{n+1} = x_n + 2n + 1$$

that we examined previously, for $n \geq 0$. Notice that all the graphs are the right halves of parabolas because the solution we constructed is $x_n = n^2 + C$ for $n \geq 0$. We call such a picture of a typical sample of solutions the *solution field* associated with the difference equation. In the solution field, intercepts on the vertical axis represent values of the initial conditions x_0. The horizontal axis is the independent variable n.

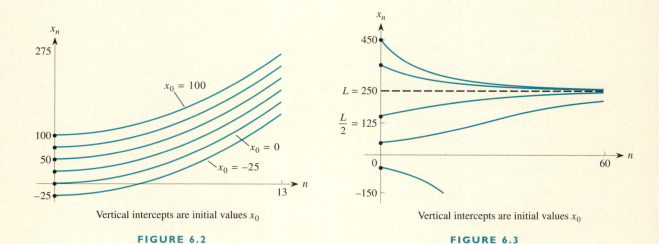

Vertical intercepts are initial values x_0

FIGURE 6.2

Vertical intercepts are initial values x_0

FIGURE 6.3

The solution field associated with the logistic difference equation

$$x_{n+1} = (1 + a)x_n - bx_n^2 \qquad \text{or} \qquad \Delta x_n = ax_n - bx_n^2$$

with $a = 0.05$ and $b = 0.0002$ is shown in Figure 6.3. The particular logistic model used has a limit to growth of $L = \frac{a}{b} = 250$, so its point of inflection occurs at a height of 125. The different behaviors depend on the initial condition x_0. If $x_0 > L$, then the solution decays from x_0 to the limiting value L. If x_0 is between $\frac{1}{2}L$ and L, then the solution grows from x_0 toward L in a purely concave down manner. If x_0 is between 0 and $\frac{1}{2}L$, then the solution grows from x_0 toward L in the S-shape typical of a logistic curve. If x_0 is negative, then the solution decreases toward $-\infty$.

The solution field gives us a graphical connection between the difference equation and its solution that adds to our understanding of both. Let's see how it can help us to understand the behavior described in the previous paragraph. We rewrite the equation for the logistic model by factoring out bx_n to get

$$\Delta x_n = ax_n - bx_n^{\;2} = bx_n\!\left(\frac{a}{b} - x_n\right)\!.$$

Recall that $b > 0$. The logistic difference equation has two solutions that are constants, one at the maximum sustainable population level $L = \frac{a}{b}$ and the other when $x_n = 0$. We see them graphically as the two horizontal lines in the solution field. These lines divide the plane into three distinct regions and we analyze the behavior of the solution within each region. First, whenever x_0 is greater than $L = \frac{a}{b}$, the value of $\frac{a}{b} - x_0$, will be negative, so Δx_0 is negative and therefore x_1 is less than x_0; that is, the solution starts off by decreasing. In fact, whenever $x_n > L$ for any value of n, Δx_n is negative, and so the solution decreases toward the limiting value $L = \frac{a}{b}$.

Second, whenever x_0 is between 0 and $L = \frac{a}{b}$, the expression $\frac{a}{b} - x_0$ will be positive, so Δx_0 is positive and therefore x_1 is larger than x_0. Similarly, if x_n is between 0 and L for any value of n, the difference equation assures us that Δx_n is positive, so x_{n+1} is larger than x_n and the solution increases toward the limiting value L.

Third, whenever x_0 is negative, the positive term $\frac{a}{b} - x_0$ is multiplied by the negative term bx_0, so Δx_0 is negative and the solution begins by decreasing. In general, if x_n is negative for any n, Δx_n is negative and the solution will continually decrease toward $-\infty$. We show all three cases in Figure 6.4.

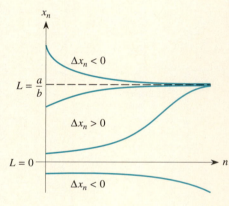

FIGURE 6.4

When creating a solution field for a difference equation, care must be taken to get a selection of initial conditions that produces solutions displaying all possible behavior patterns. Had we taken only values of $x_0 > L$, say, in the logistic equation, then we might have concluded erroneously that *all* solutions decay toward L.

Figure 6.5 shows a solution field associated with the *nonlinear* difference equation

$$x_{n+1} = x_n^{\;2} - 2.$$

The solutions shown correspond to initial values of $x_0 = -2.1, -2, -1, 0, 2$, and 2.01. As suggested by the solution for $x_0 = 2.01$, the solution diverges to ∞ very rapidly for any $x_0 > 2$. (Try some values numerically.)

Similarly, for any $x_0 < -2$, $x_1 = x_0^2 - 2 > 2$, and so the solution also diverges to ∞ very rapidly. In fact, the divergence in both cases is so extreme that, in order to display this behavior, we had to choose initial values very close to 2 and -2.

If the initial condition x_0 is any number between -2 and 2, then x_0^2 is between 0 and 4, and so $x_1 = x_0^2 - 2$ is also between -2 and 2. In fact, every subsequent term in the solution must be between -2 and 2. This means that all solutions remain within a narrow horizontal band, although many of the solutions (see Figure 6.6) clearly oscillate up and down within that band. The particular behavior pattern depends very much on the choice of x_0 and can be quite complicated as suggested by the more detailed view of the solution field shown in Figure 6.6 based on five different initial values between -2 and 2. It is because of complexities such as this that we will focus primarily on *linear* difference equations where the behavior patterns are typically much simpler.

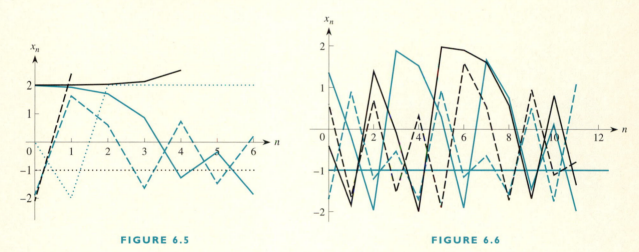

FIGURE 6.5   FIGURE 6.6

However, we use $x_{n+1} = x_n^2 - 2$ to illustrate one more idea. Suppose that $x_0 = -1$. Then $x_1 = x_0^2 - 2 = -1$ and, in fact, all subsequent values of the sequence will be identically equal to -1. You can see this graphically in the horizontal line at the height of -1 in the solution field in Figure 6.5. We call such a solution that remains the same for all values of n an **equilibrium solution,** or **equilibrium level,** for the difference equation. There is another equilibrium at $x_n = 2$ since, if $x_0 = 2$, all subsequent terms of the solution are identically equal to 2. (Think about the logistic model with $x_0 = L = \frac{a}{b}$; that is also an equilibrium. Similarly, the drug concentration model and the heating and cooling models also had equilibrium solutions.)

You can usually find the equilibrium solutions for a difference equation, if there are any, by realizing that at any equilibrium, x_{n+1} is equal to x_n. Simply equate x_{n+1} and x_n in the difference equation and solve the resulting algebraic equation. If there are any solutions, these solutions are equilibrium solutions. If there are no solutions to this equation, then the difference equation does not have an equilibrium solution. For instance in our previous example,

$$x_{n+1} = x_n^2 - 2,$$

we let $x_{n+1} = x_n = L$, the limiting value, which gives

$$L = L^2 - 2$$
$$L^2 - L - 2 = 0$$
$$(L - 2)(L + 1) = 0.$$

The last equation gives the two equilibrium levels of $L = 2$ and $L = -1$.

E X A M P L E   2

Construct the general solution to the difference equation

$$x_{n+1} - (n + 1)x_n = 0$$

and determine all equilibrium levels, if there are any.

Solution   Notice that this is a first order, linear homogeneous difference equation with variable coefficient. We can rewrite it as

$$x_{n+1} = (n + 1)x_n.$$

Therefore,

$$\begin{aligned}
&\text{if } n = 0: &&x_1 = (1)x_0 \\
&\text{if } n = 1: &&x_2 = (2)x_1 = (2)(1)x_0 \\
&\text{if } n = 2: &&x_3 = (3)x_2 = (3)(2)(1)x_0 \\
&\text{if } n = 3: &&x_4 = (4)x_3 = (4)(3)(2)(1)x_0 \\
&\text{if } n = 4: &&x_5 = (5)x_4 = (5)(4)(3)(2)(1)x_0
\end{aligned}$$

and so, in general,

$$x_n = (n)(n - 1)(n - 2) \cdots (3)(2)(1)x_0$$

for any value of n. The number x_0 plays the role of the arbitrary constant here. We can rewrite this solution using *factorial notation* (see Appendix C) as

$$x_n = n!\, x_0.$$

To see if there are any equilibrium levels for this difference equation, we let $x_{n+1} = x_n = L$ to get

$$L = (n + 1)L.$$

The only solution to this equation is $L = 0$. That is, there is an equilibrium corresponding to $x_0 = 0$ in which all terms x_n are zero.

The approach used here for constructing a solution will not work for most difference equations. It applies only if the equation is especially simple. Consequently, we will develop a more systematic method for constructing the solutions to such equations in the next section.

EXERCISES

Characterize each of the difference equations in Exercises 1–10 as being:

- *first order or second order*
- *linear or nonlinear*
- *homogeneous or nonhomogeneous*

1. $x_{n+1} = 1.5x_n$

2. $x_{n+1} = 1.5x_n^2$

3. $x_{n+1} + x_n = n^2$

4. $y_{n+1} - \frac{1}{2}y_n = 0$

5. $z_{n+1} - 5z_n = 3n$

6. $w_{n+1} = 5w_n + 3$

7. $x_{n+2} - 2x_n = 2$

8. $y_{n+3} = \frac{1}{4}y_{n+2} - 2^n$

9. $q_{n+1} = \sqrt{q_n} + 3$

10. $q_{n+1} - n^3 q_n = 1$

11. Calculate the first 10 terms of the solution to the difference equation

$$x_{n+1} = \frac{n}{n+1}x_n + \frac{1}{5^n},$$

with $x_0 = 1$. Plot the points and predict the behavior of the solution as n continues to increase.

12. Determine the equilibrium solutions for each of the following difference equations.

a. $x_{n+1} = 1.5x_n - 6$

b. $x_{n+1} = 0.2x_n + 8$

c. $x_{n+1} = x_n^2 - 6$

d. $x_{n+1} = x_n^2 + 6$

e. $x_{n+1} = x_n^3 - 7x_n^2 + 13x_n$

13. a. Calculate the first 10 terms of the solution to the difference equation

$$x_{n+1} = x_n^2 - 2$$

for $x_0 = 1$ and plot the resulting points. What behavior pattern do you observe?

b. Repeat part (a) if the initial condition is $x_0 = 1.5$. Do you get the same behavior?

c. Repeat part (a) if the initial condition is $x_0 = 1.9$. Do you get the same behavior?

d. Does starting just below the equilibrium solution $x_n = 2$ with $x_0 = 1.9$ give a solution that converges to this equilibrium? What happens if you take a starting value closer to the equilibrium, say $x_0 = 1.99$?

14. Consider the difference equation $\Delta x_n = a x_n (1 - x_n)(2 - x_n)$.
    a. Find all equilibrium solutions.
    b. The equilibrium levels divide the solution field for this difference equation into four distinct regions. Sketch a graph of the solution field by analyzing the behavior of the solution inside each of these regions.

15. Consider the solution field in the accompanying figure.

    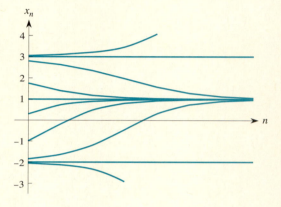

    a. What are the equilibrium levels?
    b. Write a possible difference equation for x_n that could have this solution field.

16. Construct a table showing the terms x_1, x_2, ..., x_{10} of the solution sequences for the difference equation

    $$x_{n+1} = x_n + 4n + 5$$

    corresponding to initial conditions of $x_0 = -10$, $x_0 = 0$, $x_0 = 10$, and $x_0 = 20$. Then use the table to determine a formula for the solution for any initial value x_0.

17. Construct a table showing the terms x_1, x_2, ..., x_{10} of the solution sequences for the difference equation

    $$x_{n+1} = x_n + 3n^2 + 3n + 1$$

    corresponding to initial conditions of $x_0 = 0$, $x_0 = 10$, and $x_0 = 20$. Then use the table to determine a formula for the solution for any initial value x_0.

18. a. Verify algebraically that $x_n = C(2^n) + 6(3^n)$ is the solution to the difference equation

    $$x_{n+1} = 2x_n + 6(3^n)$$

    for any constant C.
    b. Identify how each term in the solution is related to a part of the difference equation.

19. a. Verify algebraically that $x_n = C(5^n) + 4(2^n)$ is the solution to the difference equation

    $$x_{n+1} = 5x_n - 12(2^n)$$

    for any constant C.
    b. Identify how each term in the solution is related to a part of the difference equation.

20. a. Verify that $x_n = C(3^n) + 2n^2 - 2n - 3$ is the solution to the difference equation

    $$x_{n+1} = 3x_n - 4n^2 + 8n + 6$$

    for any constant C.
    b. Identify how the terms in the solution are related to parts of the difference equation.

21. Consider the difference equation $\Delta x_n = f(n)$. (Note that the differences are equal to a function of n only; they do not involve x_n at all.) For each of the functions f shown below, describe the behavior of the solution x_n: Is it increasing or decreasing? Is it concave up or concave down? Does it approach a limiting value as $n \to \infty$?

a.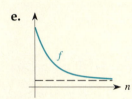

b.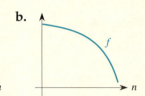

c.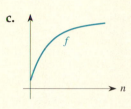

d.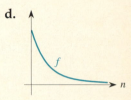

e.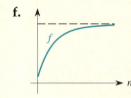

f.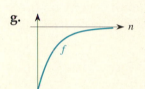

g.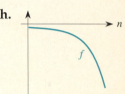

h.

22. Explain why a first order, linear, homogeneous difference equation cannot have two different equilibrium levels.

23. Consider the exponential function $f(x) = (1.5)^x$ from $x = 0$ to $x = 1$, which grows slowly at first and then more rapidly. The average rate of change over the interval 0 to 1 is defined to be the slope of the line segment connecting the endpoints of the curve:

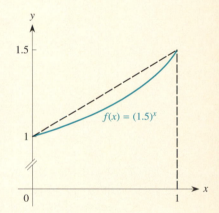

$$\frac{\Delta y}{\Delta x} = \frac{1.5 - 1}{1 - 0} = 0.5.$$

Indicate approximately the point on the curve where the rate of change is the same as the average rate of change over the entire interval. Explain why you decided on the point you indicated. Then repeat this for the interval from $x = 0$ to $x = 2$.

24. Consider the power function $f(x) = x^2$ from $x = 0$ to $x = 1$, which grows slowly at first and then more rapidly.

a. Find the average rate of change of this function over the interval from 0 to 1.

b. Indicate approximately the point on the curve where the rate of change is the same as the average rate of change over the entire interval. Explain why you decided on the point you indicated. Then repeat this for the interval from $x = 0$ to $x = 2$.

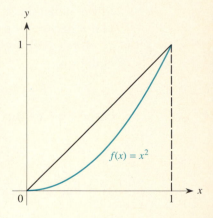

6.2 *Constructing Solutions of First Order Difference Equations*

In the last section, we considered the basic terminology regarding difference equations and their solutions. We now introduce a powerful method that allows us to actually construct the solution to most first order, linear difference equations of the form

$$x_{n+1} - ax_n = b_n \quad \text{or} \quad x_{n+1} = ax_n + b_n.$$

The method we use to construct the solution for a first order, linear, nonhomogeneous difference equation is based on the following principles regarding the *complete solution*.

> The **complete solution** of the first order, linear, nonhomogeneous difference equation with constant coefficient
>
> $$x_{n+1} - ax_n = b_n$$
>
> is made up of two parts:
>
> **1.** The *general solution* $C \cdot a^n$ to the associated homogeneous difference equation
>
> $$x_{n+1} - ax_n = 0.$$
>
> **2.** Any *particular solution* of the nonhomogeneous difference equation.

Recall that the *general solution* of a first order difference equation always involves an arbitrary constant C. When an initial condition x_0 is specified, we can determine the corresponding value for C to get a *specific solution*. A linear, nonhomogeneous difference equation may have many different solutions, but all we need is one *particular solution*. The *complete solution* of a first order, linear, nonhomogenous difference equation consists of the sum of the general solution of the associated homogeneous difference equation and a particular solution.

For example, in Section 5.4, we created a solution to the difference equation

$$D_{n+1} = aD_n + B$$

based on Newton's Laws of Heating and Cooling. The solution is

$$D_n = \frac{B}{1-a}(1 - a^{n+1}) = \frac{B}{1-a} - \frac{aB}{1-a}a^n.$$

This solution is composed of two parts, a constant term $\frac{B}{1-a}$ and a multiple of a^n. Notice that the corresponding homogeneous difference equation is

$$D_{n+1} = aD_n,$$

the difference equation for the exponential growth and decay model whose solution is Ca^n, a multiple of a^n.

Similarly, consider the difference equation

$$x_{n+1} - 3x_n = 12(5^n).$$

The associated homogeneous difference equation is

$$x_{n+1} - 3x_n = 0,$$

which has the immediate solution

$$x_n = C(3^n)$$

for some arbitrary constant C. The *nonhomogeneous term* b_n is $12(5^n)$, and the problem we face is to find a particular solution to the nonhomogeneous equation.

EXAMPLE I

Verify that

$$x_n = C(3^n) + 6(5^n)$$

is the complete solution of the difference equation

$$x_{n+1} - 3x_n = 12(5^n).$$

Solution   To check that this is indeed the solution, we calculate

$$x_{n+1} = C(3^{n+1}) + 6(5^{n+1}) = 3 \cdot 3^n C + 6 \cdot 5(5^n)$$

and substitute this expression for x_{n+1} and the given expression for x_n into the left-hand side of the difference equation to find

$$x_{n+1} - 3x_n = [3 \cdot 3^n C + 6 \cdot 5(5^n)] - 3[3^n C + 6 \cdot 5^n]$$
$$= 30 \cdot 5^n - 18 \cdot 5^n$$
$$= (30 - 18)(5^n)$$
$$= 12(5^n),$$

as required.

Now that we know that this expression does represent the complete solution, let's see how the facts stated previously suggest a method for actually finding the solution. The associated homogeneous difference equation is

$$x_{n+1} - 3x_n = 0,$$

and its solution is given by $3^n C$, where C is any arbitrary constant. Further, the term $6(5^n)$ represents a single particular solution of the nonhomogeneous difference equation. We can check this fact by calculating

$$x_{n+1} - 3x_n = 6(5^{n+1}) - 3 \cdot 6(5^n)$$
$$= 6(5 \cdot 5^n) - 18(5^n) = 12(5^n),$$

which is the right-hand side of the difference equation. The complete solution consists of a combination of these two pieces.

What is the source of the particular solution $6(5^n)$? Suppose we compare $6(5^n)$ with the nonhomogeneous term $12(5^n)$ in the original difference equation. Clearly, the particular solution is just a multiple of this nonhomogeneous term. This situation is usually the case. The solution of the nonhomogeneous equation typically has the same form as the term on the right-hand side of the difference equation.

We use this observation to find the particular solution $6(5^n)$. Suppose we suspect that the particular solution might be some multiple of the nonhomogeneous term, but we aren't sure what that multiple is. We would therefore expect it to be a sequence of the form $x_n = A(5^n)$, for some unknown multiple A. (Notice that we essentially ignore the constant multiple 12 in the actual nonhomogeneous term.) Thus,

$$x_{n+1} = A(5^{n+1})$$

and when we substitute these expressions into the left-hand side of the original difference equation, $x_{n+1} - 3x_n = 12(5^n)$, we get

$$x_{n+1} - 3x_n = A(5^{n+1}) - 3(A \cdot 5^n)$$
$$= A(5 \cdot 5^n - 3 \cdot 5^n)$$
$$= A \cdot 5^n(5 - 3)$$
$$= 2A \cdot 5^n.$$

From the original difference equation, we see that we must have

$$2A \cdot 5^n = 12 \cdot 5^n.$$

Since 5^n is never zero, it follows that $2A = 12$, and so $A = 6$. Therefore the particular solution is $6 \cdot 5^n$, as stated.

E X A M P L E   2

Solve the linear, nonhomogeneous difference equation

$$x_{n+1} + 4x_n = 15 \cdot 3^n.$$

Solution   The solution of the associated homogeneous difference equation

$$x_{n+1} + 4x_n = 0, \quad \text{or} \quad x_{n+1} = -4x_n$$

is given by

$$x_n = C(-4)^n.$$

Further, the nonhomogeneous term $15 \cdot 3^n$ is a constant multiple of 3^n, so we expect that a particular solution is also a multiple of 3^n. Hence we assume that $x_n = A(3^n)$, for some constant A to be determined. We therefore have

$$x_{n+1} = A(3^{n+1}) = A \cdot 3 \cdot 3^n.$$

When we substitute these expressions for x_n and x_{n+1} into the left-hand side of the difference equation, we find

$$
\begin{aligned}
x_{n+1} + 4x_n &= A \cdot 3 \cdot 3^n + 4(A \, 3^n) \\
&= A(3 \cdot 3^n + 4 \cdot 3^n) \\
&= A \, 3^n(3 + 4) \\
&= 7A \, 3^n.
\end{aligned}
$$

From the original nonhomogeneous difference equation, we therefore have

$$7A(3^n) = 15(3^n).$$

So

$$A = \frac{15}{7}.$$

Thus the particular solution is $(15/7)3^n$ and the complete solution of the difference equation is

$$x_n = (-4)^n C + \frac{15}{7} \cdot 3^n,$$

where C is some arbitrary constant. Verify this result by substituting the expressions for x_n and x_{n+1} into the difference equation and showing that

$$x_{n+1} + 4x_n = 15 \cdot 3^n.$$

The particular solution x_n of a nonhomogeneous difference equation usually has the same mathematical form as the nonhomogeneous term b_n of the difference equation. If this term is an exponential function k^n, then we assume that x_n is also a multiple of k^n. If the nonhomogeneous term is a polynomial in n, then we assume that x_n is also a polynomial in n of the same degree. The process we use to find the particular terms that work is called the *method of undetermined coefficients*, and the expression we try for x_n is called the *trial solution*.

In summary, to solve the first order, linear, nonhomogeneous difference equation

$$x_{n+1} = ax_n + b_n$$

we typically perform the following steps.

1. Write the solution $C(a^n)$ of the associated homogeneous difference equation

$$x_{n+1} = ax_n.$$

2. Find any particular solution of the nonhomogeneous difference equation using the method of undetermined coefficients.
3. The complete solution is the sum of these two expressions.
4. If an initial condition x_0 is given, use it to determine the value of the arbitrary constant C and thus produce the specific solution that satisfies the initial condition.

E X A M P L E   3

a. Find the complete solution of the difference equation

$$x_{n+1} - 2x_n = n + 4.$$

b. Find the specific solution if $x_0 = 10$.

Solution   a. The solution of the associated homogeneous difference equation

$$x_{n+1} - 2x_n = 0, \quad \text{or} \quad x_{n+1} = 2x_n$$

is

$$x_n = C(2^n).$$

Further, the right-hand side of the difference equation, $n + 4$, is a polynomial of degree one, so we assume that the particular solution we seek is also a polynomial of degree one. Therefore we write

$$x_n = An + B,$$

where A and B are two constants whose values must be determined. It then follows that

$$x_{n+1} = A(n + 1) + B.$$

When we substitute these expressions into the left-hand side of the difference equation and then collect like terms, we obtain

$$\begin{aligned} x_{n+1} - 2x_n &= [A(n + 1) + B] - 2[An + B] \\ &= An + A + B - 2An - 2B \\ &= -An + (A - B). \end{aligned}$$

Looking back at the original nonhomogeneous difference equation, we must have

$$-An + (A - B) = n + 4.$$

We now equate the coefficients of n to find that

$$-A = 1$$
$$A = -1.$$

Also, when we equate the constant terms, we find that

$$(A - B) = 4$$

or, since $A = -1$,

$$-1 - B = 4$$
$$B = -5.$$

Therefore the particular solution is $An + B = -n - 5$, and so the complete solution to the difference equation is

$$x_n = C(2^n) - n - 5.$$

b. Using the initial condition $x_0 = 10$, we have when $n = 0$,

$$x_0 = C \cdot 2^0 - 0 - 5 = C - 5.$$

So

$$C - 5 = 10$$
$$C = 15.$$

Therefore the specific solution satisfying this initial condition is

$$x_n = 15(2^n) - n - 5.$$

Now let's summarize the rules needed to apply the given procedure to the most common forms possible for the nonhomogeneous term. First, the method of undetermined coefficients applies when the difference equation is *linear* and x_n has a *constant coefficient a*:

$$x_{n+1} = ax_n + b_n \qquad \text{or} \qquad x_{n+1} - ax_n = b_n.$$

Second, we use the form of b_n to decide what type of expression to use as the trial solution: If b_n on the right consists of

an exponential term k^n,	then use Ak^n
a linear term $an + b$,	then use $An + B$
a quadratic term $an^2 + bn + c$,	then use $An^2 + Bn + C$
a sum or difference of such terms,	then use a sum or difference of such terms.

For instance, if the right-hand side is $2(5^n) - 7(3^n)$, then we would use the trial solution $x_n = A \cdot 5^n + B \cdot 3^n$. If the right-hand side is $8n^2 - 12$, then the trial solution is $x_n = An^2 + Bn + C$. (Notice that even though no linear term is part of $8n^2 - 12$, we must include the term Bn as part of the quadratic trial solution because the trial solution represents *any* possible quadratic function.) If the right-hand side is $4(5^n) + 8n$, then the trial solution is $x_n = A(5^n) + Bn + C$. (Notice that we must include the C term since the trial solution represents *any* possible linear function.)

E X A M P L E   4

Solve the difference equation

$$x_{n+1} - 5x_n = 8n^2 - 12.$$

Solution   The solution of the associated homogeneous difference equation is $x_n = 5^nD$, where D is any arbitrary constant. Since the right-hand side of the nonhomogeneous equation is a quadratic function of n, we use

$$x_n = An^2 + Bn + C$$

as the trial solution. Consequently,

$$
\begin{aligned}
x_{n+1} &= A(n + 1)^2 + B(n + 1) + C \\
&= A(n^2 + 2n + 1) + Bn + B + C.
\end{aligned}
$$

When we substitute these expressions into the left-hand side of the difference equation and collect like terms, we obtain

$$
\begin{aligned}
x_{n+1} - 5x_n &= [A(n^2 + 2n + 1) + Bn + B + C] - 5[An^2 + Bn + C] \\
&= An^2 + 2An + A + Bn + B + C - 5An^2 - 5Bn - 5C \\
&= -4An^2 + (2A - 4B)n + (A + B - 4C).
\end{aligned}
$$

We therefore must have

$$-4An^2 + (2A - 4B)n + (A + B - 4C) = 8n^2 - 12.$$

Equating the coefficients of like terms, we find the following.

> Quadratic terms: $-4A = 8$,   so $A = -2$.
>
> Linear terms: $2A - 4B = 0$,   so $4B = 2A$.
>
> Since $A = -2$, we have $B = -1$.
>
> Constant terms: $A + B - 4C = -12$,   so
>
> $4C = 12 + A + B$.
>
> Since $A = -2$ and $B = -1$, we have $C = \dfrac{9}{4}$.

Consequently, the particular solution of the nonhomogeneous equation is

$$-2n^2 - n + \frac{9}{4},$$

and the complete solution of the difference equation is

$$x_n = 5^nD - 2n^2 - n + \frac{9}{4}.$$

There is one exception to the rules summarized above Example 4. It occurs when a term in b_n on the right-hand side of the nonhomogeneous equation matches a term that arises in the solution to the associated homogeneous difference equation. For instance, in the difference equation

$$x_{n+1} - 3x_n = 8 \cdot 3^n,$$

the solution to the associated homogeneous equation is

$$x_n = C \cdot 3^n,$$

which represents every possible multiple of 3^n, since C is an arbitrary constant. If we tried to apply the given rules, we would use $x_n = A \cdot 3^n$ as the trial solution for the nonhomogeneous equation, where A is some constant. However, this term is no different from the homogeneous solution. In fact, if you did try this, you would be led to an algebraic contradiction because we are not introducing a new term for the trial solution.

Think About This

Substitute $x_n = A(3^n)$ into the difference equation and attempt to solve for A.

In such a case, there is a simple "fix": Multiply the usual trial solution by the variable n to produce an appropriate new form for the particular solution. In this case, it is

$$x_n = An(3^n).$$

We illustrate this procedure in the following example.

E X A M P L E 5

Solve

$$x_{n+1} - 3x_n = 8 \cdot 3^n.$$

Solution The homogeneous solution consists of $x_n = C3^n$, for some arbitrary constant C. Since 3^n also occurs on the right-hand side in the nonhomogeneous term, we try $x_n = An3^n$ for the trial solution. Thus we have

$$x_{n+1} = A(n+1)3^{n+1}.$$

When we substitute into the left-hand side of the original difference equation, we find

$$
\begin{aligned}
x_{n+1} - 3x_n &= A(n+1)3^{n+1} - 3A \cdot n \cdot 3^n \\
&= A(n+1) \cdot 3 \cdot 3^n - 3A \cdot n \cdot 3^n \\
&= A \cdot 3^n[(n+1) \cdot 3 - 3n] \\
&= A \cdot 3^n[3n + 3 - 3n] \\
&= A \cdot 3^n[3].
\end{aligned}
$$

We therefore get

$$3A \cdot 3^n = 8 \cdot 3^n$$
$$3A = 8$$
$$A = \frac{8}{3}.$$

Hence the particular solution of the nonhomogeneous equation is $\frac{8}{3}n(3^n)$, and consequently the complete solution to the difference equation is

$$x_n = C3^n + \frac{8}{3}n(3^n).$$

Verify that this is indeed the complete solution by substituting the expressions for x_n and x_{n+1} into the original difference equation and showing that

$$x_{n+1} - 3x_n = 8 \cdot 3^n.$$

In all of these examples, we have been looking for the solution of a difference equation in "closed form." That is, we have sought a solution for which we could write a precise formula. As we pointed out in Chapter 5, it is nice to have such a result, but not every difference equation can be solved in this manner. In fact, it is relatively rare that a given equation can be solved in closed form. As we have said, no one has been able to find a closed form solution for the logistic difference equation.

Fortunately, there are other ways to generate a solution to any difference equation. In particular, the difference equation itself provides us with a way to *calculate* each term in the sequence for the solution. That is, once we select the initial value x_0, we can use the difference equation itself to produce each successive term, one after another. This is precisely what we did previously for solutions to the logistic equation. In fact, we could use this constructive approach to generate the terms of the solutions to the preceding difference equations instead of determining the closed form solutions as we did in the examples.

In the rest of this chapter, we will consider a variety of applications that lead to first order linear difference equations because we now have the means to solve most of the difference equations that arise.

EXERCISES

Find the complete solution for the difference equations in Exercises 1–12.

1. $x_{n+1} = 1.5x_n$

2. $x_{n+1} = -1.5x_n$

3. $x_{n+1} = -2x_n + 20 \cdot 5^n$

4. $y_{n+1} = \frac{1}{2}y_n - 12 \cdot 2^n$

5. $z_{n+1} = 5z_n + 12 \cdot 3^n$ **6.** $w_{n+1} = w_n + 3 \cdot \left(\frac{1}{2}\right)^n$

7. $x_{n+1} = 2x_n - 20 \cdot 4^n$ **8.** $y_{n+1} = \frac{1}{4}y_n - 5 \cdot 2^n$

9. $z_{n+1} = 3z_n + 12n + 2$ **10.** $x_{n+1} = 2x_n + n^2 - 12n + 20$

11. $x_{n+1} - 4x_n = 12 \cdot 4^n$ **12.** $x_{n+1} - x_n = 6n$

13–20. In Exercises 3–10, find the specific solution corresponding to the initial condition that, when $n = 0$, the initial term in the solution sequence is 1. How do these answers change if the initial condition is 2 instead of 1?

21. a. Find the complete solution of the difference equation

$$x_{n+1} = \frac{1}{4}x_n + 6\left(\frac{1}{3}\right)^n.$$

b. Assuming that the arbitrary constant $C = 1$, calculate the first 10 terms of the solution and plot them. What are the values for x_{15}, x_{20}, and x_{25}? What do you think happens to the solution as n continues to increase?

c. Repeat part (b) if $C = 100$.

22. The following difference equations were developed in Example 3 of Section 5.3 to model the level of contaminant in a lake under different scenarios. Solve each one and check that you obtain the same solutions cited in Example 3.

a. $C_{n+1} = 0.9C_n$, $C_0 = 500$

b. $C_{n+1} = 0.9C_n + 100$, $C_0 = 500$

c. $C_{n+1} = 0.9C_n + 100 + 25n$, $C_0 = 500$

d. $C_{n+1} = 0.9C_n + 100(1.20)^n$, $C_0 = 500$

e. $C_{n+1} = 0.9C_n + 100(0.75)^n$, $C_0 = 500$

6.3   *Applications of First Order, Nonhomogeneous Difference Equations*

So far, we have developed the mechanics for constructing solutions of first order, linear, nonhomogeneous difference equations of the form

$$x_{n+1} = ax_n + b_n \quad \text{or} \quad x_{n+1} - ax_n = b_n.$$

In this section, we demonstrate how such difference equations arise in various fields and how the techniques we developed in the last section let us produce the solutions directly and easily.

We begin with several situations that we already have encountered. Our intention is to show how these new methods provide us with powerful tools that replace either the brute force techniques or the gimmicks to which we were limited in the last chapter.

E X A M P L E   1

In Section 5.4, we developed the difference equation

$$D_{n+1} = 0.75D_n + 16, \qquad \text{or} \qquad D_{n+1} - 0.75D_n = 16$$

to model the level of a drug in the bloodstream, based on the assumptions that the kidneys remove 25% of the drug in any four-hour period and the patient takes an additional 16 mL of the drug every four hours. Find the specific solution to this difference equation and determine the maintenance level for the drug.

Solution   The equation is a first order, linear, nonhomogeneous difference equation. The solution to the corresponding homogeneous equation

$$D_{n+1} = 0.75D_n$$

is $C(0.75)^n$, where C is any arbitrary constant. For the nonhomogeneous case, the right-hand side is a constant, so we assume that the trial solution is also a constant, $D_n = A$. Consequently, $D_{n+1} = A$ as well, so that when we substitute D_n and D_{n+1} into the left-hand side of the difference equation, we get

$$D_{n+1} - 0.75D_n = A - 0.75A = 0.25A.$$

We therefore must have

$$0.25A = 16$$
$$A = 64.$$

The complete solution is therefore

$$D_n = C(0.75)^n + 64.$$

Substituting $n = 0$ into this solution of the difference equation, we find

$$D_0 = C(0.75)^0 + 64 = C + 64.$$

Since the initial dose is $D_0 = 16$, we find that $C = -48$. Therefore, the specific solution subject to the initial condition $D_0 = 16$ is

$$D_n = 64 - 48(0.75)^n.$$

Notice that this is the identical expression we created for the solution in Section 5.4 by observing the behavior of the solution as an inverted decaying exponential function. Notice also that the term $(0.75)^n$ approaches 0 as n increases, and so we see that the limiting value for D_n is 64 mL; we have seen that this is the maintenance level for the drug (see Figure 6.7).

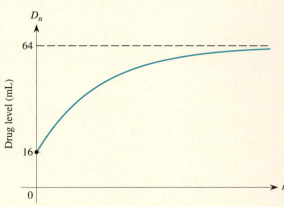

FIGURE 6.7

You may recall that, in Section 5.4 we raised the question of what happens if a person takes an initial dose higher than the maintenance level and then takes repeated doses of 16mL thereafter. We are now in a position to analyze this situation precisely because we can easily solve the difference equation with a different initial condition, say $D_0 = 100$ mL instead of 16 mL. The complete solution is still $D_n = C(0.75)^n + 64$. In this case when $n = 0$,

$$D_0 = C(0.75)^0 + 64 = C + 64 = 100,$$

and so $C = 36$. The specific solution corresponding to this initial condition $D_0 = 100$ is

$$D_n = 64 + 36(0.75)^n.$$

It is clear that the exponential term must die out as n increases. However, notice that the coefficient of this term is positive, so all values for D_n are greater than 64. Thus the solution decays from an initial value of 100 mL toward the maintenance level of 64 mL (see Figure 6.8).

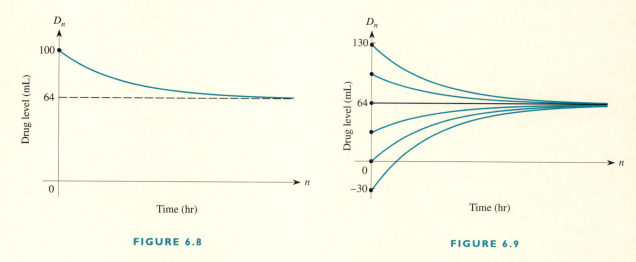

FIGURE 6.8   FIGURE 6.9

Figure 6.9 illustrates the solution field associated with the difference equation

$$D_{n+1} = 0.75D_n + 16$$

for the drug level model. We find the equilibrium solution for the difference equation by substituting $D_{n+1} = D_n = L$ into the difference equation. We thus get

$$L = 0.75L + 16$$

$$0.25L = 16.$$

So the equilibrium level is

$$L = \frac{16}{0.25} = 64,$$

which is the maintenance level for the drug.

Now let's consider the behavior of the solutions. For initial values greater than the maintenance level, the solutions decay in a concave up pattern toward that equilibrium level. For initial values between 0 and the maintenance level, all solutions increase steadily toward the maintenance level in a concave down pattern. Although the model makes no practical sense when D_0 is negative, it still is possible to generate solutions to the difference equation. Notice that when D_0 is negative, the solutions still grow toward the same equilibrium value.

The reason for the behavior patterns for the different solutions is quite clear if we look at the formula for the solutions. In Example 1, we found the complete solution to be

$$D_n = C(0.75)^n + 64.$$

When we consider any initial condition D_0, we have

$$D_0 = C(0.75)^0 + 64 = C + 64,$$

and so the arbitrary constant $C = D_0 - 64$. Therefore the specific solution to this difference equation is

$$D_n = (D_0 - 64)(0.75)^n + 64.$$

Realize that the term $(0.75)^n$ is a decaying exponential function. Whenever $D_0 > 64$, the coefficient of $(0.75)^n$ is positive. Thus the solution consists of 64 plus a decaying exponential term; it starts at a level higher than 64 and decays toward the equilibrium level at 64. On the other hand, whenever $D_0 < 64$, including negative values for D_0, the coefficient $(D_0 - 64)$ is negative. The solution then consists of 64 plus a negative multiple of a decaying exponential function. The solution starts below the equilibrium level 64 and rises toward it over time.

EXAMPLE 2

When their son Jeremy was born, Steve and Marian established a college savings plan to pay for his education. They agreed to put $1000 each year into the plan, which pays 5% interest every year. What will be the value of the plan on Jeremy's eighteenth birthday?

Solution   The balance in this account grows at the rate of 5% per year, and the balance is augmented each year by $1000. Therefore, if B_n represents the balance after n years, it satisfies the difference equation

$$B_{n+1} = 1.05B_n + 1000 \quad \text{or} \quad B_{n+1} - 1.05B_n = 1000,$$

which is a first order, linear, nonhomogeneous difference equation. The corresponding homogeneous difference equation is

$$B_{n+1} = 1.05B_n,$$

whose solution is

$$B_n = C(1.05)^n,$$

where C is an arbitrary constant. Because the right-hand side of the nonhomogeneous difference equation is the constant 1000, we assume a constant A for the trial solution. Thus, we have $B_n = A$ and $B_{n+1} = A$, and so

$$B_{n+1} - 1.05B_n = A - 1.05A = -0.05A.$$

From the difference equation, we therefore see that

$$-0.05A = 1000$$
$$A = -20,000.$$

As a result, the complete solution to the difference equation is

$$B_n = C(1.05)^n - 20,000.$$

When we use the initial condition $B_0 = 1000$, we find that

$$C(1.05)^0 - 20,000 = C - 20,000 = 1000,$$

and so $C = 21,000$. Hence the specific solution is

$$B_n = 21,000(1.05)^n - 20,000.$$

The balance on Jeremy's eighteenth birthday therefore will be

$$B_{18} = 21,000(1.05)^{18} - 20,000 = \$30,539.$$

Sum of the Terms in a Geometric Sequence

We first encountered the sum of the terms in a finite geometric sequence with common ratio r in Section 5.7, where we obtained the formula for the sum of the terms $1, r, r^2, \ldots, r^n$ as

$$1 + r + r^2 + \cdots + r^n = \frac{1 - r^{n+1}}{1 - r}$$

by applying a clever trick. Unfortunately, this trick does not apply to any other situation, so it is of limited use. Alternatively, we can derive this formula using our methods for solving difference equations. While more complicated than the earlier approach, this method does provide us with a means for addressing more complicated situations later. Suppose we start with

$$S_n = 1 + r + r^2 + \cdots + r^n.$$

So

$$S_{n+1} = 1 + r + r^2 + \cdots + r^n + r^{n+1}$$
$$= (1 + r + r^2 + \cdots + r^n) + r^{n+1}$$
$$= S_n + r^{n+1}.$$

This equation is equivalent to

$$S_{n+1} - S_n = r^{n+1},$$

which is a first order, linear, nonhomogeneous difference equation in S_n. The associated homogeneous difference equation is

$$S_{n+1} = S_n = 1 \cdot S_n,$$

whose solution is simply

$$S_n = C(1)^n = C$$

for an arbitrary constant C. To find a particular solution to the nonhomogeneous difference equation, we observe that the right-hand term is just r^{n+1}, so we assume the trial solution also is a multiple of r^{n+1}; that is,

$$S_n = A \cdot r^{n+1}.$$

Therefore

$$S_{n+1} = A \cdot r^{n+2} = Ar \cdot r^{n+1}.$$

When we substitute for S_n and S_{n+1} in the left-hand side of the difference equation, we get

$$Ar \cdot r^{n+1} - Ar^{n+1} = Ar^{n+1}(r-1).$$

We therefore must have

$$Ar^{n+1}(r-1) = r^{n+1}.$$

If $r \neq 0$, we can divide both sides by the common term r^{n+1}, which gives

$$A(r-1) = 1$$

$$A = \frac{1}{r-1},$$

provided that $r \neq 1$. The nonhomogeneous solution is therefore

$$\frac{1}{r-1} r^{n+1},$$

and hence the complete solution is

$$S_n = \frac{1}{r-1} r^{n+1} + C.$$

If we now set $n = 0$ and use the fact that the initial term $S_0 = 1$, we find that

$$S_0 = \frac{1}{r-1} r^1 + C$$

$$= \frac{r}{r-1} + C = 1.$$

We can determine the value for C by combining terms over a common denominator:

$$C = 1 - \frac{r}{r-1}$$

$$= \frac{(r-1) - r}{r-1}$$

$$= \frac{-1}{r-1}.$$

Thus the sum of the terms $1, r, r^2, \ldots, r^n$ of any geometric sequence (provided that $r \neq 1$) is given by

$$S_n = \frac{1}{r-1} r^{n+1} - \frac{1}{r-1}$$

$$= \frac{r^{n+1} - 1}{r-1}$$

$$= \frac{1 - r^{n+1}}{1-r},$$

which is the same formula we saw before.

Sums of Integers

In Section 4.6, we developed the formulas for the sum of the first n integers and the sum of their squares:

$$\sum_{k=1}^{n} k = 1 + 2 + 3 + \cdots + n = \frac{n(n+1)}{2}$$

$$\sum_{k=1}^{n} k^2 = 1^2 + 2^2 + 3^2 + \cdots + n^2 = \frac{n(n+1)(2n+1)}{6}.$$

We now show how these results can be proved for all values of n using difference equations. The important thing is not the particular result, but rather the demonstration of how a variety of different and powerful techniques can be brought to bear on the same problem.

We begin with the sum of the first n integers. Let

$$S_n = 1 + 2 + 3 + \cdots + n.$$

Now consider the sum of the first $n + 1$ terms:

$$S_{n+1} = 1 + 2 + 3 + \cdots + n + (n+1)$$

$$= [1 + 2 + 3 + \cdots + n] + (n+1)$$

$$= S_n + (n+1).$$

This last equation is equivalent to

$$S_{n+1} - S_n = n + 1,$$

which is a first order, linear, nonhomogeneous difference equation. The solution of the corresponding homogeneous difference equation

$$S_{n+1} - S_n = 0$$

is $A(1)^n = A$, where A is any arbitrary constant. To solve the nonhomogeneous equation, we observe that the right-hand side, $n + 1$, is a polynomial of degree 1 in n, so we might be tempted to try $S_n = Bn + C$, where B and C are two constants to be determined. Using this expression for S_n, we get

$$S_{n+1} = B(n + 1) + C.$$

So the left-hand side of the difference equation yields

$$\begin{aligned} S_{n+1} - S_n &= [B(n + 1) + C] - [Bn + C] \\ &= Bn + B + C - Bn - C \\ &= B. \end{aligned}$$

We therefore must have

$$B = n + 1.$$

Since B is supposed to be a constant, it cannot be equal to the variable term $n + 1$, which has a different value for each value of n. Something is wrong!

The problem here is that the homogeneous solution is just a constant, A, and the trial solution we used for the nonhomogeneous solution $S_n = Bn + C$ also includes a constant term, C; but we did not introduce a new term. (See Example 5 of Section 6.2, where we first considered this type of situation.) As we did there, we must compensate for the duplication of terms by multiplying the polynomial in the trial solution by n. We therefore consider instead

$$S_n = n(Bn + C) = Bn^2 + Cn,$$

and so

$$S_{n+1} = B(n + 1)^2 + C(n + 1).$$

Substituting these expressions into the left-hand side of the difference equation, we get

$$\begin{aligned} S_{n+1} - S_n &= [B(n + 1)^2 + C(n + 1)] - [Bn^2 + Cn] \\ &= [B(n^2 + 2n + 1) + Cn + C] - Bn^2 - Cn \\ &= 2Bn + B + C. \end{aligned}$$

Therefore we must have

$$2Bn + B + C = n + 1.$$

When we equate coefficients of like terms, we find

$$2B = 1 \quad \text{and} \quad B + C = 1.$$

Thus

$$B = \frac{1}{2}$$

and

$$C = 1 - B = \frac{1}{2}.$$

As a result, the complete solution to the difference equation is

$$S_n = A + n\left(\frac{1}{2}n + \frac{1}{2}\right).$$

To determine the arbitrary constant A, we use the fact that $S_1 = 1$. We substitute $n = 1$ into the expression for the solution to find

$$S_1 = A + (1)\left(\frac{1}{2} + \frac{1}{2}\right) = A + 1,$$

and since $S_1 = 1$, we immediately get

$$A = 0.$$

Consequently, the specific solution is

$$S_n = n\left(\frac{1}{2}n + \frac{1}{2}\right) = \frac{1}{2}n(n + 1) = \frac{n(n + 1)}{2},$$

which is the same formula we saw before for the sum of the first n integers.

We ask you to derive the comparable formula for the sum of the squares of the first n integers in an exercise in this section.

EXERCISES

1. Suppose that the periodic dosage used in Example 1 is doubled to 32 mL.
   a. Write the difference equation for this scenario.
   b. Find the complete solution of the difference equation.
   c. What is the effect on the maintenance level?

2. If the periodic dosage in Example 1 is halved to 8 mL, what is the effect on the maintenance level?

3. Suppose that the interest rate on the savings plan in Example 2 was only 4% per year. What would be the balance in the account after 18 years?

4. Suppose that the interest rate in Example 2 was 6%. What would be the balance after 18 years?

5. Suppose that Steve and Marian wanted the account in Example 2 to reach $100,000 in 18 years. Assuming the interest rate is 5% per year, how much must they contribute each year?

6. Suppose that Steve and Marian set up the savings plan in Example 2 in such a way that they will increase their contribution each year by $100 after starting with the initial $1000. Solve the corresponding difference equation

$$B_{n+1} = 1.05B_n + 100n + 1000$$

and find the value after 18 years.

7. Repeat Exercise 6 assuming Steve and Marian increase their annual contribution by $250 each year.

8. Suppose that the annual contribution to the savings plan in Example 2 increases 10% per year after starting with $1000. Set up the appropriate difference equation and solve it to find the value of the plan after 18 years.

9. An IRA (individual retirement account) allows a person to put aside a maximum of $2000 each year as a tax-deferred investment for retirement. Suppose that Craig intends to put the maximum of $2000 into an IRA account paying 6% per year starting at age 25. What will be the value of this account when he retires at age 65?

10. The population of the United States is growing at an annual rate of 0.7% and the 1990 population was 248 million. If one million immigrants enter the country each year, set up the appropriate difference equation and solve it to find the U.S. population in the year 2000.

11. (Continuation of Exercise 10.) Suppose that the number of immigrants permitted to enter the United States increases a quarter million each year. Set up the appropriate difference equation and solve it to find what the population will be in the year 2000.

12. In Exercise 7 of Section 5.3 and in Exercise 15 of Section 5.5 you were asked to construct a set of difference equations to model the growth of the fish population in a lake and the deer population in a wildlife refuge. Identify which of these difference equations are linear or nonlinear and which are homogeneous or nonhomogeneous. For those that are linear, find the complete solution.

13. Psychologists have developed a learning model to account for the way that an individual learns any new body of knowledge or develops any new skill. (Picture, for example, learning 250 new words for a vocabulary test by going through a set of flashcards repeatedly.) If K_n represents the amount of knowledge that a person has after n time periods, psychologists have found that the gain or change in knowledge ΔK_n is proportional to $100 - K_n$, the amount that the person does not know. So K_n satisfies the difference equation

$$\Delta K_n = m(100 - K_n),$$

where m is a constant of proportionality and K_0 is the initial amount of the person's knowledge or initial level of skill.

   a. Is the constant of proportionality in the difference equation positive or negative? Is it less than 1 or greater than 1?
   b. Find the complete solution to this learning model difference equation.
   c. Sketch the graph of this solution to the learning curve model. Does it behave as you would expect?

    d. What would the model predict if you took $K_0 > 100$? Can you give any reasonable interpretation for this situation?

    e. Suppose that Paul is trying to improve his free-throw shooting percentage in basketball. He practices free throws for an hour every day. When he starts, he makes 20% of his shots; on his tenth day of practice, he makes 40%. Use your result from part (b) to find the formula to predict Paul's percentage of success on any day. How long will it take him to increase his success rate the next 20%—from 40% to 60%?

14. Consider the difference equation $\Delta x_n = f(n)$. (Note that the differences are a function of n only: They do not involve x_n at all.) For each of the functions f shown below, the solution x_n eventually behaves like a function with which you are familiar. In each case, identify the type of function that the solution eventually looks like.

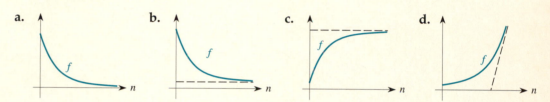

15. Chlorine is added to swimming pools to control the growth of microorganisms. If too little chlorine is present, green slime appears; if too much chlorine is present, it burns the eyes of swimmers. Chlorine is usually added to a pool each day since about 15% of the amount present each day dissipates due to reactions with bacteria and the effects of sunlight. The optimal concentration that should be maintained is between 1 and 2 parts per million (ppm). Determine the smallest and largest "doses" of chlorine, in ppm, that should be added to a pool each day to maintain the concentration between the desired levels. Based on your results, what would be *your* recommendation for the appropriate daily dose?

16. Derive the formula

$$S_n = 1^2 + 2^2 + 3^2 + \cdots + n^2 = \frac{n(n + 1)(2n + 1)}{6}$$

by considering S_{n+1} and using it with S_n to form a first order, nonhomogeneous difference equation and then solving the difference equation to get the desired expression for S_n, for all n.

17. Consider the sum of the first n cubes

$$1^3 + 2^3 + 3^3 + \cdots + n^3.$$

What form might you suspect a formula for this sum will take? Outline two different methods you could apply to determine what this formula actually is. (Note that we are not asking you to do the actual work, only to think through how you would proceed.)

6.4 *Fitting Quadratic Functions to Data*

A financial analyst is studying the growth in revenues of K-Mart, one of the nation's largest retailers. She finds that the annual revenues, in billions of dollars, between 1984 and 1991 are as shown in the following table.

Year	Revenue
1984	21.1
1985	22.5
1986	23.8
1987	25.5
1988	27.4
1989	29.5
1990	32.0
1991	34.6

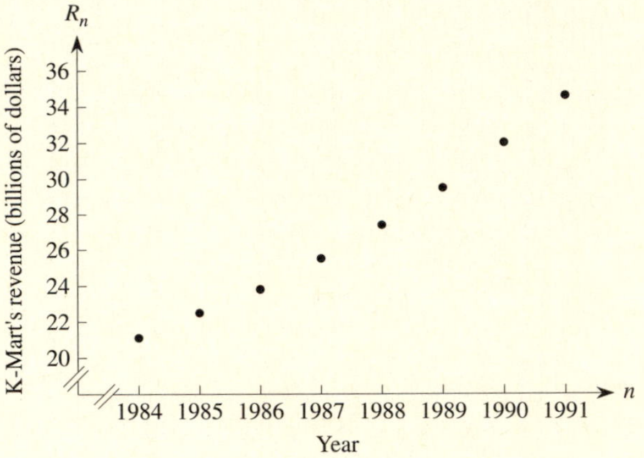

FIGURE 6.10

She constructs the scatterplot for this data, as shown in Figure 6.10, and decides that the pattern is apparently nonlinear. It is clearly increasing and is concave up, so the annual revenues are growing ever more rapidly. She verifies this conclusion numerically by examining the sequence of differences ΔR_n for the values of the yearly revenues R_n and concludes that ΔR_n is certainly not constant.

She then examines the ratios of successive terms for the revenues in the hopes of demonstrating that the growth pattern is exponential. However, she finds that the successive ratios are not constant (you should verify this), and so concludes that the pattern in the data is not exponential. She verifies this by examining the residuals after the data values have been transformed and $\log R_n$ is plotted against n. There appears to be a pattern in the residuals suggesting that an exponential model is not a good fit to the data. In a similar way, she transforms the data values and compares $\log R_n$ to $\log n$ and inspects the resulting residuals. She then concludes that a power function is also not a particularly good fit to the data.

At this point, the financial analyst suspects that the revenue data for K-Mart may follow a polynomial growth pattern. How does she determine if there is indeed a polynomial pattern to the data? If there is a polynomial pattern, how does she determine what this polynomial is?

We have previously seen that a linear model fits a set of data precisely if the successive differences ΔR_n are all constant. Thus, if the differences are *approximately* constant, a linear fit is appropriate. We have also seen that the second differences $\Delta^2 R_n$ can provide similar information about whether a quadratic model fits a set of values. If the second differences are all constant, then the points lie on a parabola. If the second differences are *approximately* constant, then a quadratic polynomial should provide a good fit to the data. Let's examine the various differences for the K-Mart revenues. For the first two years, with $R_0 = 21.1$ and $R_1 = 22.5$, we have $\Delta R_0 = 22.5 - 21.1 = 1.4$. We display all such calculations in the following table.

Year	Revenue	ΔR_n	$\Delta^2 R_n$
1984	21.1		
		1.4	
1985	22.5		-0.1
		1.3	
1986	23.8		0.4
		1.7	
1987	25.5		0.2
		1.9	
1988	27.4		0.2
		2.1	
1989	29.5		0.4
		2.5	
1990	32.0		0.1
		2.6	
1991	34.6		

As we indicated above, the first differences are clearly not constant; in fact, they seem to be growing fairly steadily, at least after the first entry, 1.4. However, notice that the second differences seem to be relatively constant. We therefore conclude that the data values fall into a roughly quadratic pattern. We then apply the regression technique of Chapter 3 to the linearized data given by ΔR_n compared to the number of years n since 1984 (see Figure 6.11). The corresponding regression line for ΔR_n versus n is

$$\Delta R_n = 0.23n + 1.24.$$

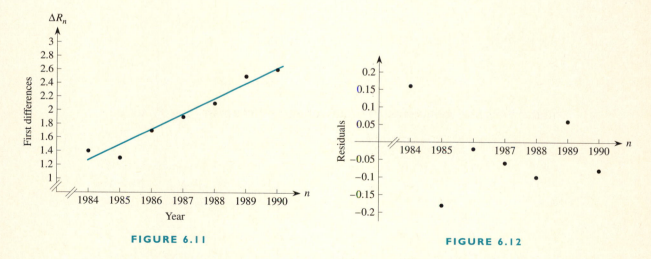

FIGURE 6.11   **FIGURE 6.12**

The associated residual plot is shown in Figure 6.12. We notice that the residuals are quite small and there appears to be no pattern to them, which suggests that a linear function is a reasonably good fit to the transformed data. Further, the corresponding correlation coefficient $r = 0.976$ indicates a high degree of positive linear correlation between ΔR_n and n.

Now, since $\Delta R_n = R_{n+1} - R_n$, the above equation becomes

$$R_{n+1} - R_n = 0.23n + 1.24,$$

which is a first order, linear, nonhomogeneous difference equation whose solution, R_n, is the best relationship between R_n and n. The solution of the associated homogeneous difference equation is

$$R_n = C(1^n) = C,$$

where C is an arbitrary constant. Since the nonhomogeneous term also includes a constant, our trial solution for the nonhomogeneous solution is

$$R_n = n(An + B) = An^2 + Bn.$$

As a consequence,

$$\begin{aligned} R_{n+1} &= A(n + 1)^2 + B(n + 1) \\ &= A(n^2 + 2n + 1) + B(n + 1). \end{aligned}$$

We now substitute these quantities into the left-hand side of the difference equation to get

$$A(n^2 + 2n + 1) + B(n + 1) - An^2 - Bn = 2An + A + B.$$

So from the difference equation, we get

$$2An + A + B = 0.23n + 1.24.$$

When we equate coefficients of the linear term, we find that

$$2A = 0.23$$
$$A = 0.115.$$

When we equate the constant terms, we find that

$$A + B = 1.24$$
$$B = 1.24 - A = 1.125.$$

Hence the complete solution to the difference equation is

$$R_n = 0.115n^2 + 1.125n + C.$$

If we now use the initial condition $R_0 = 21.1$ for the revenues in 1984 (when $n = 0$), we find that $C = 21.1$. Therefore the desired quadratic relationship is

$$R_n = 0.115n^2 + 1.125n + 21.1$$

for all $n \geq 0$, where n is the number of years since 1984 (see Figure 6.13).

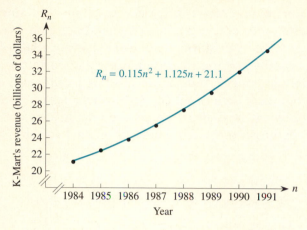

FIGURE 6.13

It is worth noting that the method built into graphing calculators for finding the best fit quadratic is different from the one discussed here, so the results obtained with the two approaches might be slightly different. For instance, applying a graphing calculator's quadratic fit routine to the K-Mart data, we get

$$R_n = 0.119n^2 + 1.083n + 21.175,$$

which is slightly different from the result we previously found.

Also, the method used to construct a quadratic function that best fits a set of data is very different from the ways that we construct linear, exponential, or power function fits to a set of data. As a result, it is not possible to calculate a correlation coefficient r for such a fit. Instead, statisticians have developed a different measure for the goodness of fit known as the *coefficient of multiple determination*, denoted by R^2. It is discussed in most statistics texts.

So far, we have used ideas on difference equations to provide a mathematical framework for finding the quadratic function that best fits a set of data. It is sometimes possible to reverse this approach and use the data fitting techniques to construct the solution of a difference equation, as illustrated in the following example.

E X A M P L E

Determine the specific solution to the difference equation

$$x_{n+1} = x_n + 2n + 5, \quad \text{for } x_0 = 3.$$

Solution The associated homogeneous difference equation, $x_{n+1} = x_n$, has $x_n = C$ as its general solution. Normally, we would use $x_n = An + B$ for the trial solution to the nonhomogeneous difference equation. But since the constant term is already accounted for in the homogeneous solution, we use the quadratic function $x_n = n(An + B) = An^2 + Bn$ instead as our trial solution.

Rather than actually solving the difference equation in this way, however, let's consider the following approach. Recall that a difference equation along with an initial condition defines all the subsequent terms in the solution sequence. Our objective is to find a formula for that sequence. Since the initial condition is $x_0 = 3$, the difference equation yields the following values.

When $n = 0$: $x_1 = x_0 + 2(0) + 5 = 8$

When $n = 1$: $x_2 = x_1 + 2(1) + 5 = 15$

When $n = 2$: $x_3 = x_2 + 2(2) + 5 = 24,$

and so on. Eventually, the first 10 terms of the solution sequence are

$$3, 8, 15, 24, 35, 48, 63, 80, 99, 120, 143,$$

as shown in Figure 6.14. We know from our discussion about the trial solution to the nonhomogeneous difference equation and the solution of the associated homogeneous equation that the solution should be a quadratic function of n. So, if we apply the calculator's quadratic fitting routine to the set of values above from the beginning of the solution sequence, we immediately obtain

$$x_n = n^2 + 4n + 3,$$

which is a perfect fit to the data, as shown in Figure 6.15. Thus, this is the specific solution to the difference equation.

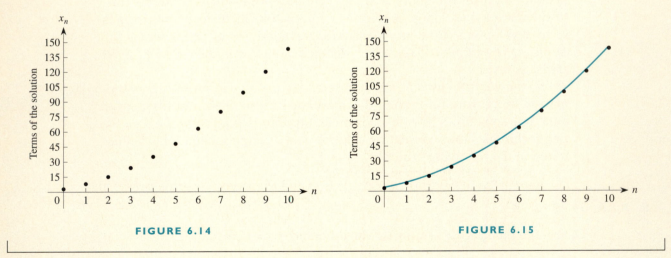

FIGURE 6.14   FIGURE 6.15

Think About This

Verify algebraically that $x_n = n^2 + 4n + 3$ is the specific solution to the difference equation by substituting it into the equation.

Unfortunately, the method we used here works only for a difference equation of the form

$$x_{n+1} = x_n + P(n),$$

where $P(n)$ is a polynomial of degree k in n. In any such case, the corresponding solution will be a polynomial of degree $k + 1$, using the same reasoning as in the Example. However, if the difference equation has just a slightly different form, say

$$x_{n+1} = 2x_n + 2n + 5$$

instead of $x_{n+1} = x_n + 2n + 5$, then the homogeneous solution is $C(2^n)$ and the specific solution will consist of the sum of this exponential term and a linear function that arises from the nonhomogeneous term $2n + 5$. Our curve fitting methods then do not apply because the desired function involves both exponential and linear terms.

Torricelli's Law on Fluid Leaks[1]

An ongoing environmental hazard is posed by old fuel tanks that develop holes through which their contents leak out and consequently contaminate any underlying groundwater. We now investigate the mathematics related to such leakages.

Let's assume that a gasoline tank develops a small hole near the bottom. The fluid will leak out until its level drops to the height of the hole. Consider the rate at which the liquid escapes. Clearly, when the tank is almost full, the rate will be quite high because there is considerable pressure pushing the liquid out the hole. However, as the level of the fluid drops, the pressure diminishes, and so the rate of leakage likewise diminishes. In Figure 6.16 we show the graph of the height of the liquid as a function of time. Notice that it is a decreasing function that is con-

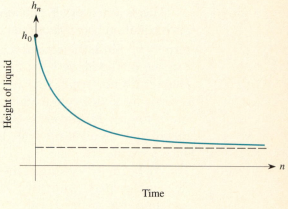

FIGURE 6.16

cave up—the level drops rapidly at first and then slows as time passes and the pressure diminishes. The level of the liquid eventually drops to the height of the hole, which is a horizontal asymptote.

In general, the height of the liquid in an upright circular cylinder is governed by Torricelli's Law, which states that the change in volume V of

---

[1] This development is adapted from *Physical Demonstrations in the Calculus Classroom* by Tom Farmer and Fred Gass in the *College Mathematics Journal*, March 1992.

the liquid draining from the cylinder is proportional to the square root of the height h_n of the liquid. That is,

$$\Delta V_n = k\sqrt{h_n},$$

where k is a constant of proportionality. Note that because the height h_n is decreasing, ΔV_n will be negative, and so the constant k also must be negative. Further, since we assume that the tank is a circular cylinder, its volume $V_n = \pi r^2 h_n$ depends only on the height h_n because the radius r is constant. Thus a change in the volume ΔV_n corresponds to a change in the height Δh_n, or

$$\Delta V_n = \pi r^2 \Delta h_n.$$

When we equate the two expressions for ΔV_n, we have

$$\pi r^2 \Delta h_n = k\sqrt{h_n}$$

$$\Delta h_n = \frac{k}{\pi r^2}\sqrt{h_n} = K\sqrt{h_n},$$

where $K = \dfrac{k}{\pi r^2}$ is a new constant that is also negative. This gives

$$h_{n+1} - h_n = K\sqrt{h_n}.$$

This is a first order, *nonlinear* difference equation because of the nonlinear term $\sqrt{h_n}$. Since the equation is not linear, the methods we have used in this chapter do not apply, and we have no direct way to determine an explicit formula for h_n. However, that does not mean that there is no way to obtain such a result.

We begin by looking at the solution field associated with this difference equation for a fixed value of K, as shown in Figure 6.17. Note that it makes sense to consider only initial values $h_0 \geq 0$ since the height of the liquid at time 0 cannot be negative (and since we cannot take the square root of a negative quantity in the difference equation). Further, the solution field reflects the behavior pattern we desire—the solutions die out as time goes by. Thus we have visual support that the assumptions we used to construct the difference equation lead to a reasonable model for the situation. We can use the difference equation to calculate as many terms of its solution as we

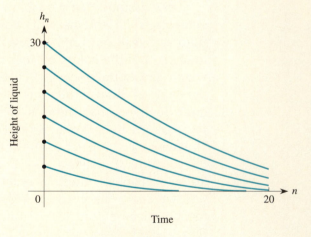

FIGURE 6.17

want for any desired initial condition h_0 and any particular value of the constant of proportionality K.

Now, suppose that we approach this problem the way Torricelli did—by conducting an actual experiment and using the experimental data to obtain a formula for the height h_n. Picture a circular cylinder filled with water that has a small hole near the bottom so that the water is leaking out. We want to measure the water level as time passes. We can use an empty plastic soda bottle that is basically cylindrical in shape if we ignore the top and the base. We ignore the top by not filling the bottle too full; we ignore the base by regarding the "bottom" to be 6 cm, say, above the actual bottom (see Figure 6.18). We drill a 4 mm hole at a height of 3 cm above our "baseline" and fill the bottle to a height of 10 cm above this baseline. As the water leaks out, we record the time in seconds and the height of the liquid in centimeters. In one such experiment, for example, we obtained the following measurements with heights read to the nearest tenth of a centimeter.

Time n (sec)	h_n (cm)
0	10
10	8.7
20	7.5
30	6.4
40	5.4
50	4.4
60	3.5
68	3.0

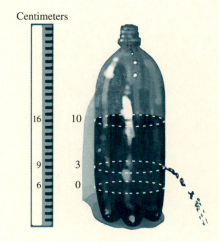

FIGURE 6.18

We can use these data values to estimate the value of the constant of proportionality K. Using the first two measurements, we have

$$h_0 = 10 \quad \text{and} \quad \Delta h_0 = 8.7 - 10 = -1.3.$$

Since $\Delta h_0 = K\sqrt{h_0}$, we get

$$-1.3 = K\sqrt{10},$$

$$K = \frac{-1.3}{\sqrt{10}} = -0.411.$$

Our next thought, based on the graph in Figure 6.16, is that the pattern could be a decaying exponential. We can check this by examining the sequence of successive ratios to see whether they are constant. (Do it, but be careful to omit the final measurement at time $t = 68$ because it is not uniformly spaced with respect to the previous measurements.) Since they are not approximately constant, we might be tempted to try a polynomial function. To check this, we construct a table of differences of the data values (omitting the final measurement at $t = 68$).

Time	h_n	Δh_n	$\Delta^2 h_n$
0	10.0		
		-1.3	
10	8.7		0.1
		-1.2	
20	7.5		0.1
		-1.1	
30	6.4		0.1
		-1.0	
40	5.4		0
		-1.0	
50	4.4		0.1
		-0.9	
60	3.5		

Observe that the sequence of values for the differences Δh_n are clearly not constant; if anything, they appear to be increasing roughly linearly. Further, the second differences, $\Delta^2 h_n$, are nearly constant; the discrepancies could be blamed on slight errors or rounding in measurement. Consequently, we assume that the second differences are constant and so conclude that the data values fall into a quadratic pattern. We find the best quadratic function for h_n in terms of n in the least squares sense using a calculator or computer program. The corresponding quadratic function is

$$h_n = 0.0004414n^2 - 0.13368n + 10.0007.$$

(We note that it is usually not reasonable to carry so many digits in an expression where the original data was measured to one decimal place accuracy. However, if we rounded the above expression for h_n to one or even two decimal places, we would not have a quadratic function.) The resulting predictions for the heights of the liquid based on this function, rounded to two decimal places, are shown below. We see that the predicted values agree quite closely with the actual data measurements, as shown by the table of values and in Figure 6.19.

n	h_n	Predicted
0	10.0	10.00
10	8.7	8.71
20	7.5	7.50
30	6.4	6.39
40	5.4	5.36
50	4.4	4.42
60	3.5	3.57
68	3.0	2.95

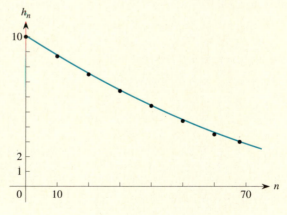

FIGURE 6.19

EXERCISES

1. The table below shows the number of 18- to 24-year-olds in the United States in recent years. Find the quadratic function that best fits this data set using the methods of this section and use it to predict the number of people in this age range in 1995.

Year	1970	1975	1980	1985	1990
Population (millions)	24.71	28.76	30.35	29.48	26.14

2. According to the theory of relativity, the mass of an object increases as its velocity increases, so that $M = f(v)$. Suppose that the mass of an object is 1 unit when it is at rest ($v = 0$). The table below gives values for the mass of this object at different speeds expressed as fractions of c, the speed of light, about 186,280 miles per second. Find the best quadratic fit to this set of data using the methods of this section.

Velocity (fraction of c)	0	0.1	0.2	0.3	0.4	0.5
Mass	1	1.0050	1.0206	1.0483	1.0911	1.1547

3. While approaching the Verrazano Bridge in New York City, Lori noticed that the main cable looks like a parabola. As her car crawled across the bridge in heavy traffic, she estimated the following heights (in feet) of the cable above the road level starting from one of the vertical support columns.

Distance (ft) from support column	0	1000	2150	3000	4000	4300
Estimated height (ft)	500	150	20	100	400	500

Find an equation of the parabola that best fits her estimates using techniques from this section. (Think how to set up the coordinate axes.)

4. A Pythagorean triple is a set of three integers, a, b, and c, that satisfy the Pythagorean identity $a^2 + b^2 = c^2$ and hence represent the sides of a right triangle. The following is a sequence of Pythagorean triples (a_n, b_n, c_n):

n	a_n	b_n	c_n
1	3	4	5
2	5	12	13
3	7	24	25
4	9	40	41

(These are not the only Pythagorean triples.)

Notice that we can write, for any n, $a_n = 2n + 1$ and $c_n = b_n + 1$.
   a. Construct a difference table to determine the pattern in the b_n terms.
   b. Solve the associated difference equation to find a formula for b_n.
   c. Show that the resulting triple, (a_n, b_n, c_n), forms a Pythagorean triple for any integer n.

5. Find the specific solution to the difference equation

$$x_{n+1} = x_n + 4n - 10, \quad x_0 = 7$$

   a. by solving the difference equation algebraically using the method of undetermined coefficients.
   b. by fitting a function to the first 10 terms of the solution sequence.

6. Find the specific solution to the difference equation

$$x_{n+1} = x_n + 3n^2 - 5n + 2, \quad x_0 = 1$$

   a. by solving the difference equation algebraically.
   b. by fitting a function to the first 10 terms of the solution sequence.

7. Using ideas from Section 4.6 on the differences of terms in a sequence, explain why you should expect that the following difference equation has a linear solution.

$$x_{n+1} = x_n + a, \quad \text{or} \quad x_{n+1} - x_n = a,$$

   where a is a constant.

8. Explain why you should expect that the following difference equation has a quadratic solution.

$$x_{n+1} = x_n + an + b \quad \text{or} \quad x_{n+1} - x_n = an + b,$$

   where a and b are constants.

9. Water is being poured, at a constant rate, into vases having the shapes shown. Sketch a graph showing the level of the water as a function of time, paying careful attention to concavity. What is the significance of each of the points of inflection, if any?

a.

b.

c.

d.

10. Use the first three measurements from our experiment in this section on liquid flowing out of the soda bottle to construct algebraically the quadratic function that passes through those three points. How does it compare to the quadratic function we found based on fitting a function to all the data points?

11. Suppose that water is being poured into a cylindrical pail at a constant rate R_0 while it leaks out of a small hole at the bottom.

   a. Write a difference equation for this situation.
   b. Clearly, if the rate R_0 at which water is poured in is great enough, the level of liquid will rise despite the leakage; if the rate is too slow, then the level will drop because of the leakage. Determine the rate at which water must be poured into the pail that will maintain a constant height of liquid.

12. Suppose, incorrectly, that oil leaks out of a full 200 gallon tank at the constant rate of 8 gallons per hour for 5 hours. Suppose also that the hole then suddenly increases in size, so that the oil leaks out at the rate of 20 gallons per hour for the next 5 hours. Then it is partially patched so that the oil leaks out at the rate of 2 gallons per hour for 5 more hours until the tank is completely repaired and the leakage stops.

   a. Sketch the graph of the *rate* of oil leaking out of the tank as a function of time.
   b. Use the graph to determine the total amount of oil that leaks out.
   c. Sketch the graph of the *number* of gallons of oil that leak out as a function of time.
   d. Find the equations of the line segments you drew in part (c).
   e. Sketch the graph of the amount of oil remaining in the tank as a function of time.
   f. Find the equations of the line segments you drew in part (e).

6.5 *Financing a Car or a Home*

Suppose that a car dealership offers you a "special" financing package on your dream car with a 4.9% annual interest rate. How good is the deal? Should you buy the car because of the offer? Let's examine the mathematics behind this type of financial arrangement.

Essentially, the company is loaning you a certain amount of money, the amount that you are *financing*. You will repay this debt with a series of equal monthly payments for a fixed number of months. These payments are used for two purposes:

1. They pay off the interest that is accumulating on the loan.
2. They pay off part of the original loan.

This process is known as *amortization*.

There are two important questions that you, as a consumer, should ask:

1. How much will the monthly payments be?
2. How much will this financing arrangement actually cost?

We find out how to answer both questions in this section.

To analyze this situation, let D_0 represent the original debt balance or loan (say \$10,000) and let D_n represent the balance owed after n months. Clearly, as you pay off the loan, the values for the outstanding balance, D_n, will decrease as n increases. Let p be the fixed monthly payment and let I be the fixed *monthly* interest rate. That is, if the annual percentage rate is 4.9%, then the monthly rate I is one-twelfth of 4.9%, or $\frac{0.049}{12} = 0.0040833$.

Finally, suppose the debt is amortized over a period of M months; that is, it will take M months to pay off the loan completely.

Starting with an initial debt of D_0, let's see what happens from one month to the next. During the first month:

1. The debt increases by the interest due on the outstanding debt balance, or $I \cdot D_0$.
2. The debt decreases by the monthly payment p.

Thus by the end of the first month,

$$\text{New debt} = \text{Old debt} + \text{interest} - \text{payment}$$
$$D_1 = D_0 + I \cdot D_0 - p$$
$$= D_0(1 + I) - p.$$

Similarly, by the end of the second month,

$$D_2 = D_1 + I \cdot D_1 - p$$
$$= D_1(1 + I) - p$$

and, in general, after $n + 1$ months,

$$D_{n+1} = D_n + I \cdot D_n - p$$
$$= D_n(1 + I) - p,$$

or

$$D_{n+1} - (1 + I)D_n = -p. \tag{1}$$

This is a first order, linear, nonhomogeneous difference equation with constant coefficient that we can solve easily with the methods at our disposal. The corresponding homogeneous difference equation

$$D_{n+1} = (1 + I)D_n$$

has the solution

$$D_n = C(1 + I)^n,$$

where C is an arbitrary constant. Further, since the term, $-p$, on the right-hand side of the nonhomogeneous difference equation (Equation 1) is a constant, the particular solution must also be a constant. If we try

$$D_n = A,$$

then, applying our usual methods, we quickly find that

$$A = \frac{p}{I}.$$

So the complete solution is

$$D_n = C(1 + I)^n + \frac{p}{I}.$$

If we now substitute the initial value for D_0, we find that

$$D_0 = C(1) + \frac{p}{I}$$

from which we get

$$C = D_0 - \frac{p}{I}.$$

Therefore the amount owed on this loan after any number of months n is

$$D_n = \left(D_0 - \frac{p}{I}\right)(1 + I)^n + \frac{p}{I}$$

$$= D_0(1 + I)^n - \frac{p}{I}(1 + I)^n + \frac{p}{I}$$

$$= D_0(1 + I)^n - \frac{p}{I}[(1 + I)^n - 1]. \tag{2}$$

Now that we have this formula for the remaining debt on a loan after n time periods, we can address the two questions raised at the start of the section. We begin by considering the second question: How much will this loan cost? In the process of paying off the loan, we make precisely M payments of p dollars each. Therefore the total amount paid is Mp. This total includes D_0, the original amount that was financed, plus all the interest accrued and paid. Thus

$$Mp = D_0 + \text{interest,}$$

which means that the amount of interest paid on the loan is

$$\text{Interest} = Mp - D_0.$$

We now turn to the first question: What is the monthly payment p? We have to think through what is happening with such an amortization situation. We pay an amount p each month for M months until the entire loan plus all interest is paid off. Therefore the debt after M months, D_M, must be reduced to 0. Thus using Equation (2), we find that the debt after $n = M$ months is

$$D_M = D_0(1 + I)^M - \frac{p}{I}[(1 + I)^M - 1] = 0,$$

where we want to solve for the monthly payment p. Consequently, we find that

$$D_0(1 + I)^M = \frac{p}{I}[(1 + I)^M - 1]$$

so that

$$\frac{p}{I} = \frac{D_0(1 + I)^M}{(1 + I)^M - 1}$$

and therefore

$$p = \frac{D_0 I(1 + I)^M}{(1 + I)^M - 1} \tag{3}$$

is the monthly payment.

In summary,

Financing cars or homes

$$I = \text{Fixed monthly interest rate} = \frac{\text{yearly interest rate}}{12}$$

$$p = \text{Fixed monthly payment}$$

$$M = \text{Total number of months}$$

$$D_0 = \text{Initial loan balance}$$

$$\text{Debt after } n \text{ months} = D_0(1 + I)^n - \frac{p}{I}[(1 + I)^n - 1]$$

$$\text{Monthly payment } p = \frac{D_0 I(1 + I)^M}{(1 + I)^M - 1}$$

E X A M P L E

You plan to buy a new car with a $10,000 loan amortized over 30 months at an annual interest rate of 4.9% per year. What will your monthly payment be and what will the loan cost you over the full 30 months?

Solution We are told that

$$D_0 = 10,000$$

$$M = 30$$

$$I = \frac{0.049}{12}.$$

If we substitute these results into Equation (3) for the monthly payments, we find that

$$p = \frac{10,000\left(\frac{0.049}{12}\right)\left(1 + \frac{0.049}{12}\right)^{30}}{\left(1 + \frac{0.049}{12}\right)^{30} - 1}$$

$$= \$354.85$$

is the monthly payment.

Further, because there are 30 of these $354.85 monthly payments until the original $10,000 loan is fully paid off, the total amount you pay is

$$Mp = 30(354.85) = 10,645.50.$$

Therefore the loan costs you $Mp - D_0$, which is the difference between the total amount paid and the original loan balance $D_0 = \$10,000$, or $645.50

We note that the identical mathematics applies to many other situations, including financing a home mortgage. We explore some other applications in the exercises for this section.

EXERCISES

1. Use the values given in the Example in this section to sketch a graph of the amount owed D_n as a function of time.

2. In amortization situations, the fixed monthly payment consists of one portion that pays off the accrued interest and the other portion that pays off part of the loan principal. Use the values from the Example in this section to sketch the following.

   a. A graph of the amount of interest paid each month as a function of time.
   b. A graph of the amount paid against the principal each month as a function of time.

3. a. Repeat the Example in this section using $10,000 loans for 30 months at 3.9%, at 5.9%, and at 6.9% annual finance rates.
   b. Sketch the graph of the monthly payment p as a function of the interest rate I based on the results of the Example in this section and your results from part (a). What does the concavity tell you?
   c. What kind of simple function best fits the data values graphed in

part (b)? How close are the values obtained by using this approximate formula to the actual values for the monthly payment based on the formula derived in this section if the interest rate $I = 8\%$? If $I = 10\%$?

4. **a.** Repeat the Example in this section for loan periods of 24 months, 36 months, and 48 months.

   b. Sketch the graph of the monthly payment p as a function of the length of the loan M based on the results of the Example and your results from part (a). What does the concavity tell you?

   c. What kind of simple function best fits the data values graphed in part (b)? How close are the values obtained by using this approximate formula to the actual values for the monthly payment based on the formula derived in this section if the loan period is $M = 30$ months? If $M = 60$ months?

5. **a.** Repeat the Example in this section for loan amounts of $8000, $9000, and $12,000 for a 30 month loan at 2.9% annual interest.

   b. Sketch the graph of the monthly payment p as a function of the loan amount D_0 based on the results of the Example and your results from part (a). What does the concavity tell you?

   c. What kind of simple function best fits the data values graphed in part (b)? How close are the values obtained by using this approximate formula to the actual values for the monthly payment based on the formula derived in this section if the loan amount $D_0 = \$15,000$?

6. A family plans to get a $100,000 mortgage at 7% annual interest over 30 years in order to purchase a new home. What is the monthly payment? What is the total cost of the loan?

7. If interest rates drop to $6\frac{1}{2}\%$ per year just before the family in Exercise 6 takes out its mortgage, how much of a difference will this make in their monthly payment and the total cost?

8. The same family thinks that it can put up more cash and only needs a $90,000 mortgage at the $6\frac{1}{2}\%$ interest rate. How much of a difference does this make in their monthly payment and the total cost compared to the values in Exercise 7?

9. Interest rates increase to $7\frac{1}{2}\%$ per year just before the family in Exercise 6 takes out its mortgage. The family decides that it cannot carry any greater monthly payment than the one you calculated in Exercise 6. How much more money would the family have to put up initially to reduce the size of the loan to a level where the monthly payment is the same?

10. The following table gives the average weekly wages for individuals in private industry over the last few decades. Determine the best fit to this data using each of a linear, exponential, and power function model. What is your best prediction for weekly wages in 1995? In 2000? Describe any reasons for believing that these predictions may be inaccurate.

Year	1970	1975	1980	1985	1988	1989
Average weekly wage (dollars)	120	164	235	299	322	335

6.6 *Sorting Methods in Computer Science*

One of the most important procedures used in computer science or data processing is *sorting* a set of data. Suppose you have a list of names or numbers. It is usually desirable to arrange such a list in either alphabetical or numerical order so that you can access particular entries in the list, insert new entries into the list, or delete old entries from it.

When the list is relatively short, sorting the values is a fairly simple process that can be performed quickly and easily. However, if the list contains thousands of entries, the process is no longer simple, even with the aid of a computer. Because of this, computer scientists have devised a variety of techniques to sort a collection of data into a particular order. The effectiveness of any such method depends on how fast it sorts a set of data. We compare different sorting methods by analyzing how long they take to perform a sorting routine. One way to do this is to keep track of the number of operations that must be performed by the computer in implementing each method. In this section, we study three of the simplest and most common sorting methods.

The Bubble Sort

Suppose we have two numbers a and b, and want to list them in increasing numerical order. To do so, we compare a to b. If a is smaller than b, we do nothing; the numbers are in the desired order. If a is larger than b, then we interchange them to put b before a. Thus there are at most two steps involved in this sorting process: a comparison and, if necessary, a swap or interchange.

Next suppose we have three numbers a, b, and c. We start with a and compare it first to b and, if necessary, we interchange a and b so that the smaller of a and b is at the head of the list. We then compare the first number to c and, if necessary, swap them. In this way, we are sure that the first number in the list is the smallest of the three. Notice that this requires two comparisons and a maximum of two swaps. However, we still do not know if the second number or the third number in the list is the next smallest. Therefore it is necessary to compare the second entry to the third and, if necessary, interchange them. Thus, to sort a group of three numbers, it is necessary to make three comparisons and up to three swaps.

Incidentally, in the process of making a swap, say $a = 43$ and $b = 5$, we keep things simple by changing their names as well. That is, the number in the first position will always be called a, whether it is 43 initially or 5 after the swap. Similarly, the number in the second position will be called b whether it is the original 5 or the final 43.

Next, suppose we have four numbers a, b, c, and d. We first want to position the smallest of the four numbers at the start of the list. To do so, we compare the first entry a to each of the remaining entries b, c, and d and swap any of them, if necessary. Doing this involves three comparisons and a maximum of three swaps. Having assured ourselves that the smallest entry is first, we then want to get the next smallest entry into the second position. To do so, we compare the second entry b to each successive entry c and d; when necessary, we interchange them. This involves another two comparisons and possibly two swaps. Finally, we want the third entry c to be the third smallest of the group. To do this, we have to compare c to d and swap them if necessary. Therefore, there are a total of $3 + 2 + 1 = 6$ comparisons necessary when sorting a set of four numbers.

Similarly, if we have five entries a, b, c, d, and e, we first compare a to b, c, d, and e (four comparisons); we then compare b to c, d, and e (three more); we then compare c to d and e (two more); and finally we compare d to e (one last comparison). Therefore, there are a total of $4 + 3 + 2 + 1 = 10$ comparisons.

Before proceeding to the general case, let's look at an actual example. Suppose we want to sort the five numbers 40, 30, 50, 10, 20 using the *bubble sort* (where the smallest value "rises to the front"). On the first pass, we want to move the smallest entry to the first position. So we begin by comparing 40 to the entries following; since 30 is smaller than 40, we swap them to get

$$30, 40, 50, 10, 20.$$

We then compare 30, which is now in the lead position, to entries from the third position on; clearly, 10 is smaller than 30, so we swap them to get

$$10, 40, 50, 30, 20.$$

Finally, we compare 10 to the number in the last position and no swap is made.

Next, we seek to get the second smallest number into the second position, so we compare 40 to the entries following. Since 30 is smaller than 40, we swap them to get

$$10, 30, 50, 40, 20,$$

and then compare 30 to the last value. Since 20 is smaller than 30, we swap them to get

$$10, 20, 50, 40, 30.$$

We then seek to get the third smallest number into the third position. We compare 50 to the entries following and find we have to swap 50 and 40 to get

$$10, 20, 40, 50, 30.$$

We then compare 40 to the last value, and so we have to swap 40 and 30 to produce

$$10, 20, 30, 50, 40.$$

Finally, we want the fourth smallest entry in the fourth position, so we compare 50 to the last entry and swap 50 and 40 to produce

$$10, 20, 30, 40, 50.$$

In the process of doing this bubble sort, we had to make the following number of comparisons

4 (for first entry) + 3 (for second entry) + 2 (for third entry) + 1 (for fourth entry),

or a total of 10 comparisons and a total of 7 swaps.

Now let's look at the general case where we have a set of n entries to be sorted. Let C_n denote the total number of comparisons needed. (This also may involve as many as C_n swaps.) Now suppose that one additional entry is added to this list, so that the number of comparisons needed is C_{n+1}. How does C_{n+1} relate to C_n? The first n entries need C_n comparisons. To determine the correct location for the new entry, we must compare it to each of those n entries. Thus we need an additional n comparisons. The total number of comparisons C_{n+1} needed to sort $n + 1$ entries is therefore

$$C_{n+1} = C_n + n,$$

or

$$C_{n+1} - C_n = n.$$

This is a first order, linear, nonhomogeneous difference equation that we can solve using our usual methods. The associated homogeneous difference equation has as its solution

$$C_n = A(1)^n = A,$$

where A is some arbitrary constant. Since the right-hand side is a linear function of n, we would normally use the trial solution

$$C_n = Bn + D.$$

However, since the homogeneous solution already contributes a constant term, it is necessary to adjust this trial solution, and therefore we use

$$C_n = n(Bn + D) = Bn^2 + Dn.$$

Consequently,

$$C_{n+1} = B(n + 1)^2 + D(n + 1)$$
$$= B(n^2 + 2n + 1) + Dn + D,$$

so

$$C_{n+1} - C_n = [B(n^2 + 2n + 1) + Dn + D] - [Bn^2 + Dn]$$
$$= 2Bn + B + D.$$

Substituting into the original difference equation yields

$$2Bn + B + D = n.$$

When we equate the coefficients of the linear terms, we get

$$2B = 1$$

$$B = \frac{1}{2}.$$

When we equate the constant terms, we get

$$B + D = 0, \qquad \text{so} \qquad D = -B = -\frac{1}{2}.$$

Therefore the complete solution to the difference equation is

$$C_n = A + \frac{1}{2}n^2 - \frac{1}{2}n.$$

Further, as we saw before, when there are $n = 2$ items in the list, we need only $C_2 = 1$ comparison. Because

$$C_2 = A + \frac{1}{2}(2^2) - \frac{1}{2}(2)$$

$$= A + \frac{1}{2}(4) - 1$$

$$= A + 2 - 1 = A + 1,$$

we must have

$$A + 1 = 1$$

$$A = 0.$$

Thus, the number of comparisons needed to sort n objects using the bubble sort is

$$C_n = \frac{1}{2}n^2 - \frac{1}{2}n = \frac{1}{2}n(n - 1).$$

Let's check the accuracy of this result. We saw above that it requires 10 comparisons to sort a group of five entries. Using the solution of the difference equation, we see that

$$C_5 = \frac{1}{2}(5)(5 - 1)$$

$$= \frac{1}{2}(5)(4) = 10.$$

Next, let's interpret the meaning of the formula $C_n = \frac{1}{2}n^2 - \frac{1}{2}n$. Notice that when n is large, the number of comparisons needed to sort n items is essentially $\frac{1}{2}n^2$ since the linear term is relatively negligible. Thus to sort 1000 items involves roughly $\frac{1}{2}(1000)^2 = 500{,}000$ comparisons, but to sort 2000 items requires roughly 2,000,000 comparisons. The amount of work, and hence the time needed, grows much faster than the number of items in the list. In fact, it grows large as fast as $\frac{1}{2}n^2$ does, when n is large.

So far, we have considered only the number of comparisons, C_n, needed to sort n objects using the bubble sort. In addition, we could also consider the number of swaps, S_n, needed to sort them. However, that becomes a considerably more difficult problem because the number of swaps needed depends on the particular arrangement of the items being sorted. Different arrangements of the same items will likely require different numbers of swaps. For instance, suppose the items are the numbers 1 and 2; if the arrangement is 1, 2 then zero swaps are required while if the arrangement is 2, 1 then one swap is needed. As a consequence, we cannot expect a closed form expression for the number of swaps needed; the problem is a probabilistic or random one depending on the arrangement.

Let's experiment with this process by using a few cards from a deck of cards to represent the items being sorted. Suppose we start with $n = 1$ card, an ace; there is only one arrangement possible and hence we require $S_1 = 0$ swaps. Next, suppose we take $n = 2$ cards, an ace and a two, mix them randomly, and deal them out in a row. Depending on the arrangement, we might require either $S_2 = 0$ swaps or $S_2 = 1$ swap. Suppose our particular arrangement requires 1 swap.

Next, suppose we take $n = 3$ cards, an ace, a two, and a three, mix them, deal them out in a row, and sort them into order. Say the results of the deal are 2–3–1. On the first pass, we interchange the 2 and the 1 to get 1–3–2; on the next pass, we get 1–2–3 and so require $S_3 = 2$ swaps for this one particular arrangement. (*Note:* There are actually 6 arrangements possible for the three cards.) Continuing in this way, we next take four cards, and find, for instance, that the particular arrangement we have requires $S_4 = 5$ swaps.

The important question here is: How fast does the number of swaps needed grow as n increases? In particular, does the number grow linearly? quadratically? exponentially? Recall that the number of comparisons needed when sorting n objects is $C_n = \frac{1}{2}n(n-1)$ and that, at worst, each comparison is accompanied by a swap. Typically the number of swaps S_n is smaller than C_n, for any n. As a result, we conclude that, at worst, S_n must grow quadratically.

Think About This

For $n = 3$ and $n = 4$, find the arrangements that require the maximum number of swaps.

Think About This

Using a deck of cards, conduct the experiment described above for one random arrangement of $n = 1, 2, 3, 4, 5, 6, 7$ and then 8 cards. In each case, keep track of the number of swaps you performed. Then use these values as data on which to analyze the growth rate for the process. Decide if the growth rate is a linear, quadratic, or power function. If so, find a formula that approximates the growth as a function of n.

The following table shows the time needed (in seconds) to sort the indicated number of randomly generated quantities with the bubble sort using a very fast microcomputer.

Number of entries n	200	400	600	800	1000	1200	1400	1600	1800	2000
Sorting time (sec)	0.12	0.51	1.26	2.16	3.36	4.83	6.56	8.61	10.77	13.38

Notice how quickly the sorting time increases as the number of entries n increases. If the set of numbers to be sorted is much larger or if sorting must be done very frequently, the bubble sort is just too inefficient to be practical. We will ask you to fit a function to this data as one of the exercises at the end of the section.

The Simple Insertion Sort

Another sorting algorithm, known as the *simple insertion sort,* is occasionally used in computer science. It is based on an approach often used by card players when they are being dealt a fresh hand. As they pick up each new card, they insert it into their hand in the appropriate order, say from lowest on the left to highest on the right.

Suppose we have a list of numbers to be sorted into ascending order. We begin by taking the first entry as the starting point of a new, sorted list. We then consider the next item in the original list and either insert it ahead of the first number in the sorted list if it is smaller or after the first number if it is larger. We then take the third item in the original unsorted list and insert it into the correct place in the new, sorted list. We continue this procedure, taking each entry in turn from the original list and inserting it into the appropriate spot in the new sorted list, until all of the entries have been sorted.

There are two operations involved in performing the simple insertion sort. One is determining the appropriate position in which to insert each entry based on a series of comparisons; the other is the actual insertion. As with the bubble sort, the number of comparisons needed is a function of n, the number of items in the list. In some sample tests of this sorting routine, the following data were obtained on the number of comparisons C_n needed as a function of n.

n	100	200	300	400	500	600
C_n	2595	10,267	22,618	40,382	62,786	90,385

We now find the growth pattern for this function using our data analysis techniques. To increase the size of the data set and so improve accuracy, there is one other point we can use: When $n = 0$, $C_0 = 0$. If you examine the difference between consecutive values, you will see that these differences seem to be growing steadily, so the pattern is certainly not linear. We saw previously that the number of comparisons involved in the bubble sort grows quadratically, so we might be tempted to try a quadratic fit here also. Using the regression features of the calculator and the above set of results, we find that

$$C_n = 0.249n^2 + 0.943n + 12.60.$$

If you examine this fit visually with your function grapher, you will see that it looks extremely accurate. In fact, it is shown in computer science that the average number of comparisons needed with the simple insertion sort is

$$\frac{n^2}{4} + \frac{3n}{4} - \sum_{i=1}^{n} \frac{1}{i}$$

where the adjustment term $\sum_{i=1}^{n} \frac{1}{i}$ grows very slowly as n increases.

The Quicksort

Computer scientists have developed far more effective sorting routines that involve fewer comparisons and fewer swaps than the bubble sort or the simple insertion sort. We outline one of them, known as the *quicksort*, here. Suppose we have the ten numbers

$$50, 20, 80, 30, 60, 90, 10, 70, 100, 40.$$

We begin with the first entry, 50, which is known as the *pivot* and which we highlight by marking it with a double underline. We then go through the remaining terms after the pivot working from the left looking for the first entry that is larger than the pivot (it is 80) and then from the right looking for the first entry smaller than the pivot (it is 40). We underline these entries and see

$$\underline{\underline{50}}, 20, \underline{80}, 30, 60, 90, 10, 70, 100, \underline{40}.$$

We then interchange the two indicated entries to get

$$\underline{\underline{50}}, 20, \underline{40}, 30, 60, 90, 10, 70, 100, \underline{80}.$$

We continue this process to find the next entry from the left that is larger than the pivot (it is 60) and the next entry from the right that is smaller than the pivot (it is 10):

$$\underline{\underline{50}}, 20, 40, 30, \underline{60}, 90, \underline{10}, 70, 100, 80.$$

We then interchange these two entries to get

$$\underline{\underline{50}}, 20, 40, 30, \underline{10}, 90, \underline{60}, 70, 100, 80.$$

The next scan of the list shows no further entries that need to be interchanged. The result has been to split the original list into two smaller sublists, one consisting of all entries smaller than the pivot, the other consisting of all entries larger than the pivot. We then move the pivot, 50, into its correct place in this new list (fixing it using color) where it separates the two sublists:

$$20, 40, 30, 10, \mathbf{50}, 90, 60, 70, 100, 80.$$

The quicksort process is then applied to each of the two sublists: The first entry in each (20 and 90) is now considered a new pivot

$$\underline{\underline{20}}, 40, 30, 10, \mathbf{50}, \underline{\underline{90}}, 60, 70, 100, 80$$

and we go through each sublist from both the left and the right to identify the first entry larger than each pivot and the first entry smaller than each pivot:

$$\underline{\underline{20}}, \underline{40}, 30, \underline{10}, \mathbf{50}, \underline{\underline{90}}, 60, 70, \underline{100}, \underline{80}.$$

When we interchange them, we get

$$\underline{\underline{20}}, \underline{10}, 30, \underline{40}, \mathbf{50}, \underline{\underline{90}}, 60, 70, \underline{80}, \underline{100}.$$

The next scan shows that there are no further entries that need be switched. We then move the two pivots into their correct places in each sublist to get:

$$10, \mathbf{20}, 30, 40, \mathbf{50}, 60, 70, 80, \mathbf{90}, 100.$$

For this set of numbers, the entries are now in order. (If they were not, we would have as many as four new sublists, each usually much shorter than the original, to which we apply the same process.)

The following table shows the comparable times, in seconds, needed for the quicksort process to sort the same sets of data previously studied using the bubble sort. Notice how much faster the data values are arranged in order.

Number of entries (n)	200	400	600	800	1000	1200	1400	1600	1800	2000
Sorting time (sec)	0.006	0.014	0.023	0.031	0.041	0.050	0.060	0.070	0.080	0.092

We ask you to find a function to fit this data in Exercise 6 of this section. Notice, though, that when n becomes reasonably large, the sorting time becomes nearly a linear function of n.

We mentioned previously that the number of comparisons needed with the bubble sort or the simple insertion sort both grow quadratically as a function of n. Computer scientists write this using *big O* notation as $O(n^2)$ to indicate that a quantity grows *on the order of* n^2—any lower degree terms are ignored because, when n is very large, only the highest power term has a significant effect. Similarly, even the coefficient of the highest power term is ignored because, when n is very large, the coefficient does not affect the behavior of the function significantly. Thus, $O(n^2)$ means that the function grows roughly as fast as n^2. If a function is $O(n^3)$, then it grows roughly as fast as n^3. It can be shown that the number of comparisons needed with the quicksort is of order $O(n \log n)$, so that the number grows roughly as fast as

$n \log n$ grows when n is very large. Since $n \log n$ grows much more slowly than n^2 for large n, as shown in Figure 6.20, we see that the quicksort is the superior choice among the three methods we have considered.

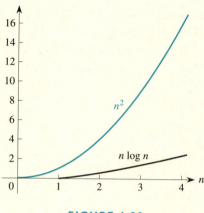

FIGURE 6.20

Computer scientists have developed other methods, including the *mergesort,* the *heapsort,* and the *shellsort,* though we will not go into any of them here. Each has its advantages, but none are significantly better, in general, than the quicksort. Ideally, we would like a sorting method in which the number of comparisons grows as $O(n)$ since a linear function grows even more slowly than $n \log n$ (check this out on your calculator), but no method that effective has been developed.

EXERCISES

1. Using the bubble sort, how many comparisons would be needed to sort a list of 100 numbers? A list of 1000 numbers? A list of 10,000 numbers?

2. Use this list of numbers: 5, 20, 8, 12, 2, 15. Sort the numbers from least to greatest by hand using the following methods.

   a. Use the bubble sort. Keep track of the total number of comparisons needed and the total number of swaps.
   b. Use the simple insertion sort. Keep track of the total number of comparisons needed. How does this compare to the results in part (a)?
   c. Use the quicksort process. Keep track of the total number of comparisons needed and the total number of swaps.

3. Compare the formula for the number of comparisons needed by the bubble sort, $\frac{1}{2}n(n-1)$, with the regression formula for the number needed by the simple insertion sort, $0.249n^2 + 0.943n + 12.60$. In particular, for what values of n does the simple insertion sort seem to require fewer

comparisons? For what values of n does the bubble sort seem to require fewer comparisons?

4. Describe which arrangements of n objects require 0 swaps using the bubble sort? Which arrangements require $\frac{1}{2}n(n-1)$ swaps? Which arrangements require precisely 1 swap?

5. The following set of values shows the times in seconds needed to bubble sort a list of n quantities. Determine the best quadratic fit to the data.

Number of entries (n)	200	400	600	800	1000	1200	1400	1600	1800	2000
Sorting time (sec)	0.12	0.51	1.26	2.16	3.36	4.83	6.56	8.61	10.77	13.38

6. The following set of values shows the time in seconds needed to quicksort the same list as used in Exercise 5. Determine the best quadratic fit to the data. How does the coefficient of the quadratic term compare to the one you found in Exercise 5? What does this tell you about the relative efficiency of the bubble sort and the quicksort? How well do other functions fit this set of data?

Number of entries (n)	200	400	600	800	1000	1200	1400	1600	1800	2000
Sorting time (sec)	0.006	0.014	0.023	0.031	0.041	0.050	0.060	0.070	0.080	0.092

7. In this section we found that the number of comparisons needed for the bubble sort to arrange n objects is $C_n = \frac{1}{2}n(n-1)$ by solving a difference equation. Since C_n satisfies $\Delta C_n = C_{n+1} - C_n = n$, you know that C_n must be a quadratic function of n. (Think about the form for the particular solution.) Assume that $C_n = an^2 + bn + c$ and solve for a, b, and c by using the fact that $C_1 = 0$, $C_2 = 1$, and $C_3 = 3$. Since the result will be slightly different, explain which approach likely produces a better result.

8. Follow the procedure suggested in this section and find the number of comparisons needed to sort lists of two, three, and four names into alphabetical order using the bubble sort. Is there any difference in the number of comparisons needed to sort a list of numbers and a list of words?

9. For small n, we have $n \log n < n$ and, for large n, we have $n \log n > n$. Find the first value of n for which $n \log n$ is greater than n.

6.7   *Modeling the Stock Market*

In the 16 years from the beginning of 1980 to the beginning of 1996, the Dow-Jones average of 30 industrial stocks on the New York Stock Exchange increased 517% from 839 to 5177. We get the percent increase from $(5177 - 839)/839 = 5.170$, which is the same as 517%. We can find the equivalent compounded growth rate, assuming a roughly exponential growth pattern, by solving

$$839a^{16} = 5177$$

for a. This leads to a value of $a \approx 1.12$, so the average annual growth rate was approximately 12% per year over the 16-year period. Using this, we might conclude that the Dow-Jones average is modeled by the exponential function

$$D(t) = 839(1.12)^t,$$

where t is the number of years since 1980.

The problem with this conclusion is that we know that the stock market has not risen steadily. It rises and falls from one day to the next, although the overall pattern has been one of roughly exponential growth over this 16-year period. We can get a somewhat better idea of the actual growth pattern by looking at the following set of data values that represent the Dow-Jones average at the beginning of each year:

Year	1980	1981	1982	1983	1984	1985	1986	1987	1988
Dow	839	964	875	1047	1259	1212	1547	1896	1939

1989	1990	1991	1992	1993	1994	1995	1996
2169	2753	2634	3169	3301	3758	3824	5177

The associated scatterplot is shown in Figure 6.21.

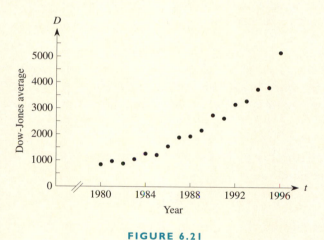

FIGURE 6.21

In Exercise 8 of Section 3.5, you were asked to find the best linear, exponential, or power function fit to this set of data. You presumably found that the best choice is the exponential function

$$D(t) = 787.6(1.119)^t,$$

where t is the number of years since 1980. The corresponding correlation coefficient is $r = 0.9911$, which indicates an extremely high positive correlation. Note that the growth factor here is surprisingly similar to the result we obtained above for $D(t)$ by simply using the initial and final readings on the Dow-Jones average. If anything, we should have more confidence in this exponential regression fit to the behavior of the Dow since it uses 17 data points instead of just the two end points. Nevertheless, it does not at all reflect the daily increases and decreases that we know occur in any stock average.

To account for this unpredictable variation from one day to the next, we must look at the mathematical model more closely. We can think of the exponential growth model we constructed using the 17 data points as based on the first order difference equation

$$D_{n+1} = 1.119D_n,$$

with a growth factor of 1.119 and a growth rate of 11.9% per year, where n is measured in years since 1980. To refine this considerably, suppose that the stock exchange operates for a total of 250 trading days each year. Let a represent the daily growth rate and $1 + a$ the daily growth factor corresponding to the annual growth factor of 1.119. After one year or 250 days,

$$D_1 = 1.119D_0 = (1 + a)^{250}D_0.$$

We conclude that for the 11.9% growth rate,

$$(1 + a) = (1.119)^{1/250} = 1.00045,$$

and so the daily growth rate a is $0.00045 = 0.045\%$. (Notice that simply dividing 11.9% by 250 leads to

$$\frac{0.119}{250} = 0.000476,$$

which is not the same because it ignores the effects of daily compounding.) The difference equation model based on the daily growth rate is

$$S_{m+1} = 1.00045S_m,$$

where m is the number of *days* since the beginning of 1980. The solution to this difference equation,

$$S_m = 839(1.00045)^m,$$

is an exponential function expressed on a daily rather than a yearly basis. It is still growing steadily from one day to the next.

What we need is a way to introduce some degree of fluctuation into the daily stock market result. This fluctuation should be either positive (when the market increases more than the exponential model would predict) or negative (when the market increases less than the exponential model predicts or when the market decreases). Further, this fluctuation should be a random quantity since we do not know in advance whether the market will increase or decrease, let alone by how much. Realize that we will not be able to get an explicit solution to this problem because of the randomness of the situation; what we seek is a way to produce a mathematical model that behaves reasonably like the actual Dow-Jones average.

To simplify things, suppose we assume that the *maximum* swing in the Dow-Jones average on any given day is between -30 and $+30$ points from the exponential curve. (That is equivalent to a maximum of somewhat more than half of one percent of the value of the Dow when it is near 5000.) We write $RAND(-30, 30)$ to represent a random number between -30 and 30; this means that every number in this interval is as likely to be picked as any other. Thus we assume that a swing of -30 is just as likely to happen as a swing of $+2.4$. (We will discuss the reasonableness of this assumption later.)

Using this, we can rewrite the difference equation as

$$S_{m+1} = 1.00045S_m + RAND(-30, 30).$$

Such a difference equation with a random term is known as a *stochastic difference equation;* the word *stochastic* signifies that there is a random element present.

If the new term were a fixed constant, we could find an explicit solution to this linear, nonhomogeneous difference equation using our usual methods. However, it is definitely not fixed; it is not even a predictable quantity. As such, there is no hope of obtaining an exact expression for the solution. Despite this, we can examine the behavior of the solution of such a difference equation using the computer. Realize that each time the program runs, the computer selects a different set of random numbers for each "day," so the overall results will be different. The question is: How different? If the results are dramatically different, then we cannot expect to get any useful guidance from such an approach. On the other hand, if the results of different runs are relatively similar, then we might be able to infer something useful about the performance of a quantity such as the Dow-Jones average.

In Figures 6.22 and 6.23, we show the results of two separate simulations of the Dow-Jones average. In each case, we superimpose the theoretical exponential growth curve based on the values entered in the computer simulation for comparison.

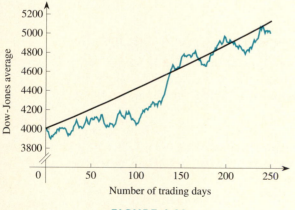

FIGURE 6.22

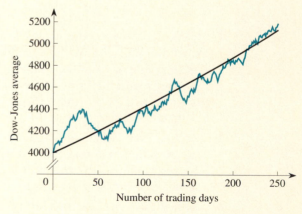

FIGURE 6.23

Notice that, in each instance, while the individual fluctuations are quite different, the overall pattern fairly closely mirrors the theoretical curve. Our simulation thus gives a reasonably accurate portrayal of the behavior of the stock market. We note that such random simulations are one of the fundamental tools used by sophisticated financial analysts who attempt to predict the behavior of the stock market.

The approach we just described is limited by the fact that we have intentionally restricted ourselves to swings between -30 and $+30$ on any given day. If you follow the stock market, you know that this is reasonable on *most* days. However, it is certainly possible for the Dow to increase 60 points or even drop 120 points on a given day, although these occurrences are relatively rare. It would be nice to work with larger intervals than just -30 to $+30$, but if we extend them too far, say to -75 to $+75$, it is unrealistic to expect the Dow-Jones average to fluctuate that much often.

The problem is that we used a uniform scheme to generate the random values; each possible value had the same chance as any other of being generated. Other schemes are possible; the most common is to generate numbers from a bell-shaped distribution (known as the *normal distribution*), which is centered about 0 and is shown in Figure 6.24. In such a case, values near the center are much more likely to be selected than values farther out in either direction. In fact, the farther you go to either side, the less likely it is for the corresponding value to be generated. Therefore at a more sophisticated level, it would make sense to use such a bell-shaped distribution, with values between -100 and $+100$, say, instead of the equally likely random values we described previously. However, such a model is beyond the scope of this book.

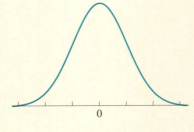

FIGURE 6.24

EXERCISES

1. If you have a program available to simulate the daily performance of the stock market, try it with four different runs. Use the same beginning and ending values, say 4000 and 5000, the same maximum daily swing, say 30 points, and the same number of days of trading, say 250. How close do the four sets of random outcomes come to reflecting the same behavior pattern?

2. Repeat Exercise 1 using a smaller value for the maximum daily swing, say 10 points. Do you think that the simulated results give a better or a poorer match to the theoretical exponential growth curve than you observed in Exercise 1?

3. Repeat Exercise 2 using a larger daily swing, say 50 points.

4. Repeat Exercises 1–3 using the same values again, but over the course of 300 trading days. How well do the simulated results mirror the theoretical ones? Do they seem more accurate, less accurate, or comparable?

5. Construct a stochastic difference equation to model the daily behavior of the stock if the maximum daily fluctuation is plus or minus half of one percent of the value of the Dow on that day.

6.8 *Fitting Logistic Curves to Data*

The logistic model is one of the most useful mathematical models we have for realistically describing growth processes in many different fields. In a typical application, we start with a set of data representing actual values or measurements of some quantity, and we want to determine the particular logistic curve that best fits that data. This means determining the values for the coefficients a and b in the logistic difference equation

$$\Delta y_n = a y_n - b y_n^2.$$

As we have seen before with other functional patterns, it is not reasonable to expect that the data we have will fit precisely on any given logistic curve. Rather, we again face the problem of obtaining the best fit.

Recall our strategy in Chapter 3 when we fit exponential, power, and logarithmic functions to a set of data. We first *linearized* the data through an appropriate transformation (using logarithms) and then applied linear regression analysis to the transformed data. Once we found the equation of the regression line that best fit the transformed data in the least-squares sense, we detransformed the linear equation to obtain the nonlinear function that best fits the original data. In this section, we will develop comparable methods for treating data falling into a logistic pattern.

As with any curve-fitting situation, we begin with a scatterplot. Suppose that the data points fall into a pattern such as the one shown in Figure 6.25 and we want to fit a logistic curve to it.

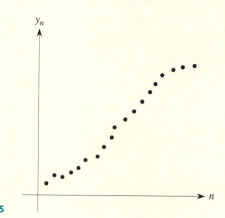

FIGURE 6.25

To do so, we must transform the data so that the resulting points are in a linear pattern. If a set of data $(0, y_0), (1, y_1), (2, y_2), \ldots$ appears to be logistic, then we know that y_n satisfies the difference equation

$$\Delta y_n = a y_n - b y_n{}^2 = y_n(a - b y_n).$$

Dividing both sides by y_n, we obtain

$$\frac{\Delta y_n}{y_n} = a - b y_n.$$

Thus, if y_n follows a logistic pattern, the quantity $\dfrac{\Delta y_n}{y_n}$ is actually a linear function of y_n. Consequently, if we plot $\dfrac{\Delta y_n}{y_n}$ against y_n and find that it falls into a roughly linear pattern, then we should find the best linear fit for the transformed data pairs, $\left(y_n, \dfrac{\Delta y_n}{y_n}\right)$. Once we have that equation, we use the regression coefficients for that line as the logistic coefficients a and $-b$.

We illustrate these ideas in the following examples.

EXAMPLE 1

The German biologist R. Carlson studied the growth of yeast under controlled conditions and obtained the following set of measurements for the weight of yeast as a function of time, n:

Time (n)	1	2	3	4	5	6	7	8	9	10
Weight (y_n)	9.6	29.0	71.1	174.6	350.7	513.3	594.4	640.0	655.9	661.8

Find the parameters a and b for the best logistic fit to this growth situation.

Solution By inspecting either the entries in the preceding table or the corresponding scatterplot of the points shown in Figure 6.26, we see that they suggest a logistic pattern.

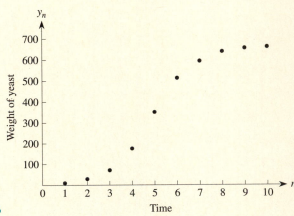

FIGURE 6.26

To linearize the data, we add two extra columns to the previous table, one for the differences Δy_n of the successive terms and the other for the ratio, $\dfrac{\Delta y_n}{y_n}$, as shown in the following table.

Time (n)	Weight (y_n)	Δy_n	$\Delta y_n/y_n$	
1	9.6	19.4	2.021	(= 19.4/9.6)
2	29.0	42.1	1.452	(= 42.1/29.0)
3	71.1	103.5	1.456	
4	174.6	176.1	1.009	
5	350.7	162.6	0.464	
6	513.3	81.1	0.158	
7	594.4	45.6	0.077	
8	640.0	15.9	0.025	
9	655.9	5.9	0.009	
10	661.8			

We next plot the values of $\dfrac{\Delta y_n}{y_n}$ versus y_n (not n), as shown in Figure 6.27, and find that they do indeed seem to fall into a roughly downward-sloping linear pattern. The corresponding regression equation is

$$\frac{\Delta y_n}{y_n} = 1.66 - 0.00271 y_n,$$

where $a = 1.66$ and $b = -0.00271$. Moreover, the corresponding correlation coefficient is $r = -0.967$, which suggests a very high degree of negative correlation between $(\Delta y_n)/y_n$ and y_n. Multiplying both sides of the regression equation by y_n, we find

$$\Delta y_n = a y_n - b y_n^2 = 1.66 y_n - 0.00271 y_n^2.$$

This is the logistic difference equation on which the logistic curve that best fits the data on the growth of the yeast is based.

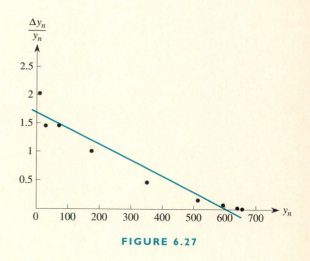

FIGURE 6.27

EXAMPLE 2

In Section 3.3, we found that the growth in the U.S. population from 1780 to 1900 very closely followed an exponential growth pattern with growth rate of

32.1% per decade. In fact, corresponding to the best-fit exponential curve, we had a correlation coefficient of $r = 0.998$. However, we pointed out that this exponential pattern does not apply during the twentieth century for various reasons. Let's now examine the growth in the U.S. population over the entire period since 1780, as shown in the following table and the accompanying scatterplot in Figure 6.28.

Year	Population	Ratio
1780	2.8	1.39
1790	3.9	1.36
1800	5.3	1.36
1810	7.2	1.33
1820	9.6	1.34
1830	12.9	1.33
1840	17.1	1.36
1850	23.2	1.35
1860	31.4	1.27
1870	39.8	1.26
1880	50.2	1.25
1890	62.9	1.21
1900	76.0	1.21
1910	92.0	1.15
1920	105.7	1.16
1930	122.8	1.07
1940	131.7	1.14
1950	150.7	1.19
1960	179.3	1.13
1970	203.3	1.11
1980	226.5	1.10
1990	248.7	

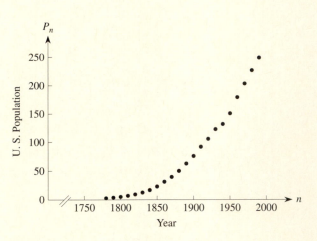

FIGURE 6.28

Both from the table and from the scatterplot, we notice that the rate of population growth has slowed considerably during the present century. Notice also that the successive ratios have slowly decreased, so that the rate of population growth has slowed from over 20% per decade at the start of the twentieth century to 10% per decade. Let's try to fit a logistic curve to this data. To do so, we must calculate the differences $\Delta P_n = P_{n+1} - P_n$ and then the ratios

$$\frac{\Delta P_n}{P_n} = \frac{P_{n+1} - P_n}{P_n}.$$

For instance, over the first two decades, we have $\Delta P_0 = 3.9 - 2.8 = 1.1$ and $(\Delta P_0)/P_0 = 1.1/2.8 = 0.393$. The following table shows all the results.

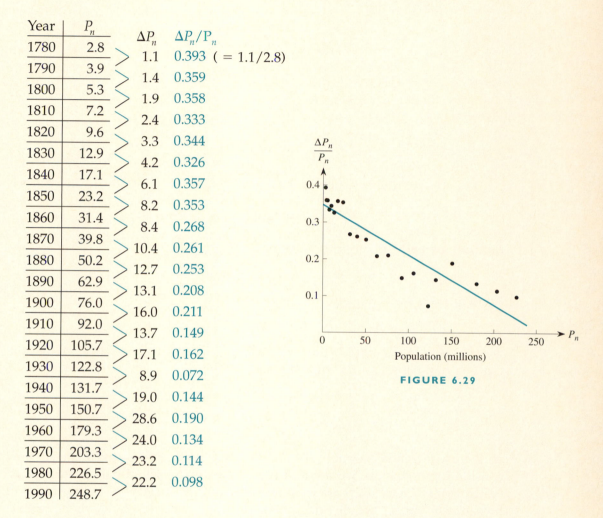

Year	P_n	ΔP_n	$\Delta P_n/P_n$
1780	2.8		
		1.1	0.393 (= 1.1/2.8)
1790	3.9		
		1.4	0.359
1800	5.3		
		1.9	0.358
1810	7.2		
		2.4	0.333
1820	9.6		
		3.3	0.344
1830	12.9		
		4.2	0.326
1840	17.1		
		6.1	0.357
1850	23.2		
		8.2	0.353
1860	31.4		
		8.4	0.268
1870	39.8		
		10.4	0.261
1880	50.2		
		12.7	0.253
1890	62.9		
		13.1	0.208
1900	76.0		
		16.0	0.211
1910	92.0		
		13.7	0.149
1920	105.7		
		17.1	0.162
1930	122.8		
		8.9	0.072
1940	131.7		
		19.0	0.144
1950	150.7		
		28.6	0.190
1960	179.3		
		24.0	0.134
1970	203.3		
		23.2	0.114
1980	226.5		
		22.2	0.098
1990	248.7		

FIGURE 6.29

Notice from the table that the ratios $\dfrac{\Delta P_n}{P_n}$ fall into an overall decreasing pattern when we plot them against the values of P_n, as shown in Figure 6.29. Also, the scatterplot suggests a roughly linear pattern. However, there does seem to be a fair amount of variation about the regression line. The associated correlation coefficient, $r = -0.90$, represents a high degree of negative correlation for 21 data pairs. Incidentally, we should expect negative correlation between the transformed values and the actual population values since the slope of the regression line is negative.

The corresponding plot of the residuals, as shown in Figure 6.30 (on the next page), indicates that the fit is reasonably accurate, although there may be some pattern to the residuals. This suggests that the logistic model may not fully explain all the variation in the data.

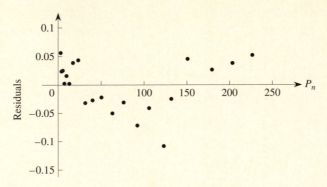

FIGURE 6.30

Nonetheless, the equation of the regression line for the $\left(P_n, \dfrac{\Delta P_n}{P_n}\right)$ data pairs is

$$\frac{\Delta P_n}{P_n} = -0.001299 P_n + 0.338.$$

Multiplying both sides of this equation by P_n gives

$$\Delta P_n = -0.001299 P_n^{\,2} + 0.338 P_n.$$

Consequently, in the logistic difference equation

$$\Delta P_n = a P_n - b P_n^{\,2},$$

we conclude that $a = 0.338$ and $b = 0.001299$. Based on the logistic model, the limiting population for the United States will be

$$L = \frac{a}{b} = \frac{0.338}{0.001299}$$

$$= 260.2 \text{ million},$$

if all current trends continue. We note that this prediction is unrealistically low considering how close we are to it now, how far the actual data values in the scatterplot in Figure 6.28 are from leveling out to a horizontal asymptote, and how much "room" we obviously have in the United States for more people. However, recall the residual plot. It suggests that the logistic model does not seem to account for all the variation in the data. It provides a relatively good fit to the growth of the population, but it is not an outstanding fit.

Note that some calculators have a routine for performing logistic regression analysis. However, the method they use involves an approach based on methods from calculus instead of the difference equation model we have considered. Thus they produce results that appear quite different.

produce x_1; then to apply the function f to x_1 to get x_2, and so on indefinitely. Such a process is known as **iteration** and the successive values produced are called **iterates.**

Suppose $r > 1$ so that we have exponential growth. Then, for any initial value x_0, the sequence of iterates diverges to infinity as the process continues because each successive iteration involves multiplying the previous value by r which is greater than 1. On the other hand, suppose that $0 < r < 1$ and we have exponential decay. Then, for any x_0, the sequence of iterates converges to 0 because each successive value is multiplied by a number between 0 and 1. Similarly, if $-1 < r < 0$, the sequence of iterates converges to 0 in an oscillatory manner. Also, if $r < -1$, the sequence of iterates diverges in an oscillatory manner. If $r = 1$, the sequence remains constant and is always equal to the initial value x_0. If $r = -1$, the sequence bounces back and forth between x_0 and $-x_0$ forever. Finally, if $r = 0$, the sequence becomes zero immediately and stays at zero thereafter. We illustrate these possibilities in Figure 6.31.

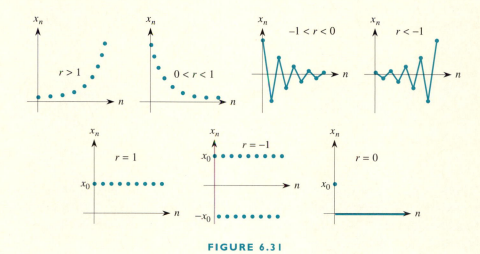

FIGURE 6.31

In summary:

If $|r| > 1$, the sequence diverges.
If $|r| < 1$, the sequence converges to 0.
If $r = 1$, the sequence is constant and is equal to x_0.
If $r = -1$, the sequence oscillates back and forth from x_0 to $-x_0$.

Since the particular behavior of the iterates depends on the value of the coefficient r, we think of r as a *parameter* that can take on any of a variety of values. Further, the *limit* of the sequence of iterates also depends on the value of r. We illustrate this in Figure 6.32, where the values of r are displayed along the horizontal axis and

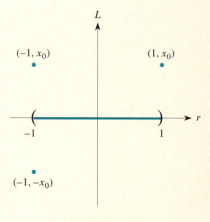

FIGURE 6.32

the limits L of the sequences of iterates are shown vertically. The resulting graph consists of the following:

1. A line segment along the horizontal axis on the interval $-1 < r < 1$ since the limit L of the sequence of iterates is 0 when $-1 < r < 1$.
2. The point $(1, x_0)$ since the limit L of the sequence of iterates is x_0 when $r = 1$.
3. The pair of points $(-1, x_0)$ and $(-1, -x_0)$ since the process oscillates from x_0 to $-x_0$ when $r = -1$.

Outside this interval the sequence diverges, and so there is no limit. The limiting value $L = 0$ is called an *attractor* for the sequence of iterates because they are "pulled" to this specific value as the process proceeds.

Similar results occur with functions other than $f(x_n) = rx_n$. For instance, try the following experiment with your calculator in radian mode. Select any initial value x_0, enter $\sin x_0$, and then repeatedly press the *SIN* key followed by previous *ANSWER*. What do you observe about the sequence of iterates being produced? What do you think will happen if you continue this process long enough? What happens if you start with a different value for x_0? What do you think is the attractor for the difference equation

$$x_{n+1} = f(x_n) = \sin x_n?$$

To see what is happening, think of the problem of finding the equilibrium solution of the difference equation, if it exists. To make things simpler, suppose we write x instead of x_n so that finding the equilibrium is equivalent to solving the equation

$$x = f(x) = \sin x.$$

We visualize the solution graphically as the point of intersection of the line $y = x$ and the curve $y = \sin x$, as shown in Figure 6.33. From this, it is clear that the only solution occurs at $x = 0$ since this is the one point where the line and the sine curve cross.

Now let's see how the original problem

$$x_{n+1} = f(x_n) = \sin x_n$$

fits into this picture. When we start with an initial value x_0, we have a point on the x-axis and we calculate $\sin x_0$, which represents the height of the sine curve, as shown in Figure 6.33. We then use this height $x_1 = \sin x_0$ as the next point in the sequence along the x-axis. To locate x_1 on the x-axis, we move horizontally from the point $(x_0, y_0) = (x_0, x_1)$ on the sine curve until we reach the diagonal line $y = x$ (at which point both x and y are equal to this new x_1, from which we drop down to the x-axis. Once

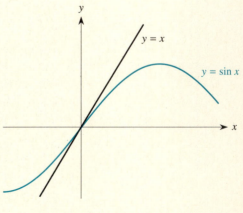

FIGURE 6.33

we have this point x_1 on the x-axis, we move vertically, back up, to the sine curve to find the new height $x_2 = \sin x_1$. Then we move horizontally across to the diagonal line, drop back down to the x-axis, and repeat the process indefinitely. In effect, what we are doing is the following: Start at x_0 on the x-axis, move up to the sine curve, move across to the line, move down to the sine curve, move across to the line, move to the curve, and so on, as shown in Figure 6.34. You should trace this sequence of steps on the graph to be sure you understand what is happening.

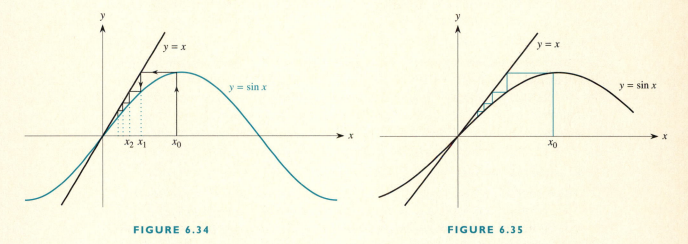

FIGURE 6.34   FIGURE 6.35

The diagram corresponding to this process is known as a *cobweb*. From Figure 6.35, it should be clear that the sequence of points generated is moving inexorably down toward the origin, so that the attractor is 0.

Think About This

Apply the same iterative process on your calculator to the cosine function. What do you find is the attractor here? Can you draw a cobweb diagram to explain what is happening?

Some truly remarkable and unexpected results occur if we try the same process with the logistic difference equation

$$\Delta x_n = a x_n - b x_n^2,$$

which is equivalent to

$$x_{n+1} = (1 + a)x_n - b x_n^2$$

or

$$x_{n+1} = f(x_n) = r x_n - b x_n^2, \qquad (4)$$

where we have written $r = 1 + a$. We can simplify things considerably if we consider the special case where $b = r$, so that

$$x_{n+1} = f(x_n) = r x_n(1 - x_n), \qquad (5)$$

which involves a single parameter r. (*Note:* Equation (4) can always be transformed into Equation (5), so the two forms are actually equivalent difference equations. We ask you to do this as an exercise at the end of the section.)

In our earlier discussions on the logistic model, we saw that the limit of the process, L, is the ratio

$$L = \frac{a}{b} = \frac{r-1}{b}.$$

In this case where $b = r$, we get

$$L = \frac{r-1}{r}.$$

However, some strange things happen on the way to the attractor for this equation.

First, if the initial value x_0 is 0, then from Equation (5), we see that all successive iterates are 0. Second, if any iterate x_n is either 0 or 1, then all successive values will remain 0. Third, if any iterate x_n is negative, then the succeeding value x_{n+1} will also be negative; further, x_{n+1} will be more negative than x_n because of the term $(1 - x_n)$. (Try this with $x_n = -\frac{1}{2}$ to convince yourself.) Consequently, the successive terms will diverge toward $-\infty$. Fourth, if any iterate x_n is larger than 1, then the succeeding value x_{n+1} will be negative and the sequence will thereafter diverge to $-\infty$. As a result, we see that the only meaningful cases involve starting with x_0 between 0 and 1 and remaining there for all values of n thereafter.

Now let's see what effect the parameter r has on the process. If we select any value of r between 0 and 1, the successive iterates become smaller and smaller and converge to a limit of 0. In terms of population growth, this just means that the population becomes extinct. If r is between 1 and 3, the sequence of iterates converges to the limit $L = \frac{r-1}{r}$, as we would expect. Thus for $r = 1$, the limit is 0 and for $r = 3$, the limit is $\frac{2}{3}$.

However, if $r > 3$, things become very different. Suppose we take $r = 3.2$ with $x_0 = 0.5$. We then obtain the following iterates

$$x_1 = 0.8 = 3.2(0.5)(1 - 0.5)$$
$$x_2 = 0.512$$
$$x_3 = 0.79954$$
$$x_4 = 0.51288$$
$$x_5 = 0.79947$$
$$x_6 = 0.51302$$
$$x_7 = 0.79946$$
$$x_8 = 0.51304$$
$$x_9 = 0.79946 = x_7$$
$$x_{10} = 0.51304 = x_8$$

and all succeeding values keep oscillating between these two values, x_7 and x_8. Thus we get a result similar to what we had earlier with the exponential growth model whose solution likewise oscillated between two values when $r = -1$. If we use $r = 3.3$ instead of 3.2, the sequence eventually oscillates between 0.82360 and 0.47943. Note that these pairs of limits or attractors hold no matter what initial value x_0 we use provided it is between 0 and 1. The only thing that changes is how long it takes to reach the limit values.

However, things get even more complicated than this. Suppose we take $r = 3.5$, also with $x_0 = 0.5$. The resulting values for x_n are

$$0.50000, \ 0.87500, \ 0.38281, \ 0.82693,$$
$$0.50091, \ 0.87500, \ 0.38281, \ 0.82694,$$
$$0.50088, \ 0.87500, \ 0.38281, \ 0.82694,$$
$$0.50088, \ 0.87500, \ldots$$

Thus we see that after a few transient values, the sequence settles down to oscillate among *four* different limit values in a cycle. This is termed a *period-4 cycle* and the previous case with two limiting values is known as a *period-2 cycle*.

Are there other cycles? If you try $r = 3.55$, you will find a period-8 cycle. In fact, it is possible to obtain cycles of period 2^k, for $k = 1, 2, 3, 4, \ldots$. Moreover, the intervals of values for the parameter r that produce these cycles become smaller and smaller. We summarize these intervals as follows.

Region I: $0 < r \le 1$ x_n approaches 0 ($L = 0$)
Region II: $1 < r \le 3$ x_n approaches a limit $L = (r - 1)/r$
Region III: $3 < r < 3.56994\ldots$ period doubling region where x_n oscillates between 2, 4, 8, ... different values.

The length of the cycles depend on r as follows.

- If r is in the interval $3 < r \le 3.44949 \ (= 1 + \sqrt{6})$, there are period-2 cycles.
- If r is in the interval $3.44949 < r \le 3.54409$, there are period-4 cycles.
- If r is in the interval $3.54409 < r \le 3.56441$, there are period-8 cycles.
- If r is in the interval $3.56441 < r \le 3.56876$, there are period-16 cycles.
- If r is in the interval $3.56876 < r \le 3.56969$, there are period-32 cycles.
- If r is in the interval $3.56969 < r \le 3.56989$, there are period-64 cycles.

Notice that the critical values are spaced ever more closely. Does this continue forever? Yes, in the sense that there are ever more period-doubling cycles. However, all of these cycles occur until r reaches a critical value of about 3.56994. Until this point, the results can be predicted. Beyond that value for r, the results become totally unpredictable. The values for x_n

change among an infinite number of possible values in a way that cannot be predicted. We describe such behavior as being *chaotic*. In such a case, the successive iterates are extremely sensitive to small changes in the initial value x_0. That is, arbitrarily small changes in the initial value will lead to totally different sequences of values that bear no resemblance to one another.

Instances of such chaotic phenomena have been observed in many different areas. For instance, in long-range weather predictions, all existing mathematical models are subject to this type of uncertainty. The smallest change in initial values results in dramatic changes in the final predictions. Scientists sometimes call this the *butterfly effect* to reflect the fact that the fluttering of the wings of a butterfly on one continent could conceivably affect the weather on a different continent. In a similar way, efforts at predicting the precise location of the moons of some of the outer planets have run into the same problem. Seemingly minor perturbations, based on the interaction of the gravitational fields of other planets and moons, cause unpredictable effects. The same problem also arises in medicine when cardiologists study the results of stimuli applied to the heart.

Note that *chaotic behavior* is not the same as *random behavior*. As we saw in our discussion of radioactive decay, a random process is one where we cannot predict individual outcomes, but can predict overall patterns in the long run. A chaotic process is one where it is not possible to predict either individual outcomes or overall patterns. The results are purely chaotic.

Incidentally, the region of chaotic behavior is itself limited; it extends only as far as $r = 4$. Beyond that, the successive iterates will diverge to ∞. However, within the regions of chaotic behavior, we often see many (short) intervals of normal periodic behavior as well.

We can understand the different types of behavior using cobweb diagrams that display the iteration process graphically. First, realize that the function $y = f(x) = rx(1 - x) = rx - rx^2$, for different positive values of r, represents a family of parabolas that are opening downward. Moreover, each of these parabolas has roots at $x = 0$ and $x = 1$. In addition, the height of the parabola depends on the value of the parameter r. In particular, by symmetry, we can conclude that the turning point of each parabola occurs when $x = 0.5$, so that the maximum height that the parabola achieves is

$$y_{max} = r(0.5)(1 - 0.5) = 0.25r.$$

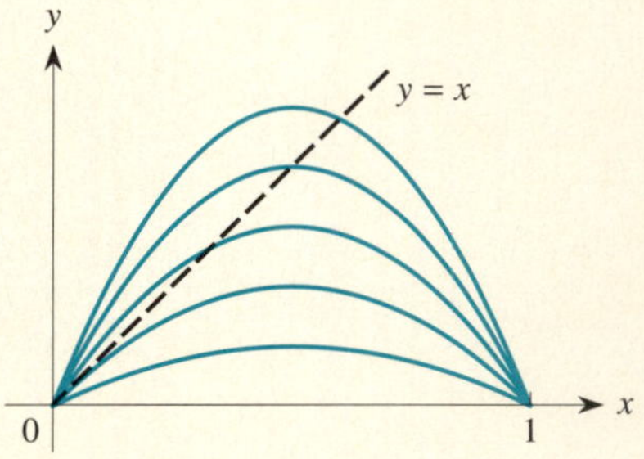

FIGURE 6.36

Thus the larger the parameter r, the higher the arch of the parabola, as shown in Figure 6.36.

As with the sine function we discussed before, the limiting behavior of the iteration process with the logistic function involves the diagonal line $y = x$ that is also drawn in Figure 6.37. However, the particular behavior depends very much on the value of r. Notice that some of the parabolas, corresponding to small values of r, lie entirely below the line for $x > 0$. The parabolas corresponding to larger values of r intersect the line at a point between $x = 0$ and 1. These points of intersection, which depend on the value of r, give rise to the limiting values of the iteration process.

Figure 6.37(a) shows a value of r between 0 and 1, so the parabola has a very small arch. Notice that the successive values generated by the cobweb construction converge toward the origin, since it is the only point of intersection between the line and the parabola. This is very similar to what happened earlier with the sine function.

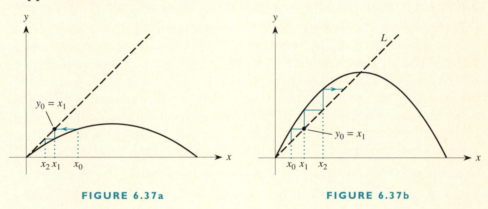

FIGURE 6.37a   FIGURE 6.37b

Figure 6.37(b) shows what happens with a slightly larger value of r, now between 1 and 3, so the arch of the parabola is somewhat higher. In this case, the cobweb process converges to a single value at the limiting point $L = \dfrac{r - 1}{r}$ where the line crosses the parabola.

Figure 6.37(c) shows the results with an even larger value of r. In this case, the cobweb process oscillates between two limiting values and bounces back and forth between them forever.

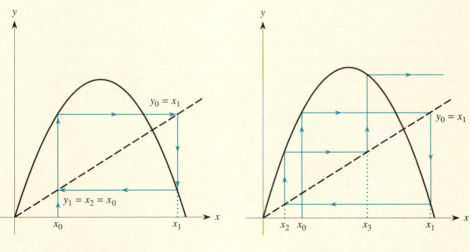

FIGURE 6.37c   FIGURE 6.37d

Finally, Figure 6.37(d) shows what happens with a still larger value for r. We see that the cobweb process quickly diverges. In fact, the reason that the sequence diverges is that the height corresponding to x_0 is greater than 1 and, as we saw earlier, all values for the iterates must remain between 0 and 1 if we are to have convergence. However, let's consider the worst case scenario for this situation. This corresponds to selecting x_0 as the point at the vertex, since this produces the maximum possible value, namely $0.25r = \frac{r}{4}$. Thus if $\frac{r}{4}$ is larger than 1, then we should expect a divergent sequence. Alternatively, if $\frac{r}{4}$ is smaller than 1, which means $r < 4$, then we should expect convergence. This is precisely the condition we specified previously.

We can visualize the various types of behavior in a different way using what is known as a *bifurcation diagram*, which is an extension of Figure 6.32. Again, we indicate values of r in the horizontal direction and limiting values L for the logistic sequence vertically (see the top graph of Figure 6.38). You may want to interpret this figure as resulting from a vertical line moving from left to right for different values of r.

For r between 0 and 1, the limit is always 0 and we get the horizontal segment at the left. Between $r = 1$ and $r = 3$, the limit is $L = \frac{r-1}{r}$, and so, when we think of the limit L as a function of r, it will be a slowly rising curve as r increases to the right. Beyond $r = 3$, we get into the period-doubling region and there is an initial split (bifurcation) to show the pair of limiting values. Each of these in turn then splits further, as r gets somewhat larger, into sets of four values for the period-4 cycles; then, further to the right, these four branches bifurcate into eight branches, and so forth. This is seen better in the enlarged graph for r between 2.9 and 4, shown in the bottom graph of Figure 6.38. Eventually, we get to the region of chaos where there is no predictability. Beyond $r = 4$, there is nothing shown because there are no limits—the sequences all diverge no matter what initial value is used.

There are many computer programs available that allow you to explore some of the implications of the period-doubling and the chaos effects. We suggest that, if you have access to one, you should take the opportunity to see these effects "live." For instance, you might want to experiment with the ideas behind the bifurcation diagram for a class of generalized logistic models based on the difference equation

$$x_{n+1} = rx_n(1 - x_n^p),$$

where p is any desired power, or you can experiment with functions such as

$$x_{n+1} = rx_n^2 - 1$$

or

$$x_{n+1} = \frac{rx_n}{(1 + x_n)^p}$$

that have been studied by other people, or with any other function of your choice. We will encounter other examples of chaos later in this book.

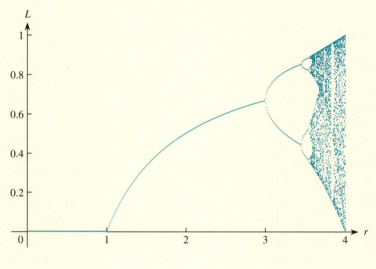

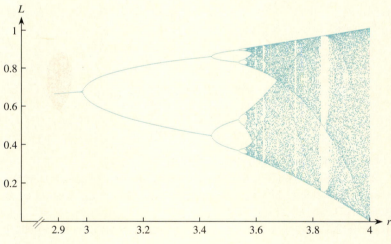

FIGURE 6.38

EXERCISES

1. Repeat the calculations in the text based on the logistic equation with $r = 3.2$ using $x_0 = 0.2$ instead of 0.5; using $x_0 = 0.8$. What do you observe about the limiting values?

2. What are the limiting values for the period-2 cycle using $r = 3.1$? How do they compare to the results in Exercise 1?

3. What are the limiting values for the period-2 cycle using $r = 3.4$? How do they compare to the results in Exercises 1 and 2?

4. Repeat the calculations done with the logistic equation with $r = 3.5$ using $x_0 = 0.2$ instead of 0.5. How do the results compare to the limiting values found in the text?

5. What are the limiting values for the period-4 cycle using $r = 3.48$? How do they compare to the results from Exercise 4?

6. In Exercise 1, you found that the period-2 cycle limiting values were approximately 0.513 and 0.799, based on the difference equation $x_{n+1} = 3.2x_n (1 - x_n)$. Use this equation to obtain an equation for x_{n+2} in terms of x_n, then let L represent the limiting value, and solve for L when $x_{n+2} = x_n = L$.

7. Derive the form of the logistic Equation (5) discussed in this section by introducing the substitution $y_n = \frac{b}{r} x_n$ into the original logistic difference equation $x_{n+1} = rx_n - bx_n^2$. After you have simplified the resulting equation in terms of y_n, simply replace y_n by x_n.

CHAPTER SUMMARY

In this chapter, you have learned the following:

- What the order of a difference equation is.
- What it means for a difference equation to be linear or nonlinear.
- What it means for a difference equation to be homogeneous or nonhomogeneous.
- How the solution of a first order difference equation depends on the initial condition x_0.
- What the solution field of a difference equation tells us about the behavior of the solutions.
- How the complete solution of a first order, linear, nonhomogeneous difference equation is composed of two parts: the solution to the associated homogeneous difference equation and a particular solution of the nonhomogeneous difference equation.
- How to find a particular solution using the method of undetermined coefficients.

- How to solve and interpret the solutions of difference equations arising from modeling a variety of phenomena.
- How to fit quadratic functions to data.
- How to model the stock market using a difference equation with a random term.
- How to fit logistic curves to data.
- How chaotic behavior arises from iterating the logistic function for different values of a parameter.

REVIEW EXERCISES

1. Consider the nonlinear difference equation $x_{n+1} = x_n^3 - x_n^2 - 5x_n$.
   a. Find all equilibrium levels for the difference equation.
   b. Construct a table containing the first 10 terms in each solution starting with the initial conditions $x_0 = -3, -2, -1, \ldots, 4$.
   c. Sketch the solution field for the difference equation based on the table you constructed in part (b).
   d. Describe the behavior of the solutions in each of the four regions separated by the equilibrium levels.

2. Aunt Jean deposited $35,000 in a savings account when her nephew Brett was born. The interest on the account is 6% per year, compounded monthly. What is the value of the account when Brett celebrates his 21st birthday?

3. Suppose that every month Aunt Jean deposits an additional $100 into Brett's account as described in Exercise 2. How much would be in the account when Brett is 21?

4. Jerry has secured a loan of $100,000 at 9% per year compounded monthly. If he can afford to pay $1000 per month, how much toward the principal is he paying in the first three months? Make a chart showing the part of the monthly payment that is interest and the part that is payment on the principal for each of the three months. Write a difference equation for the principal after $n + 1$ months.

5. How much money can you borrow at 8% interest per year, compounded monthly, if you wish to pay it off in monthly payments, have the loan retired in 5 years, and can afford $400 per month?

6. Initially 100 pounds of a pollutant are dumped in a lake and 8% of it is washed away each year. Find the expression for the amount of pollutant in the lake after n years. Determine the amount after three years. When will the level of the pollutant fall below 10 pounds?

7. Initially 100 pounds of a pollutant are dumped in a lake and 8% of it is washed away each year. Every year thereafter 100 pounds of pollutant are dumped into the lake. Find an expression for the level of pollutant after n years. Find the level of pollutant after 4 years. Is there a limiting value for the level of the pollutant? Explain.

8. Chlorine is added to the water supply in a reservoir at the rate of 30 units a day. It is estimated that 20% of this amount disappears each day through evaporation or filters. Is there a maintenance level of chlorine in the reservoir?

9. Determine the first five terms of the sequence d_n and find the solution of the difference equation

$$d_{n+1} = 0.8d_n + 20(1.3)^n.$$

10. Joe decided to create his own sequence with the following scheme. He started with the initial value $x_0 = 2$ and calculated the following term

$$x_1 = 1 + \frac{1}{x_0} = 1 + \frac{1}{2} = \frac{3}{2}.$$

He then repeated this to obtain

$$x_2 = 1 + \frac{1}{x_1} = 1 + \frac{1}{3/2} = \frac{5}{3},$$

and so forth.

a. What are the first 10 terms of Joe's sequence?
b. Toward what limiting value do the terms appear to be converging?
c. Repeat this process with any other nonzero initial value for x_0.
d. Write a first order, nonlinear difference equation that represents this process.
e. What are the equilibrium levels for the difference equation?
f. Sketch a graph of the solution field for this difference equation.

7

MODELING PERIODIC BEHAVIOR

7.1 *Introduction to the Trigonometric Functions*

One of the most common behavior patterns in nature is a *periodic oscillatory effect*—one that repeats over and over again. For example, think about the ocean level that varies at a beach between low tide and high tide approximately every 12 hours. If low tide occurs at midnight, then high tide will be around 6 A.M. and low tide will occur again around noon, and so on indefinitely. This periodic oscillatory behavior is shown in Figure 7.1. The word *periodic* refers to the fact that this phenomenon repeats indefinitely; the **period** is the time it takes to complete one full cycle.

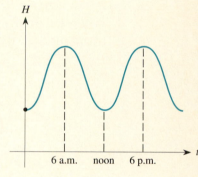

FIGURE 7.1

In a similar way, consider the number of hours of daylight each day in a particular city, say San Diego. The minimum number of hours of daylight occurs on the winter solstice, December 21; it increases slowly until the maximum daylight occurs on the summer solstice, June 21, and then decreases back to the same minimum the following December 21. This identical oscillatory behavior repeats, year after year, giving rise to the same type of graph shown in Figure 7.1.

In this chapter, we will consider how to model such periodic phenomena. This type of behavior pattern is quite different from linear, exponential, or polynomial behavior. We therefore need a different class of functions, the **trigonometric functions,** to describe it mathematically.

Historically, the trigonometric functions are based on ratios of sides in any right triangle. The two most common ratios are the *sine* and *cosine* functions. The key ideas about the sine and cosine are as follows.

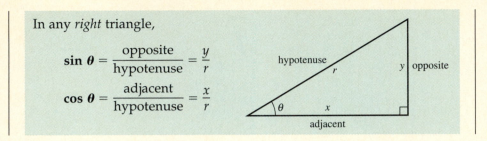

In any *right* triangle,

$$\sin \theta = \frac{\text{opposite}}{\text{hypotenuse}} = \frac{y}{r}$$

$$\cos \theta = \frac{\text{adjacent}}{\text{hypotenuse}} = \frac{x}{r}$$

Before going on, some of you may want to review some right-angle trigonometry in Appendix A.

For our study of the trigonometric functions, we begin by considering the **unit circle**—a circle of radius 1 centered at the origin—as shown on the left in Figure 7.2. A point $P(x, y)$ lies on this circle if

$$x^2 + y^2 = 1.$$

We consider separately the horizontal distance x to the left or right of the y-axis and the vertical height y above or below the x-axis. Each of these quantities is actually a function of θ, the angle at the center of the circle. Let's start with the height y. Consider the point $A(1, 0)$ at the extreme right of the circle as our starting point; it has a height $y = 0$ and an angle of inclination $\theta = 0$. When we think of y as a function of θ, the pair $(\theta, y) = (0, 0)$ corresponds to the point A' at the origin in the graph on the right in Figure 7.2.

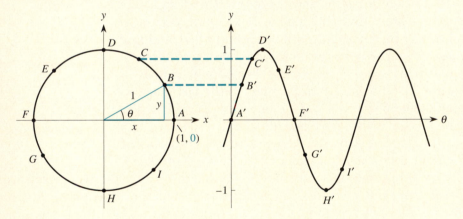

FIGURE 7.2

As we proceed, keep clear the two pictures to which we refer in the coordinate plane—the circle on the left where we think of the height y as a function of x, and the associated graph that we construct on the right showing y as a function of the angle θ. In the circle, the horizontal axis measures x while in the graph, the horizontal axis measures θ. The position of any point, say B, on the circle can be specified either (1) in terms of its x- and y-coordinates or (2) in terms of its angle of inclination θ and its vertical distance y above or below the horizontal axis of the circle. It is these (θ, y) pairs that we graph on the right.

The Sine Function

Let's consider all the points P on the circle. We have labeled several specific points as $A, B, C, D, E, \ldots, I$ in Figure 7.2. Each point corresponds to a particular angle θ, and we plot the heights y corresponding to these angles θ in the graph at the right in Figure 7.2. Notice that as point P traces out the circle, starting at point A, passing through the points B, C, D, E, and so forth, the height y on the circle rises from 0 (at A) up to a maximum of 1 (at D), then decreases down past 0 (at F) to a minimum height of -1 (at H), and then back up to 0 as P finally returns to the starting point A. However, this motion of P can continue; P can keep tracing around the circle again and again, and so the identical pattern of heights can recur repeatedly, every 360°. As a result, the oscillatory pattern we see in the right-hand graph is periodic; it repeats forever with a period of 360°. The curve we produced on the right is part of the graph of the *sine function*, $f(\theta) = \sin \theta$, for any angle θ.

You can see this development dynamically using your graphing calculator. Set the mode for radians (which we discuss in a moment), for parametric graphing (we will discuss this topic in a later chapter), and for simultaneous plotting. See the instructions for your particular calculator if necessary. Go to the "Y = " menu and enter

$$X1 = \cos T$$
$$Y1 = \sin T$$
$$X2 = T$$
$$Y2 = \sin T$$

For the viewing window, set T between 0 and 7, $\Delta T = 0.1$, x between -1.5 and 7, and y between -2 and 2. When the graphs are drawn, you will see the circle (somewhat flattened because of the screen dimensions) and the sine curve traced out simultaneously. Watch how the heights of points on the circle precisely match the heights on the sine curve for any angle T. You may also want to trace this behavior with your fingers on the graphs in Figure 7.2.

We summarize the above ideas on the sine function as follows:

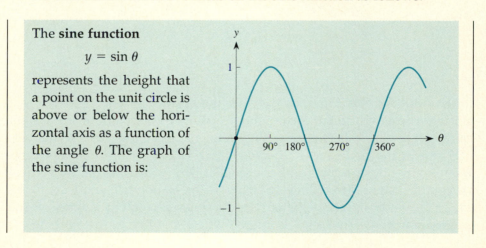

The **sine function**

$$y = \sin \theta$$

represents the height that a point on the unit circle is above or below the horizontal axis as a function of the angle θ. The graph of the sine function is:

Notice that the graph of the sine function oscillates between a maximum height of 1 and a minimum height of -1. Also, the basic shape repeats every 360° so that the behavior pattern you see from 0° to 360° occurs again from $\theta = 360°$ to $\theta = 720°$, and again from 720° to 1080°, and so forth. Similarly, the same pattern occurs between $-360°$ and 0°, $-720°$ and $-360°$, and so on. Thus the sine function is a *periodic function* and its **period** is 360°.

In addition, the sine curve reaches its maximum height of 1 where $\theta = 90°$, and again where $\theta = 450°$ (90° + 360°), 810° (90° + 2 × 360°), ..., as well as where $\theta = -270°$ (90° − 360°), $-630°$ (90° − 2 × 360°), ... Similarly, the sine curve reaches its minimum height of -1 where $\theta = 270°$, 630°, ..., and where $\theta = -90°, -450°, -810°, ...$ Also, notice that the sine curve crosses the horizontal axis at the origin where $\theta = 0°$, again where $\theta = 180°$ (corresponding to the extreme left-hand point on the unit circle), still again where $\theta = 360°$, and so forth indefinitely. In fact, the sine function has zeros at every integer multiple of 180°.

The Cosine Function

In an analogous way, let's consider the horizontal distance x from the vertical axis of the unit circle to points $P(x, y)$ on the circle, as shown on the left in Figure 7.3. We now treat x as a function of θ, so that in the graph of (θ, x) points on the right in Figure 7.3, the angle θ is measured along the horizontal axis and the distance x is measured vertically.

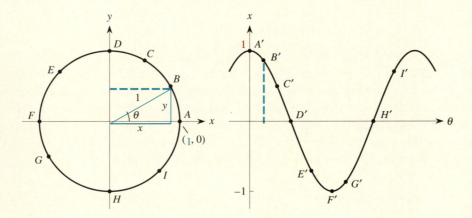

FIGURE 7.3

Our initial point $A(1, 0)$ on the circle lies at a distance of $+1$ to the right of the vertical axis, so the corresponding point A' on the graph at the right has a height of $x = 1$. The succeeding points B and C are closer to the vertical axis of the circle and so, because the x-values are smaller, the corresponding heights on the graph are smaller. The point D is on the vertical axis, so its distance from the vertical axis is 0. The points E and F are to the left of the vertical axis of the circle, so the corresponding heights on the graph are negative. In fact, F is located where $\theta = 180°$ at the extreme left point of the circle at a distance of −1 from the vertical axis, and so the corresponding point F' on the graph is a minimum. As the angle θ continues to increase, the points on the circle approach the vertical axis from the left, and so the corresponding points on the graph now rise toward 0. Eventually, the tracing point P on the circle passes the vertical axis and approaches the initial point A where $\theta = 360°$. The horizontal distance that P is from the axis changes from negative to positive and approaches the distance 1 that we started with, and the graph on the right crosses the θ-axis and rises to its initial starting height of 1.

Further, we can allow θ to continue beyond $\theta = 360°$, and the previous pattern will repeat exactly. The graph we are constructing on the right in Figure 7.3 is also a periodic, oscillatory curve. This curve, which corresponds to the horizontal distances from the vertical axis of the unit circle to points on the circle is the graph of the *cosine function*, $g(\theta) = \cos \theta$. The cosine function, like the sine function, is periodic and repeats every 360°, so its period also is 360°. The maximum value of the cosine function is 1, which occurs where $\theta = 0°, 360°, 720°, \ldots$, as well as where $\theta = -360°, -720°, \ldots$. The minimum value of the cosine function is −1, which occurs where $\theta = \pm 180°, \pm 540°, \ldots$. The cosine function has zeros when $\theta = \pm 90°, \pm 270°, \pm 450°, \ldots$.

We summarize the key facts about the cosine function as follows:

The **cosine function**, $y = \cos \theta$, represents the horizontal distance that a point on the unit circle is to the right or left of the vertical axis as a function of the angle θ. The graph of the cosine function is:

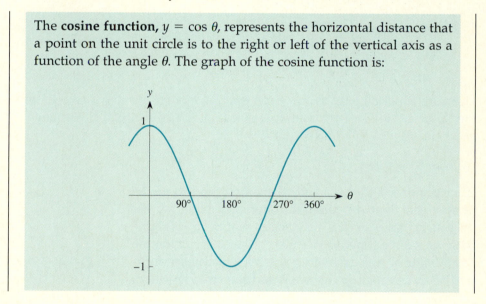

You may find the preceding construction of the cosine curve somewhat easier to visualize using the following trick. Rotate the circle at the left of Figure 7.3 through an angle of 90° counterclockwise. As a result, each of the horizontal distances, x, now is transformed into an equivalent height above or below the new horizontal axis, and it is these heights that produce the heights of the points on the cosine curve in the right-hand graph in Figure 7.3.

Because of the way that the sine and cosine functions can be defined in terms of the unit circle, they are sometimes called *circular functions*.

When you look at the graphs of the sine and the cosine functions and consider how we defined them, it should be clear that these two functions are closely related. The unit-circle definition suggests what is perhaps the most important relationship between them. From Figure 7.2, we see that the vertical height y to a point on the unit circle is equal to sin θ. Similarly, the horizontal distance x from the vertical axis to the same point is equal to cos θ. Since $x = \cos \theta$ and $y = \sin \theta$ must satisfy the equation of the unit circle,

$$x^2 + y^2 = 1,$$

it follows that

$$(\cos \theta)^2 + (\sin \theta)^2 = 1.$$

By convention, we write

$$(\cos \theta)^2 \quad \text{as} \quad \cos^2 \theta \quad \text{and} \quad (\sin \theta)^2 \quad \text{as} \quad \sin^2 \theta.$$

Since the above relationship is based on the Pythagorean theorem, we have

> The Pythagorean Identity:
>
> $$\sin^2 \theta + \cos^2 \theta = 1$$
>
> for *any* angle θ.

Think About This

Use your function grapher to graph the function $f(X) = \sin^2 X + \cos^2 X$ for any interval of X values. What does it look like? (You will likely have to enter this as $(\cos X)\hat{\ }2 + (\sin X)\hat{\ }2$.)

Radian Measure

If we are to use the sine and cosine functions to model phenomena that are periodic over time, such as the heights of tides or the number of hours of daylight, then we need a function of time t rather than a function of an angle θ. Also, we often use a trigonometric function of some other quantity, say a distance x. Therefore it is essential that we avoid dependence on an angle θ measured in degrees in our definitions of these functions. To accomplish this, we introduce an alternative unit for measuring an angle known as the *radian*. In the unit circle shown in Figure 7.4, we start on the horizontal axis at the point $A(1,0)$, move counterclockwise along the circle, and measure off a distance equal to the radius of the circle (1 unit). This produces an angle α whose size we define to be **one radian.**

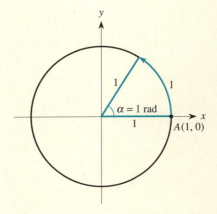

FIGURE 7.4

In degrees, this angle turns out to be approximately 57°.

We can get a precise measurement for one radian in terms of degrees as follows. The length of the arc that defines one radian is equal to the radius of the circle, which is 1. However, since $r = 1$, the total circumference of the circle is $2\pi r = 2\pi$. Further, the angle α represents a fraction of the full 360° in the circle. As a result, we can set up the proportion

$$\frac{1 \text{ radian}}{360°} = \frac{1}{2\pi}.$$

If we cross-multiply, we get

$$2\pi \text{ radians} = 360°,$$

or

$$\pi \text{ radians} = 180°.$$

Alternatively,

$$1 \text{ radian} = \left(\frac{180}{\pi}\right)° = 57.29578° \approx 57.3°.$$

We also can find that

$$1° = \frac{\pi}{180} \text{ radians.}$$

Since $180° = \pi$ radians, we can obtain the radian measure for some common angles:

$$90° = \frac{180°}{2} = \frac{\pi}{2} \text{ radians} = \frac{\pi}{2}$$

$$60° = \frac{180°}{3} = \frac{\pi}{3} \text{ radians} = \frac{\pi}{3}$$

$$45° = \frac{180°}{4} = \frac{\pi}{4} \text{ radians} = \frac{\pi}{4}$$

$$30° = \frac{180°}{6} = \frac{\pi}{6} \text{ radians} = \frac{\pi}{6}$$

To summarize:

$$\pi \text{ radians} = 180°$$

In particular,

$$30° = \frac{\pi}{6}$$

$$45° = \frac{\pi}{4}$$

$$60° = \frac{\pi}{3}$$

$$90° = \frac{\pi}{2}$$

These results will arise throughout our study of the trigonometric functions, and you will have to know them thoroughly.

For these standard angles, we have the following sine values (which are discussed in Appendix A):

$$\sin 30° = \sin \frac{\pi}{6} = 0.5 = \frac{1}{2}$$

$$\sin 45° = \sin \frac{\pi}{4} = 0.707 = \frac{\sqrt{2}}{2}$$

$$\sin 60° = \sin \frac{\pi}{3} = 0.866 = \frac{\sqrt{3}}{2}$$

$$\sin 90° = \sin \frac{\pi}{2} = 1$$

Be sure that you can use your calculator to obtain the value for the sine or cosine of any argument, both in degrees and radians. We strongly recommend that you permanently set your calculator mode to radians; we will work with radians almost exclusively from this point on.

If we perform the same construction in any circle with radius r—that is, if we measure off an arc whose length is precisely equal to the radius r—then the corresponding angle would be the same 1 radian, or about 57.3°. Thus an angle measured in radians is the same no matter what the size of the circle. More importantly, radians are not tied directly to angles in a triangle the way that degrees are. Using radians, we can consider any variable whatsoever and apply the sine and cosine functions to it. Thus, we can use a variable representing time, height, or any other desired quantity as the independent variable with either the sine or the cosine function.

The Behavior of the Sine and Cosine Functions

Let's now consider the important aspects of the behavior of the sine and cosine functions. In general, there are several questions we want to answer for any function:

1. Where is it increasing?
2. Where is it decreasing?
3. Where is it concave up?
4. Where is it concave down?
5. Where are its points of inflection?
6. Where are its zeros?
7. Where does it achieve its maximum value, and what is that maximum value?
8. Where does it achieve its minimum value, and what is that minimum value?
9. Is it periodic? If so, what is its period?

We can answer all these questions about the sine and cosine functions by examining their graphs and applying ideas we have encountered earlier in this book. However, for other functions that may not be as well known, some of these questions require the use of calculus in order to answer them.

Let's consider the behavior of the sine function. It is evident from its graph that the sine curve increases for θ between 0 and $\frac{\pi}{2}$, decreases from $\frac{\pi}{2}$ to $\frac{3\pi}{2}$, increases from $\frac{3\pi}{2}$ to 2π, and then repeats this cycle thereafter. You also can see this clearly from the unit circle definition of the sine. Further, the sine curve is concave down for θ between 0 and π, concave up for θ between π and 2π, and then repeats this cycle thereafter. Consequently, we see that the sine curve has points of inflection at $\theta = 0, \pm\pi, \pm 2\pi, \ldots$, where its concavity changes.

In addition, the sine function has zeros when $\theta = 0, \pm\pi, \pm 2\pi, \ldots$. A special characteristic of the function $f(\theta) = \sin\theta$ is that its roots and its points of inflection are identical, which does not happen for most other common functions. Finally, we see that the sine function achieves its maximum value of 1 at $\theta = \frac{\pi}{2}$. It reaches this same maximum value at $\theta = \frac{5\pi}{2}, \frac{9\pi}{2}, \ldots$, as well as at $\theta = -\frac{3\pi}{2}, -\frac{7\pi}{2}, \ldots$. The sine function achieves its minimum value of -1 at $\theta = \frac{3\pi}{2}$. It also reaches this minimum value at $\theta = \frac{7\pi}{2}, \frac{11\pi}{2}, \ldots$, as well as at $\theta = -\frac{\pi}{2}, -\frac{5\pi}{2}, \ldots$.

We ask you to describe the behavior of the cosine function in an exercise in this section.

EXERCISES

1. Janis trims her fingernails every Saturday morning. Sketch the graph of the length of her nails as a function of time. Can this process be modeled by a periodic function? If it is periodic, what is the period?

2. Harry gets a haircut on the first of every month. Sketch the graph of the length of his hair as a function of time. Can this process be modeled by a periodic function? If it is periodic, what is the period?

3. Convert the following angles from degrees into radians:

    a. 15°   **b.** 75°   **c.** 120°   **d.** 150°   **e.** 225°   **f.** 315°

4. Convert the following angles from radians into degrees.

    a. $\frac{3\pi}{4}$ **b.** $\frac{4\pi}{5}$ **c.** $\frac{2\pi}{3}$ **d.** 1.5   **e.** 2.5   **f.** 3

5. Given $f(\theta) = 5 \sin \theta$, evaluate:

    a. $f(30°)$ b. $f(45°)$ c. $f(60°)$ d. $f(120°)$

    e. $f(-15°)$ f. $f(873°)$ g. $f\left(\dfrac{\pi}{4}\right)$ h. $f\left(\dfrac{\pi}{3}\right)$

    i. $f\left(\dfrac{\pi}{12}\right)$ j. $f\left(-\dfrac{\pi}{6}\right)$ k. $f(5.27)$ l. $f(-25.614)$

6. In this section we posed nine questions about the behavior of any function. Answer these questions for the cosine function.

7. The population growth patterns of two species are interrelated when one species preys on the other. This occurs in northern Canada where lynxes are the predators and hares are the prey. The accompanying figure is based on records kept by the Hudson's Bay Trading Company on the number of animals of each species caught by fur trappers from 1845 through 1935. The graphs indicate that both populations change in roughly periodic cycles.

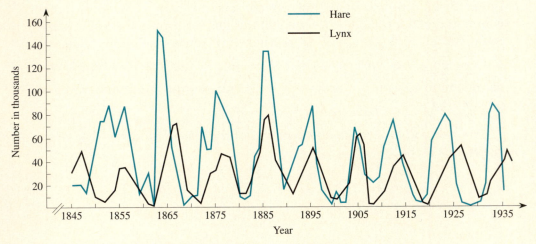

    a. Estimate the period of the cycle for the lynxes.
    b. Estimate the period of the cycle for the hares.
    c. Estimate the years in which the lynx population reached its maximum and minimum values.
    d. Estimate the years in which the hare population reached its maximum and minimum values.
    e. Can you find any relationship between the lengths of the periods in parts (a) and (b) and the times in parts (c) and (d)?
    f. Estimate the years when the hare population passed its points of inflection. How do these compare to any of the times you found in parts (c) and (d)?

8. Plot the functions $y = \sin x$ and $y = \cos\left(x - \dfrac{\pi}{2}\right)$. Explain why you see only one graph. (If you see two graphs, check that your calculator MODE is set for radians.)

9. Consider the functions $y = \cos x$ and $y = \sin$ (#$%#$). What could #$%#$ represent so that the two graphs are identical? Is there only one correct answer to this question? Explain.

10. One of the fundamental problems in calculus is finding the area of a plane region determined by a curve. For instance, what is the area of the region that lies between the x-axis and one arch of the sine curve? We can estimate this area as follows. First, we know that the sine curve is symmetric about $x = \frac{\pi}{2}$, so the total area under the arch is twice the area on the left side of $x = \frac{\pi}{2}$; that is, between $x = 0$ and $x = \frac{\pi}{2}$ (see the figure). Furthermore, either on the accompanying graph or from your function grapher, observe that the sine curve appears to be below the line $y = x$ and above the line from the origin to point M, the peak of the sine curve.

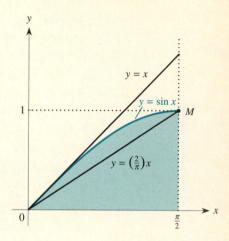

a. What is the area of the larger right triangle? (This is an overestimate of the desired area.)

b. What is the area of the smaller right triangle? (This is an underestimate of the desired area.)

c. How would you estimate the area of the region under the sine curve from 0 to $\frac{\pi}{2}$?

d. What is your estimate for the total area under one arch of the sine curve?

11. Use the reasoning suggested by Exercise 10 to estimate the area under the sine curve from $x = 0$ to $x = \frac{\pi}{4}$. Is your result a better estimate than the estimate made in Exercise 10? How do you know?

12. Use the graph of $y = \cos x$ to help you sketch the graph of its reciprocal function $y = \frac{1}{\cos x}$. (This is known as the *secant function*.)

13. Consider the nonlinear difference equation $x_{n+1} = \sin x_n$. Select the initial value $x_0 = \frac{\pi}{4}$ and use the difference equation to generate the next 10 terms, $x_1, x_2, \ldots, x_{10}$ by entering your value for x_0 in your calculator and successively taking the sine of the previous result. What do you notice about the terms you obtain? Do they appear to be converging to some limit? Continue the process until you think you know what the limit is. Now repeat the process with *any* other initial value x_0. To what value does this new sequence appear to be converging?

14. (Continuation of Exercise 13) Since the values for $\sin x$ must be between -1 and 1, you know that no matter what initial value you pick for x_0 in Exercise 13, the succeeding term x_1 must be between -1 and 1. Use the fact from Exercise 10 that the graph of the sine function always is below the line $y = x$ when $x > 0$ and above this line when $x < 0$ to decide how each value for x_{n+1} must compare to the preceding value x_n. Does this give you any reason to conclude that the limiting value you found in Exercise 13 must be correct? Why? You can determine this limiting value precisely by assuming (falsely) that the limit is reached for some sufficiently large value of n, call it M, so that $x_{M+1} = \sin x_M = x_M$. Finding the

limit then reduces graphically to finding the point of intersection between the curves $y = \sin x$ and $y = x$. What is it?

15. Repeat Exercise 13 with the difference equation $x_{n+1} = \cos x_n$. What limiting value do you seem to get? Is it the same as that in Exercise 13?

16. Determine graphically what the limiting value for Exercise 15 must be by using the same line of reasoning outlined in Exercise 14.

7.2 *Trigonometric Functions and Periodic Behavior*

In the introduction to Section 7.1, we noted that the trigonometric functions can be used as the mathematical model to represent certain periodic behavior. For example, the number of hours of daylight H each day of the year in San Diego can be modeled by the function

$$H(t) = 12 + 2.4 \sin\left(\frac{2\pi}{365}(t - 80)\right),$$

where t is the number of days from January 1 of any given year (on January 1, $t = 1$). We will see how a similar formula can be found for any city later in this section. For now, let's see what the different numbers in the formula actually mean in terms of the number of hours of daylight.

First, we can use a calculator—be sure that it is set to radian mode—to find the number of hours of daylight on February 15, the 46th day of the year:

$$H(46) = 12 + 2.4 \sin\left(\frac{2\pi}{365}(46 - 80)\right)$$

$$= 12 + 2.4 \sin\left(\frac{2\pi}{365}(-34)\right) = 10.67 \text{ hours.}$$

The number of hours on March 21, the 80th day, is

$$H(80) = 12 + 2.4 \sin\left(\frac{2\pi}{365}(80 - 80)\right)$$

$$= 12 + 2.4 \sin\left(\frac{2\pi}{365} \cdot 0\right) = 12 \text{ hours.}$$

March 21 is the spring equinox, and so we should expect 12 hours of daylight and 12 hours of darkness. The number of hours of daylight on June 21, the 172nd (and longest) day, is

$$H(172) = 12 + 2.4 \sin\left(\frac{2\pi}{365}(172 - 80)\right) = 14.40 \text{ hours.}$$

Think About This

How many hours of daylight are there on December 21, the 355th (and shortest) day?

Of course, the maximum and minimum number of hours of daylight, as well as the number on any given date, all depend on the location itself and are therefore based on a somewhat different function from H; just think about how long a "day" is during the winter or the summer in the far north, the so-called "land of the midnight sun."

The graph of H, the number of hours of daylight in San Diego, is shown in Figure 7.5. Notice that it has the same *shape* as the graph of the sine or cosine function. However, it does not oscillate about the horizontal axis, but rather lies considerably above it. In fact, it oscillates about the horizontal line $y = 12$, which represents the average number of hours of daylight over the course of a full year. Also, its maximum and minimum heights are no longer $+1$ and -1; instead, the graph varies from a minimum of 9.60 hours to a maximum of 14.40 hours.

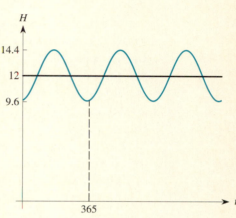

FIGURE 7.5

Notice some additional differences: The graph is shifted sideways (the curve does not "start" at the vertical axis where $t = 0$ and $H = 12$). Also, the period is 365 days, rather than the usual 2π radians, or 360°. (Incidentally, the ancient Babylonians believed that the length of a year was 360 days. This is the reason we divide a circle into 360 degrees.)

There are four ways in which this particular function differs from the standard, or base, function $\sin \theta$ that we discussed in Section 7.1. They are:

1. a vertical shift
2. an oscillation other than from -1 to $+1$
3. the length of a cycle
4. the "starting" point of the cycle (a horizontal shift).

Understanding how to incorporate these variations is critical for applying the trigonometric functions to describe periodic phenomena. We therefore focus on them in detail. The equation for the number of hours of daylight in San Diego is

$$H = 12 + 2.4 \sin\left(\frac{2\pi}{365}(t - 80)\right).$$

Consider the more general *sinusoidal function*

$$S(x) = D + A \sin(B(x - C)),$$

where $A, B, C,$ and D are all constants and x is the independent variable. In the San Diego situation, $D = 12$, $A = 2.4$, $B = \frac{2\pi}{365}$, and $C = 80$. Let's investigate how each of these four *parameters* affects the graph of the basic sine curve. To do so, we consider each parameter separately.

The Vertical Shift

To see the significance of the D term, consider the function

$$S(x) = D + \sin x.$$

We know that $y = \sin x$ oscillates repeatedly between -1 and $+1$. What is the effect of adding a constant D? From our discussion in Section 4.4, we know that D raises or lowers the basic sine curve by the amount D. The graph of $S(x) = 2 + \sin x$ has the same shape as the basic sine function, but is shifted up 2 units; it oscillates about the horizontal line $y = 2$, between 1 and 3 units above the x-axis (see Figure 7.6). Similarly, the graph of $S(x) = -5 + \sin x$ oscillates about the horizontal line $y = -5$, between -6 and -4.

Thus the effect of the constant D in

$$S(x) = D + \sin x$$

is to produce a sinusoidal curve that oscillates about the horizontal line $y = D$, between $D - 1$ and $D + 1$. If D is positive, the curve is shifted upward; if D is negative, the curve is shifted downward. The number D is known as the *vertical shift*. In the formula for the number of hours of daylight in San Diego, the vertical shift is 12.

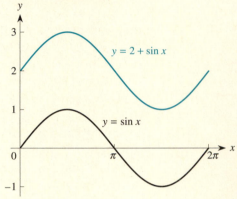

FIGURE 7.6

The Amplitude

We next investigate the effect of the multiplicative constant A in

$$S(x) = A \sin x.$$

For example, if $A = 2$, we get $S(x) = 2 \sin x$, whose graph is shown in Figure 7.7 where it is compared to the basic curve for the sine function, $y = \sin x$ (for which $A = 1$). For comparison, we also show the graph of $T(x) = \frac{1}{2} \sin x$. Notice that while the basic sine function oscillates between -1 and $+1$, the transformed function $S(x)$ oscillates between -2 and $+2$ and the transformed function $T(x)$ oscillates between $-\frac{1}{2}$ and $+\frac{1}{2}$. In general, the effect of multiplying the sine function by a constant A is to increase its vertical height by the factor $|A|$.

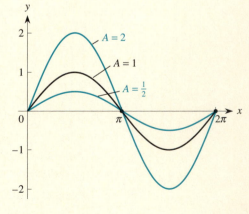

FIGURE 7.7

To see why the absolute value is necessary, consider the graph of

$$S(x) = -4 \sin x,$$

which has the same shape as the basic sine curve, but oscillates from -4 to $+4$. We naturally think of it as being four times as high as the base curve, not -4 times as high. (Draw simultaneously the graphs of $y = \sin x$ and $y = -4 \sin x$ using your function grapher.) The graph of $-4 \sin x$ is flipped over compared to the graph of $\sin x$. This is the effect of the negative multiple. (Notice, though, that this curve has the same period ($2\pi = 360°$) and the same zeros ($x = 0, \pm\pi, \pm2\pi, \dots$) as the basic sine curve.

The quantity $|A|$ is known as the **amplitude** of the sine function. In our expression for the number of hours of daylight in San Diego, the amplitude is 2.4.

What happens when we combine the two transformations? For instance, let us graph

$$S(x) = 2 + 3 \sin x.$$

We know that the effect of multiplying the sine function by 3 is to increase its vertical size by a factor of 3. Further, adding the constant 2 to the function simply raises the entire curve 2 units vertically. Consequently, the combined effect is to produce a sinusoidal function which oscillates about the horizontal line $y = 2$, from 3 units below to 3 units above that line; that is, from -1 to $+5$, as shown in Figure 7.8.

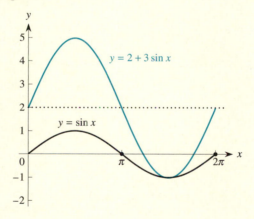

FIGURE 7.8

Incidentally, $2 + 3 \sin x$ is not the same as $5 \sin x$; the coefficients cannot be combined because 2 and $3 \sin x$ are not like terms. Graph both to see that they produce very different results.

Use your function grapher to examine the graphs of several functions of the form $D + A \sin x$ for different values of A and D. Predict and then observe how the different values are reflected in the corresponding sinusoidal curve.

The Frequency and the Period

We next consider the effect of the parameter B, which multiplies the term $(x - C)$ in

$$S(x) = D + A \sin (B(x - C)).$$

For simplicity, we concentrate on B individually and assume that $B > 0$. Consider how the function

$$S(x) = \sin 2x$$

compares to the basic curve, $y = \sin x$, as shown in Figure 7.9. Notice that the resulting sinusoidal curve $y = \sin 2x$ completes two full cycles between $x = 0$ and $x = 2\pi$ compared to the one complete cycle for the basic sine curve.

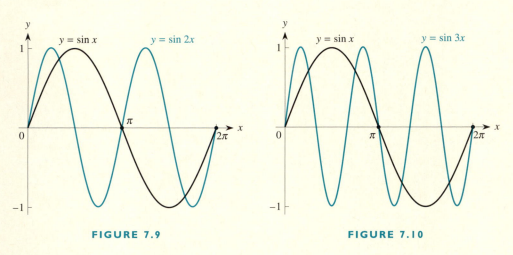

FIGURE 7.9   FIGURE 7.10

Similarly, the graph of

$$S(x) = \sin 3x$$

shown in Figure 7.10 completes three full cycles across the interval from 0 to 2π. Based on these two results, we conclude that the graph of

$$S(x) = \sin (n \cdot x),$$

for any positive integer n, will complete n full cycles between $x = 0$ and $x = 2\pi$. Try this with your function grapher for values of n such as 5 or 8.

What happens if the positive multiple B is not an integer? To investigate this, let's consider the cases where $B = \frac{1}{2}$ and $B = 2.5$. The corresponding graphs for $\sin \frac{1}{2}x$ and $\sin (2.5x)$ are shown in Figures 7.11 and 7.12, respectively. Notice that in Figure 7.11, the function $y = \sin \frac{1}{2}x$ completes half of a complete cycle between 0 and 2π; it actually requires an interval of values for x from 0 to 4π to complete the full cycle. In Figure 7.12, we see that the function $y = \sin (2.5x)$ completes 2.5 full cycles between 0 and 2π. (Trace over the graph with your finger and count the cycles.) It therefore completes one full cycle in $\frac{1}{2.5} = 0.4 = \frac{2}{5}$ of this interval.

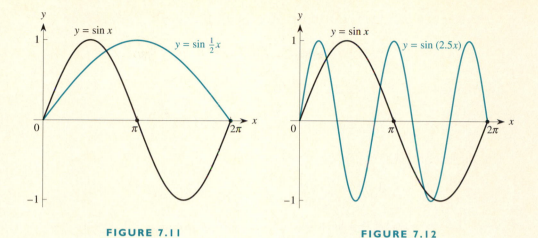

FIGURE 7.11   FIGURE 7.12

The parameter B in $\sin Bx$ is called the **frequency** of the sinusoidal function. It tells us the number of complete cycles that occur between $x = 0$ and $x = 2\pi = 360°$. For instance, the function $\sin 6x$ completes six full cycles across this interval while the function $\sin \frac{3}{8}x$ completes $\frac{3}{8}$ of one cycle.

As with any periodic function, the *period* of a sinusoidal function $\sin Bx$ is the length of the interval needed to complete one full cycle. For $\sin 2x$, the period is π radians $= 180°$ since a full cycle is completed in any interval of x-values of length π (see Figure 7.9). For $\sin 3x$, the period is $\left(\frac{1}{3}\right)2\pi = \frac{2\pi}{3}$, or $\left(\frac{1}{3}\right)360° = 120°$ (see Figure 7.10). For $\sin \frac{1}{2}x$, the period is

$$\frac{1}{1/2}(2\pi) = 4\pi$$

radians, or $720°$ (see Figure 7.11). In general, the period of $y = \sin Bx$ is

$$\text{period} = \frac{2\pi}{B} = \frac{2\pi}{\text{frequency}}.$$

In particular, for $\sin (2.5x)$, the period is

$$\text{period} = \frac{2\pi}{2.5} = \frac{2\pi}{5/2} = \frac{2}{5}(2\pi)$$

because this is the length of the interval across which this sinusoidal function completes one full cycle (see Figure 7.12). This agrees with our earlier statement that the function $\sin (2.5x)$ completes one full cycle in $\frac{2}{5}$ of the interval from 0 to 2π. In general, the period of any periodic function is the length of the interval needed to complete one full cycle. Alternatively, if we start with the period, then

$$\text{frequency} = \frac{2\pi}{\text{period}}.$$

In our expression for the number of hours of daylight in San Diego,

$$H = 12 + 2.4 \sin\left(\frac{2\pi}{365}(t - 80)\right),$$

the frequency of the sinusoidal curve is

$$\text{frequency} = \frac{2\pi}{365} = 0.0172.$$

The period of the sinusoidal curve is

$$\text{period} = \frac{2\pi}{\text{frequency}} = \frac{2\pi}{0.0172} = 365 \text{ days}.$$

As we would expect, the period is one year. In summary,

$$\text{period} = \frac{2\pi}{B} = \frac{2\pi}{\text{frequency}}.$$

The Phase Shift

Finally, we consider the role of the parameter C in

$$S(x) = D + A \sin (B(x - C)).$$

Let's start with $A = 1$, $B = 1$, and $D = 0$. Figure 7.13 compares the graph of $S(x) = \sin\left(x + \frac{\pi}{4}\right)$ to the basic curve $y = \sin x$. Note that the two curves appear identical, but $S(x)$ seems to be shifted to the left (backward) by $\frac{\pi}{4}$, or $\frac{1}{8}$ of 2π (which is one-eighth of a full cycle). Similarly, Figure 7.14 shows the graph of $T(x) = \sin\left(x - \frac{\pi}{3}\right)$. It appears to have been shifted to the right (forward) by $\frac{\pi}{3}$, or $\frac{1}{6}$ of 2π (which is one-sixth of a full cycle).

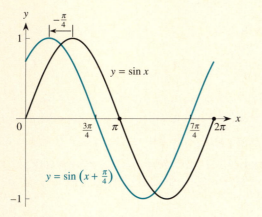

FIGURE 7.13

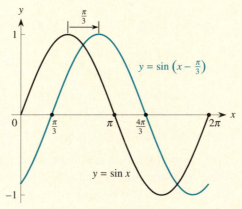

FIGURE 7.14

In general, the parameter C serves to shift the curve to the left or the right by the amount C. If C is positive in the term $(x - C)$, as in $\sin\left(x - \frac{\pi}{3}\right)$, the curve is shifted to the right; if C is negative in $(x - C)$, as in $\sin\left(x + \frac{\pi}{4}\right)$, the curve is shifted to the left. This parameter is called the **phase shift** for the sinusoidal curve. This is the same idea as the horizontal shift we discussed in Section 4.4. In our expression for the daylight function for San Diego,

$$H = 12 + 2.4 \sin\left(\frac{2\pi}{365}(t - 80)\right),$$

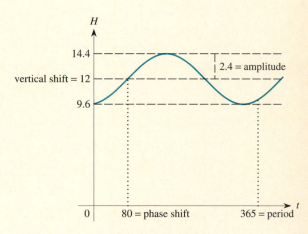

the phase shift is 80 days. The phase shift is responsible for adjusting the curve to the right to reflect the fact that March 21, the spring equinox (the day when there are equal numbers of hours of daylight and darkness) is the 80th day of the year. On this date, the graph for the sinusoidal function crosses the "middle" or average level of $D = 12$ hours.

We summarize all these results for the San Diego daylight function in Figure 7.15.

FIGURE 7.15

E X A M P L E   1

The water at a boat dock is 7 feet deep at low tide and 11 feet deep at high tide. On a certain day, low tide occurs at 4 A.M. and high tide at 10 A.M. Find an equation for the height of the tide, y, as a function of time t.

Solution

We use the given information to sketch the graph of a sinusoidal curve in Figure 7.16.

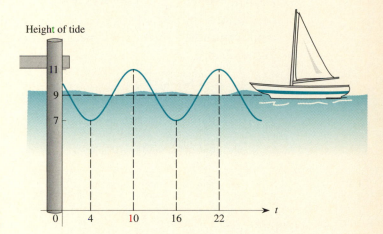

FIGURE 7.16

Since the tide ranges from a minimum height of 7 feet to a maximum height of 11 feet above sea bottom, the curve oscillates about the middle value of 9 feet, which is the vertical shift. Therefore the amplitude of the sinusoidal curve is 2. Further, the time interval between the minimum and maximum heights of the water level is 6 hours; consequently, a complete tide cycle takes 12 hours, and so the period is 12. As a result,

$$\text{frequency} = \frac{2\pi}{\text{period}} = \frac{2\pi}{12} = \frac{\pi}{6}.$$

Finally, because the tide level increases from 4 A.M. to 10 A.M., the curve passes across the middle height of 9 feet halfway between 4 A.M. and 10 A.M. (or at 7 A.M.), which gives the phase shift. (From the graph, we see that even though the tide function also crosses the 9-foot level at 1 A.M., the function is decreasing there, and so this does not give the phase shift.) Therefore the height y of the water at any time t is modeled by

$$y = 9 + 2\sin\left(\frac{\pi}{6}(t - 7)\right).$$

Identical ideas about vertical shift, amplitude, period, frequency, and phase shift apply to cosine functions of the form

$$y = D + A\cos(B(x - C)),$$

whose behavior also is described as *sinusoidal*. The only difference is in finding the phase shift. For a sine function, the phase shift corresponds to the first point to the right of the origin where the curve increases and crosses the level of the vertical shift. For a cosine function, the phase shift corresponds to the first point to the right of the origin where the curve reaches its maximum.

As an illustration, consider a spring hanging vertically from the ceiling with a weight attached at the bottom, as shown in Figure 7.17. If you displace the weight, either by pulling it down or pushing it up, and then release it, the weight will bob up and down with smaller and smaller oscillations until it settles to a stop in the original rest position, called its equilibrium. We show this vertical displacement y as a function of time t in Figure 7.18.

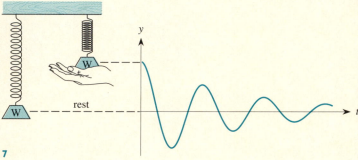

FIGURE 7.17   **FIGURE 7.18**

What kind of function is it? The oscillatory effect certainly suggests a sinusoidal function, either a sine or a cosine, but the amplitude is not constant. The overall effect of the decreasing amplitude suggests an exponential decay function. Two possible formulas for curves that combine these two behavior patterns are

$$y = Ab^t \sin ct \qquad \text{or} \qquad y = Ab^t \cos ct$$

with $b < 1$. You can think of the decaying exponential function as a variable amplitude. Use your function grapher to verify that a function of this type has the desired shape for different choices of A, b, and c.

EXAMPLE 2

Figure 7.19 shows the results of recording the vertical oscillations of an object attached to a particular spring as a function of time from $t = 0$ to $t = 2\pi$. Construct a function that models this behavior pattern.

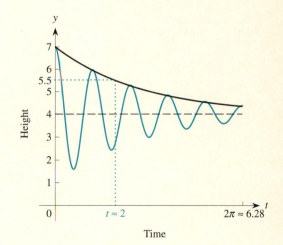

Solution First, notice that the initial height of the object is approximately $y = 7$. Starting from there, the object drops at first. Its height decreases from a maximum, which suggests a cosine function rather than a sine function.

Second, notice that the final or equilibrium height for the object is about 4, so the object seems to be oscillating about a height of 4. Since the maximum height is 7, or 3 above this equilibrium level, the form for the function seems to be

$$y = f(t) = 4 + 3b^t \cos ct,$$

where $f(0) = 4 + 3 = 7$. Further, we estimate from the graph that, between $t = 0$ and $t = 2\pi \approx 6.28$, there are about five complete diminishing cycles. So the frequency for the cosine function is approximately 5, giving the equation

$$y = f(t) = 4 + 3b^t \cos 5t.$$

FIGURE 7.19

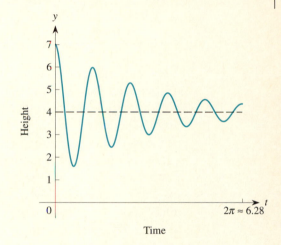

FIGURE 7.20

Finally, consider the exponential decay curve $g(t) = 3b^t$ that is superimposed over the successive peaks of the decaying sinusoidal function in Figure 7.20. It starts with an initial height of 7 and decays to a final level of 4. Using a ruler, we can estimate that it has dropped halfway (down to a height of 5.5) at about $t = 2$, and so we use $t = 2$ as an approximation for the "half-life" of the pure exponential decay function $g(t) = 3b^t$. Thus

$$g(2) = 3b^2 = 1.5.$$

This equation gives

$$b^2 = 0.5$$
$$b = \sqrt{0.5} \approx 0.71.$$

We can now estimate that the desired function is given by

$$y = f(t) = 4 + 3(0.71)^t \cos 5t.$$

Verify that this function matches the required pattern for the oscillation shown in Figure 7.19 by using your function grapher to graph f between 0 and 2π.

Incidentally, we have implicitly assumed throughout this section that all periodic processes actually follow a sinusoidal pattern (either a sine or a cosine curve) precisely. In practice, this may be expecting a lot. Think about the length of Janis' fingernails in Exercise 1 from Section 7.1. The nail length is a periodic function, but it is not sinusoidal. Even if we observe that the overall pattern for some periodic process is smooth and appears to be that of a sine curve, we have no guarantee that the behavior is exactly sinusoidal. Nevertheless, sinusoidal behavior is our best model for such types of periodic phenomena and consequently is the model we should choose when faced with such behavior.

Finally, just as we can construct linear, exponential, power, and other functions to fit a set of data, we often are faced with the problem of having a set of data that exhibits a periodic pattern and we want to find the periodic function that best fits the data. We ask you to explore several such cases in the exercises.

EXERCISES

1. Decide which of the following functions are periodic. For those that are periodic, what is the period? (Assume that each graph continues in the same pattern indefinitely to the left and right.)

a.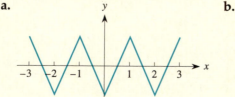

b.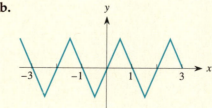

c.

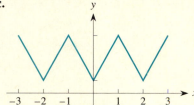

d.

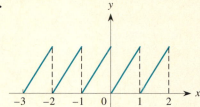

e.

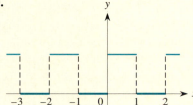

f.

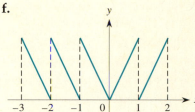

g.

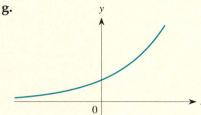

h.

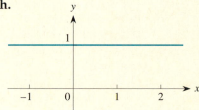

i.

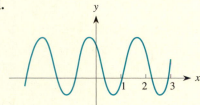

j.

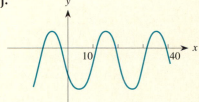

2. Find the number of hours of daylight in San Diego on March 1, on May 12, on July 4.

3. The number of hours of daylight in Montreal is given by

$$H(t) = 12 + 3.6 \sin\left(\frac{2\pi}{365}(t - 80)\right),$$

where t is the number of days after January 1.

   a. What is the amplitude of this function?
   b. What is the period of this function?
   c. What is the length of daylight on the shortest day of the year?
   d. What is the length of daylight on the longest day of the year?

4. The shortest day of the year in Fairbanks, Alaska, has 3.70 hours of daylight. Find an equation for the number of hours of daylight there on any day of the year.

5. Write an equation giving the number of hours of darkness in San Diego as a function of the day of the year.

6. Consider Example 1 again on the height of the tide at a dock. Suppose that low tide still occurs at 4 A.M., but that high tide actually occurs at 10:30 A.M. Find an equation for the height of the tide as a function of time t.

7. Suppose that the low water level at a pier is 4 feet and the high water level is 12 feet. If low tide occurs at 2 A.M. and high tide at 10:32 A.M., find an equation for the water level as a function of time.

8. The Bay of Fundy in eastern Canada is known for the highest tides in the world. The tides there rise and fall by as much as 50 feet. If the tidal cycle takes 11 hours, find a sinusoidal function that best models the tides in the bay. For convenience, assume that low tide corresponds to a height of zero.

9. The thermostat in Sylvia's home in Baltimore is set at 66°. Whenever the temperature drops to 66° (roughly every half-hour), the furnace comes on and stays on until the temperature reaches 70°.

    a. Write a trigonometric function that models this situation.
    b. Gary's thermostat in upper New York State is set the same way. How would the model you created in part (a) change to reflect Gary's climate?
    c. Jodi, who lives in central Florida, likewise has her thermostat set to come on at 66°. How would you change the model you created for parts (a) and (b) to reflect her climate?
    d. Do you think that a trigonometric function is necessarily a good model? Explain why or why not. Think about the *rates* at which the temperature increases and decreases.

10. Ocean waves move in a roughly sinusoidal pattern. As a rule of thumb, the length of a wave (crest to crest, say) on the open seas is about 20 times the height of the wave (trough to crest). (This rule does not apply near coastlines where waves are much choppier and their intervals shorter.)

    a. Write a formula for ocean waves that are 4 feet high in moderately calm seas.
    b. Write a formula for ocean waves that are 15 feet high in rough seas.

11. Meryl is a normal individual with a pulse rate of 72 beats per minute and a blood pressure of 120 over 80. Thus her heart is beating 72 times each minute and her blood pressure is oscillating between a low (diastolic) reading of 80 and a high (systolic) reading of 120. Assume that the oscillation in Meryl's blood pressure can be modeled by a sinusoidal function.

    a. What is the period of this sinusoid?
    b. What is the frequency of this sinusoid?
    c. What is the equation of this sinusoid?

12. Your Thanksgiving turkey is taken from a refrigerator at 40°F and placed in an oven set at 350°F. Suppose that the temperature of the bird is 130° after 60 minutes. In practice, you know that an oven cycles on and off as some of the heat escapes. Estimate how often the cycles occur and a reasonable set of ranges for the actual temperatures inside the oven. Use these estimates to construct a trigonometric function to model the temperature of the oven. About how large a variation is possible in the temperature of the turkey after 60 minutes? after 100 minutes?

13. Sketch by hand the graphs of the following functions. For each, draw the basic curve $y = \sin x$ or $y = \cos x$ on the same set of axes for comparison. (Do not use your function grapher.)

    a. $y = 3 \sin 4x$ **b.** $y = 3 \sin \left(\frac{1}{2}x\right)$ **c.** $y = 2 \sin 3x$

    d. $y = 4 \cos 2x$ **e.** $y = -3 \cos 2x$ **f.** $y = 4 + 2 \sin x$

    g. $y = \sin \left(x - \frac{\pi}{4}\right)$ **h.** $y = 3 \sin \left(2x - \frac{\pi}{6}\right)$ **i.** $y = 4 + 2 \cos\left(x + \frac{\pi}{3}\right)$

14. Identify the following trigonometric functions from their graphs:

a.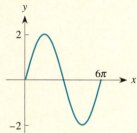

b.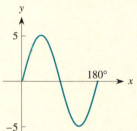

c.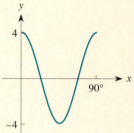

d.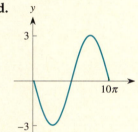

e.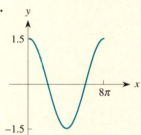

f.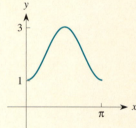

g.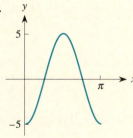

h.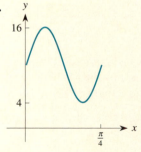

i.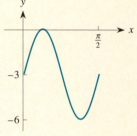

j.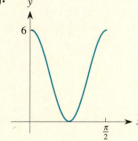

k.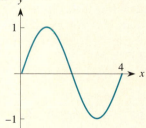

l.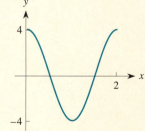

15. Identify a possible formula for each of the sinusoidal functions whose values are given in the following table. Note that there are many possible correct answers.

x	f_1	f_2	f_3	f_4	f_5	f_6
−6	0.279	2.279	0.559	0.537	−0.141	0.721
−5	0.959	2.959	1.918	0.544	−0.598	0.041
−4	0.757	2.757	1.514	−0.989	−0.909	0.243
−3	−0.141	1.859	−0.282	0.279	−0.997	1.141
−2	−0.909	1.091	−1.819	0.757	−0.841	1.909
−1	−0.841	1.159	−1.683	−0.909	−0.479	1.841
0	0.000	2.000	0.000	0.000	0.000	1.000
1	0.841	2.841	1.683	0.909	0.479	0.159
2	0.909	2.909	1.819	−0.757	0.841	0.091
3	0.141	2.141	0.282	−0.279	0.997	0.859
4	−0.757	1.243	−1.514	0.989	0.909	1.757
5	−0.959	1.041	−1.918	−0.544	0.598	1.959
6	−0.279	1.721	−0.559	−0.537	0.141	1.279

16. The following table gives the outdoor temperatures in Chicago during one 24-hour period:

midnight	2 A.M.	4 A.M.	6 A.M.	8 A.M.	10 A.M.
53	48	47	49	53	59

noon	2 P.M.	4 P.M.	6 P.M.	8 P.M.	10 P.M.	midnight
66	71	68	65	58	54	53

If you were to fit a sinusoidal function to this set of data, what is the vertical shift? the amplitude? the period? the frequency? What is the equation of the trig function you would use?

17. The following table shows the average daytime high temperature each month in San Diego.

   a. Construct a sinusoidal function that best fits this data.
   b. How does the phase shift for this function compare to the phase shift we used in the text for the number of hours of daylight in San Diego? In particular, can you explain in practical terms why the sinusoidal function for air temperature should lag behind the function for hours of daylight?

Month	Jan	Feb	Mar	Apr	May	June	July	Aug	Sept	Oct	Nov	Dec
Avg. daily high temp. (°F)	65.2	64.4	65.9	67.8	68.6	71.3	75.6	77.6	76.8	74.6	69.9	66.1

18. The following table gives the average daytime high temperature in Dallas on different days of the year (roughly every two weeks) based on historical weather records.

Day	1	15	32	46	60	74	91	105	121	135	152	196	213
Avg. daily high temp. (°F)	55	53	56	59	63	67	72	77	81	84	89	98	99

227	244	258	274	288	305	319	335	349
98	94	90	85	80	72	66	61	58

Assuming the temperature behavior is periodic from year to year, determine a sinusoidal function that models the average daytime high temperature in Dallas.

19. The following table shows the average number of tornados reported in the United States per month based on historical records.

Month	Jan	Feb	Mar	Apr	May	June	July	Aug	Sept	Oct	Nov	Dec
Tornados	16	24	60	111	191	179	96	66	41	26	31	22

Determine a sinusoidal function that models the monthly number of tornados as a function of time.

20. Astronomers recently reported the discovery of the first known planets outside the solar system. They found three worlds orbiting around a pulsar, a rotating star that emits radiation with extremely constant frequency. For this pulsar, the astronomers detected slight variations in the intensity of the radiation as shown in the accompanying figure. This variation would be the effect of a planet in orbit about the pulsar.

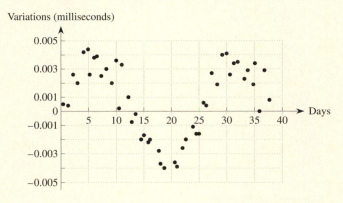

a. From the figure, estimate the length of the year for the planet.
b. Use the results on Kepler's Law from Example 1 in Section 3.3 (assuming the same coefficient applies) to calculate the distance from this planet to its star.
c. Assuming that the orbit of this new planet is circular, how fast is it moving in its orbit about the pulsar?
d. For comparison, Earth takes 365 days to complete one revolution about the sun at a distance of about 93 million miles. How fast is Earth moving in its orbit about the sun?

21. Graph the following functions and determine whether they are periodic. For those that are periodic, what is the period?

a. $|x|$ **b.** $x + \sin x$ **c.** $|\sin x|$

d. $|\cos x|$ **e.** $\sin |x|$ **f.** $\sin^2 x$

g. $\cos^2 x$ **h.** $\sin^3 x$ **i.** $\cos^4 x$

22. In Example 2 of Section 4.2, we modeled the Gateway Arch by fitting different polynomials to data taken from a graph. The shape of the arch, as shown in the accompanying figure, may also suggest an arch of a trigonometric function. Use the same set of measurements of heights on the arch as a function of the horizontal distance and construct a sinusoidal function that is a good fit to the data. How does the fit with the sinusoidal function compare to the fits with the quadratic and quartic functions that we found in Section 4.2?

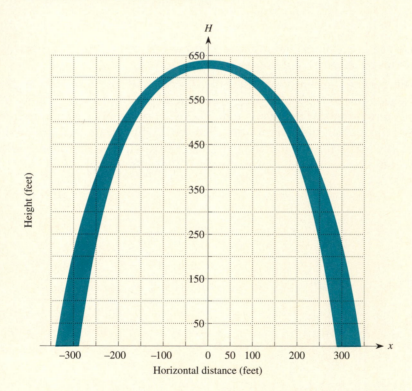

23. The accompanying figure shows the McDonald's arches superimposed over a grid. Decide on an appropriate scale for the grid and use it to estimate a set of measurements for heights on one arch as a function of the horizontal distance and then determine a sinusoidal function that is a good fit to the arch.

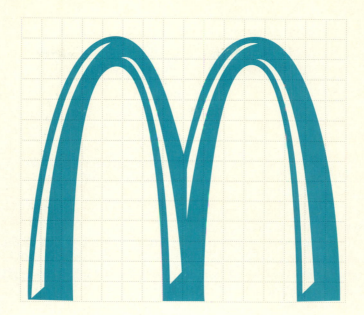

24. You are taking a ride on a Ferris wheel that is 200 feet in diameter that has a bottom point 10 feet above the ground. Suppose that the wheel rotates twice every minute and, from your friend's viewpoint on the ground, is rotating in a clockwise direction.

    a. Sketch your height y above the ground as it depends on the horizontal distance x from a vertical axis through the center of the Ferris wheel.

    b. Sketch your height y above the ground as a function of time t.

    c. Find a formula for your height y as a function of t. Does it agree with your rough sketch in part (b)?

    d. Find a formula for the horizontal distance x as a function of t.

    e. Find all intervals of t values for which you are moving forward. Indicate these intervals on the graph of the function in part (d). What do you observe?

    f. Suppose your friend moves around to the opposite side of the wheel so that it now appears to him to be moving counterclockwise. From this new vantage point, the direction of the x-axis is reversed. How do your answers to parts (d) and (e) change?

    g. Find a formula relating your height y above the ground and the horizontal distance x from the vertical axis through the center of the wheel.

25. Many people believe that a person's life is determined by three independent cycles, called *biorhythms*. One cycle, with a period of 23 days, represents the physical or health dimension of a person, $H(t) = \sin\left(\frac{2\pi t}{23}\right)$, where time t is measured in days starting at birth. A second cycle, with a period of 28 days, represents the emotional or sensitivity aspects of a person, $E(t) = \sin\left(\frac{2\pi t}{28}\right)$. A third cycle, with a period of 33 days, represents the mental or intellectual aspects of an individual, $M(t) = \sin\left(\frac{2\pi t}{33}\right)$.

   a. Suppose Tony was born on January 1. Consider the 60-day period immediately following his twentieth birthday. What set of values for t are appropriate?
   b. Which days would you recommend as being suitable for Tony to compete in a track-and-field meet?
   c. Which days would you recommend as being good days for Tony to ask his girlfriend to marry him?
   d. Which days could you suggest as days on which Tony could hope to have a major exam at school?
   e. Are there any days when you would recommend that Tony simply not get out of bed?
   f. Are there any days when all the signs are highly positive?

26. As part of a study on the possibility of global warming at a National Science Foundation math modeling workshop at Pellissippi State College, the accompanying scatterplot was produced. It suggests that the average global temperature values appear to oscillate about the best-fit line, $T = 0.0042t + 14.67$, where t represents years since 1880. Use the scatterplot to estimate the parameters for a sinusoidal function that oscillates above and below the indicated best-fit line. What is the equation of the resulting function? Use your function grapher to draw the graph of that function. Does it have the correct shape? What is your prediction for the average global temperature in the year 2000 based on the combination of the given linear function and the sinusoidal function you created?

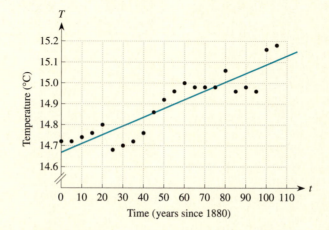

27. Use the graph of $y = \sin x$ to sketch the graph of $y = \frac{1}{\sin x}$. (This function is known as the *cosecant function*.)

7.3 *Relationships Between Trigonometric Functions*

In many applications of trigonometry, particularly in calculus, it often is necessary to transform one trigonometric function into another. To do so, we need the concept of a *trigonometric identity*. An **identity** is a relationship that is true for *all values of the variable.* For example, the Pythagorean identity

$$\sin^2 x + \cos^2 x = 1 \tag{1}$$

that we discussed in Section 7.1 is an identity because it holds for any value of x. However, suppose we ask if $\sin x + \cos x$ is also equal to 1. We show a portion (one complete cycle) of the graph of $y = \sin x + \cos x$ in Figure 7.21. Notice that the function is not identically equal to 1 because its graph is not a horizontal line of height 1. While there are several specific values of x for which $\sin x + \cos x$ is equal to 1, such as $x = 0$ or $x = \frac{\pi}{2}$, this relationship does not hold for all x. So $\sin x + \cos x = 1$ is not an identity.

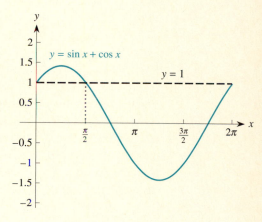

FIGURE 7.21

Now consider again the Pythagorean identity in Equation (1). One way that we use it is to transform sines into cosines with

$$\cos^2 x = 1 - \sin^2 x,$$

or

$$\cos x = \pm\sqrt{1 - \sin^2 x}$$

and cosines into sines with

$$\sin^2 x = 1 - \cos^2 x,$$

or

$$\sin x = \pm\sqrt{1 - \cos^2 x}.$$

Each of these equations holds for all values of the variable x, so each one is an identity.

The Reflection Identities

There are many other useful relationships among the trigonometric functions that we will explore here. The following are two properties of the sine and cosine functions:

$$\sin(-x) = -\sin x \tag{2}$$

$$\cos(-x) = \cos x \tag{3}$$

for any x. These two relationships, known as the **reflection identities,** or the **odd-even identities,** are very easy to see graphically. If you picture the graph of the cosine function, you should realize that it is symmetric about the vertical y-axis (see Figure 7.22). That is, for any positive value of x, the height of the cosine function is the same to the left of the y-axis (at $-x$) as it is at the same distance to the right of the y-axis (at x). Thus

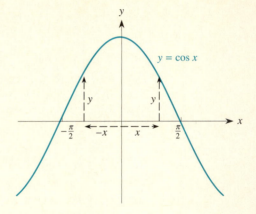

$$\cos(-x) = \cos x$$

for any value of x. This is the same type of behavior we discussed in Section 2.4 for power functions with even powers such as $f(x) = x^2$ and $g(x) = x^4$. For this reason, the cosine function is called an *even function*.

FIGURE 7.22

On the other hand, the sine curve is not symmetric about the y-axis. Rather, if you move a distance of x to the left and consider the height to the sine curve, it is equivalent, but opposite in sign, to the height you would find if you move the same distance x to the right of the y-axis (see Figure 7.23). Thus

$$\sin(-x) = -\sin x$$

for any value of x. This is the same type of behavior we have with the linear function $f(x) = x$ or the cubic function $g(x) = x^3$. As a result, the sine function is called an *odd function*.

These ideas of even and odd functions will come up again in Section 7.7 where you will see that there really are relationships between polynomial functions and trigonometric functions.

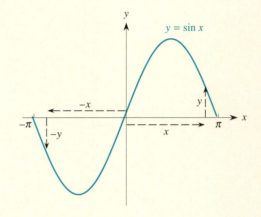

FIGURE 7.23

The Double-Angle Identities

Let's consider some additional relationships involving the sine and cosine. We begin with $\sin 2x$. Suppose we want to express $\sin 2x$ in an equivalent form that does not have the frequency 2 showing explicitly. Is it possible that $\sin 2x$ and $2 \sin x$ are equivalent? Graph the two functions and you will see that they cannot be the same. The first, $\sin 2x$, is a sinusoidal curve with an amplitude of 1 and a frequency of 2, so it completes two full cycles between $x = 0$ and $x = 2\pi$; the second function, $2 \sin x$, is a sinusoidal curve with an amplitude of 2 and a frequency of 1, so it completes one full cycle between 0 and 2π.

It turns out that the actual relationship between these two functions is

$$\sin 2x = 2 \sin x \cdot \cos x. \tag{4}$$

You can verify this graphically using your function grapher. When you graph the two functions $\sin 2x$ and $2 \sin x \cdot \cos x$ simultaneously, you will see only one graph being drawn. Tracing will show that the one graph matches both functions. This means that, for any value of x whatsoever, the two functions give the identical value. You can also check this out numerically: Pick any value for x and evaluate $\sin 2x$ and $2 \sin x \cdot \cos x$. The results will be identical for every value of x that you choose, thus supporting the fact that Equation (4) is indeed an identity.

Similarly, we have the identity

$$\cos 2x = \cos^2 x - \sin^2 x. \tag{5}$$

Again, we urge you to examine this relationship graphically to see that the two expressions are indeed identical for all values of x. We can rewrite Equation (5) in several alternative, but equivalent, forms by making use of Equation (1). Thus

$$\cos 2x = \cos^2 x - \sin^2 x = (1 - \sin^2 x) - \sin^2 x$$

$$\cos 2x = 1 - 2 \sin^2 x. \tag{5a}$$

Similarly, we can rewrite Equation (5) as

$$\cos 2x = \cos^2 x - \sin^2 x = \cos^2 x - (1 - \cos^2 x)$$

$$\cos 2x = 2 \cos^2 x - 1 \tag{5b}$$

Verify each of these identities visually by using your function grapher and numerically by substituting in several different values for x. The identities in Equations (4), (5), (5a), and (5b) are known as the **double-angle identities** for the sine and cosine.

The Sum and Difference Identities

The double-angle identities in Equations (4) and (5) actually are special cases of more general identities known as the **sum and difference identities** for sine and cosine. The sum identities are

$$\sin (x + y) = \sin x \cdot \cos y + \cos x \cdot \sin y \tag{6}$$

$$\cos (x + y) = \cos x \cdot \cos y - \sin x \cdot \sin y. \tag{7}$$

When $y = x$, these equations reduce to the double-angle formulas in Equations (4) and (5).

If we now replace y with $-y$ in the two sum identities in Equations (6) and (7) and then apply Equations (2) and (3), we immediately have the **difference identities** for the sine and cosine:

$$\sin (x - y) = \sin x \cdot \cos (-y) + \cos x \cdot \sin (-y)$$

$$= \sin x \cdot \cos y - \cos x \cdot \sin y \tag{8}$$

$$\cos (x - y) = \cos x \cdot \cos (-y) - \sin x \cdot \sin (-y)$$

$$= \cos x \cdot \cos y + \sin x \cdot \sin y. \tag{9}$$

EXAMPLE 1

Reduce $\sin 3x$ to an equivalent expression involving only sines and not including any multiple angles.

Solution   We can write

$$\begin{aligned}
\sin 3x = \sin(2x + x) &= \sin 2x \cdot \cos x + \cos 2x \cdot \sin x \\
&= (2\sin x \cdot \cos x) \cdot \cos x + (1 - 2\sin^2 x) \cdot \sin x \\
&= 2\sin x \cdot \cos^2 x + \sin x - 2\sin^3 x \\
&= 2\sin x \cdot (1 - \sin^2 x) + \sin x - 2\sin^3 x \\
&= 3\sin x - 4\sin^3 x.
\end{aligned}$$

EXAMPLE 2

Reduce $\sin 4x$ to an equivalent expression that has no multiple angles.

Solution   Following the approach in Example 1, we could write

$$\begin{aligned}
\sin 4x &= \sin(3x + x) \\
&= \sin 3x \cdot \cos x + \cos 3x \cdot \sin x.
\end{aligned}$$

This approach involves expanding $\cos 3x$, but we have not worked this out yet. (You will be asked to do this in an exercise in this section.) Alternatively, we could start with

$$\begin{aligned}
\sin 4x = \sin(2x + 2x) &= \sin[2 \cdot (2x)] \\
&= 2\sin 2x \cdot \cos 2x \\
&= 2(2\sin x \cdot \cos x)(\cos^2 x - \sin^2 x) \\
&= 4\sin x \cdot \cos^3 x - 4\sin^3 x \cos x.
\end{aligned}$$

The Half-Angle Identities

Often we are faced with the problem of starting with powers of the sine or cosine function and having to eliminate all powers by rewriting such expressions in terms of sines and cosines with multiple angles. You will see in Section 7.6 that one of the fundamental ideas in applied mathematics is based on expressing any periodic function in terms of such sinusoidal functions. In order to eliminate the powers, we must make use of two additional identities. Starting with the double-angle identity in Equation (5a),

$$\cos 2x = 1 - 2\sin^2 x,$$

we have

$$2 \sin^2 x = 1 - \cos 2x$$

$$\sin^2 x = \frac{1}{2}(1 - \cos 2x). \tag{10}$$

Similarly, if we start with the double-angle identity in Equation (5b),

$$\cos 2x = 2 \cos^2 x - 1,$$

then we get

$$\cos^2 x = \frac{1}{2}(1 + \cos 2x). \tag{11}$$

The identities in Equations (10) and (11) are called the **half-angle identities.** Verify them graphically using your function grapher. We illustrate the use of these identities in the following example.

EXAMPLE 3

Rewrite $\cos^4 x$ by eliminating all exponents.

Solution   Using Equation (11), we have

$$\cos^4 x = (\cos^2 x)^2 = \left[\frac{1}{2}(1 + \cos 2x)\right]^2$$

$$= \frac{1}{4}[1 + 2 \cos 2x + \cos^2(2x)]$$

Since this expression involves the second power of $\cos 2x$, we apply Equation (11) again to get

$$\cos^4 x = \frac{1}{4}\left[1 + 2 \cos 2x + \frac{1}{2}\{1 + \cos(2 \cdot 2x)\}\right]$$

$$= \frac{1}{4}\left[1 + 2 \cos 2x + \frac{1}{2} + \frac{1}{2}\cos 4x\right]$$

$$= \frac{1}{4} + \frac{1}{2}\cos 2x + \frac{1}{8} + \frac{1}{8}\cos 4x$$

$$= \frac{3}{8} + \frac{1}{2}\cos 2x + \frac{1}{8}\cos 4x$$

For reference purposes, we list together all of the essential trigonometric identities involving the sine and cosine functions. It is essential that you memorize the Pythagorean, reflection, and double-angle identities. You should also become very familiar with the others, or you will be unable to recognize them when the need arises. These identities will appear repeatedly both in this course and in later mathematics and associated courses.

Trigonometric Identities

Pythagorean identity: $\sin^2 x + \cos^2 x = 1$ (1)

Reflection identities: $\sin(-x) = -\sin x$ (2)

$\cos(-x) = \cos x$ (3)

Double-angle identities: $\sin 2x = 2\sin x \cdot \cos x$ (4)

$\cos 2x = \cos^2 x - \sin^2 x$ (5)

$= 2\cos^2 x - 1 = 1 - 2\sin^2 x$

Sum identities: $\sin(x + y) = \sin x \cdot \cos y + \cos x \cdot \sin y$ (6)

$\cos(x + y) = \cos x \cdot \cos y - \sin x \cdot \sin y$ (7)

Difference identities: $\sin(x - y) = \sin x \cdot \cos y - \cos x \cdot \sin y$ (8)

$\cos(x - y) = \cos x \cdot \cos y + \sin x \cdot \sin y$ (9)

Half-angle identities: $\sin^2 x = \frac{1}{2}(1 - \cos 2x)$ (10)

$\cos^2 x = \frac{1}{2}(1 + \cos 2x)$ (11)

The Law of Cosines

In addition to the preceding identities, there is one other very useful relationship involving the cosine function. This identity relates the three sides of any triangle and any one of the three angles. Consider the triangle shown in Figure 7.24 with sides a, b, and c, where the angle opposite side c is C. The sides and angle are related by the *law of cosines*.

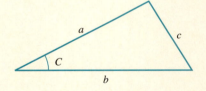

FIGURE 7.24

Law of Cosines

$$c^2 = a^2 + b^2 - 2ab\cos C$$

Notice that the triangle need not be a right triangle; the law of cosines applies to any triangle whatsoever. The law of cosines allows us to determine (1) the length of the side opposite a known angle if the other two sides are known, or (2) any angle if the three sides of the triangle are known. For example, if $a = 5$, $b = 7$, and the angle between them is $C = 60°$, then

$$c^2 = a^2 + b^2 - 2ab\cos C$$
$$= 5^2 + 7^2 - 2(5)(7)\cos 60°$$
$$= 25 + 49 - 70\left(\frac{1}{2}\right) = 39.$$

Thus the third side is $c = \sqrt{39} = 6.245$.

EXAMPLE 4

Alpha Centauri is the closest star to the sun. Use the law of cosines to find out just how far away it is.

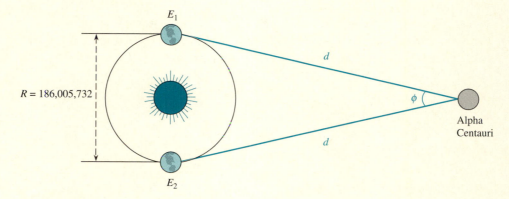

FIGURE 7.25

Solution Figure 7.25 shows two positions of the Earth six months apart in its orbit about the sun. It is known that the distance R between the two points E_1 and E_2 is 186,005,732 miles. Alpha Centauri is represented by point A and we wish to determine the distance d. The angle ϕ (the Greek letter phi) represents the angle subtended by the two points on the Earth's orbit at Alpha Centauri and is called the *parallax*. Astronomers have worked out an ingenious way to find the parallax from measurements made at E_1 and E_2; for Alpha Centauri, the parallax $\phi = 7.3691680 \times 10^{-6}$ radians.

Using the law of cosines, we have

$$R^2 = d^2 + d^2 - 2d^2 \cos \phi = 2d^2(1 - \cos \phi),$$

and so

$$d^2 = \frac{R^2}{2(1 - \cos \phi)}$$

$$d = \frac{R}{\sqrt{2(1 - \cos \phi)}}.$$

If we use the given values for $R = 186,005,732$ and $\phi = 7.3691680 \times 10^{-6}$, we find that the distance to Alpha Centauri is

$$d = \frac{186,005,732}{\sqrt{2(1 - \cos (7.3691680 \times 10^{-6}))}} = 25,242,154,000,000 \text{ miles.}$$

Think About This

Interstellar distances are usually given in terms of light-years—the distance that light travels in one year. Use the fact that the speed of light is about 186,280 miles per second to convert the answer in Example 4 to light-years. What does the distance to Alpha Centauri in light-years mean in terms of what we are actually seeing when we observe Alpha Centauri through a telescope?

Some of you might find that your calculators give vastly different answers when evaluating d in Example 4; in fact, some of you might get an error message indicating division by zero because the calculator may round the argument of the cosine function, 7.3691680×10^{-6}, to 0. We will discuss a way around this situation in Section 7.7.

EXERCISES

1. Using ideas on amplitude and frequency, explain why $\cos 3x$ cannot be identically equal to $3 \cos x$.

2. Using ideas on amplitude, explain why it is reasonable that $\cos 2x = 2 \cos^2 x - 1$. (Realize that such an argument cannot constitute a proof.)

Examine each of the equations in Exercises 3–14 graphically to see if the given relationship may be an identity. If it is not an identity, attempt to locate graphically or numerically at least one point that lies on both curves. If it seems to be an identity, prove it algebraically.

3. $\sin^3 x + \cos^3 x = 1$

4. $\cos 3x = \cos^3 x - \sin^3 x$

5. $\dfrac{\sin 2x}{\sin x} = 2 \cos x$

6. $(1 - \cos \theta)(1 + \cos \theta) = \sin^2 \theta$

7. $\sin 3x = 3 \sin x$

8. $\dfrac{\cos^2 \theta}{1 + \sin \theta} = 1 - \sin \theta$

9. $\dfrac{1 - \cos \alpha}{\sin \alpha} = \dfrac{\sin \alpha}{1 + \cos \alpha}$

10. $\cos 3\beta = 3 \cos^3 \beta - 1$

11. $\sin^2 3x + \cos 6x = \cos^2 3x$

12. $\cos^2 2x = 3(1 - \sin 2x)$

13. $\sin(\cos x) = \cos(\sin x)$

14. $\sin(\cos x) = \sin x \cdot \cos x$

15. Rewrite $\sin^4 x$ by eliminating all exponents.

16. Rewrite $\cos^6 x$ by eliminating all exponents.

17. Rewrite $\sin^2 x \cdot \cos^2 x$ by eliminating all exponents.

18. Express $\cos 3x$ in terms of powers of $\sin x$ and $\cos x$, but with no multiple angles.

19. Express $\cos 4x$ in terms of powers of $\sin x$ and $\cos x$, but with no multiple angles.

20. Express $\cos 5x$ in terms of powers of $\sin x$ and $\cos x$, but with no multiple angles.

21. Examine the results of Exercises 18–20 as well as the formula for $\cos 2x$ and see if you can determine any patterns in the terms.

22. By setting $y = x$ in the sum identities in Equations (6) and (7), show that you get the double-angle identities in Equations (4) and (5).

23. **a.** Sketch the graph of $\sin\left(x + \frac{\pi}{2}\right)$. What familiar function do you get from this phase shift?

b. Use the sum identity for the sine function to show that $\sin\left(x + \frac{\pi}{2}\right)$ is actually equal to that function.

24. **a.** Repeat Exercise 23 for $\sin(x + \pi)$.

b. Repeat Exercise 23 for $\cos\left(x + \frac{\pi}{2}\right)$.

25. In the half-angle identities in Equations (10) and (11), let $y = 2x$, so that $x = \frac{1}{2}y$. Rewrite each of these identities in terms of y to see why they are called half-angle identities.

Exercises 26 and 27 refer to Figure 7.24 in the text.

26. Find c if $a = 5$, $b = 3$, and $C = 20°$.

27. Find c if $a = 7$, $b = 4$, and $C = 25°$.

28. Suppose that a star is 75,000,000,000,000 miles away from the sun. What would its parallax angle be?

29. Find the distances to the following nearby stars based on their given parallax angles:

Barnard's Star	5.3523430×10^{-6} radians
Sirius	3.8979020×10^{-6} radians
Epsilon-Eridani	2.9573635×10^{-6} radians

30. **a.** Refer to the functions shown in Exercise 1 of Section 7.2 and decide which ones are odd, even, or neither.

b. Refer to the functions given in Exercise 21 of Section 7.2 and decide which ones are odd, even, or neither.

31. **a.** Explain why you can calculate the value of

$$1 + \sin x + \sin^2 x + \sin^3 x + \sin^4 x + \cdots$$

as

$$\frac{1}{1 - \sin x}.$$

Are there any values of x for which this does not work?

b. What formula would you get for the sum of the terms

$$1 + \sin x + \sin^2 x + \cdots + \sin^n x$$

for some n?

32. Use the result of Exercise 31(b) with different values of n to calculate the value of

$$1 + \sin x + \sin^2 x + \sin^3 x + \sin^4 x + \cdots$$

for $x = \frac{\pi}{6}$ correct to three decimal places. Now suppose you wanted to do this with $x = \frac{\pi}{3}$ instead. Do you expect that you will need approximately the same number of terms, more terms, or fewer terms to get the same three decimal place accuracy?

7.4 *Solving Trigonometric Equations: The Inverse Functions*

We have seen that the number H of hours of daylight in San Diego as a function of the day of the year t is given by

$$H(t) = 12 + 2.4 \sin\left(\frac{2\pi}{365}(t - 80)\right).$$

Suppose we now raise this question: When will there be 13 hours of daylight? That is, on which day t of the year will $H = 13$? (See Figure 7.26). To find this, we must solve the equation

$$H(t) = 12 + 2.4 \sin\left(\frac{2\pi}{365}(t - 80)\right) = 13$$

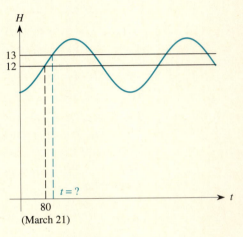

for t. We subtract 12 from both sides to get

$$2.4 \sin\left(\frac{2\pi}{365}(t - 80)\right) = 1.$$

When we divide both sides by 2.4, we have

$$\sin\left(\frac{2\pi}{365}(t - 80)\right) = \frac{1}{2.4} = 0.417.$$

FIGURE 7.26

Our problem now is to extract the variable t from the argument of the sine function.

Compare this problem to the situation we repeatedly have faced of extracting the variable from an exponential function, such as 10^x. We solved that problem by using logarithms to undo the exponential function. It works because the exponential and log functions are inverses of one another.

Similarly, we can undo the sine function by using the *inverse sine function*. On your calculator, this typically is done by pressing either INV or 2nd followed by SIN (to get SIN^{-1}). When you do this in radian mode, you will find that

$$\text{Inverse sine of } 0.417 = 0.430.$$

Therefore,

$$\frac{2\pi(t - 80)}{365} = 0.430.$$

Multiplying both sides by 365, we get

$$2\pi(t - 80) = (0.430)(365) = 156.95$$

$$t - 80 = \frac{156.95}{2\pi} = 24.98 \approx 25.$$

Hence,

$$t = 25 + 80 = 105.$$

That is, there will be 13 hours of daylight in San Diego on approximately the 105th day of the year (April 15). Check this on your calculator by evaluating $H(105)$. Our answer also agrees with common sense because we know that there are 12 hours of daylight on March 21, the spring equinox, and the amount of daylight increases about 2 to 3 minutes per day at that time of the year.

Actually, if you think about it a little and examine Figure 7.27, you will realize that this answer is not complete since the function H oscillates sinusoidally. Looking at the graph in Figure 7.27, you should expect another date later in the year when there likewise will be 13 hours of daylight. To determine when this happens, let's use a little common sense and some facts about the sine function. We found that the solution $t = 105$ corresponds to April 15, which is 25 days *after* the spring equinox on March 21. Using the symmetry of the sine curve, we should expect that there will also be 13 hours of daylight 25 days *before* the fall equinox on September 21; that is, August 27 should also have 13 hours of daylight. Check this by substituting the corresponding value for t into the formula for H. Finally, because of the periodicity of the sine function, there will be 13 hours of daylight in San Diego on April 15 and August 27 of every year.

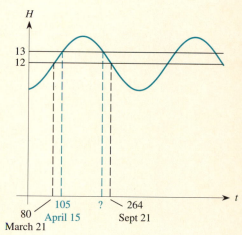

FIGURE 7.27

The Inverse Sine Function

Let's now examine more carefully just what the inverse sine function is all about. Recall that a continuous function f has an inverse f^{-1} when it is either strictly increasing or strictly decreasing; or equivalently if it satisfies the horizontal line test. Obviously, the sine function does not fulfill either of these conditions because of its shape. The only way to obtain an inverse for the sine function is to restrict its domain, just as we did in Section 2.7 when we considered only the right side of the parabola $y = x^2$. We thus use only a small portion of the sine curve $y = \sin \theta$ where the function is strictly increasing. By convention, the restricted domain used is from $\theta = -\frac{\pi}{2}$ to $\theta = \frac{\pi}{2}$ (see Figure 7.28).

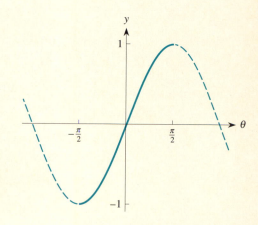

FIGURE 7.28

For simplicity, let's work temporarily in degrees. Suppose we have the equation

$$\sin \theta = 0.825,$$

and θ is some unknown angle. From the graph of the sine function, we should expect that a value for θ will be closer to 90° than it is to 0°. The inverse sine function, written **Arcsin y** or **Sin^{-1} y**, allows us to solve for the correct angle θ. That is, if

$$\sin \theta = 0.825$$

then

$$\theta = \text{Arcsin } (0.825) = 55.59°.$$

Equivalently, if we use radians instead of degrees, then

$$\theta = \text{Arcsin } (0.825) = 0.9702 \text{ radians.}$$

Thus the inverse sine function undoes the sine function to extract any angle θ between $-90°$ and $90°$ or equivalently, between $-\frac{\pi}{2}$ and $\frac{\pi}{2}$ radians. Check this by taking the sine of the angle 55.59° or 0.9702 radians. It is helpful to think of Arcsin y as *the angle whose sine is y*.

In general, if $y = f(x)$ and $x = g(y) = f^{-1}(y)$ are inverse functions, then we know that

$$f^{-1}(f(x)) = x \qquad \text{for all } x \text{ in the domain of } f$$
$$f(f^{-1}(y)) = y \qquad \text{for all } y \text{ in the domain of } f^{-1}.$$

In the case of the sine function and its inverse, we have:

$$\text{Arcsin } (\sin \theta) = \theta, \qquad \text{provided that } -\frac{\pi}{2} \le \theta \le \frac{\pi}{2}$$

$$\sin (\text{Arcsin } y) = y, \qquad \text{provided that } -1 \le y \le 1.$$

As we saw previously when we wanted to find the date for which there would be 13 hours of daylight in San Diego, the inverse sine function returned only $t = 105$. This answer is unique because we must restrict the domain in order to have an inverse function. To see this, suppose we try a variety of different values for y and see the effects of the inverse sine function.

If $y = 0.2$,	Arcsin $(0.2) = 11.5°$.
If $y = 0.6$,	Arcsin $(0.6) = 36.9°$.
If $y = 0.95$,	Arcsin $(0.95) = 71.8°$.
If $y = -0.4$,	Arcsin $(-0.4) = -23.6°$.
If $y = -0.88$,	Arcsin $(-0.88) = -61.6°$.
If $y = 2.5$,	ERROR

E X A M P L E

Find all angles for which $\cos \theta = 0.92$.

Solution We immediately find that

$$\theta = \text{Arccos} \ (0.92) \approx 23°.$$

Because of the periodicity of the cosine function, we also know that the value of 0.92 will repeat every 360°, so that the solutions include

$$\theta = 23°, 23° + 360°, 23° + 2(360°), 23° + 3(360°), 23° + 4(360°), \ldots$$

or, in general,

$$\theta = 23° + n(360°)$$

for any integer $n \geq 0$.

Further, from the graph of the cosine function we know that there must be another angle just before 360° whose cosine is 0.92. In particular, since our first solution is 23°, we realize that the next angle will be 23° short of 360°, or 337°. Thus, the solutions also include

$$\theta = 337°, 337° + 360°, 337° + 2(360°), 337° + 3(360°), 337° + 4(360°), \ldots$$

or, in general,

$$\theta = 337° + n(360°)$$

for any integer $n \geq 0$. Equivalently, using radian measure, the solutions are

$$\theta = 0.401 + 2n\pi \text{ radians} \quad \text{and} \quad \theta = 5.882 + 2n\pi \text{ radians},$$

for any integer $n \geq 0$.

Moreover, because the cosine function is symmetric about the vertical axis $\theta = 0$ (it is an even function), we know that the same patterns repeat with negative angles:

$$\theta = -23° - n(360°) \quad \text{and} \quad \theta = -337° - n(360°) \quad \text{for any integer } n \geq 0.$$

Equivalently, with radian measure, $\theta = -0.401 - 2n\pi$ and $\theta = -5.882 - 2n\pi$, for any integer $n \geq 0$.

In summary,

Properties of the Inverse Cosine Function $y = \text{Arccos} \ x$

1. Arccos x is defined only for values of x (the domain) between -1 and 1.
2. The principal values for y (the range) are between 0 and π radians.
3. Arccos $(\cos y) = y \qquad 0 \leq y \leq \pi$.
4. $\cos (\text{Arccos} \ x) = x \qquad -1 \leq x \leq 1$.

Finally, as with the graph of the inverse sine function, the graph of the inverse cosine function $y = \text{Arccos } x$ is the mirror image of the cosine graph about the line $y = x$, as shown in Figure 7.31. Notice that the inverse cosine is defined only for x between -1 and 1 and the inverse cosine values are between 0 and π.

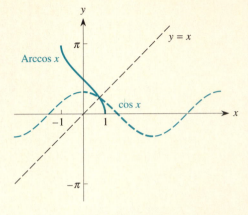

FIGURE 7.31

EXERCISES

1. Several of the following trigonometric equations do not have solutions. By inspection, decide which ones do not (give reasons) and then find the solutions to the remaining equations.

   a. $\sin \theta = 4$ **b.** $\sin \theta = 0.4$ **c.** $3 \sin \theta = 4$
   d. $4 \sin \theta = 3$ **e.** $4 \sin \theta = -3$ **f.** $5 \sin 2x = 3$
   g. $5 \cos 2x = -3$ **h.** $3 \cos x = 2$ **i.** $5 \sin 2x = 3 \cos x$

2. On what days of the year will San Diego have 11 hours of daylight? 10 hours of daylight? 9 hours of daylight?

3. The height of water at a dock is given by the formula $h(t) = 10 + 4 \sin \frac{\pi t}{6}$, where t is measured in hours since midnight.

   a. When does high tide occur?
   b. When does low tide occur?
   c. When does the water level reach 8 feet? 10 feet? 11 feet? 12 feet?

4. An air conditioner is being used to cool a room. The temperature T oscillates up and down according to the formula $T(t) = 69 + 3 \sin \frac{\pi t}{10}$, where t is measured in minutes.

   a. At about what temperature is the thermostat set? (when does the air conditioner kick in?)
   b. At about what temperature does the air conditioner kick out?
   c. When does the room temperature reach 70°? 67°?

5. One of the dangers at places that have very high tides, such as Canada's Bay of Fundy, is the rate at which the tide can come in and potentially trap unwary visitors. Use the formula you devised for a sinusoidal function that models the heights of the tides at Fundy in Exercise 8 of Section 7.2. Determine how long it takes for the water level to rise 5 feet

   a. from a point of low tide.
   b. from a point at the average tide level.

6. A 25-foot ladder is leaning against the side of a building and begins to slip. Write a formula for the angle θ that the ladder makes with the ground as a function of the distance x from the foot of the ladder to the building. Use your function grapher to draw the graph of this function. What are appropriate values for the domain of the function? Is the graph concave up or concave down? When is it maximum?

7. A *lunar cycle*, new moon to full moon and back, takes 28 days. According to legend, a werewolf comes out only when the moon is full. If you interpret this to mean that the visible part of the moon is 99% or more illuminated, then for how many hours can a werewolf function during each 28-day cycle?

8. Find the angle C if $a = 7, b = 11$, and $c = 5$ in the accompanying figure.

9. Find the angle opposite the side of a triangle whose length is 6 if the lengths of the other two sides are 10 and 8 in the accompanying figure.

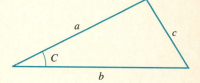

7.5 *The Tangent Function*

In this section we introduce a third trigonometric function, the *tangent*, which is defined by the ratio

$$\tan x = \frac{\sin x}{\cos x}$$

for any value of x for which $\cos x \neq 0$.

Just as with the sine and cosine functions, the tangent function also can be written as a ratio of sides in a right triangle in terms of an angle θ, as shown in Figure 7.32.

$$\tan \theta = \frac{\text{side opposite } \theta}{\text{side adjacent } \theta}$$

$$\tan \theta = \frac{\text{opposite}}{\text{adjacent}}$$

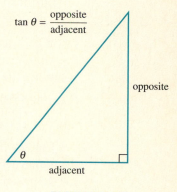

FIGURE 7.32

As with any function, our first concern is to determine the graph of the tangent function to help us understand its behavior. The graph of the tangent function is shown in Figure 7.33 on the next page. From it, we see that the tangent function is periodic with period π; the tangent graph completes one full cycle between $x = -\frac{3\pi}{2}$ and $-\frac{\pi}{2}$, another between $-\frac{\pi}{2}$ and $\frac{\pi}{2}$, a third between $\frac{\pi}{2}$ and $\frac{3\pi}{2}$, and so forth. Each of these segments is called a *branch* of the graph. Also, the tangent function has zeros at $x = 0, \pm\pi, \pm2\pi, \ldots$. Further, the tangent graph has vertical asymptotes at $x = \pm\frac{\pi}{2}, \pm\frac{3\pi}{2}, \pm\frac{5\pi}{2} \ldots$, so the tangent function is not defined at these points (which separate the branches). In addition, the tangent is an increasing function for all intervals between asymptotes. Also as seen from the graph in Figure 7.33, the tangent curve is first concave down and then concave up on each branch. Finally, all points of inflection occur at the points where the tangent curve crosses the x-axis: at $x = 0, \pm\pi, \pm2\pi, \ldots$.

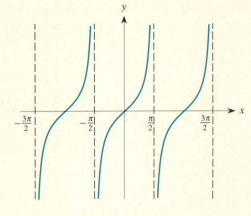

FIGURE 7.33

To see why the tangent function has this behavior, we go back to the defining relationship:

$$\tan x = \frac{\sin x}{\cos x}.$$

First, the tangent function has a zero wherever the numerator is zero. This corresponds to points where the sine is zero, namely at $x = 0, \pm\pi, \pm2\pi, \ldots$, which clearly agrees with what we see on the graph. Second, the tangent function is undefined and has a vertical asymptote wherever its denominator, cos x, is zero. This happens at $x = \pm\frac{\pi}{2}, \pm\frac{3\pi}{2}, \ldots$, and again this agrees with what we see on the graph of the tangent function. Third, between $x = 0$ and $\frac{\pi}{2}$, we know that the sine function is positive and increasing toward 1 while the cosine function is positive and decreasing toward 0. As a result, their ratio involves a positive numerator that is getting larger and a positive denominator that is getting smaller and approaching zero. Thus the tangent is a positive function that increases toward ∞.

Similarly, between $x = -\frac{\pi}{2}$ and $x = 0$, the sine function is negative and increasing toward 0 while the cosine function is positive and increasing toward 1. Thus, their ratio is negative and increases toward 0. Finally, because the tangent function is periodic with period π, we see that it will repeat the behavior we have outlined here, leading to the graph shown in Figure 7.33. For reference purposes, you should know that $\tan 0 = \frac{\sin 0}{\cos 0} = \frac{0}{1} = 0$ and $\tan\frac{\pi}{4} = \tan 45° = 1$.

Just as there are trigonometric identities relating the sine and cosine, there are identities involving the tangent function. The defining equation for the tangent function,

$$\tan x = \frac{\sin x}{\cos x},$$

is by far the most important. We also can derive an analog to the

Pythagorean identity as follows. Since

$$\sin^2 x + \cos^2 x = 1$$

for any x, we can divide both sides by $\cos^2 x$ when $\cos x \neq 0$ and get

$$\frac{\sin^2 x}{\cos^2 x} + \frac{\cos^2 x}{\cos^2 x} = \frac{1}{\cos^2 x}$$

and hence

$$\tan^2 x + 1 = \frac{1}{\cos^2 x}.$$

Likewise, there are double-angle, sum, and difference formulas, and so on, for the tangent function, but we will not concern ourselves with them here. If you would like to see them, you can find them in any trigonometry textbook.

The Inverse Tangent

Suppose that we are faced with an equation such as $\tan \theta = 1.5$. To find the angle θ that satisfies this equation, we use an inverse tangent function, called **Arctan x,** that gives the angle whose tangent value is x. Therefore,

$$\theta = \text{Arctan } 1.5 = 0.9828 \text{ radian} = 56.31°.$$

As with the inverse sine and inverse cosine functions, we have to restrict the domain in order to define the inverse tangent function. It is defined for all real values, and the inverse tangent values are only between $-\frac{\pi}{2}$ and $\frac{\pi}{2}$ where the tangent function is strictly increasing. Accordingly, a calculator returns a value only between $-\frac{\pi}{2}$ and $\frac{\pi}{2}$ (or between $-90°$ and $90°$) for the inverse tangent. If a given situation is such that other values of θ are needed, you will have to determine them by using what you know about the graph of the tangent function.

E X A M P L E

You enter a movie theater that has a screen 20 feet high positioned 5 feet above your eye level. If you sit too far back in the theater, the screen appears too small; if you sit too close, the picture will seem distorted because your viewing angle is too small. Suppose you sit 40 feet back from the screen.

a. What is your viewing angle θ? (See Figure 7.34 on the next page.)
b. Does the viewing angle increase or decrease if you move farther back?

Solution **a.** Your viewing angle θ is the angle subtended by the screen. There is no direct way to get an expression for θ, so we have to get one in a somewhat indirect manner. To do so, we introduce the angle α shown in Figure 7.34 representing the angle from eye level vertically upward to the bottom of the screen. This gives us two right triangles. In the smaller triangle, we have

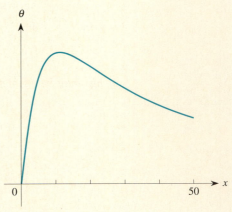

$$\tan \alpha = \frac{5}{40} = \frac{1}{8}$$

so

$$\alpha = \text{Arctan} \frac{1}{8}.$$

FIGURE 7.34

In the larger triangle, we have

$$\tan (\theta + \alpha) = \frac{25}{40} = \frac{5}{8}$$

so

$$\theta + \alpha = \text{Arctan} \frac{5}{8}.$$

Consequently, the desired expression for θ is

$$\theta = (\theta + \alpha) - \alpha = \text{Arctan} \frac{5}{8} - \text{Arctan} \frac{1}{8}$$

$$= 0.5586 - 0.1244 = 0.4342 \text{ radian} \quad \text{or} \quad 24.9°.$$

b. To determine what happens to this viewing angle θ as you move farther back, we could just replace the 40-foot distance with somewhat larger values, say 41 or 45 feet. Alternatively, we can treat the distance to the screen as a variable, x, and consider the angle θ as a function of x. We then have

$$\theta = \text{Arctan} \frac{25}{x} - \text{Arctan} \frac{5}{x}.$$

(Note that these two terms *cannot* be combined algebraically.) If we now graph this function on the interval from 0 to 50, say, as shown in Figure 7.35, we can determine the behavior of this function for θ more thoroughly. The function increases rapidly starting at $x = 0$, rises to a maximum viewing angle when x is approximately 11 feet, and then slowly decreases as x increases thereafter. Therefore if you move farther back from the screen, the viewing angle will decrease.

FIGURE 7.35

Finally, we note that it is possible to introduce three other trigonometric functions, the **cotangent, secant,** and **cosecant,** which are just reciprocals of the sine, cosine, and tangent functions. These new functions are defined as

$$\cot \theta = \frac{1}{\tan \theta}, \qquad \sec \theta = \frac{1}{\cos \theta}, \qquad \csc \theta = \frac{1}{\sin \theta}.$$

These functions were important in the past because they simplified the hand calculations required in working with certain trigonometric problems. However, with calculators, it is just as easy to work with the actual reciprocals shown as it is to use $\cot \theta$, $\sec \theta$, and $\csc \theta$ (which also have associated formulas and properties). Consequently, the cotangent, secant, and cosecant functions gradually are being laid to rest, and so we will not consider them further.

EXERCISES

1. **a.** Verify graphically that

$$\tan \theta + \frac{1}{\tan \theta} = \frac{1}{\sin \theta \cdot \cos \theta}$$

for all θ for which the denominators are nonzero.

   b. Show analytically that it is an identity. (*Hint:* Transform $\tan \theta$ into equivalent expressions in $\sin \theta$ and $\cos \theta$.)

2. Use appropriate trigonometric identities to show that

$$\tan x - \frac{1}{\tan x} = \frac{-2}{\tan 2x}$$

3. Use the identity in Exercise 2 to construct a double-angle formula for the tangent function.

Examine each of the equations in Exercises 4–12 graphically to see if the given relationship may be an identity. If it is not an identity, attempt to locate graphically or numerically at least one point that lies on both curves. If it seems to be an identity, prove it algebraically.

4. $1 + \dfrac{1}{\tan^2 \theta} = \dfrac{1}{\sin^2 \theta}$

5. $\tan 2\theta = 2 \tan \theta$

6. $\tan^2 x - \sin^2 x = (\tan x \cdot \sin x)^2$

7. $\tan 2x = \dfrac{2 \tan x}{1 - \tan^2 x}$

8. $\tan \dfrac{\alpha}{2} = \dfrac{1 - \cos \alpha}{\sin \alpha}$ (*Hint:* Let $\dfrac{\alpha}{2} = \theta$)

9. $\tan^2 x = 1 + 2 \tan x$

10. $1 - \cos 2x = \tan x \cdot \sin 2x$

11. $\tan (\sin x) = (\tan x)(\sin x)$

12. $\cos (\tan x) = \tan (\cos x)$

13. What is wrong with the following "proof"?

$$\cos(\tan x) = \cos\left(\frac{\sin x}{\cos x}\right) = \sin x.$$

14. The Statue of Liberty is 46 meters tall and stands on a base that is also 46 meters tall. Find an expression for the angle subtended by the statue from ground level as a function of distance from the base of the statue. Use this function to estimate graphically the distance when the angle is maximum. Approximately what is this maximum angle?

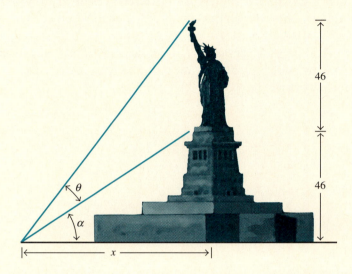

15. Assuming we ignore all forces other than gravity, the path of a baseball is a parabola with equation

$$y = -\frac{16}{v_0^2 \cos^2 \alpha} x^2 + (\tan \alpha)x + 3,$$

where x and y are measured in feet. In this equation, v_0 represents the initial velocity with which the ball leaves the bat and α is the initial angle of inclination of the path of the ball as it leaves the bat.

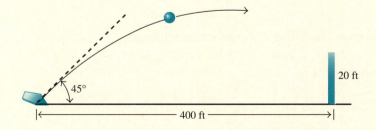

a. If a batter hits a baseball at an initial angle of $\alpha = 45°$ with an initial velocity of 80 miles per hour (= 117 feet per second), will the ball clear a 20-foot-high fence that is 400 feet away from home plate and thus result in a home run?

    b. At an initial velocity of 117 feet per second, for what range of angles α will the ball clear the fence?

    c. At an initial angle of $\alpha = 45°$, what is the minimum velocity v_0 for which the ball will clear the fence?

16. Solve the following trigonometric equations.

    a. $4 \tan \theta = 5$ **b.** $5 \tan \theta = 4$ **c.** $5 \sin \theta = 4 \cos \theta$.

17. The lines $y = x, y = 2x, y = 3x$, and $y = 4x$ all pass through the origin. Find the angle each line makes with the x-axis.

18. **a.** In the general equation of a line through the origin, $y = mx$, can you give an interpretation of the meaning of the slope m in terms of trig functions?

    b. What is the significance of the slope m in $y = mx + b$ from this point of view?

19. Use the graph of $y = \tan x$ to sketch the graph of $y = \frac{1}{\tan x}$.

20. In the exercises at the end of Section 7.1, we examined the behavior of solutions of nonlinear difference equations involving the sine and cosine functions. Consider the comparable problem here with the nonlinear difference equation $x_{n+1} = \frac{1}{2}\tan(x_n)$ for any initial value x_0. Does the resulting sequence appear to converge to a limit? If so, what is it? If not, can you explain why geometrically? What would happen if you use the difference equation $x_{n+1} = \tan(x_n)$ instead?

7.6 Approximating Periodic Functions with Sine and Cosine Terms

By this time, you may feel that most periodic phenomena can be represented by sine or cosine functions. Unfortunately, this is not quite the case. For instance, Exercise 1 of Section 7.1 asked that you construct a graph representing the length of Janis's fingernails if she cuts them every Saturday morning. The result should look something like Figure 7.36. This is certainly a periodic function with a period of seven days, but it is not sinusoidal; rather, it is made up of a series of diagonal lines.

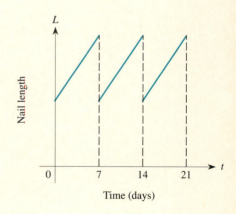

FIGURE 7.36

Nevertheless, it turns out that any periodic function, including those that are not continuous such as the function in Figure 7.36, can be *approximated* by appropriate combinations of sine and cosine terms. That is, these two functions can be thought of as a set of fundamental building blocks out of which any other periodic function can be constructed.

To see how this works, let's look at a still simpler periodic function known as a *square wave*, which often arises in electronics and whose graph is shown in Figure 7.37. Such a function can be used to model the on-off states of an electrical device or the transmission of digital signals. Notice that the function we have drawn has period 2π. We do this for simplicity because the period we use for the sine and cosine functions is 2π. If a given periodic function f has a period other than 2π, we can adapt the ideas developed here by transforming the function with appropriate shifts and stretches.

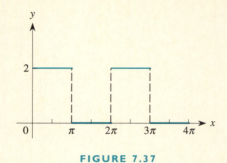

FIGURE 7.37

Since one period of the square wave function is made up of two distinct line segments, we can write a formula for this function if we separately express the equation for each portion of the graph along with the domain for each segment. The first line segment is the horizontal line $y = 2$ extending from $x = 0$ to $x = \pi$. The second segment is the line $y = 0$ from $x = \pi$ to $x = 2\pi$. We then think of this pattern as repeating indefinitely to form the complete function. We therefore get

$$f(x) = \begin{cases} 2 & 0 \le x < \pi \\ 0 & \pi \le x < 2\pi \end{cases}$$

as the basic equation of the square wave; it is repeated for each successive cycle. Because this function is defined differently for different intervals of the domain, this type of function is known as a **split domain function.**

Now consider the graphs of the following sequence of trigonometric functions,

$$f_1(x) = 1 + \frac{4}{\pi} \sin x$$

$$f_2(x) = 1 + \frac{4}{\pi}\left(\sin x + \frac{\sin 3x}{3}\right)$$

$$f_3(x) = 1 + \frac{4}{\pi}\left(\sin x + \frac{\sin 3x}{3} + \frac{\sin 5x}{5}\right)$$

$$f_4(x) = 1 + \frac{4}{\pi}\left(\sin x + \frac{\sin 3x}{3} + \frac{\sin 5x}{5} + \frac{\sin 7x}{7}\right),$$

shown in Figure 7.38.

Use your function grapher (in radian mode) to examine these successive functions to get a more dynamic view of what happens. Notice that each successive trigonometric function is a better fit to the square wave. Further, there is a clear pattern so that if you want to achieve a better fit, it should be evident what additional terms need to be included. Try this out on your function grapher as well. See how close a match you can achieve by including more terms. Also, extend the view beyond $x = 2\pi$ to see that the identical pattern repeats with period 2π.

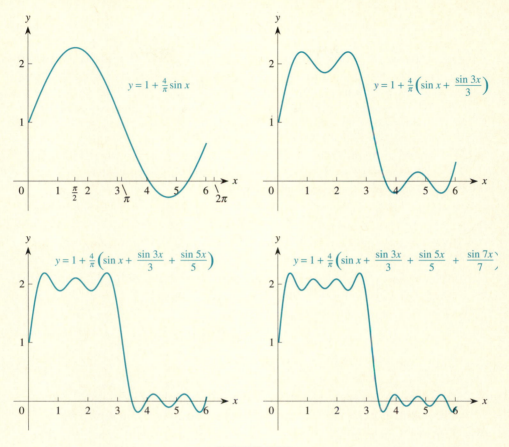

FIGURE 7.38

This type of approximation of a periodic function by a combination of sinusoidal waves is known as a **Fourier approximation** after the French mathematician Joseph Fourier, who first investigated them in the early 1800s. There is a systematic way of determining the specific combination of sinusoidal terms to use for any given periodic function, but that involves using calculus. At this stage, all we can do is experiment with some simple yet interesting and important cases.

As a final application, note that all sound waves are made up of combinations of sinusoidal waves that model the vibration of air molecules. A new high-tech industry is emerging that uses this fact to combat noise pollution. Microprocessors are used to monitor loud noises and to generate the identical combination of sinusoidal waves, but ones that are out of phase with the actual sounds by $\pi = 180°$. If done accurately enough and fast enough, the result is to cancel out the original noises. Prototypes have been installed on four buses in New York City and, in tests, have reduced the level of operating noise from the buses by about 90%.

EXERCISES

1. **a.** Sketch a possible graph of sales as a function of time over several years for a lawn and garden shop. Keep in mind the seasonal nature of the business.

    b. Many such shops introduce a second line, often merchandise related to the December holidays, to keep busy during the slow season for lawn and garden. Sketch a possible graph of sales of holiday items as a function of time. Then sketch a graph showing the sum of both types of sales over the course of several successive years.

2. Use your function grapher to graph

$$y = \sin x - \frac{\sin 3x}{9} + \frac{\sin 5x}{25}.$$

This should come out as an approximation to a "triangle wave." Suppose you want the approximation to come out a little closer to "true triangle-ness." Take a guess at an additional term you could add to the sum to improve the approximation and check by graphing your new function.

3. Use your function grapher to graph

$$y = \sin x + \frac{\sin 2x}{2} + \frac{\sin 3x}{3} + \frac{\sin 4x}{4}.$$

This should come out as an approximation to a "sawtooth wave." Suppose you want the approximation to come out a little closer to "true sawtoothed-ness." Take a guess at an additional term you could add to the sum to improve the approximation and check by graphing your new function.

4. Use your function grapher to sketch the graph of $\sin x - \sin 2x$. What is its period?

5. Use your function grapher to graph

$$y = \sin x + \frac{\sin 3x}{3} + \frac{\sin 5x}{5}.$$

Can you decide what function you are approximating? How does this differ from what we did in the text?

6. **a.** Show that if $f(x) = \sin 2x + \sin 3x$, then $f(x + 2\pi) = f(x)$.

    b. Graph $\sin 2x$ and $\sin 3x$ on the same set of axes. Can you tell from these graphs what the period of $f(x)$ is?

7. **a.** If $f(x) = \sin x - 3 \cos 2x + \frac{1}{2}\sin 3x - \frac{1}{4}\cos 4x$, use appropriate trig identities to simplify $f(x + \pi)$.

    b. Graph $f(x) - f(x + \pi)$ using your function grapher. Is the result surprising? Why?

8. The accompanying graph shows a sawtooth function on the interval 0 to 2π. It passes through the points $(0, 0)$, $\left(\frac{\pi}{2}, 1\right)$, $\left(\frac{3\pi}{2}, -1\right)$ and $(2\pi, 0)$. Use this information to construct a formula for this split-domain function.

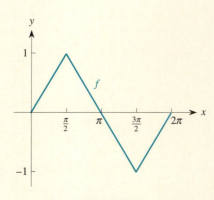

EXERCISE 8

9. Tides are affected by a number of factors including the rotation of the Earth on its axis, the rotation of the Earth about the sun, and the rotation of the moon about the Earth. The method for modeling tides that is used by the National Oceanic and Atmospheric Administration is based on 23 different "constituents," expressed in cosine terms, which are characteristic of (and hence individually calculated for) any given harbor. If the time t is measured in hours from midnight on January 1 and all angles are given in degrees, then the following are the most significant constituents for Bridgeport, Connecticut, in order of decreasing amplitude:

Mean height = 3.61 feet

$C_1 = 3.226 \cos (28.984t - 145.4)$ (effect of the moon assuming circular orbit)

$C_2 = 0.705 \cos (28.439t + 19.7)$ (adjustment for elliptical orbit of the moon)

$C_3 = 0.538 \cos (30t - 343.7)$ (effect of the sun if the Earth's orbit is circular)

$C_4 = 0.287 \cos (15.041t - 96.1)$ (effect of the moon due to inclination of the Earth)

$C_5 = 0.221 \cos (29.528t - 141.4)$ (a more subtle effect)

a. Use your function grapher to plot the graph of the tide function for Bridgeport based on the sum of these constituents for the first day of January; for the first week of January. What is the apparent shape?

b. How high was the tide at 10 A.M. on January 1? At 10 A.M. on January 5?

c. Estimate when high tide occurred on January 1. How high was the tide? When did low tide occur? How low was it?

d. Estimate when high tide occurred on February 1. How high was the tide? When did low tide occur? How low was it?

e. Which answers do you think are more accurate, in part (c) or part (d)? Give reasons.

10. Suppose that the average daily temperature in New York City ranges from a low of 25° on February 1 to a high of 85° on August 1. Sketch a very rough graph of the average daily temperature over the course of an entire year. Write a possible formula for this graph.

11. Suppose that on any given day, the actual temperature in New York City typically is somewhere between 10° below the daily average and 10° above it. Sketch a very rough graph of the temperature in New York at any time t. Can you write a possible formula that models this temperature?

7.7 *Approximating Sine and Cosine with Polynomials*

Have you ever wondered what happens when you press either the SIN or COS key on your calculator and the value for the function appears? How does the calculator actually find the values of these functions?

Approximations Using Curve Fitting

In this section, we consider one approach that has been used to compute these values. We begin by examining the following table of values rounded to three decimal places for sin x, where x is measured in radians.

x	$\sin x$	x	$\sin x$
0	0		
0.1	0.100	−0.1	−0.100
0.2	0.199	−0.2	−0.199
0.3	0.296	−0.3	−0.296
0.4	0.389	−0.4	−0.389
0.5	0.479	−0.5	−0.479
0.6	0.565	−0.6	−0.565
0.7	0.644	−0.7	−0.644

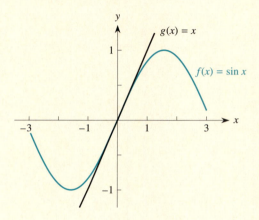

FIGURE 7.39

Notice that when x is reasonably close to 0, sin x is quite close to x itself. This is reflected in the graphs of both $f(x) = \sin x$ and $g(x) = x$ (see Figure 7.39). When x is close to 0, either positive or negative, the two graphs are virtually indistinguishable from one another. Of course, as the value of x gets farther away from 0, the sine curve eventually bends away from the line $y = x$. However, from the graph, it is clear that

$$\sin x \approx x \qquad \text{if } x \text{ is close to 0.}$$

To see this from a different viewpoint, use your calculator to zoom in on the graph of $f(x) = \sin x$ near $x = 0$. The closer you look, the more the sine curve looks like a straight line through the origin. What is the slope of this line? Look at the following table:

x	$\sin x$	$\Delta(\sin x)$
−0.010	−0.009999	
		−0.005
−0.005	−0.004999	
		−0.004999
0	0	
		0.004999
0.005	0.004999	
		0.005
0.010	0.009999	

Notice that the differences in the values of the sine function are nearly constant, so the numbers fall into an approximately linear pattern with slope

$$m = \frac{\Delta y}{\Delta x} = \frac{0.004999}{0.005} \approx 1,$$

which is the slope of the line $y = x$.

Alternatively, if we apply linear regression to the given set of $(x, \sin x)$ values, we find that the line that best fits the data is $y = 0.99988x$ with a correlation coefficient $r = 0.99999999$, and so a linear fit with slope of about 1 is close to perfect. Thus we again see that when x is close to 0,

$$\sin x \approx x.$$

As an example, we use this approximation to estimate the value of

$$\sin \frac{1}{8} = \sin(0.125) \approx 0.125.$$

A more accurate value, correct[1] to five decimal places, is 0.12467. In a similar fashion,

$$\sin\left(-\frac{1}{12}\right) = \sin(-0.08333) \approx -0.08333$$

compared to the correct value, which is -0.08324. Of course, if we move too far away from $x = 0$, the accuracy of the approximation breaks down. For instance, we would not want to approximate $\sin(0.75) = 0.6816$ with the approximate value of 0.75.

Before going on, examine how the two functions x and $\sin x$ compare to one another using your function grapher. The visual interpretation helps make these ideas clear.

This idea of approximating a function such as $y = \sin x$ with a simpler function (often a linear function) is an essential principle in mathematics. We use this principle to approximate trigonometric function values because it is just not possible to calculate them directly using algebraic methods.

With this in mind, let's consider how to improve on the linear approximation for $\sin x$ that we used previously. As we have seen, the problem is that the sine curve, by its very nature, bends away from the approximating line $y = x$. Therefore, in order to approximate the values of $\sin x$ when x is somewhat farther away from 0, we need a simple curve (at least one that is simpler to work with than the sine function) that bends in a similar fashion. For computational purposes, the simplest curves to work with are polynomials.

We suggested above that you zoom in on the sine curve for values of x that are very close to zero, when the sine curve looks linear. Now suppose that you zoom out a little and see what happens for values of x from -3 to 3, as shown in Figure 7.39. As you begin to see the first pair of turning points in the sine curve, observe that the overall shape is quite suggestive of a cubic curve. Of course, if you zoom out a little farther, then more turning points appear and the cubic-like appearance disappears.

Similarly, suppose you look at the cosine curve near $x = 0$, say from -2 to 2. Its overall shape looks like a parabola opening downward. Therefore we can approximate the cosine function with a quadratic function as long

---

[1] Throughout this section, "correct" means that the answer given has been rounded to the number of decimal places shown.

as x remains close to 0. There are several ways to find an equation for such a quadratic. One is to apply our methods of data analysis to fit a quadratic to some set of values for cos x. We ask you to do this as an exercise in this section.

Approximations Using Trigonometric Identities

Instead, we will approach this problem using several trig identities. One of our identities from Section 7.3 is the double-angle formula

$$\cos 2\theta = 1 - 2\sin^2\theta.$$

Let $x = 2\theta$ so that $\frac{x}{2} = \theta$ and the expression for cos 2θ becomes

$$\cos x = 1 - 2\sin^2\left(\frac{x}{2}\right) = 1 - 2\left(\sin\frac{x}{2}\right)^2.$$

When θ is close to 0, we have $\sin\theta \approx \theta$, which means

$$\sin\frac{x}{2} \approx \frac{x}{2},$$

and consequently the given expression becomes

$$\cos x \approx 1 - 2\left(\frac{x}{2}\right)^2 = 1 - \frac{x^2}{2}$$

for x close to 0. Graph this quadratic function on your function grapher along with the cosine curve and see how close they are for values of x between -2 and 2, for example. The two curves, for values of x from -3.5 to 3.5, are shown in Figure 7.40. For x between -1 and 1, there appears to be very close agreement.

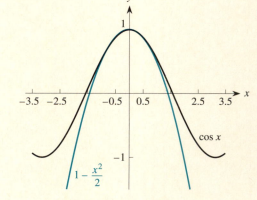

Furthermore, we can use the other double-angle formula

$$\sin 2\theta = 2\sin\theta\cos\theta$$

with $x = 2\theta$ so that $\frac{x}{2} = \theta$. We get

$$\sin x = 2\sin\frac{x}{2}\cos\frac{x}{2}.$$

Using the approximations for

$$\sin x \approx x \quad \text{and} \quad \cos x \approx 1 - \frac{x^2}{2},$$

we find

$$\sin\frac{x}{2} \approx \frac{x}{2} \quad \text{and} \quad \cos\frac{x}{2} \approx 1 - \frac{\left(\frac{x}{2}\right)^2}{2} = 1 - \frac{x^2}{8}$$

FIGURE 7.40

Thus an approximation for sin x that improves upon sin x ≈ x is given by

$$\sin x \approx 2\left(\frac{x}{2}\right)\cdot\left(1-\frac{x^2}{8}\right)$$

$$= x\left(1-\frac{x^2}{8}\right)$$

$$= x - \frac{x^3}{8}.$$

This is an equation of a cubic function that approximates the sine function near $x = 0$. We show the graph of this cubic along with the sine function on the interval from -3 to 3 in Figure 7.41. Notice how the two curves seem almost identical for x between -1 and 1, are reasonably close between -2.5 and -1 and again between 1 and 2.5, and apparently begin to diverge from one another when we move even farther away from 0. Notice also how much better this cubic function seems to be as an approximation to sin x than our original linear approximation $y = x$. Check out how close the cubic is to the sine function at $x = 0.5$, $x = 1$, and $x = 1.5$ using your calculator.

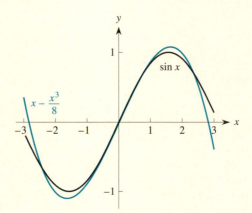

FIGURE 7.41

We can continue this process to produce still better approximations to both the cosine and the sine functions by using the same trig identities. For instance, using $\cos 2\theta = 1 - 2\sin^2\theta$ and the fact that $\sin x \approx x - \frac{x^3}{8}$, we get

$$\cos x = 1 - 2\sin^2\left(\frac{x}{2}\right)$$

$$= 1 - 2\left(\sin\frac{x}{2}\right)^2$$

$$\approx 1 - 2\left(\frac{x}{2} - \frac{(x/2)^3}{8}\right)^2.$$

After some algebraic manipulation, this eventually simplifies to

$$\cos x \approx 1 - \frac{x^2}{2} + \frac{x^4}{32} - \frac{x^6}{2048}.$$

The two graphs in Figure 7.42 illustrate that this sixth degree polynomial $P_6(x)$ is an almost perfect match to the cosine function from around $x = -1.5$ to $x = 1.5$. It is quite accurate from around -2 to 2, and thereafter the accuracy diminishes. For comparison, in Figure 7.43, we show three graphs, the basic cosine curve, our initial quadratic approximation $P_2(x) = 1 - \frac{x^2}{2}$, and this sixth degree polynomial approximation, $P_6(x)$. The higher degree polynomial is clearly a much better fit: It follows the bends of the cosine curve and stays close to it over a longer interval of x-values.

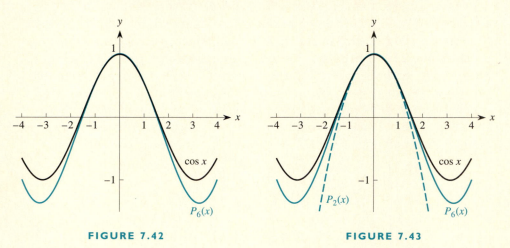

FIGURE 7.42   **FIGURE 7.43**

Approximations Using Taylor Polynomials

We could continue this process to get better and better polynomial approximations to both $\sin x$ and $\cos x$ by using polynomials of higher and higher degrees. Unfortunately, each improvement is based on using a series of approximations—we used the fact that $\sin x \approx x$ to generate the approximation $\cos x \approx 1 - \frac{x^2}{2}$, and so on. The approximation errors in this process will mount up and give us less than the best possible approximation at each successive stage. For instance, we first found

$$\sin x \approx x = Q_1(x)$$

and then used it to find

$$\sin x \approx x - \frac{x^3}{8} = Q_3(x).$$

It turns out, as you will see in calculus, that the *best* possible cubic curve to approximate the sine curve near $x = 0$ is

$$\sin x \approx x - \frac{x^3}{6} = T_3(x).$$

The graph of this cubic and the underlying sine curve are shown in Figure 7.44. The sine curve and the two different cubic approximations are shown in Figure 7.45; notice that the curve corresponding to $T_3(x) = x - \dfrac{x^3}{6}$ remains closer to the sine curve over a longer interval than $Q_3(x)$ does. Notice also that $T_3(x)$ bends in such a way that it remains very close to the sine curve over a relatively large portion of the first arch of the sine curve.

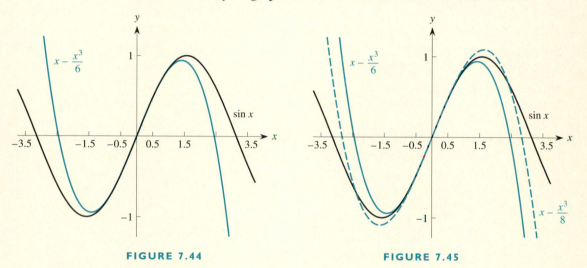

FIGURE 7.44   FIGURE 7.45

Use your function grapher to view what happens in a more dynamic way. In particular, examine the three curves near $x = 0$ to see that T_3 is actually closer to the sine curve than Q_3 is.

Best polynomial approximations, such as T_3 for the sine function, are known as *Taylor polynomial approximations* after the English mathematician Brook Taylor, who investigated them in the early 1700s. They form an important component of calculus, and similar Taylor approximations are available for other common functions such as the cosine function and exponential and logarithmic functions.

To illustrate the accuracy of the approximation for sin x using $T_3(x)$, the *best cubic* polynomial for values of x near 0,

$$\sin(0.125) \approx T_3(0.125)$$

$$= (0.125) - \frac{(0.125)^3}{6}$$

$$= 0.12467,$$

which agrees with the true value of sin (0.125) to five decimal places. Similarly, if we move farther away from 0,

$$\sin(0.7) \approx 0.7 - \frac{(0.7)^3}{6} = 0.643,$$

compared to the actual value of sin (0.7) = 0.644. This is correct to the nearest hundredth, so the approximation still is fairly accurate. However, from the graphs in Figure 7.44, it is evident that the two curves eventually diverge from one another. Thus if we take x too far away from 0, the accuracy of the Taylor approximation diminishes. For instance,

$$\sin 1 \approx 1 - \frac{1^3}{6} = 0.83333$$

compared to the correct value of $\sin 1 = 0.84147$;

$$\sin 1.5 \approx 1.5 - \frac{(1.5)^3}{6} = 0.93750$$

compared to the correct value of 0.99749;

$$\sin 2 \approx 2 - \frac{2^3}{6} = 0.66667$$

compared to the correct value of 0.90930;

$$\sin \pi \approx \pi - \frac{\pi^3}{6} = -2.02612$$

compared to the correct value $\sin \pi = 0$.

   Just as we get a better estimate by using a cubic in place of a linear approximation, we can improve the previous approximation with a still higher degree Taylor polynomial, which is

$$\sin x \approx x - \frac{x^3}{6} + \frac{x^5}{120} = T_5(x).$$

The graph of this *fifth-degree* polynomial $T_5(x)$ is superimposed over the graph of the sine function in Figure 7.46. Try the comparison with your function grapher. Notice that the two curves remain extremely close together over a much longer interval than was the case with the cubic approximation. The turning points in the polynomial tend to match the first turning points in the sine curve, and so we have visual evidence that the polynomial should be a good approximation to the sine curve. To verify this numerically, let's see how much of an improvement we get using the examples from before:

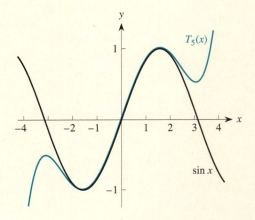

FIGURE 7.46

$$\sin 0.7 \approx T_5(0.7) = 0.7 - \frac{(0.7)^3}{6} + \frac{(0.7)^5}{120} = 0.64423$$

compared to the actual value of $\sin 0.7 = 0.64422$;

$$\sin 1 \approx 1 - \frac{1^3}{6} + \frac{1^5}{120} = 0.84167$$

compared to the actual value of $\sin 1 = 0.84147$;

$$\sin 1.5 \approx 1.5 - \frac{(1.5)^3}{6} + \frac{(1.5)^5}{120} = 1.00078$$

compared to the actual value of 0.99749;

$$\sin 2 \approx 2 - \frac{2^3}{6} + \frac{2^5}{120} = 0.93333$$

compared to the actual value of 0.90930;

$$\sin \pi \approx \pi - \frac{\pi^3}{6} + \frac{\pi^5}{120} = 0.52404$$

compared to the actual value $\sin \pi = 0$.

We see that this approximation is better still in the sense that we get more accurate estimates for the values of sin x over still larger intervals of x-values around 0. It turns out that we can continue this process using higher degree Taylor polynomials and get even better approximations. However, before doing that, let's examine the sequence of Taylor polynomial approximations we have so far. They are

$$\sin x \approx x$$

$$\sin x \approx x - \frac{x^3}{6}$$

$$\sin x \approx x - \frac{x^3}{6} + \frac{x^5}{120}.$$

First of all, notice that each successive polynomial involves just one additional term. Second, each of the polynomial terms involves only *odd* powers, and we know that the sine function is an *odd* function. This observation means that both the sine function and the approximating Taylor polynomials are symmetric about the origin. Third, notice that successive coefficients alternate in sign, so the next term will be subtracted from the previous terms. Fourth, there is a definite pattern in the coefficients. In particular, notice that $6 = 3 \times 2 \times 1 = 3!$ and $120 = 5!$, using factorial notation (see Appendix C), so that

$$\sin x \approx x$$

$$\sin x \approx x - \frac{x^3}{3!}$$

$$\sin x \approx x - \frac{x^3}{3!} + \frac{x^5}{5!}$$

Think About This

What do you expect for the next higher degree Taylor polynomial approximation to sin x? How accurate is this next approximation to sin x for the values of $x = 0.7, 1, 1.5, 2$, and π?

Using Properties of Sin x

We could continue this process and construct approximating polynomials of higher and higher degree. It turns out that it is not necessary to do this if we are clever in using some of the basic properties of the sine function. First of all, recall that

$$\sin(-x) = -\sin x.$$

This property means we can obtain approximations for $\sin x$ when x is negative simply by using the corresponding positive value for x and reversing the sign of the estimate.

Second, we know that the sine function is periodic with period 2π. Therefore if x is any number greater than 2π, then the value of $\sin x$ will be the same as the value of $\sin x_0$, where x_0 is some number between 0 and 2π radians. Consequently, we only need to obtain an approximation that is accurate as far out as 2π. Anything beyond that can be handled by reducing the value of x to an appropriate x_0 by "removing" all multiples of 2π.

Furthermore, think about the sine curve (see Figure 7.47). The portion of it between $x = \pi$ and $x = 2\pi$ has precisely the same shape as the portion from $x = 0$ to $x = \pi$. Now picture flipping this portion of the curve about the x-axis. You have side by side above the x-axis two identical arches. Thus, if we have a value of x between π and 2π (where $\sin x$ is negative), there is a matching value between 0 and π, namely at $x - \pi$, where the sine function has the same value, but with a positive sign. That is,

$$\sin x = -\sin(x - \pi).$$

Consequently, all we need do is obtain an approximation that is accurate for x between 0 and π. Any larger number between π and 2π can be reduced to an appropriate one between 0 and π.

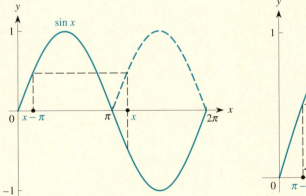

FIGURE 7.47

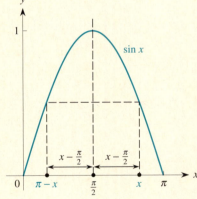

FIGURE 7.48

Now picture the sine curve from $x = 0$ to $x = \pi$. The two halves are symmetric, as shown in Figure 7.48. Therefore if we have a value of x between $\frac{\pi}{2}$ and π, its sine is the same as that of a corresponding value between 0 and $\frac{\pi}{2}$. In particular, suppose x is between $\frac{\pi}{2}$ and π so that it lies at a distance of $x - \frac{\pi}{2}$ to the right of the vertical line at $\frac{\pi}{2}$. We need to determine the point in the symmetric position on the left side of the vertical line; it is located at the same distance to the left of the line, namely

$$\frac{\pi}{2} - \left(x - \frac{\pi}{2}\right) = \pi - x.$$

Consequently, if x is any number between $\frac{\pi}{2}$ and π, then the associated number having the same sine value is $\pi - x$ and so

$$\sin x = \sin (\pi - x).$$

Using these symmetry arguments, all we need is an approximation to $\sin x$ that is sufficiently accurate for x between 0 and $\frac{\pi}{2}$, and the previous fifth-degree Taylor polynomial $T_5(x)$ gives two-decimal accuracy for any value of x in this interval. If we want more than two-decimal accuracy, then we would have to use a higher degree polynomial, say the seventh-degree Taylor polynomial that we asked you to produce in the previous *Think About This* exercise. Try the seventh-degree polynomial for various values of x in this interval. Does it provide four-decimal accuracy? Is that adequate? If not, what would you do?

Taylor Polynomial Approximations to Cos x

We can use the same type of analysis with the cosine function. The successive Taylor polynomial approximations are

$$\cos x \approx 1 - \frac{x^2}{2!}$$

$$\cos x \approx 1 - \frac{x^2}{2!} + \frac{x^4}{4!}$$

$$\cos x \approx 1 - \frac{x^2}{2!} + \frac{x^4}{4!} - \frac{x^6}{6!}.$$

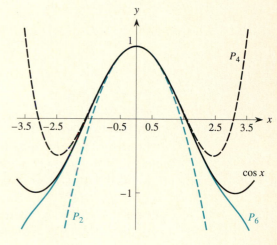

The graphs of these polynomials, as well as that of the cosine curve, are shown in Figure 7.49. Notice that each successive polynomial fits the cosine curve more accurately over a larger and larger interval centered about the y-axis. You should examine these successive approximations to the cosine function using your function grapher.

FIGURE 7.49

Think About This

What types of powers occurring in these Taylor approximations to the cosine function can be related to any special geometric property of the cosine function? Predict what the next higher term will be. Devise a scheme to reduce any value of x to an equivalent one that allows us to use the smallest possible interval of x-values. How accurate is the fourth-degree Taylor polynomial on this interval? How accurate is the sixth-degree polynomial?

Taylor polynomial approximations can be constructed for most other common functions, including exponential and logarithmic functions, which we ask you to explore in the exercises. We urge you to investigate these approximations, either with a graphing calculator or an appropriate computer program. For the various functions, use the calculator or computer to draw the graph of the indicated function and superimpose over it the successive Taylor polynomials of higher degrees. Does each successive polynomial appear to give better agreement with the function? Is there any instance where the successive polynomials do not necessarily improve the accuracy? You will have to wait until you take calculus to learn why these phenomena occur.

As a final note, let's look at the ideas we have developed in this section from a somewhat different perspective. Up to now, we have interpreted Taylor polynomials as a means of *approximating* one function by a polynomial. An alternative interpretation is that we have been *constructing* a function (or a portion of a function) out of simpler functions. In particular, we have been constructing the trigonometric functions using polynomials as the fundamental building blocks. More specifically, we have used *linear combinations* of power terms—that is, sums of constant multiples of power terms—as these fundamental building blocks. This idea of using linear combinations of basic mathematical elements to construct more complicated mathematical structures is a continuing theme throughout mathematics.

EXERCISES

1. Use the Taylor polynomial approximation to $f(x) = \cos x$ of degree 2 to estimate the value of the cosine function for $x = 0, 0.1, 0.2, 0.3, 0.4, 0.5,$ and 0.6. Compare each estimate to the correct value.

2. Construct a table using the values you calculated in Exercise 1 that has a column containing the error in the approximation (the difference between the estimate and the correct value). Analyze the column of differences. Do they appear to grow approximately linearly? exponentially? quadratically? cubically?

3. Use the Taylor polynomial approximation to $f(x) = \sin x$ of degree 3 to construct a table of estimates for the values of the sine function when $x = 0, 0.1, 0.2, 0.3, 0.4, 0.5,$ and 0.6. Calculate the errors and analyze them in the same way as in Exercise 2.

4. Construct a table containing Taylor polynomial approximations of degrees $n = 1$ and $n = 3$ to $f(x) = \sin x$ for $x = 0°, 10°, 20°, \ldots, 60°$. (Remember to

convert the angles into radians first, or the approximation formulas make no sense!) For each value, compare the linear and cubic approximations to the correct value. Is there any consistent pattern you notice regarding how the two approximations relate to the correct value? Can you account for this pattern based on the graphs of the approximation curves and the sine curve? How can you use this pattern to improve the approximation?

5. The cosine function looks like a quadratic for x close to 0. Use the values of $\cos x$ for $x = 0, \pm0.1, \pm0.2$, and ±0.3 to fit the best possible quadratic using the statistical features of your calculator or computer software. Then complete the following table to compare the accuracy of this regression equation to that of the quadratic Taylor approximation $\cos x \approx 1 - \frac{x^2}{2}$.

x	0.05	0.15	0.25	0.40	0.50
$\cos x$					
$1 - \dfrac{x^2}{2}$					
Regression equation					

6. Repeat Exercise 5 with cubic approximations to the sine function.

7. Repeat Exercise 6 for the sine function using $x = 0, \pm0.05, \pm0.10$, and ±0.15. Since the points used now are all closer to $x = 0$, is your resulting cubic a better fit to the sine function near the origin than the cubic you found in Exercise 6?

8. Construct a table of values of $\sin x$ for $x = -\frac{4\pi}{25}, -\frac{3\pi}{25}, -\frac{2\pi}{25}, \ldots, \frac{4\pi}{25}$. If your calculator has as one of its statistical features the ability to find the best cubic polynomial fit to a set of data, find that polynomial for this set of sine values. How close does it come to the cubic Taylor polynomial approximation $\sin x \approx x - \frac{x^3}{6}$?

9. Use the Taylor polynomial approximation of degree $n = 5$ to $f(x) = \sin x$ to find a polynomial approximation of degree $n = 5$ to $g(x) = \sin(-x)$. Is the result surprising?

10. Use the Taylor polynomial approximation of degree $n = 5$ to $f(x) = \sin x$ to find a polynomial approximation of degree $n = 10$ to $g(x) = \sin(x^2)$. Graph both $g(x)$ and your approximation to it on $[-\pi, \pi]$. Based on this graph, over what interval does your polynomial seem to be a good approximation to $g(x)$?

11. a. Use the Taylor polynomial approximation of degree $n = 5$ to $f(x) = \sin x$ to find a polynomial approximation of degree $n = 5$ to $h(x) = \sin 2x$.
    b. What do you get if you multiply the polynomial approximation of degree $n = 3$ to $f(x) = \sin x$ by the polynomial approximation of degree $n = 4$ to $g(x) = \cos x$?
    c. Graph $h(x)$ and twice the product of the two approximations given in part (b). What do you observe?

12. Write out the Taylor polynomial approximation of degree 3 to $f(x) = \sin x$ and the approximation of degree 4 to $g(x) = \cos x$. Square each expression and add them. What do you get? What do you think will happen if you use higher degree approximations?

13. In the text, we used the identity $\cos 2\theta = 1 - 2\sin^2 \theta$ with $\theta = \frac{x}{2}$ to construct the approximation

$$\cos x \approx 1 - \frac{x^2}{2} + \frac{x^4}{32} - \frac{x^6}{2048}.$$

Instead, use the alternative form of the identity $\cos 2\theta = \cos^2 \theta - \sin^2 \theta$ and any lower degree polynomials to find a different approximation to $\cos x$.

14. Repeat Exercise 13 with the third form of the double-angle identity $\cos 2\theta = 2\cos^2 \theta - 1$ to construct still another approximation formula for $\cos x$.

15. The function $f(x) = \frac{\sin x}{x}$ is not defined at $x = 0$.

    a. Use values of $x = 0.1, 0.01, 0.001, 0.0001, 0.00001, \ldots$ to investigate the behavior of this function close to $x = 0$. What limiting value appears to exist?

    b. Use the linear Taylor polynomial approximation to $\sin x$ to explain why the limiting value you found in part (a) appears to make sense.

16. In calculus, you will have to determine the value of

$$\frac{\sin (x + \Delta x) - \sin x}{\Delta x},$$

where Δx is a very small quantity. Estimate the value of this quotient, first using linear approximations to both sine expressions and then using cubic approximations. With the cubic approximation, suppose that Δx is actually 0; what does the resulting expression suggest?

17. In Example 4 of Section 7.3, we used the law of cosines to find the distance to the star Alpha Centauri as

$$d = \frac{R}{\sqrt{2(1 - \cos \phi)}} = 25{,}242{,}154{,}000{,}000 \text{ miles},$$

where $R = 186{,}005{,}732$ and $\phi = 7.3691680 \times 10^{-6}$. We pointed out that some calculators will give an error message because of rounding errors that can lead to 0 in the denominator. Use a quadratic Taylor polynomial to approximate the term in the denominator, and calculate the value for the distance to Alpha Centauri. How different are the two answers?

18. The exponential function $f(x) = e^x$ with base $e = 2.71828\ldots$ arises extensively throughout mathematics and the sciences. As with the trig functions, its values are calculated using Taylor polynomial approximations:

$$e^x \approx 1 + x$$

$$e^x \approx 1 + x + \frac{x^2}{2!}$$

$$e^x \approx 1 + x + \frac{x^2}{2!} + \frac{x^3}{3!},$$

and so forth. Use these and any further approximations you feel are needed to approximate the values of

a. $e^{0.1}$ b. $e^{-0.1}$

c. Use the given polynomials and any additional approximations to e^x that you need to estimate the value of e reasonably accurately. What degree polynomial will produce two-decimal accuracy? What degree produces three-decimal accuracy? four-decimal accuracy?

19. In Chapter 2, we encountered the logarithmic function with base e, $\ln x = \log_e x$. The successive Taylor polynomial approximations to $f(x) = \ln x$ are

$$\ln x \approx x - 1$$

$$\ln x \approx (x - 1) - \frac{(x - 1)^2}{2}$$

$$\ln x \approx (x - 1) - \frac{(x - 1)^2}{2} + \frac{(x - 1)^3}{3}$$

$$\ln x \approx (x - 1) - \frac{(x - 1)^2}{2} + \frac{(x - 1)^3}{3} - \frac{(x - 1)^4}{4},$$

and so forth. (These polynomials are expressed in powers of $(x - 1)$ rather than powers of x because $\ln x$, like all log functions, is not defined at $x = 0$). Use these and any further approximations you feel are needed to approximate the values of

a. $\ln 1.1$ b. $\ln 0.75$

c. Use the above polynomials and any additional approximations to $\ln x$ that you need to approximate the value of $\ln 1.5$ to two-decimal accuracy. What degree polynomial produces three-decimal accuracy? What do you conclude about the rate of convergence of the sequence of approximations when $x = 1.5$? How does this rate of convergence compare to what happens when $x = 1.1$ or $x = 0.75$?

20. The successive Taylor polynomial approximations to $f(x) = \dfrac{1}{x}$ are

$$\frac{1}{x} \approx 1 - (x - 1)$$

$$\frac{1}{x} \approx 1 - (x - 1) + (x - 1)^2$$

$$\frac{1}{x} \approx 1 - (x - 1) + (x - 1)^2 - (x - 1)^3,$$

and so forth. (These polynomials are expressed in powers of $(x - 1)$ because $1/x$ is not defined at $x = 0$.) Use these and any further approximations you need to approximate the values of

a. $\dfrac{1}{1.2}$ b. $\dfrac{1}{0.75}$

(Many of you may wonder why anyone would want to approximate $1/x$ when you can just divide. Recall that calculators and computers are able to do only addition and subtraction. They perform multiplication by repeated addition. To perform division, they must reduce the operation to a simpler one involving only multiplication, addition, and subtraction.)

7.8 *Properties of Complex Numbers*

One of the most startling developments in the history of mathematics was the introduction of complex numbers to allow people to solve quadratic equations. For example, from the quadratic formula, the roots of $x^2 + 4 = 0$ are $x = 2i$ and $x = -2i$, and the roots of $x^2 - 2x + 10 = 0$ are $x = 1 - 3i$ and $x = 1 + 3i$. These roots are based on the imaginary number $i = \sqrt{-1}$.

In our explorations on the nature of the roots of polynomials in Section 4.3, we found that quadratic, cubic, and consequently higher degree polynomials have a surprisingly high proportion of complex zeros. We now need to develop a way to visualize these zeros that will give us a deeper understanding of the processes that lead to such polynomial equations.

Any complex number $z = a + bi$ is composed of two parts, a *real part*, a, and an *imaginary part*, bi. For instance, in $z = 4 + 7i$, the real part is 4 and the imaginary part is $7i$. We occasionally denote these by $a = \text{Re}(z)$ and $b = \text{Im}(z)$, respectively. Note that a and b are both real numbers; it is the combination $a + bi$ that is a complex number. In the special case when $b = 0$, the complex number reduces to a real number. In another special case where $a = 0$, the complex number reduces to an imaginary number, bi.

The arithmetic of complex numbers, for the most part, is quite straightforward, and we review it briefly in Appendix B. Notice that, since $i = \sqrt{-1}$, it immediately follows that

$$i^2 = (\sqrt{-1})^2 = -1$$
$$i^3 = (i^2)(i) = (-1)(i) = -i$$
$$i^4 = (i^2)(i^2) = (-1)(-1) = 1$$
$$i^5 = i^4 \cdot i = 1 \cdot i = i.$$

Thus all higher powers simply cycle through these four "values": $i, -1, -i,$ and 1. That is:

$$i = i$$
$$i^2 = -1$$
$$i^3 = -i$$
$$i^4 = 1$$

It is extremely helpful to be able to visualize complex numbers geometrically, and we can do this using the *complex plane*. It is a two-dimensional coordinate system designed to display a complex number by using a number on the horizontal axis for the real part and a number on the vertical axis for the imaginary part. In Figure 7.50, we plot the complex numbers $2 + 5i$, $1 - 3i$, and $-2 + i$ in the complex plane. Any real number, such as 4 (which is $4 + 0i$) or -6 (which is $-6 + 0i$), lies on the horizontal axis.

Any imaginary number, such as i (which is $0 + i$) or $-3i$ (which is $0 - 3i$), lies on the vertical axis.

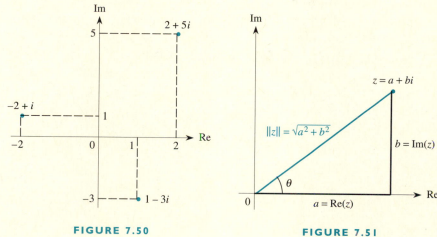

FIGURE 7.50   FIGURE 7.51

Suppose that $z = a + bi$ is an arbitrary complex number. In addition to plotting the corresponding point in the complex plane, we now connect the point to the origin with a line segment and think of it as the hypotenuse of a right triangle, as shown in Figure 7.51. Notice that the base of the triangle is equal to a, the real part of z, and the height of the triangle is equal to b, the size of the imaginary part of z. Using the Pythagorean theorem, we see that the length of the hypotenuse is $\sqrt{a^2 + b^2}$, and we interpret this length as representing the size, or magnitude, of the complex number, z. We call the size the **modulus** of z and write it as

$$\|z\| = \sqrt{a^2 + b^2}.$$

For instance, if $z = 4 + 3i$, then its modulus is

$$\|z\| = \sqrt{4^2 + 3^2} = \sqrt{25} = 5.$$

Similarly, the complex numbers $4 - 3i$, $-4 + 3i$, and $-4 - 3i$ all have the same modulus of 5. Sketch them to see that this is indeed the case.

We again consider the complex number $z = a + bi$ and the associated right triangle in the complex plane, but now we focus on the angle θ. Notice from Figure 7.51 that

$$\tan \theta = \frac{b}{a}.$$

We immediately have the two further relations

$$\cos \theta = \frac{a}{\|z\|} \quad \text{and} \quad \sin \theta = \frac{b}{\|z\|},$$

which lead to

$$a = \|z\| \cos \theta \quad \text{and} \quad b = \|z\| \sin \theta.$$

Consequently, we can write the original complex number z in the equivalent *trigonometric form*

$$z = a + bi = \|z\| \cos \theta + i \cdot \|z\| \sin \theta$$
$$= \|z\| (\cos \theta + i \cdot \sin \theta).$$

The **trigonometric form** for the complex number $z = a + bi$ is

$$z = \|z\| \cos \theta + i \cdot \|z\| \sin \theta,$$

where

$$\|z\| = \sqrt{a^2 + b^2} \quad \text{and} \quad \tan \theta = \frac{b}{a}.$$

In our example $z = 4 + 3i$, where $\|z\| = 5$, we then have

$$z = 4 + 3i = 5(\cos \theta + i \cdot \sin \theta),$$

where $\tan \theta = \frac{3}{4}$ or $\theta = \text{Arctan} \frac{3}{4} = 0.644$ radian or $36.87°$.

Think About This

Use the given value for the angle θ and show that the trigonometric form for the complex number z is identical to the original expression $4 + 3i$.

Powers of Complex Numbers

The trigonometric form for a complex number z allows us to interpret z as being located at a certain distance, the modulus, from the origin and rotated through a certain angle, θ, from the horizontal. This model gives us a way to gain some special insights into complex numbers. Consider again the complex number $z = 4 + 3i$; it has a modulus of 5 and an associated angle of $\theta = 36.87°$. Now let's square it, so we get

$$z^2 = (4 + 3i)^2 = 4^2 + 2(4)(3i) + (3i)^2$$
$$= 16 + 24i + 9(i^2)$$
$$= 16 + 24i - 9 \qquad \text{since } i^2 = -1$$
$$= 7 + 24i.$$

This algebraic result provides no special understanding of how z^2 is related to z. Alternatively, think about the trigonometric form for z^2. Its modulus is

$$\|z^2\| = \sqrt{7^2 + 24^2} = \sqrt{49 + 576} = \sqrt{625} = 25,$$

which is the square of the modulus of the original complex number z. Next, the angle ϕ associated with z^2 is defined by

$$\tan \phi = \frac{24}{7}$$

and so

$$\phi = \text{Arctan}\,\frac{24}{7} = 73.74°,$$

which is exactly twice the angle $\theta = 36.87°$ associated with z. For this instance, z^2 is related to z by first squaring the modulus and then doubling the angle. Does this rule hold in general?

Let's look at two other simple cases. We know that $i = 0 + 1i$ is located at a distance of 1 from the origin with an associated angle of $\frac{\pi}{2}$ measured in the usual positive direction from the horizontal axis. Therefore the modulus of i is 1 and the associated angle is $\frac{\pi}{2}$. Notice that when we square i to get $i^2 = -1 = -1 + 0i$, we obtain a different complex number at the same distance 1 from the origin but rotated through an additional angle of $\frac{\pi}{2}$ in the positive direction, thus ending up at a final angle of π. Therefore we have the same pattern; the modulus of i^2 is the square of the modulus of i and the associated angle is double the angle associated with i.

Similarly, $i^3 = -i = 0 - 1i$ is also located at a distance of 1 from the origin and is rotated through an additional $\frac{\pi}{2}$ to end up at a final angle of $\frac{3\pi}{2}$. The modulus of i^3 is the cube of the modulus of i and the associated angle, $\frac{3\pi}{2}$, is three times the angle $\frac{\pi}{2}$ associated with i.

Based on the last few observations about the geometric significance of powers of z for these special cases, we might wonder what happens with the powers z^n of any complex number z. For any z, how do $z^2, z^3, \ldots$, relate graphically to z? Let's start with any complex number $z = a + bi$ in the equivalent trigonometric form $z = \|z\|\,(\cos\theta + i \cdot \sin\theta)$ and square it to find

$$z^2 = \|z\|^2\,(\cos\theta + i \cdot \sin\theta)^2$$
$$= \|z\|^2\,(\cos^2\theta + 2i \cdot \cos\theta \cdot \sin\theta + i^2 \cdot \sin^2\theta).$$

If we use the fact that $i^2 = -1$ and collect the real and imaginary terms, we obtain

$$z^2 = \|z\|^2\,[(\cos^2\theta - \sin^2\theta) + 2i \cdot \cos\theta \cdot \sin\theta].$$

Now recall the double-angle identities in Equations (5) and (4) from Section 7.3:

$$\cos 2\theta = \cos^2\theta - \sin^2\theta$$
$$\sin 2\theta = 2\sin\theta \cdot \cos\theta.$$

If you examine the terms in the parentheses of the expression for z^2 above, you will see that they are equal to $\cos 2\theta$ while the remaining term is $i \cdot \sin 2\theta$. Thus we have

$$z^2 = \|z\|^2\,(\cos 2\theta + i \cdot \sin 2\theta).$$

Geometrically, squaring any complex number always produces a new complex number whose modulus is the square of the original modulus and whose angle is two times the original angle. If the modulus of the original number is greater than 1, then z^2 is a "larger" complex number (see Figure 7.52a). If $\|z\|$ is smaller than 1, then z^2 is a "smaller" complex number (see Figure 7.52b).

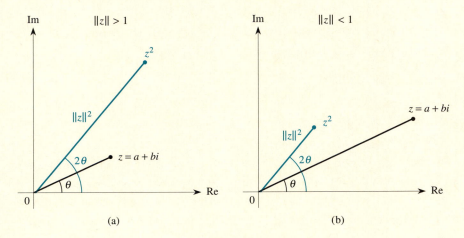

(a)   (b)

FIGURE 7.52

The key idea here is that the act of squaring a complex number always doubles the original angle θ to produce an angle of 2θ, so z^2 is rotated through twice the angle that z is.

Let's investigate what happens if we take the third power of z. Since

$$z^2 = \|z\|^2 (\cos 2\theta + i \cdot \sin 2\theta),$$

we have

$z^3 = z^2 \cdot z$

$\quad = [\|z\|^2 (\cos 2\theta + i \cdot \sin 2\theta)] \cdot [\|z\| (\cos \theta + i \cdot \sin \theta)]$

$\quad = \|z\|^3 [\cos 2\theta \cos \theta + i \cdot \cos 2\theta \sin \theta + i \cdot \sin 2\theta \cos \theta + i^2 \sin 2\theta \sin \theta].$

When we use the fact that $i^2 = -1$ and collect like terms, we get

$z^3 = \|z\|^3 [(\cos 2\theta \cos \theta - \sin 2\theta \sin \theta) + i(\cos 2\theta \sin \theta + \sin 2\theta \cos \theta)].$

We now use the sum identities for $\cos (x + y)$ and $\sin (x + y)$ (see Equations (7) and (6) from Section 7.3) to get

$$z^3 = \|z\|^3 [\cos (2\theta + \theta) + i \cdot \sin (2\theta + \theta)]$$

$$= \|z\|^3 [\cos 3\theta + i \cdot \sin 3\theta].$$

Thus we see that cubing a complex number always results in tripling the rotation of the original complex number. It either "lengthens" the complex number if the original modulus is greater than 1 or "contracts" it if the modulus is less than 1 (see Figure 7.53). If the modulus is equal to 1 and $\theta \neq 0$, all that happens is a rotation.

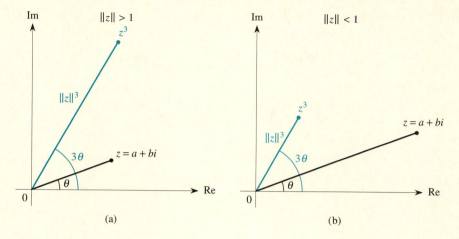

FIGURE 7.53

In the exercises at the end of this section, we ask you to show that

$$z^4 = z^3 \cdot z = \|z\|^4 \cdot [\cos 4\theta + i \cdot \sin 4\theta]$$

and that, in general for any positive integer power n,

$$z^n = z^{n-1} \cdot z$$
$$= \|z\|^n \cdot (\cos \theta + i \cdot \sin \theta)^n$$
$$= \|z\|^n \cdot [\cos n\theta + i \cdot \sin n\theta].$$

This important and useful result is known as *DeMoivre's theorem.*

DeMoivre's Theorem

If

$$z = a + bi = \|z\| (\cos \theta + i \sin \theta)$$

then

$$z^n = \|z\|^n [\cos n\theta + i \sin n\theta]$$

for any positive integer n.

Complex Conjugates

We know that complex numbers arise in complex conjugate pairs, such as $3 + 5i$ and $3 - 5i$, when we use the quadratic formula. If $z = a + bi$ is any complex number, we write its conjugate as $\bar{z} = a - bi$. Geometrically, this is shown in Figure 7.54 on the next page. It is clear that z and $\bar{z}$ have the same modulus, so $\|\bar{z}\| = \|z\|$. Also, the angle associated with $\bar{z}$ is $-\theta$ compared to the angle θ associated with z. Since

$$\cos (-\theta) = \cos \theta$$

and

$$\sin (-\theta) = -\sin \theta,$$

we find that

$$\bar{z} = a - bi = \|z\| (\cos \theta - i \cdot \sin \theta)$$

and so, in a totally similar manner to what we did previously,

$$(\bar{z})^n = \|z\|^n \cdot (\cos \theta - i \cdot \sin \theta)^n$$
$$= \|z\|^n \cdot [\cos n\theta - i \cdot \sin n\theta].$$

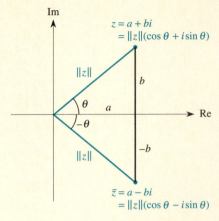

FIGURE 7.54

EXERCISES

Find the modulus and the associated angle for each of the following complex numbers.

1. $z = 4 - 3i$ **2.** $z = 5 + 12i$ **3.** $z = 12 - 5i$

4. $z = 15 + 20i$ **5.** $z = 64 - 36i$ **6.** $z = 8 - 3i$

7. $z = -5 + 7i$ **8.** $z = 3 + \sqrt{8}i$ **9.** $z = -8 - \sqrt{3}i$

10–18. Find the trigonometric form for each of the complex numbers in Exercises 1–9.

19–22. For each of the complex numbers in Exercises 1–4, find z^2 algebraically.

23–31. For each of the complex numbers in Exercises 1–9, find z^2 using DeMoivre's theorem.

32–35. For each of the complex numbers in Exercises 1–4, find z^3 algebraically.

36–44. For each of the complex numbers in Exercises 1–9, find z^3 using DeMoivre's theorem.

45. For $z = 1 + 2i$, calculate and plot z^n, for $n = 0, 1, 2, 3$, and 4.

46. Repeat Exercise 45 for $z = 0.6 + 0.8i$. What difference do you observe about the behavior of the two sets of points?

47. Show that $z^4 = z^3 \cdot z = \|z\|^4 [\cos 4\theta + i \cdot \sin 4\theta]$ for any complex number z.

48. Prove DeMoivre's theorem for any integer power n:

$$z^n = \|z\|^n [\cos n\theta + i \cdot \sin n\theta]$$

Hint: Write $z^n = z^{n-1} \cdot z$ and assume that

$$z^{n-1} = \|z\|^{n-1} [\cos (n - 1)\theta + i \cdot \sin (n - 1)\theta].$$

49. Suppose you have two complex numbers $z = a + bi$ and $w = c + di$.

   a. What is the product of z and w algebraically?
   b. What is the product of z and w using the trigonometric form for z and w?
   c. Hypothesize and prove an extension of DeMoivre's theorem that will allow you to multiply any two complex numbers in trigonometric form. (*Hint:* Your extension should reduce to DeMoivre's theorem for z^2 when $w = z$.)
   d. Apply the rule you discovered in part (b) to the product of

$$z = \frac{1}{2} + \frac{\sqrt{3}}{2}i \quad \text{and} \quad w = \frac{\sqrt{3}}{2} - \frac{1}{2}i.$$

50. a. Hypothesize an extension of DeMoivre's theorem that will allow you to divide one complex number by another in trigonometric form.
   b. Apply the rule you proposed in part (a) to the quotient of

$$z = \frac{1}{2} + \frac{\sqrt{3}}{2}i \quad \text{and} \quad w = \frac{\sqrt{3}}{2} - \frac{1}{2}i.$$

51. a. Hypothesize an extension of DeMoivre's theorem that will allow you to determine the square root of a complex number z.
   b. Apply the rule you proposed in part (a) to find the square root of

$$z = \frac{1}{2} + \frac{\sqrt{3}}{2}i.$$

   c. Algebraically square the complex number you obtained in part (b) to verify that it is actually the square root of the original number in part (a).
   d. Can you hypothesize a further extension of DeMoivre's theorem to extract any desired root of a complex number? Any desired rational power of a complex number?

52. A negative real number can be thought of as being produced by rotating the corresponding positive real number (which is located on the horizontal axis) through an angle π in the complex plane. Use this interpretation to explain why the product of two negative numbers is positive.

7.9 *The Road to Chaos*

In this section we will investigate some fascinating results that arise from iteration processes applied to complex numbers. Suppose that we start with any complex number $z_0 = \|z_0\|(\cos \theta + i \sin \theta)$ and square it to produce $z_1 = z_0^2$. Geometrically, we know that the result is a complex number whose associated angle is 2θ and whose modulus is $\|z_0\|^2$. Thus if $\|z_0\| > 1$, we get a rotation and an elongation; if $\|z_0\| < 1$, we get a rotation and a contraction; if $\|z_0\| = 1$, all we get is a rotation. Suppose we now square z_1 to produce $z_2 = z_1^2$. If $\|z_0\| > 1$, we get a further rotation (to the angle $2 \times 2\theta = 4\theta$) and a further elongation. If $\|z_0\| < 1$, then we get the same further rotation (to 4θ) and a further contraction. If $\|z_0\| = 1$, then all we get is the rotation (to 4θ).

What happens if we continue this process indefinitely to produce a sequence of complex numbers $z_0, z_1 = z_0^2, z_2 = z_1^2, z_3 = z_2^2, \ldots$. The behavior of the terms of this sequence can be predicted easily on geometric grounds by extending the reasoning we just used. If the modulus of the initial value z_0 is greater than 1, then each successive term is further and further out at a larger and larger angle and the sequence clearly diverges in a counterclockwise spiral pattern, as seen in Figure 7.55(a). If $\|z_0\| < 1$, then each successive term is closer and closer to the origin of the complex plane; the successive iterates converge to 0 in a counterclockwise spiral pattern as each one is a further rotation of the original angle θ, as shown in Figure 7.55(b). Finally, if $\|z_0\| = 1$, then all successive iterates fall on the boundary of the unit circle centered at the origin in the complex plane.

Let's focus on the possible initial values for z_0: any initial point inside the unit circle starts a sequence that spirals in to 0; any initial point on the circle itself starts a sequence that remains on that circle; any initial point outside that circle starts a sequence that spirals away toward infinity.

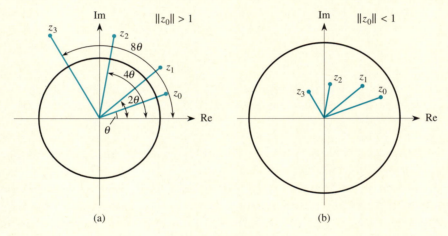

FIGURE 7.55

The Julia Set

We can display this graphically in the following way. Picture the unit circle centered at the origin in the complex plane, as shown in Figure 7.56. The circle is drawn in heavy black, the interior is shaded, and the region outside the circle is unshaded. Think of the un-shaded region as indicating any point that begins a sequence that diverges; the shaded region indicates those initial points for which the sequence converges to 0; the dark black indicates those initial points for which the sequence remains *on* the circle forever. The set of initial points for which the resulting sequences do not diverge to infinity is known as the **Julia set associated with a function;** in this case the function is

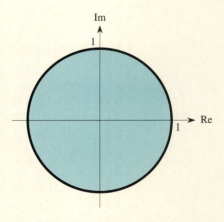

FIGURE 7.56

$f(z) = z^2$. (It is named after the French mathematician Gaston Julia who discovered the properties of these sets in the 1920s.) The Julia set associated with $f(z) = z^2$ consists of the unit circle and all points inside it.

As you will see, all it takes is a relatively small change in what we have just done to put us on the road to chaos. Rather than using the function $f(z) = z^2$, let's consider instead $f(z) = z^2 + C$, where C is any constant, either real or complex. (You may want to think of this as a family of functions for different values of C.) Thus, instead of defining $z_1 = z_0{}^2$, we take $z_1 = f(z_0) = z_0{}^2 + C$, so that $z_2 = f(z_1) = z_1{}^2 + C, \ldots$. To see what happens, let's look at a variety of cases with different values for C and with different starting values z_0.

Suppose we begin with $C = 2$. If the initial value is $z_0 = 3$, then we find

$$z_1 = z_0{}^2 + 2 = 9 + 2 = 11,$$

$$z_2 = z_1{}^2 + 2 = 121 + 2 = 123, \ldots,$$

and the sequence clearly diverges. Alternatively,

if $z_0 = 0.2$ then $z_1 = 2.04,$ $z_2 = 6.1616,$ $z_3 = 39.9653, \ldots$

if $z_0 = 1 + i$ then $z_1 = 2 + 2i,$ $z_2 = 2 + 8i,$ $z_3 = -58 + 32i, \ldots$

if $z_0 = 0.5 + 0.2i$ then $z_1 = 2.21 + 0.2i,$ $z_2 = 6.844 + 0.884i,$ $z_3 = 48.059 + 12.10i$

and seemingly all sequences diverge to infinity. Of course, we cannot make such a conclusion based on just a few examples such as this; they can provide only an indication of what may happen.

Let's see why this is indeed the case, using DeMoivre's theorem. Suppose we first take any initial value z_0 inside the unit circle. Its modulus is therefore less than 1. When we square it, the modulus for $z_0{}^2$ is smaller still. However, when we add 2 to it, the point is shifted 2 units to the right, and so z_1 must be outside and to the right of the unit circle.

Now suppose z_0 (or some subsequent iterate) is outside the unit circle. Since its modulus is greater than 1, the modulus for $z_0{}^2$ is larger still. Finally, when we add 2 to it, the point is again shifted 2 units to the right. For almost all possible values of z_0, the resulting z_1 will be outside the unit circle. There are some exceptions—say $z_0 = 1.1i$, so that $z_1 = (1.1i)^2 + 2 = -1.21 + 2 = 0.79$. However, it can be shown that, eventually, all subsequent iterations will land outside the circle and ultimately diverge to infinity. (Since each iteration involves a rotation, at some stage, one of the successive iterates will eventually land near the horizontal axis to the right and the following iterate will be outside and to the right of the unit circle.) Thus it turns out that, when $f(z) = z^2 + 2$, there are no initial points in the complex plane for which the resulting sequences do not diverge. The Julia set associated with the function $f(z) = z^2 + C$, when $C = 2$, will be completely empty since all initial points give rise to sequences which eventually diverge. Our picture of this Julia set will be entirely unshaded because there are no initial points that start convergent sequences.

Similarly, if $C = 2i$, then all sequences will diverge regardless of the initial value for z_0. The additive constant $2i$ results in a shift upward of 2 units in the imaginary direction. Pick several initial values for z_0 (real, imaginary, or complex) and see what happens when you calculate the successive iterates.

However, if $C = 0.2i$ and we start with $z_0 = 0.5 + 0.2i$, then we obtain

$$z_1 = (0.5 + 0.2i)^2 + 0.2i = 0.21 + 0.4i \qquad z_6 = -0.0394 + 0.1881i$$
$$z_2 = -0.1159 + 0.368i \qquad\qquad\qquad z_7 = -0.0338 + 0.1852i$$
$$z_3 = -0.1220 + 0.1147i \qquad\qquad\quad z_8 = -0.0332 + 0.1875i$$
$$z_4 = 0.0017 + 0.1720i \qquad\qquad\quad z_9 = -0.0341 + 0.1876i,$$
$$z_5 = -0.0296 + 0.2006i$$

and the sequence is apparently converging to some point in the complex plane.

Unfortunately, it is not practical to repeat the above process for every possible starting value z_0. Instead, we use a computer to perform such calculations for a large number of points in a grid to give a representative picture of what happens. As with the previous cases, we leave any initial point that starts a sequence that diverges to infinity unmarked to become part of our unshaded region. We put a small dot at any initial point that starts a sequence that converges to some point in the complex plane so that it will be part of the shaded Julia set.

Our resulting picture of the Julia set for the function $f(z) = z^2 + 0.2i$ is shown in Figure 7.57.

FIGURE 7.57

Realize that the picture does not indicate what the limits of any of the sequences are; it shows only those points that start sequences that have limits. Typically, if you look at points in the interior of the shaded region, it turns out that nearby starting points tend to converge to limits that are relatively close to one another. However, if you take points near the boundary, very different things can occur. Initial points that are extremely close together can produce sequences that converge to radically different limits. The result is an instance of mathematical chaos since there are no predictable patterns to the behavior. Points that are very close together, provided they are both near the boundary of the Julia set, may well lead to sequences that behave very differently. If you were to zoom in on the portion of the Julia set near these boundaries, you would see an ever more intricate design illustrating how nearby points can start sequences that either diverge or converge. This happens in a totally chaotic and unpredictable manner.

A far more striking illustration of this is shown in Figure 7.58, which is the Julia set corresponding to $C = -0.2 + 0.7i$. Notice how intricate the boundary appears. In Figure 7.59, we show the result of zooming in on the upper left corner of the Julia set in Figure 7.58. Observe how roughly similar patterns repeat; this is typical of what happens when you zoom in repeatedly on Julia sets associated with most values for C. Also, notice how much more jagged the boundary looks as more details appear in the blown-up image in Figure 7.59; this also is typical.

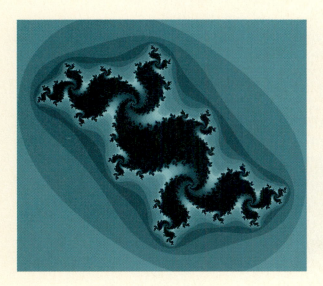

FIGURE 7.58

FIGURE 7.59

It is possible for the Julia set associated with a complex constant C to be far more complicated than we have shown so far; it can, for example, consist of a large variety of disconnected pieces. It may even consist of nothing but a collection of isolated points like a set of dust particles. You may want to experiment with some of these ideas yourself using any of the many computer programs available for displaying Julia sets for iterated functions.

The Mandelbrot Set

There is a completely different way of looking at these ideas. In the previous discussion, we considered the function $f(z) = z^2 + C$, selected a particular value for C, and then examined points in the complex plane as starting points z_0 for iterated sequences. Now let's reverse this.

Suppose instead we select a particular starting point z_0 and examine the effects of using different values for the complex constant C in $f(z) = z^2 + C$. Thus our view of the complex plane has shifted—it now represents all different constants rather than all different starting points. In particular, suppose we select $z_0 = 0$ as the starting point for all sequences. Then, for any constant C, $z_1 = 0^2 + C = C$, $z_2 = z_1^2 + C = C^2 + C$, and so forth. Clearly, if C is large (far from the origin in the complex plane), then all successive iterates will be larger still and the successive points of the sequence will diverge. However, if C is fairly small, then the successive iterates may remain close to the origin and the sequence may converge to some finite complex value.

The Julia set associated with the function $f(z) = z^2 + C$ consists of all initial points z_0 for which the sequences converge for a given constant C. Similarly, the set of all constants C for which the sequences starting from $z_0 = 0$ fail to diverge is called the **Mandelbrot set** associated with the function $f(z) = z^2 + C$ (named after Benoit Mandelbrot). For this initial point $z_0 = 0$, the Mandelbrot set (see Figure 7.60) shows those constants C for which the corresponding sequences remain close to the origin. As with a Julia set, the boundary of the Mandelbrot set is an incredibly intricate structure. If you zoom in on it, as shown in Figure 7.61, you will see remarkable shapes with no predictable patterns; however, the original overall shape in Figure 7.60 appears to repeat at all levels of magnification. The main heart-shaped portion of the Mandelbrot set is called a cardioid and will be studied in a later chapter; the portion to the left of the cardioid is actually a circle.

FIGURE 7.60

FIGURE 7.61

In our displays we show the Mandelbrot set with different shadings to indicate how quickly different sequences diverge from the starting value $z_0 = 0$. When different colors are used for this, the results are even more dramatic. You may want to examine this using one of the programs available for displaying the Mandelbrot set. All such programs allow you to see the details at different levels of magnification as you zoom in on the boundary. In theory, there is no limit to the degree of complexity of the boundary. Such a shape is known as a *fractal*.

You may also want to look for some of the many shareware programs that are available for more detailed investigation of both Julia and Mandelbrot sets. This subject is one of the most exciting areas of current mathematical research and many new and important theorems have been proven in the last few years.

EXERCISES

1. Consider iterations $x = f(x)$ based on the quadratic function $f(x) = x^2 + C$ starting with $x_0 = 0$ for $C = -0.1, -0.2, -0.3, -0.9, -1.25$, and -1.38. For each, perform enough iterations to determine if there is a single limiting value, a period-2 cycle with two limiting values that the sequence oscillates between, a period-4 cycle or a period-8 cycle. (See Section 6.9.) For the cases of a single limiting value, describe the relative rates of convergence of the sequences.

2. a. Use the quadratic formula to find a condition on those values of C for which the sequence of iterates $x = f(x) = x^2 + C$ has a real limiting value.

    b. Verify your condition in part (a) by using $C = 0.1$ starting with $x = 0.5$ and performing enough iterations to see the eventual behavior.

    c. Repeat part (b) using $C = 0.4$.

3. a. What is the limiting value you expect if $C = \frac{5}{4}$ for the sequence of function iterations based on $x = f(x) = x^2 + C$?

    b. Start the iteration process at $x_0 = 0.5 + 0.5i$ and perform enough iterations to verify that the process seems to be converging to your answer for part (a).

    c. Start the iteration process at $x_0 = 1 + i$ and perform enough iterations to determine the eventual behavior of the sequence of iterates. How could you have anticipated the result without performing the actual calculations?

4. You can think of the iteration scheme for $x = f(x) = x^2 + C$ as the difference equation $x_{n+1} = f(x_n) = x_n^2 + C$. What are the equilibrium levels for the solutions to the difference equation? Under what conditions on C will the equilibrium values be real?

5. Explain graphically the significance of C in determining whether the iteration process based on $x = f(x) = x^2 + C$ has a real limiting value by looking at the graphs of $y = x^2 + C$ and $y = x$.

6. Explain graphically why the iteration process based on the function $x = f(x) = x^3 + C$ must have at least one real limiting value.

7. Consider iterations $x = f(x)$ based on the function $f(x) = x + \sin x$.

    a. Start the iteration process at $x_0 = 2$ and perform enough iterations to allow you to recognize the limit of the resulting sequence.
    b. Repeat part (a) starting with $x_0 = 5$.
    c. Repeat part (a) starting with $x_0 = 8$. How does the limiting value compare to π?
    d. Repeat part (c) starting with $x_0 = 15$.
    e. Based on the function f, explain why all limits will be some multiple of π.

8. Consider iterations $x = f(x)$ based on the function $f(x) = x + \cos x$. Predict the possible values that can arise for the limits based on the function f. Verify whether your predictions are correct if you start with initial values $x_0 = 1, 3, 7$, and -12.

CHAPTER SUMMARY
————

In this chapter, you have learned the following:

- The behavior of the sine and cosine functions.
- What vertical shift means for the sine and cosine functions.
- What amplitude means for the sine and cosine functions.
- What frequency means for the sine and cosine functions.
- What period means for the sine and cosine functions.
- What phase shift means for the sine and cosine functions.
- How to use the sine and cosine functions to model periodic behavior.
- How to fit sine and cosine functions to data.
- The fundamental identities that relate the sine and cosine functions.
- The behavior of the inverse sine and inverse cosine functions.
- How to solve trigonometric equations using the inverse sine and inverse cosine functions.
- The behavior of the tangent function.
- The behavior of the inverse tangent function.
- How to solve trigonometric equations using the inverse tangent function.
- How to approximate periodic functions with combinations of sine and cosine functions.
- How to approximate the sine and cosine functions with polynomial functions.
- How the accuracy of a polynomial approximation depends on the degree of the polynomial.
- How to convert a complex number to its equivalent in trigonometric form.
- How to construct powers of complex numbers with DeMoivre's theorem.
- About the Julia set that is associated with a function $f(z)$.
- About the Mandelbrot set that is associated with a function $f(z)$.

REVIEW EXERCISES
————

1. The student with whom you are working finishes a problem and announces her answer is $\cos(5.70)$. You get an answer of the form $\sin(7.2708)$. Why are these answers the same?

2. Suppose θ is $60°$.

    a. Find two positive angles and two negative angles that have the same sine as θ.
    b. Write the angles from part (a) in radian form.

3. Let $\theta = 45°$.

    a. Find two positive angles and two negative angles with the same cosine as θ.
    b. Write the radian form of the angles from part (a).

Use the fact that if an arc of length s on a circle of radius r subtends an angle of θ radians, then s = rθ to solve Exercises 4–6.

4. We measure the distance between two points, P and Q, on the Earth by measuring the distance along the arc of the circle through P and Q and centered at the center of the Earth, O. The radius of the Earth is about 4000 miles. Find the distance from P to Q if the angle POQ has the following measurements:

    a. $\dfrac{\pi}{4}$ **b.** $\dfrac{\pi}{3}$ **c.** $\dfrac{5\pi}{6}$ **d.** $15°$

5. A wheel of radius 2 feet rotates at a constant rate of 180 revolutions per minute (rpm).

    a. How many radians per minute are swept out by the wheel?
    b. How far does a point on the rim of the wheel travel in one minute?

6. Find the diameter of the tires on your car. Assuming that the car is traveling at the rate of 60 miles per hour, determine the number of revolutions the tire makes every minute.

7. **a.** For each of the following sinusoidal functions, identify the vertical shift, the amplitude, the frequency, the period, and if applicable, the phase shift.

    (i) $y = 325 + 10 \sin\left(\frac{2\pi}{9}t\right)$

    (ii) $y = 63 + 3 \sin\left(\frac{2\pi}{25}t\right)$

    (iii) $y = 71 + 2 \cos\left(\frac{2\pi}{15}t\right)$

    (iv) $y = 80 + 13 \cos\left(\frac{2\pi}{24}(t - 15)\right)$

    (v) $y = 38 + 8 \sin\left(\frac{2\pi}{24}(t - 5)\right)$

    (vi) $y = 100 + 25 \sin\left(\frac{2\pi}{72}t\right)$

    (vii) $y = 100 + 25 \sin\left(\frac{2\pi}{97}t\right)$

    (viii) $y = 145 + 40 \sin\left(\frac{2\pi}{83}t\right)$

    b. Each of these functions can be a model for a common periodic phenomenon. Describe a phenomenon that each function could represent. What do the variables represent? What are the units? What are possible values for the domain and range?

8. On the same set of axes, graph the functions:

$$S(x) = 2 \sin x \qquad R(x) = 2 \sin 3x \qquad T(x) = 2 \sin 0.5x$$

Clearly mark the zeros of each function.

9. For each function give the frequency, period, amplitude, and phase shift.

    a. $y = 5 + 2 \cos \frac{3}{4}x$ **b.** $y = 5 - 2 \cos\left(\frac{3}{4}x + \pi\right)$

    c. $y = 5 - 2 \cos\left(\pi x + \frac{3}{4}\right)$ **d.** $y = 5 - 2 \cos \frac{3}{4}(\pi x - 1)$

10. Determine the values of x for which each of the functions in Exercise 9 is equal to 6.

11. Graph the functions:

    a. $y = \text{Arcsin } x + 3$ **b.** $y = \text{Arcsin } (x + 3)$
    c. $y = \text{Arcsin } (\sin (x + 3))$

12. Solve for θ.

    a. $-4 \sin \theta = 6 \cos \theta$ **b.** $2 \cos 2\theta = \sin 2\theta$
    c. $3 \tan 3\theta - 21 = 0$

13. Solve for x:

    a. $\text{Arctan } x = 1.35$ **b.** $\text{Arcsin } (\cos x) = 0.5$

14. Use the Taylor polynomial approximation of degree $n = 3$ to estimate the value of the function $f(x) = \sin 3x$ at $x = 0.2$. Sketch the graph of the function and the approximation on the same set of axes.

15. Repeat Exercise 14 using $n = 5$. Discuss any differences you observe.

16. Bernice is swinging on a playground swing whose supporting cross-bar is 11 feet above the ground and the length of the chain to her seat is 8 feet. At the end of each swing, she makes an angle of 60° with the vertical and it takes her 3 seconds to complete each full cycle.

    a. Write a sinusoidal function that can be used to model the height of the seat above ground as a function of time t.

    b. Write a sinusoidal function that can be used to model the horizontal displacement from directly under the cross-bar as a function of time t.

17. A complete lunar cycle—new moon to full moon and back to new moon—takes about 28 days. Write a sinusoidal function that can be used to model the percentage of the width of the moon showing as a function of time.

18. A bungee jumper dives off a bridge across a deep gorge. The bungee cord initially stretches to a maximum length of 200 feet before the jumper begins his first rebound. Over the course of the next 60 seconds, he bounces up and down with ever diminishing oscillations, each lasting about 6 seconds, until he comes to rest at a distance of about 160 feet below the bridge. Write the equation of a decaying oscillatory function that models the height of the bungee jumper as a function of time as measured from the instant the cord is extended to its maximum stretch.

8

GEOMETRIC MODELS

8.1 *Introduction to Coordinate Systems*

What is a coordinate system? In simple terms, a coordinate system gives us a way to *locate* and *identify* points in the plane. In the usual *rectangular* or *Cartesian coordinate system,* every point can be pictured in either of two ways. First, a point $P(x_0, y_0)$ can be thought of as lying at the corner of a unique rectangle whose opposite corner is at the origin and whose sides lie along the two coordinate axes (see Figure 8.1a). The base of this rectangle is x_0 and the height is y_0. Second, the point $P(x_0, y_0)$ can be thought of as the intersection of two perpendicular lines, one parallel to the y-axis at a distance of x_0 and the other parallel to the x-axis at a height of y_0 (see Figure 8.1b).

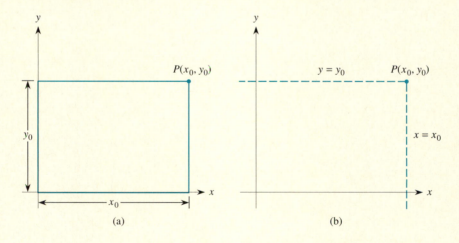

FIGURE 8.1

Mathematicians have found that there are many situations where rectangular coordinates are not the most natural or the most effective way to locate points and have developed alternative coordinate systems. One such approach involves the use of two axes that are not perpendicular, but rather meet at the origin at some angle other than a right angle; points in such a slanted coordinate system can be located at the opposing vertex of a parallelogram (see Figure 8.2).

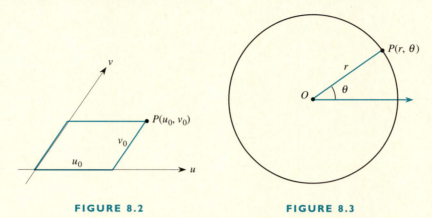

FIGURE 8.2   **FIGURE 8.3**

Another approach is to locate a point using a circle centered at the origin O instead of a rectangle. To do this, we must specify both the radius of the circle and an angle θ to indicate where on the circle the point is located. This is known as the **polar coordinate system** and is illustrated in Figure 8.3.

There are other approaches in use for particular applications that involve locating points lying on some ellipse centered at the origin (an elliptic coordinate system), on some parabola (a parabolic coordinate system), or on a hyperbola (a hyperbolic coordinate system). See Figure 8.4. In fact, the LORAN (LOng RAnge Navigation) system used by ships and planes to locate their positions is based on the fact that every point in a plane can be interpreted as lying at the intersection of two hyperbolas in a hyperbolic coordinate system.

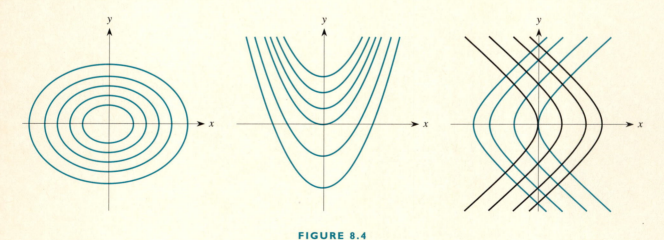

FIGURE 8.4

In this chapter, we will first develop several special characteristics of the rectangular coordinate system and then see how to represent some extremely important curves. Later we will explore other ways to represent functions, including the use of the polar coordinate system.

8.2 *Analytic Geometry*

One of the most useful and far-reaching developments in mathematics is Rene Descartes' idea of representing algebraic concepts in a geometric setting. This approach, known as *analytic geometry*, allows us to visualize the mathematics graphically to complement our approach to it algebraically through symbols. Everything we have done involving graphs of functions is an outgrowth of Descartes' ideas. In this section we examine some additional ideas involving points, lines, and circles in the plane.

We begin by considering the two points $A(x_0, y_0)$ and $B(x_1, y_1)$ in the plane. We know how to find an equation of the line through them using the point-slope form

$$y - y_0 = m(x - x_0)$$

or the slope-intercept form

$$y = mx + b,$$

where the slope of the line is

$$m = \frac{y_1 - y_0}{x_1 - x_0}.$$

Alternatively, we have the implicit form for the equation of a line,

$$ax + by = c,$$

where $\frac{c}{a}$ and $\frac{c}{b}$ represent the x- and the y-intercepts of the line, respectively.

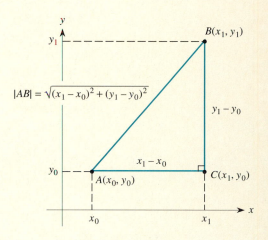

Distance Between Points

We now ask: What is the distance between the points $A(x_0, y_0)$ and $B(x_1, y_1)$? We write this distance as $|AB|$. As seen in Figure 8.5, the points determine a right triangle ABC, where the coordinates of the point C are (x_1, y_0) since C is at the same horizontal distance as B (measured from the y-axis) and at the same vertical height as A (measured from the x-axis). Moreover, the horizontal distance from A to C is just $x_1 - x_0$, the change or difference in the x-coordinates. Similarly, the vertical distance from C to B is $y_1 - y_0$, the change in the y-coordinates. Consequently, the distance from A to B is the length of the hypotenuse of this right triangle. The Pythagorean theorem therefore gives us the *distance formula*:

Distance formula

$$|AB| = \sqrt{(x_1 - x_0)^2 + (y_1 - y_0)^2}$$

E X A M P L E   I

Find the distance from $A(2, 5)$ to $B(6, 8)$.

Solution   Applying the distance formula, we find that

$$|AB| = \sqrt{(6 - 2)^2 + (8 - 5)^2}$$
$$= \sqrt{16 + 9}$$
$$= \sqrt{25} = 5 \text{ units.}$$

Consider again the two points $A(x_0, y_0)$ and $B(x_1, y_1)$ in the plane. Suppose we now want to determine the midpoint M of the line segment connecting A to B. Let's see how this can be done.

From Figure 8.6, we see that the points A and B determine a right triangle ABC and that the points A and M determine a smaller right triangle, AMD. These two right triangles are similar and hence their corresponding sides are proportional (see Appendix C). Since M is halfway from A to B, D is halfway from A to C and the height DM is half the height CB. Therefore the x-coordinate at D (and hence also at M) is

$$x = x_0 + \frac{1}{2}(x_1 - x_0).$$

Similarly, since the height DM is half the height CB, we find the y-coordinate at M is

$$y = y_0 + \frac{1}{2}(y_1 - y_0).$$

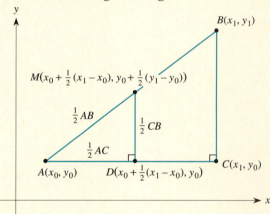

FIGURE 8.6

Midpoint M of the line segment from $A(x_0, y_0)$ to $B(x_1, y_1)$

$$x = x_0 + \frac{1}{2}(x_1 - x_0), \qquad y = y_0 + \frac{1}{2}(y_1 - y_0)$$

Notice that we can rewrite these expressions as

$$x_0 + \frac{1}{2}(x_1 - x_0) = x_0 + \frac{1}{2}x_1 - \frac{1}{2}x_0 = \frac{1}{2}(x_1 + x_0)$$

$$y_0 + \frac{1}{2}(y_1 - y_0) = y_0 + \frac{1}{2}y_1 - \frac{1}{2}y_0 = \frac{1}{2}(y_1 + y_0).$$

Thus, the coordinates of the midpoint M of a line segment are simply the averages of the x-coordinates and the y-coordinates of the endpoints, respectively.

E X A M P L E   2

Find the midpoint of the line segment joining $A(1, 11)$ and $B(3, 7)$.

Solution   The coordinates of the midpoint are

$$x = x_0 + \frac{1}{2}(x_1 - x_0)$$

$$= \frac{1}{2}(x_1 + x_0)$$

$$= \frac{1}{2}(3 + 1) = 2$$

and

$$y = y_0 + \frac{1}{2}(y_1 - y_0)$$

$$= \frac{1}{2}(y_1 + y_0)$$

$$= \frac{1}{2}(7 + 11) = 9.$$

See Figure 8.7.

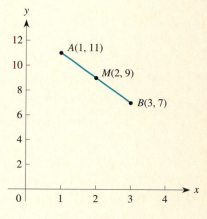

FIGURE 8.7

We might also want to determine a point at some other fraction of the distance from A to B. To do this, we simply extend the above argument to determine a point P at any distance from A to B. Suppose we want the point $\frac{1}{4}$ of the way from A to B. We then have

$$x = x_0 + \frac{1}{4}(x_1 - x_0)$$

$$y = y_0 + \frac{1}{4}(y_1 - y_0).$$

Think About This

Verify that this quarter-distance formula is correct using an argument comparable to the one used for the midpoint formula.

In general, if we want the point P at a fraction t of the distance from A to B, it will be located at

$$x = x_0 + t \cdot (x_1 - x_0)$$
$$y = y_0 + t \cdot (y_1 - y_0),$$

as shown in Figure 8.8. Incidentally, if we take $t > 1$, then we get a point on the line *beyond* B and if $t < 0$, then we get a point on the line *before* A.

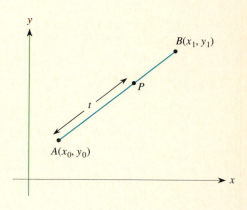

FIGURE 8.8

E X A M P L E   3

Find the point P located $\frac{3}{5}$ of the way from $A(-1, 3)$ to $B(4, 13)$ and the point Q located $\frac{7}{5}$ of the way from A to B.

Solution   For P with $t = \frac{3}{5}$, we find that

$$x = x_0 + t \cdot (x_1 - x_0) = -1 + \frac{3}{5}[4 - (-1)] = 2$$

$$y = y_0 + t \cdot (y_1 - y_0) = 3 + \frac{3}{5}(13 - 3) = 9.$$

Check by plotting $A(-1, 3)$, $B(4, 13)$, and $P(2, 9)$ to see that P is located $\frac{3}{5}$ of the way from A to B. Similarly, for Q with $t = \frac{7}{5}$, we obtain

$$x = x_0 + t \cdot (x_1 - x_0) = -1 + \frac{7}{5}[4 - (-1)] = 6$$

$$y = y_0 + t \cdot (y_1 - y_0) = 3 + \frac{7}{5}(13 - 3) = 17.$$

Plot $Q(6, 17)$ in your picture and notice that this point lies on the line through A and B and is beyond B.

We could certainly continue this example with different values of t to find other points on the line. In fact, every value of t determines a unique point on the line joining $A(x_0, y_0)$ and $B(x_1, y_1)$. Therefore the *two* equations above for x and y give us a different way of representing the line. They are known as a *parametric representation* or as *parametric equations* of the line and the quantity t is called a *parameter*.

> **Parametric equations of the line through $A(x_0, y_0)$ and $B(x_1, y_1)$**
>
> $$x = x_0 + (x_1 - x_0)t$$
> $$y = y_0 + (y_1 - y_0)t$$

Realize that this parametric form involves two interrelated equations for the line, not a single equation such as the point-slope form. However, it is possible to eliminate the parameter t and the result is to produce a single equation for the line. We ask you to do this as an exercise at the end of this section.

The Equation of a Circle

We apply the concept of distance between two points in the plane to define a circle. A **circle** is the set of all points in the plane at a fixed distance from a fixed point. The fixed distance is called the **radius** and the fixed point is called the **center**.

This definition allows us to find a general equation for any circle. Let r be the radius and let $C(x_0, y_0)$ be the center of a circle. A point $P(x, y)$ lies on this circle provided its distance from the center C is precisely r (see Figure 8.9). Using the distance formula, this is equivalent to

$$|CP| = \sqrt{(x - x_0)^2 + (y - y_0)^2} = r.$$

We can eliminate the square root in this equation by squaring both sides and thus obtain the *standard form* for the equation of a circle.

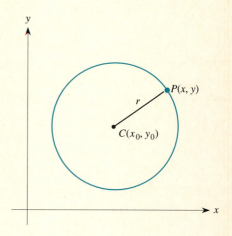

FIGURE 8.9

Equation of the circle with radius r centered at (x_0, y_0)

$$(x - x_0)^2 + (y - y_0)^2 = r^2$$

For instance, the circle of radius 7 centered at $C(6, 1)$ has equation

$$(x - 6)^2 + (y - 1)^2 = 7^2 = 49,$$

while the circle of radius 3 centered at $(-5, 0)$ is

$$(x + 5)^2 + y^2 = 9.$$

As a special case, the equation of a circle of radius r centered at the origin is

$$x^2 + y^2 = r^2.$$

Note that the equation of a circle does not represent a function. Picture any vertical line not tangent to the circle and passing through the circle—it intersects the circle twice, and so the circle fails the vertical line test. Thus for each such value of x, there correspond two values of y, and this violates the definition of a function (see Figure 8.10).

We get a similar result algebraically. For example, the circle of radius 10 centered at the origin has the equation

$$x^2 + y^2 = 100.$$

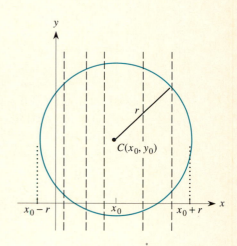

FIGURE 8.10

If we select $x = 6$, say, then

$$36 + y^2 = 100$$
$$y^2 = 64,$$

and this has solutions

$$y = 8 \quad \text{or} \quad y = -8.$$

Again we see that two values of y correspond to one value of x. Even though a circle does not represent a function, it is nonetheless a very important curve. It is, in fact, one of several important curves that do not represent functions. We will study several other such curves in the next section.

Let's now consider the equation of the circle of radius 8 centered at $(5, 2)$.

$$(x - 5)^2 + (y - 2)^2 = 8^2$$

We expand the left side and combine terms to get

$$x^2 + y^2 - 10x - 4y - 35 = 0,$$

which is an equivalent, although different, representation for the same circle. Clearly, we could do the same with the equation of any circle,

$$(x - x_0)^2 + (y - y_0)^2 = r^2.$$

Expanding the left side, we get

$$x^2 - 2x_0 x + x_0{}^2 + y^2 - 2y_0 y + y_0{}^2 = r^2$$

or equivalently,

$$x^2 + y^2 - 2x_0 x - 2y_0 y + x_0{}^2 + y_0{}^2 - r^2 = 0.$$

Since x_0, y_0, and r are constants, we can write this equation in the alternative form

$$x^2 + y^2 + Cx + Dy + E = 0,$$

where we have introduced the new constants

$$-2x_0 = C, \quad -2y_0 = D, \quad \text{and} \quad x_0{}^2 + y_0{}^2 - r^2 = E.$$

Such an equation, known as the *general equation of the circle*, represents a circle for any choice of constants C, D, and E such that the radius of the circle is positive.

Suppose we start with the equation

$$x^2 + y^2 - 10x - 4y - 35 = 0.$$

We want to see if it does indeed represent a circle and, if so, find the circle's center and radius. Our problem now becomes: How do we transform this equation to get

$$(x - 5)^2 + (y - 2)^2 = 8^2,$$

which is a representation of a circle that allows us to read off the center and the radius?

EXAMPLE 4

Find the radius and the center of the circle whose equation is

$$x^2 + y^2 + 8x - 10y - 8 = 0.$$

Solution To solve this problem, we use the technique of *completing the square* (see Appendix C) in both the x- and y-terms. We thus obtain

$$
\begin{aligned}
x^2 + y^2 + 8x - 10y - 8 &= (x^2 + 8x) + (y^2 - 10y) - 8 \\
&= [(x^2 + 8x + 16) - 16] + [(y^2 - 10y + 25) - 25] - 8 \\
&= [(x + 4)^2 - 16] + [(y - 5)^2 - 25] - 8 \\
&= (x + 4)^2 + (y - 5)^2 - 49 = 0.
\end{aligned}
$$

Then, moving the constant 49 to the right side, we get

$$(x + 4)^2 + (y - 5)^2 = 49 = 7^2$$

and recognize directly the equation of a circle with radius 7 and center $(-4, 5)$.

Incidentally, even though a circle does not fulfill the requirements of a function, if we restrict our attention to either its upper half or its lower half, then the resulting semicircle does represent a function. For the circle $x^2 + y^2 = r^2$, this is equivalent to solving for y:

$$
\begin{aligned}
y^2 &= r^2 - x^2 \\
y &= \pm\sqrt{r^2 - x^2}.
\end{aligned}
$$

The upper semicircle is the graph of the function $y = f(x) = \sqrt{r^2 - x^2}$, and the lower semicircle is the graph of the function $y = g(x) = -\sqrt{r^2 - x^2}$.

EXERCISES

Find the distance between the pairs of points in Exercises 1–6.

1. $P(2, 4)$ and $Q(5, 8)$
2. $P(2, 4)$ and $Q(7, 16)$
3. $P(-4, 1)$ and $Q(0, 4)$
4. $P(2, -5)$ and $Q(0, 4)$
5. $P(-1, 5)$ and $Q(3, 7)$
6. $P(3, 1)$ and $Q(-5, -4)$

7. Find the midpoint of the line segment joining P and Q in Exercise 1.
8. Find the midpoint of the line segment joining P and Q in Exercise 2.

9. Find the point $\frac{1}{3}$ of the way from P to Q in Exercise 3.

10. Find the point $\frac{3}{4}$ of the way from P to Q in Exercise 4.

11. Find the equation of the circle that has center $(5, 2)$ and passes through the point $P(8, -2)$.

12. Find the equation of the circle that has center $(-3, 7)$ and passes through the point $P(2, -5)$.

13. Find the equation of the circle that has $P(2, 4)$ and $Q(10, 4)$ as the endpoints of a diameter.

14. Find the equation of the circle that has $P(2, 4)$ and $Q(10, 14)$ as the endpoints of a diameter.

15. Repeat Exercises 13 and 14 using the facts that any angle inscribed in a semicircle is a right angle and that perpendicular lines have slopes that are negative reciprocals.

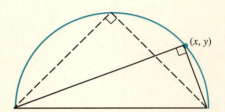

Complete the square in both the x- and y-terms for each of the following equations to obtain the standard form for the equation of a circle. Use this to determine the radius and center of the circle. Then draw the graph.

16. $x^2 + y^2 + 4x + 6y = 3$

17. $x^2 + y^2 + 4x + 6y = 12$

18. $x^2 + y^2 + 10x - 4y = 7$

19. $x^2 + y^2 + 10x - 4y = 71$

20. $x^2 + y^2 - 2x + 6y = -9$

21. $x^2 + y^2 - 2x + 6y + 6 = 0$

22. The equations $x = 3 + 2t, y = 4 - 5t$ form a parametric representation for a line.

   a. Construct a table of values for x and y corresponding to $t = -2, -1, 0, 1, \ldots, 5$.

   b. Plot these points and verify that they do seem to lie on a line.

   c. What is the slope of this line?

   d. What is a point-slope form for the equation of this line?

   e. Use the midpoint formula to find the midpoint of each of the consecutive line segments determined by the entries in your table from part (a). Then use the parametric representation of the line with $t = -1.5, -0.5, 0.5, 1.5, 2.5, 3.5,$ and 4.5. How do the results compare? Explain why.

23. Start with general parametric equations of a line,

$$x = x_0 + (x_1 - x_0)t \qquad y = y_0 + (y_1 - y_0)t,$$

and algebraically eliminate the parameter t. Identify the equation you have produced.

24. a. Find the slope of the line with parametric representation

$$x = 1 + 2t, \qquad y = 2 - 3t.$$

   b. Sketch the graph of this line.

25. Find the points where the line $x = 1 + 2t, \quad y = 2 - t$ intersects the circle $x^2 + y^2 = 25$. (*Hint*: First find values of the parameter t that satisfy the equation of the circle.)

26. The three points $P(0, 2)$, $Q(2, 4)$, and $R(4, 0)$ are noncollinear and as such determine a circle. Find an equation of this circle. (*Hint:* Substitute the coordinates of each of the points into the general equation of the circle, $x^2 + y^2 + Cx + Dy + E = 0$, and then solve the resulting system of three equations in three unknowns.)

8.3 *The Conic Sections*

When we first studied functions and their graphs, we saw that not every graph represents a function. In the last section, we saw that a circle is not a function. There are a number of other important curves that similarly are not functions, but which have valuable and interesting properties. We investigate some of these curves in this section.

The graph of any equation of the form

$$Ax^2 + By^2 + Cx + Dy + E = 0$$

is known as a *conic section*, provided that at least one of A and B is not zero. In particular, if $A = B$, we have the equation of a circle, which is therefore a conic section.

Geometrically, suppose we start with a double right circular cone (see Figure 8.11) and slice through it. If the slicing plane is horizontal, then each slice is a *circle*. However, if the slicing plane is inclined slightly from the horizontal, then the curve produced is oval in shape, rather than circular, and we have an *ellipse*. In fact, the sharper the angle of the slice, the more elongated the ellipse will be (see Figure 8.12). If the angle of slicing increases further so that it becomes parallel to the "side" of the cone, the resulting curve is a *parabola* (see Figure 8.13). If the angle of the slice increases still further, then the slicing plane intersects both the upper and lower parts of the cone and produces a pair of separated curves, known as a *hyperbola* (see Figure 8.14).

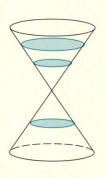

FIGURE 8.11
Slicing planes are horizontal

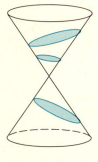

FIGURE 8.12
Slicing planes are at slight angles

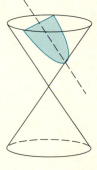

FIGURE 8.13
Slicing planes parallel to the side of the cone

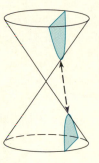

FIGURE 8.14
Slicing planes at steeper angles

In summary, there are three types of conic sections: (1) the ellipse, (2) the parabola, and (3) the hyperbola. The circle is a special case of the ellipse.

The conic sections arise in a wide variety of applications throughout mathematics and science. For instance, the orbits of the planets about the sun are ellipses. The paths of many comets and meteoroids are hyperbolic. A cross section of the metallic reflector inside a flashlight or an automobile headlight is a parabola. The path of a thrown object, such as a perfect "spiral" pass in football or a "line drive" in baseball, is also a parabola.

Although we typically work with the conic sections using formulas, we define them formally from a purely geometric perspective. This is analogous to the way we defined a circle as the set of all points at a fixed distance from a single fixed point, its center.

The Ellipse

An **ellipse** is the set of all points in the plane for which the sum of the distances to two fixed points is a constant. The two fixed points are called the **foci** of the ellipse. The midpoint of the line segment joining the foci is the **center** of the ellipse.

When the two fixed points are far apart, the resulting ellipse is very elongated. When the two foci (the singular word is *focus*) are close together, the ellipse is close to circular and, in fact, when the two foci merge into a single point, the ellipse becomes a circle.

For convenience, we assume that the center of an ellipse is at the origin and that the two foci lie on the x-axis equally distant from the y-axis. Suppose that the foci are at $F_1(c, 0)$ and $F_2(-c, 0)$ (see Figure 8.15). A point $P(x, y)$ then lies on the ellipse if the sum of the two distances $|F_1P|$ and $|F_2P|$ is some constant, k. To make things come out in a particularly nice way, it is convenient to write $k = 2a$. That is,

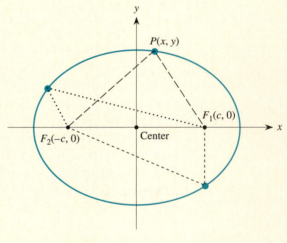

FIGURE 8.15

$$|F_1P| + |F_2P| = 2a$$

or equivalently,

$$\sqrt{(x - c)^2 + (y - 0)^2} + \sqrt{(x + c)^2 + (y - 0)^2} = 2a$$

$$\sqrt{(x - c)^2 + y^2} + \sqrt{(x + c)^2 + y^2} = 2a.$$

When this equation is simplified by eliminating both square roots (we leave this for you to do as an exercise in this section), we eventually obtain

$$\frac{x^2}{a^2} + \frac{y^2}{b^2} = 1,$$

where $b^2 = a^2 - c^2$ is a new constant. The three constants a, b, and c are related by the equation

$$a^2 = b^2 + c^2.$$

To determine where the ellipse intersects the x-axis, we set $y = 0$, and so the equation of the ellipse yields

$$\frac{x^2}{a^2} = 1.$$

Hence

$$x^2 = a^2$$

from which we find that either

$$x = a \qquad \text{or} \qquad x = -a.$$

This tells us that the distance from the center of the ellipse to the two points where the ellipse crosses the x-axis is a (see Figure 8.16). Similarly, if $x = 0$, then the equation of the ellipse yields

$$y^2 = b^2,$$

and so we find

$$y = b \qquad \text{or} \qquad y = -b.$$

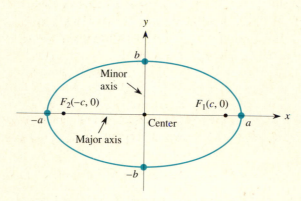

FIGURE 8.16

Thus the distance from the center of the ellipse to the two points where the ellipse crosses the y-axis is b. This yields four points, $(a, 0)$, $(-a, 0)$, $(0, b)$, and $(0, -b)$. They are called the **vertices** of the ellipse; any one of them is a **vertex.** The lines connecting opposite vertices are called the **axes** of the ellipse. The longer axis, whether horizontal or vertical, is called the **major axis** and always contains the two foci. The shorter axis is called the **minor axis.**

In summary, a represents the distance from the center to either of the two more distant vertices along the major axis of the ellipse; b represents the distance from the center to either of the two closer vertices along the minor axis; and c represents the distance from the center to either of the foci of the ellipse.

EXAMPLE 1

Describe and sketch the graph of the ellipse

$$\frac{x^2}{16} + \frac{y^2}{9} = 1.$$

Solution This ellipse is centered at the origin. Its maximum horizontal distance from the center is $a = 4$ and its maximum vertical height above the center is $b = 3$. The major axis thus extends horizontally from $x = -4$ to $x = 4$ and the minor axis extends vertically from $y = -3$ to $y = 3$. Since $a^2 = b^2 + c^2$, it follows that

$$c^2 = a^2 - b^2 = 16 - 9 = 7,$$

and so $c = \pm\sqrt{7}$. Therefore the foci are located at $(\sqrt{7}, 0)$ and $(-\sqrt{7}, 0)$. Consequently, the graph is as shown in Figure 8.17.

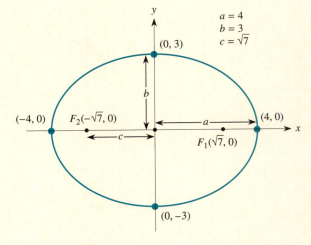

FIGURE 8.17

So far, we have assumed that the ellipse is centered at the origin and has its foci along the x-axis. We can consider the analogous situation where the foci lie along the y-axis. The resulting *standard form* equation for such an ellipse is

$$\frac{x^2}{b^2} + \frac{y^2}{a^2} = 1.$$

Notice that the constant a is still measured along the major axis while b is measured along the minor axis of the ellipse. Given the equation of an ellipse, we can identify its major axis immediately by observing which denominator is larger.

So far, we have only considered ellipses that are centered at the origin. More generally, an ellipse can be centered at any point (x_0, y_0). In such a case, we get the following *standard form* equations.

Equation of an ellipse centered at (x_0, y_0) with major axis parallel to the x-axis

$$\frac{(x - x_0)^2}{a^2} + \frac{(y - y_0)^2}{b^2} = 1$$

Equation of an ellipse centered at (x_0, y_0) with major axis parallel to the y-axis

$$\frac{(x - x_0)^2}{b^2} + \frac{(y - y_0)^2}{a^2} = 1$$

In each case,

$$a^2 = b^2 + c^2,$$

where c is the distance from the center to a focus.

E X A M P L E   2

Describe and sketch the ellipse whose equation is

$$\frac{(x - 2)^2}{4} + \frac{(y - 7)^2}{25} = 1.$$

Solution   We immediately see that this ellipse is centered at the point $(2, 7)$. Further, since 25 is larger than 4, we conclude that the major axis is parallel to the y-axis. In particular, the major axis is on the vertical line $x = 2$. The foci are on this line as well. The minor axis is on the horizontal line $y = 7$. Also $a = \sqrt{25} = 5$ and $b = \sqrt{4} = 2$. Thus the length of the major axis is 10, and the length of the minor axis is 4. Consequently, the maximum horizontal distance from the center is 2 and the maximum vertical distance from the center is 5. The ellipse therefore extends horizontally from $x = 0$ to $x = 4$ and extends vertically from $y = 2$ to $y = 12$. Moreover, since the foci are on the major axis of the ellipse, they are on the vertical line $x = 2$. To locate them, we use

$$c^2 = a^2 - b^2 = 25 - 4 = 21,$$

which gives $c = \sqrt{21}$. Thus the foci for the ellipse are located at the points $(2, 7 + \sqrt{21})$ and $(2, 7 - \sqrt{21})$. The graph of the ellipse is shown in Figure 8.18.

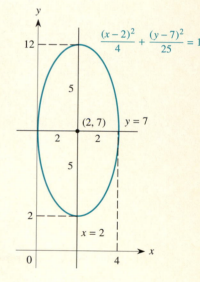

FIGURE 8.18

Suppose we again consider the ellipse in Example 2,

$$\frac{(x - 2)^2}{4} + \frac{(y - 7)^2}{25} = 1.$$

This time we multiply both sides of the equation by 100 to eliminate the fractions, which gives

$$25(x - 2)^2 + 4(y - 7)^2 = 100.$$

Expanding the terms, we get

$$25(x^2 - 4x + 4) + 4(y^2 - 14y + 49) = 100$$
$$25x^2 - 100x + 100 + 4y^2 - 56y + 196 = 100,$$
$$25x^2 + 4y^2 - 100x - 56y + 196 = 0.$$

This last equation is an equivalent equation for the same ellipse. Often we start with such an equation and have to rewrite it algebraically to uncover the key information about the ellipse. We illustrate how this is done in Example 3.

E X A M P L E   3

Verify that the equation

$$25x^2 + 9y^2 - 50x - 36y - 164 = 0$$

represents an ellipse, and find its center, vertices, and foci. Use this information to sketch the ellipse.

Solution We first collect the terms in x and y separately, factor out the coefficients of x^2 and y^2, and finally complete the squares for both x and y. Thus we obtain

$$
\begin{aligned}
25x^2 + 9y^2 - 50x - 36y - 164 &= 25x^2 - 50x + 9y^2 - 36y - 164 \\
&= [25(x^2 - 2x)] + [9(y^2 - 4y)] - 164 \\
&= 25[(x^2 - 2x + 1) - 1] + 9[(y^2 - 4y + 4) - 4] - 164 \\
&= 25[(x - 1)^2 - 1] + 9[(y - 2)^2 - 4] - 164 \\
&= 25(x - 1)^2 - 25 \cdot 1 + 9(y - 2)^2 - 9 \cdot 4 - 164 \\
&= 25(x - 1)^2 - 25 + 9(y - 2)^2 - 36 - 164 \\
&= 25(x - 1)^2 + 9(y - 2)^2 - 225.
\end{aligned}
$$

Therefore the given equation is equivalent to

$$25(x - 1)^2 + 9(y - 2)^2 - 225 = 0$$

or

$$25(x - 1)^2 + 9(y - 2)^2 = 225.$$

If we now divide both sides by 225, we find

$$\frac{(x - 1)^2}{9} + \frac{(y - 2)^2}{25} = 1,$$

which is the equation of an ellipse. From this equation we know the ellipse is centered at $(1, 2)$ and has a vertical major axis because $25 > 9$. Moreover, because $a = 5$, the major axis extends from $y = 2 - 5 = -3$ to $y = 2 + 5 = 7$. The minor axis is horizontal with $b = 3$, so it extends from $x = 1 - 3 = -2$ to $x = 1 + 3 = 4$. To find the foci, we solve

$$c^2 = 25 - 9 = 16,$$

which gives $c = 4$, and so the foci are at $(1, -2)$ and $(1, 6)$. The graph is shown in Figure 8.19.

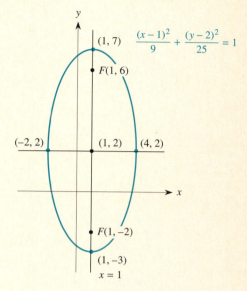

FIGURE 8.19

At the beginning of the section, we pointed out that the orbits of the planets about the sun are ellipses. More specifically, these elliptical orbits all have the sun as one of the two foci. A natural question to ask is: What is the equation of the ellipse for the orbit of the Earth? To answer this, we need two pieces of data used by astronomers to describe the orbits of the planets. The *perihelion* is the smallest distance from a planet to the sun and the *aphelion* is the greatest distance. For the Earth, the perihelion is approximately 147.1 million kilometers, or 91.38 million miles and the aphelion is approximately 152.1 million kilometers, or 94.54 million miles. These two distances help identify the location of the sun on the major axis of Earth's elliptical orbit.

E X A M P L E   4

Find an equation of the Earth's orbit about the sun.

Solution We first set up a coordinate system with the sun, the other (phantom) focus, and the major axis on the x-axis, as shown in Figure 8.20. Since the perihelion and aphelion distances are almost the same, we observe that the two foci are quite close together, and so the orbit of the Earth is very nearly circular. From Figure 8.20, we immediately see that the distance from one vertex to the other, which is equal to $2a$, is $91.38 + 94.54 = 185.92$ million miles, so

$$a = 92.96 \text{ million miles.}$$

At perihelion, the Earth is 91.38 million miles from the sun, so the distance from the center of the ellipse to the sun (a focus) must be

$$c = 92.96 - 91.38 = 1.58 \text{ million miles.}$$

Furthermore, in any ellipse,

$$a^2 = b^2 + c^2,$$

so

$$b^2 = a^2 - c^2$$
$$= 92.96^2 - 1.58^2 = 8639.07,$$

and

$$b = \sqrt{8639.07} = 92.95 \text{ million miles.}$$

Consequently, the equation of the Earth's orbit about the sun is

$$\frac{x^2}{(92.96)^2} + \frac{y^2}{(92.95)^2} = 1.$$

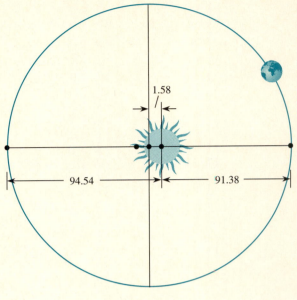

FIGURE 8.20

As we observed previously, the Earth's orbit is very nearly circular.

The accompanying table of planetary data provides some key information on the orbits of the different planets in the solar system in millions of miles. You can use it to compare the Earth's orbit to that of the other planets and you will be using some of these entries for the exercises at the end of the section.

Planet	Perihelion	Aphelion
Mercury	28.56	43.38
Venus	66.74	67.68
Earth	91.38	94.54
Mars	128.49	154.83
Jupiter	460.43	506.87
Saturn	837.05	936.37
Uranus	1700.07	1867.76
Neptune	2771.72	2816.42
Pluto	2749.57	4582.61

Reflection Property of the Ellipse   One of the most fascinating properties of an ellipse is known as the *reflection property*. Consider any line segment emanating from one of the two foci, say F_1, as shown in Figure 8.21.

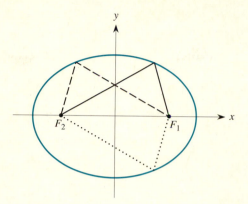

FIGURE 8.21

It eventually intersects the ellipse and then reflects. (According to physical principles, the angle of incidence with a tangent line is equal to the angle of reflection.) *Any* such reflected line segment will pass through the second focus, F_2.

This is significant because many physical phenomena, such as light and sound, travel in straight lines and reflect off solid surfaces. Thus if a lightbulb is placed at one focus of a three-dimensional shell whose cross sections containing the major axis are all ellipses, then all of its light rays will bounce off the inside surface of the shell and reflect back through the other focus. The effect is similar with sound waves. Probably the best known example of this is the whispering gallery effect in the U.S. Capitol building. The dome of the Capitol has the approximate shape of a three-dimensional ellipse and there are two foci near floor level. If you stand at one of the foci and whisper, your voice is carried, almost magically, to the second focus across the hall and can be clearly heard by anyone standing there.

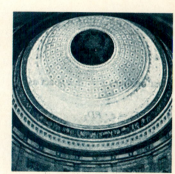

E X A M P L E   5

The distance between the foci in the "whispering gallery" of the Capitol building is 38.5 feet and the maximum height of the ceiling above ear level is 37 feet. Find the equation of an elliptical cross section of the gallery under the Capitol dome.

Solution   We set up our axes as shown in Figure 8.22. Since the distance between foci is 38.5 feet, $c = \frac{1}{2}(38.5) = 19.25$ feet.

Also, from the maximum height of the dome, $b = 37$ feet. For an ellipse, we know that

$$a^2 = b^2 + c^2 = (37)^2 + (19.25)^2,$$

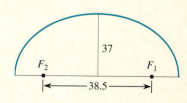

FIGURE 8.22

and so $a = 41.7$ feet. Therefore the equation of an elliptical cross section of the Capitol whispering gallery is

$$\frac{x^2}{(41.7)^2} + \frac{y^2}{37^2} = 1.$$

The Hyperbola

We have seen that the ellipse is defined geometrically by the *sum* of the distances to two fixed foci being constant. In an analogous way, we define a hyperbola in terms of the *difference* of the distances to two fixed points being constant. A **hyperbola** is the set of all points in the plane for which the *difference* between the distances to two fixed points is a constant. The two fixed points are called the **foci** of the hyperbola. The midpoint of the line segment joining the foci is the **center.**

For convenience, we place the center of a hyperbola at the origin and the foci on the x-axis at $F_1(c, 0)$ and $F_2(-c, 0)$. A point $P(x, y)$ lies on the hyperbola if

$$|F_2P| - |F_1P| = k,$$

as shown in Figure 8.23.

As with the equation of an ellipse, we let the constant $k = 2a$ for convenience. Thus

$$|F_2P| - |F_1P| = 2a,$$

so that

$$\sqrt{(x + c)^2 + y^2} - \sqrt{(x - c)^2 + y^2} = 2a.$$

When this equation is simplified by eliminating both square roots, we eventually obtain

$$\frac{x^2}{a^2} - \frac{y^2}{b^2} = 1,$$

where

$$c^2 = a^2 + b^2.$$

The graph of this hyperbola is shown in Figure 8.24. Notice that the hyperbola has two distinct *branches*, which is what we should expect in view of our original discussion of slicing through a double right circular cone. The two points where this hyperbola crosses the x-axis are called its **vertices;** they correspond to the points where $y = 0$ and so represent the points where the two branches are closest.

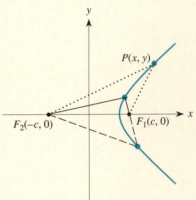

FIGURE 8.23

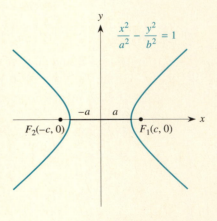

FIGURE 8.24

At the vertices, we have

$$\frac{x^2}{a^2} = 1,$$

and so

$$x = \pm a.$$

Thus the number a represents the distance from the center to a vertex while the number c represents the distance from the center to a focus. The line containing the foci is called the **axis** of the hyperbola.

Alternatively, we can place the foci for a hyperbola on the vertical axis, as shown in Figure 8.25. The equation of such a hyperbola is

$$\frac{y^2}{a^2} - \frac{x^2}{b^2} = 1.$$

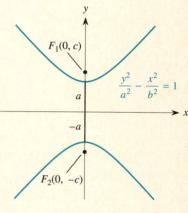

FIGURE 8.25

E X A M P L E   6

Describe the hyperbola whose equation is

$$\frac{x^2}{16} - \frac{y^2}{9} = 1.$$

Solution   The form of the equation tells us that the hyperbola is centered at the origin and its axis is horizontal. We immediately have $a = 4$ and $b = 3$, so that

$$c^2 = a^2 + b^2 = 16 + 9 = 25$$

and $c = 5$. Thus, its vertices are at $x = -4$ and $x = 4$ or the points $(-4, 0)$ and $(4, 0)$. Its foci are the points $(-5, 0)$ and $(5, 0)$.

Use your function grapher to see what the graph of this hyperbola looks like. To do this, you will have to write the equation as

$$\frac{y^2}{9} = \frac{x^2}{16} - 1$$

$$y^2 = 9\left(\frac{x^2}{16} - 1\right) = \frac{9}{16}x^2 - 9.$$

The upper and lower halves of the hyperbola are therefore given by

$$y = 3\sqrt{\frac{x^2}{16} - 1} \quad \text{and} \quad y = -3\sqrt{\frac{x^2}{16} - 1}.$$

More generally, we can consider a hyperbola as being shifted horizontally and/or vertically so that its center is at the point $P(x_0, y_0)$ rather than at the origin. We then have the following *standard form* equations.

Equation of a hyperbola centered at (x_0, y_0) with axis parallel to the x-axis

$$\frac{(x - x_0)^2}{a^2} - \frac{(y - y_0)^2}{b^2} = 1$$

Equation of a hyperbola centered at (x_0, y_0) with axis parallel to the y-axis

$$\frac{(y - y_0)^2}{a^2} - \frac{(x - x_0)^2}{b^2} = 1$$

In each case,

$$c^2 = a^2 + b^2.$$

Notice that in the equation of a hyperbola in standard form, the term with the positive coefficient determines the orientation. If the x^2-term is positive, the two branches open about the x-axis; if the y^2-term is positive, the two branches open about the y-axis.

E X A M P L E   7

Verify that

$$x^2 - y^2 + 8x - 6y = 2$$

is an equation of a hyperbola. Find the center, vertices, and foci of the hyperbola and sketch its graph.

Solution   We complete the square for both x and y to obtain

$$x^2 - y^2 + 8x - 6y = [(x^2 + 8x + 16) - 16] - [(y^2 + 6y + 9) - 9]$$
$$= (x + 4)^2 - 16 - (y + 3)^2 + 9.$$

The original equation therefore becomes

$$(x + 4)^2 - 16 - (y + 3)^2 + 9 = 2$$

or

$$(x + 4)^2 - (y + 3)^2 = 2 + 16 - 9 = 9.$$

If we divide by 9, we obtain

$$\frac{(x + 4)^2}{9} - \frac{(y + 3)^2}{9} = 1.$$

Consequently, the center of the hyperbola is $(-4, -3)$ and $a = b = 3$. Furthermore, since the x^2-term is the positive one, the axis of the hyperbola is parallel to the x-axis. That is, the vertices and the foci lie along the horizontal line $y = -3$ through the center, and the hyperbola opens to the left and the right. Since $a = 3$, the vertices are 3 units left and right of the center, or at $(-7, -3)$ and at $(-1, -3)$. Also,

$$c^2 = a^2 + b^2 = 9 + 9 = 18,$$

so that $c = \sqrt{18} = 3\sqrt{2}$. Thus the foci are located at $F_1(-4 - 3\sqrt{2}, -3)$ and $F_2(-4 + 3\sqrt{2}, -3)$. See Figure 8.26.

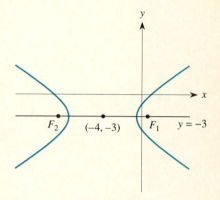

FIGURE 8.26

Notice that, based on their graphs and the vertical line test, neither ellipses nor hyperbolas can represent functions because a given x-value can lead to two distinct y-values. Of course, if x is taken beyond the limits of the ellipse, there is no corresponding y-value. Similarly, if x is taken between the two branches of a hyperbola having a horizontal axis, there again is no corresponding y-value.

The Parabola

We define a parabola in a comparable geometric manner. A **parabola** is the set of all points in the plane for which the distance to a fixed point is equal to the distance to a fixed line. The fixed point is called the **focus** of the parabola. The fixed line is the **directrix.**

For convenience, we place the focus of the parabola at the point $F(0, c)$ on the y-axis and let the directrix then be the fixed horizontal line $y = -c$, as shown in Figure 8.27. The graph shown corresponds to the case where $c > 0$. The parabola consists of all points P having the property that the distance from P to the focus F is equal to the vertical distance from P to the directrix line L. Thus, a point $P(x, y)$ lies on the parabola if the distance from P to F, $\sqrt{x^2 + (y - c)^2}$, is equal to the distance from P to the line L, which is $y + c$:

$$\sqrt{x^2 + (y - c)^2} = y + c.$$

We square both sides of this equation to get

$$x^2 + (y - c)^2 = (y + c)^2,$$

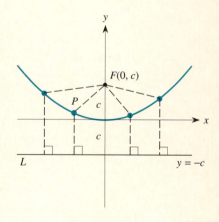

FIGURE 8.27

or equivalently, when we expand the two terms,

$$x^2 + y^2 - 2cy + c^2 = y^2 + 2cy + c^2.$$

When we subtract y^2 and c^2 from both sides of this equation, we get

$$x^2 - 2cy = 2cy.$$

Finally, we add $2cy$ to both sides and solve for y to get

$$y = \frac{x^2}{4c}.$$

This is an equation of a parabola with vertex at the origin. If $c > 0$, the parabola opens upward. If $c < 0$, the parabola opens downward. The vertical line through the vertex is called the **axis of symmetry** of the parabola.

Alternatively, had we positioned the focus on the x-axis at $F(c, 0)$ with a vertical directrix at $x = -c$, then we would have obtained

$$x = \frac{y^2}{4c}$$

as the equation for the parabola. This parabola likewise has its vertex at the origin; it opens to the right if $c > 0$ and opens to the left if $c < 0$. Finally, its axis of symmetry is now the horizontal line through the vertex.

More generally, we can describe a parabola whose vertex is at $V(x_0, y_0)$ with the following *standard form* equations.

Equation of a parabola with vertex at (x_0, y_0) and opening vertically

$$y - y_0 = \frac{(x - x_0)^2}{4c}$$

Equation of a parabola with vertex at (x_0, y_0) and opening horizontally

$$x - x_0 = \frac{(y - y_0)^2}{4c}$$

Reflection Property of the Parabola　Just as the ellipse has a remarkable—and useful—reflection property, the parabola has one that is even more commonly encountered. It can be shown that any ray coming into a parabola along a line that is parallel to the axis of symmetry of the parabola will "reflect" off the curve and pass through the focus, as shown in Figure 8.28. Alternatively, any ray emanating from the focus will reflect off the parabola and continue on a path parallel to the axis. This latter reflection property is used, for example, in a flashlight or the headlights of an automobile, where the light source is located at the focus and the beams of light bounce off a parabolic-shaped reflector to concentrate more light in a particular direction. The reflection property is used by satellite TV dishes which are constructed in such a way that every cross section containing the axis of symmetry of the dish is a parabola (see Figure 8.29). The TV signals coming down from a satellite relay in orbit arrive at the dish along rays that are parallel to the axis of the dish and its parabolic cross sections. They reflect off the dish to pass through a receptor unit positioned at the focus. There the signal is collected and then transmitted on to the television set.

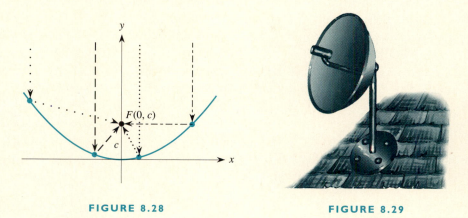

FIGURE 8.28   FIGURE 8.29

Conic Sections in General

The general equation of a conic section has the form

$$Ax^2 + By^2 + Cx + Dy + E = 0,$$

where A, B, C, D, and E are constants. When $B = A \neq 0$, we can divide both sides by A to get

$$x^2 + y^2 + \frac{C}{A}x + \frac{D}{A}y + \frac{E}{A} = 0,$$

which is the equation of a circle provided that certain conditions are satisfied which lead to a positive radius. When $B = 0$ and $A \neq 0$, the resulting equation is quadratic in x but only linear in y and so gives the equation of a parabola opening either upward or downward. Similarly, when $A = 0$ and $B \neq 0$, we get a parabola opening either left or right. For the other possibilities of A and B, ellipses and hyperbolas can be identified by comparing the signs of A and B. In particular, if A and B have the same sign, say both are positive, then the resulting curve is an ellipse provided that certain conditions are satisfied. If A and B have opposite signs, then the curve is a hyperbola. Thus, for example,

$$4x^2 + 9y^2 + 8x - 36y - 5 = 0$$

represents an ellipse while

$$4x^2 - 9y^2 + 8x - 36y - 5 = 0$$

and

$$25y^2 - 16x^2 + 10y + 8x + 3 = 0$$

both represent hyperbolas (one of which opens left and right, while the other opens up and down). It is important that you be able to identify the type of curve from the given equation.

Finally, note that all of our discussion on the conic sections has been restricted to their being in *standard position*—that is, their axes are parallel to the x- and y-axes. It is also possible to have the same shapes rotated through some angle θ about the x-axis. When this occurs, the equation for the conic section, whether it is an ellipse, a hyperbola, or a parabola, includes a term of the form xy. For the most part, we will not be concerned with such situations here. The only exception is the case with

$$xy = k,$$

where k is any constant. If we solve for y, we see that this is equivalent to the power function

$$y = \frac{k}{x} = k \cdot x^{-1}.$$

If you think about its graph from the point of view of our discussions in Chapter 2, you should realize that this represents the equation of a hyperbola rotated from standard position through an angle of 45° or $\pi/4$ (when $k > 0$), as shown in Figure 8.30.

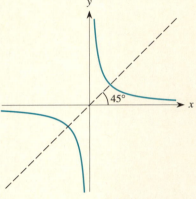

FIGURE 8.30

EXERCISES

1. For the satellite whose elliptic orbit about the Earth is shown in the accompanying diagram, indicate the location of the following points and give reasons for your answers.

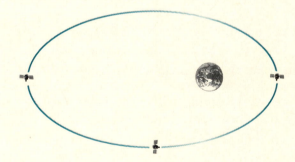

 a. The point where the gravitational force exerted by the Earth on the satellite is greatest; where it is least. (Recall Newton's law of universal gravitation: $F = f(r) = -GmM/r^2$, where r is the distance between two objects.)
 b. The point where the speed of the satellite is greatest.

c. The point where the speed of the satellite is least.

2. Suppose that the satellite in Exercise 1 fires its retro-rockets to slow down somewhat at the point in its orbit where it is closest to the Earth. Sketch the graph of the new orbit compared to the old one.

3. Suppose that the satellite in Exercise 1 is in a relatively low orbit about the Earth so that it encounters the upper fringe of the Earth's atmosphere. What will be the effect of this on the satellite's path? Sketch the graph of the resulting trajectory. What will happen eventually?

4. Suppose that the satellite in Exercise 1 fires its booster rocket to speed up at the point in its orbit where it is closest to the Earth. Sketch the graph of the new orbit compared to the old one. What happens to the orbit if the booster rockets are extremely strong or continue firing for a very long time?

5. Which of the nine planets in the solar system has the most circular orbit? the least circular orbit?

6. During the next few years, Pluto's orbit takes it inside the orbit of Neptune. Use the given values in the table of planetary data and explain why this can be so.

7. Use the fact that the perihelion and aphelion distances for Mercury are 46.0 and 69.8 million kilometers respectively to find the equation of the orbit of Mercury about the sun.

8. A salami is four inches in diameter. When the deli clerk slices it, however, the slices are at an angle of 65° to the main axis of the salami. Consequently, each slice will be in the shape of an ellipse with minor axis of length 4 inches. Find the length of the major axis of each slice.

9. The Roman Coliseum is in the shape of an ellipse whose major axis measures 620 feet and whose minor axis measures 513 feet.

   a. What is the equation of this ellipse?
   b. How far apart are the foci?
   c. How far apart are two adjacent vertices?

10. The small satellite TV dishes now on the market for home use have parabolic cross sections containing the axis of the dish. The focus is located at a point approximately 6 inches from the vertex of the parabola.

    a. Find an equation of a parabolic cross section. Assume that the dish is aimed directly upward—which only makes sense at the equator since the communications satellites are in orbit over the equator.
    b. The rim of the dish is a circle with a diameter of 18 inches at a height of about 1.5 inches above the vertex. If the dish were extended, the rim would enlarge. Find what the diameter of the rim would be if the rim would reach a height level with the focus.

11. Suppose that a satellite receiver is 36 inches across and 16 inches deep (vertex to plane of the rim). How far from the vertex must the receptor unit be located to assure that it is at the focus of the parabolic cross sections?

12. In this exercise, we ask you to investigate the significance of the quantities a and b in the equation of a hyperbola,

$$\frac{x^2}{a^2} - \frac{y^2}{b^2} = 1.$$

Suppose you zoom out far enough on the graph of the hyperbola so that what you see appears to be a pair of lines that intersect at the origin.

  a. Explain why, when x and y are both very large, either positive or negative, you can ignore the number 1 in the equation.
  b. Ignoring the 1, solve for y in terms of x to find the equations of the two lines described above.
  c. What are the slopes of the two lines that the branches of the hyperbola approach?

13. Write formulas expressing the perihelion and aphelion of an elliptic orbit in terms of the *semimajor axis* length a and the focal distance c in an ellipse.

Complete the square for x and y in each of the following equations in Exercises 14–23 to obtain the standard form for a conic section. In each case, identify the conic section and use the pertinent information to draw its graph.

14. $x^2 + 4y^2 + 2x + 8y = -1$ 15. $x^2 + 4y^2 + 2x + 8y = 11$
16. $x^2 - 4y^2 + 2x - 8y = 7$ 17. $x^2 - 4y^2 + 2x - 8y = 19$
18. $4x^2 + y^2 + 24x - 2y + 4 = 0$ 19. $4x^2 - 9y^2 - 16x - 18y = 31$
20. $9x^2 - 16y^2 - 90x + 64y = -17$ 21. $4x^2 + 4y^2 - 24x + 16y + 43 = 0$
22. $9x^2 - 4y^2 + 18x - 16y = 8$ 23. $9x^2 - 4y^2 + 18x - 16y = 6$

24. Explain why $x^2 + y^2 - 2x - 2y = -4$ is not the equation of a conic section.

25. Complete the derivation of the equation of the ellipse by simplifying the equation in the text by eliminating the two square roots. (*Hint:* First isolate one of the radicals, then square both sides, and finally eliminate the remaining radical.)

26. In this problem, we look at the mathematics behind string and wire art designs. Start with the hyperbola $xy = 1$ or $y = \frac{1}{x}$ and construct a series of lines that are tangent to the curve, as shown in the figure. For instance, the line tangent to the curve when $x = 1$ crosses the x-axis at $x = 2$ and crosses the y-axis at $y = 2$. The line tangent to the hyperbola when $x = 2$ has an x-intercept of 4 and a y-intercept of 1. If you could now erase the curve, you would still see its outline from the tangent lines. String and wire art designs use this idea to suggest a variety of curves using line segments made of the string or wire. The points on the axes are selected so as to follow the outline of a desired curve such as the hyperbola.

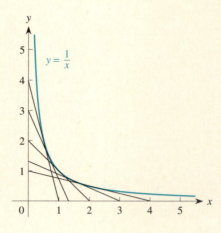

  a. Find the slope m of each of the tangent lines shown in the figure.
  b. Find a formula for the slope m as a function of the point of tangency x.

27. Consider the ellipse $\frac{x^2}{16} + \frac{y^2}{9} = 1$ and the point $P(5, 5)$. Most of the lines drawn through P either cross the ellipse at two points or miss the ellipse entirely. There are precisely two lines through P, the tangent lines, that touch the ellipse at exactly one point each.

a. Use your function grapher to estimate, by trial and error, the slope of one of those two lines, correct to two decimal places.

b. Estimate the coordinates of the point of tangency.

c. What is the second point where a line through P will be tangent to the ellipse? What is the slope of that line?

28. For the function $y = f(x) = x^2$ shown, very carefully draw lines that are tangent to the curve at $x = -2, -1, 0, 1, 2$, and 3 and estimate the slope of each. Find a formula that gives the slope m of these tangent lines as a function of the point of tangency x.

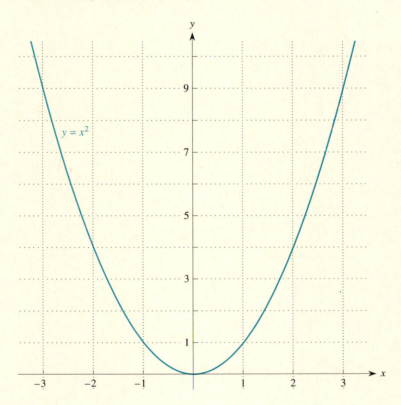

8.4 *Parametric Curves*

We have seen that there are many ways to write an equation of a line, including the point-slope form and the slope-intercept form. However, there are certain questions about a line that cannot be answered using either of these forms. For instance, as we saw in Section 8.2, if we want to locate the point that is a certain fraction of the way from $P(x_0, y_0)$ to $Q(x_1, y_1)$, it is very helpful to use the parametric form

$$x = x_0 + (x_1 - x_0)t$$
$$y = y_0 + (y_1 - y_0)t$$

for the line, where the parameter t takes on any value whatsoever. Using this form, for each possible value of t, there is a corresponding point on the line through P and Q. For example, if $x = 2 + 4t$ and $y = 5 - 3t$, then for each value of t, we get a value for x and a value for y. Thus, if $t = 1$, then $x = 6$ and $y = 2$. Each pair of values, x and y, produces a point (x, y). We show a few of these points in the following table.

t	-1	0	1	2	3	4
x	-2	2	6	10	14	18
y	8	5	2	-1	-4	-7

We plot these points and show the line containing them in Figure 8.31.

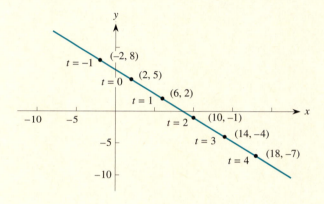

FIGURE 8.31

As another example, when you throw an object such as a football, the path it follows is a parabola of the form

$$y = ax^2 + bx + c.$$

However, for most realistic applications, the equation of the parabola by itself is of little value. It is far more important to know *when* the ball, or other object, will be at a particular point. To do this, it is necessary to introduce time as a variable and we do so by writing both x and y, the coordinates of each point along the parabola, in terms of a parameter t which represents time. In particular, if the object is released at time $t = 0$ from an initial height y_0 with an initial velocity v_0 at an initial angle α, as shown in Figure 8.32, then at any time t thereafter, it turns out that

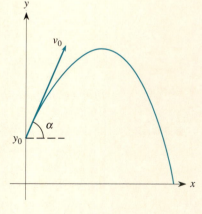

FIGURE 8.32

$$x = (v_0 \cdot \cos \alpha)t$$

and

$$y = -\frac{1}{2}gt^2 + (v_0 \cdot \sin \alpha)t + y_0.$$

Since each of these can be thought of as a function of the parameter t, we can rewrite them as

$$x(t) = (v_0 \cdot \cos \alpha)t$$

$$y(t) = -\frac{1}{2}gt^2 + (v_0 \cdot \sin \alpha)t + y_0.$$

Each value of t determines a corresponding value for x (the horizontal distance) and a value for y (the vertical height). This produces a point $P(x, y)$ on the parabola. The pair of equations for x and y in terms of the parameter t is called a *parametric representation* or the *parametric equations* for the curve.

The fact that the parametric representations for a line and for a parabola prove to be so valuable suggests that it might be useful to develop parametric representations for other curves. We will consider this possibility for the conic sections. There are two key steps: The first is to decide what would be an appropriate parameter t and the second is to find a way to express the usual variables x and y in terms of t.

Let's begin with the simplest conic section, a circle of radius r centered at the origin:

$$x^2 + y^2 = r^2.$$

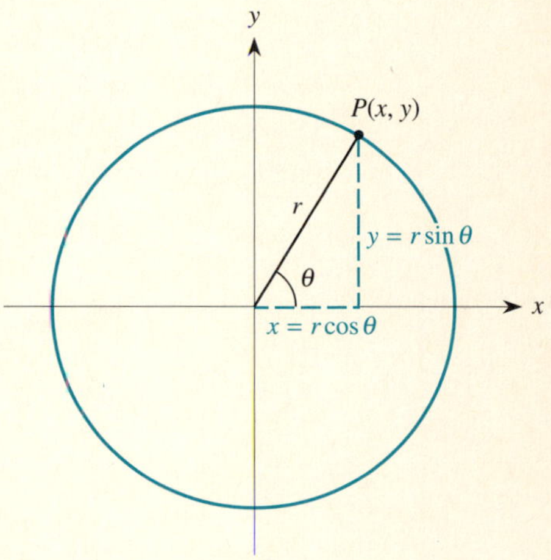

We have seen earlier that it is extremely useful to express both x and y in terms of an angle θ drawn from the center of the circle, as shown in Figure 8.33. We write

FIGURE 8.33

$$x = r \cos \theta \qquad \text{and} \qquad y = r \sin \theta.$$

This is a parametric representation for the circle with the angle θ as parameter. For each value of θ, we can calculate x and y and so have the point (x, y) on the circle. In fact, if t is any variable whatsoever (not necessarily the angle θ), then

$$x = r \cos t \qquad \text{and} \qquad y = r \sin t$$

is also a parametric representation of the circle.

If we start with a parametric representation for a curve, we can sometimes *eliminate the parameter* to construct a single equation for the curve. In this case,

$$x^2 + y^2 = (r \cos t)^2 + (r \sin t)^2 = r^2(\cos^2 t + \sin^2 t) = r^2.$$

Now let's turn to the ellipse centered at the origin with major axis along the x-axis:

$$\frac{x^2}{a^2} + \frac{y^2}{b^2} = 1.$$

Its graph is shown in Figure 8.34. The question is: What might be an appropriate parameter to introduce to help us describe this ellipse? Picture a point moving around the ellipse shown in Figure 8.34. Although we can locate each point P in terms of its x- and y-coordinates, we may also be able to locate it using the angle θ determined by P and the positive x-axis. How do we go about constructing such a parametric representation using θ as the parameter? That is, how do we express x and y as functions of θ? At first thought, you might be tempted to create a right triangle by dropping a perpendicular from point P to the x-axis just as we did for the circle. The problem with this approach is that the length of the hypotenuse would change along with θ as the point P moves around the ellipse, unlike a circle in which the lengths of the line segments from O to P remain constant. Thus the angle θ is not a good choice for the parameter. Nevertheless, our experience with the circle can provide some guidance. The parametric representation for a circle of radius r is

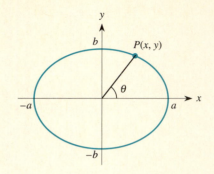

FIGURE 8.34

$$x = r \cos t$$

$$y = r \sin t.$$

If we think of the ellipse as having a "radius" of a associated with x and a "radius" of b associated with y, then we might try to write

$$x = a \cos t$$

$$y = b \sin t.$$

Let's see if this makes sense. Suppose that $P(x, y)$ is any point on the ellipse, so that x and y must satisfy

$$\frac{x^2}{a^2} + \frac{y^2}{b^2} = 1.$$

If we substitute our conjectured expressions for x and y into this equation, we find that

$$\frac{(a \cos t)^2}{a^2} + \frac{(b \sin t)^2}{b^2} = \frac{a^2 \cos^2 t}{a^2} + \frac{b^2 \sin^2 t}{b^2}$$

$$= \cos^2 t + \sin^2 t = 1.$$

Since these expressions for x and y as functions of t satisfy the equation of the ellipse, we conclude that they do form a parametric representation for the ellipse with parameter t.

One of the options available on all graphing calculators is a *parametric mode*. To use it, you need to supply an expression for X in terms of the parameter t and an expression for Y in terms of t.[1] You then have to define a

---

[1] Note that in parametric mode, T is used as the "generic" variable just as X is used as the "generic" variable in the usual function mode.

window not only in terms of x and y, but also in terms of an interval of values for the parameter t. Enter the parametric representation

$$x = 5\cos t$$
$$y = 3\sin t$$

for the ellipse

$$\frac{x^2}{25} + \frac{y^2}{9} = 1$$

with a range of values for the parameter from 0 to 2π in radians. Check that the graph is indeed that of the ellipse.

To use this parametric representation, suppose we want to know the point on the ellipse

$$\frac{x^2}{25} + \frac{y^2}{9} = 1$$

corresponding to a value of the parameter, say $t = \pi/6$. We immediately find that

$$x = 5\cos(\pi/6) = 4.330$$
$$y = 3\sin(\pi/6) = 1.5.$$

Alternatively, suppose we are told that the point $P(4, 9/5)$ lies on the ellipse. (Check that it does.) To find the value of parameter t for this point, we consider

$$x = 5\cos t = 4$$
$$y = 3\sin t = \frac{9}{5}.$$

From the first of these equations, we find that

$$\cos t = \frac{4}{5},$$

from which we see that

$$t = \text{Arccos}\,\frac{4}{5} = 0.6435 \text{ radians.}$$

Check that this value of t also satisfies the second equation for y.

It is also possible to develop a parametric representation for the hyperbola. This requires introducing two new functions known as the *hyperbolic sine* and the *hyperbolic cosine*, which we will discuss briefly in the exercises of this section.

Let's turn our attention to the parabola. At the beginning of this section, we saw how to *parameterize* the parabolic path of a projectile. We now consider the same situation in a geometric setting. It turns out that we can introduce a parameter in an extremely simple way. If the equation of the parabola is

$$y = ax^2 + bx + c,$$

then we can let

$$x = t$$

and so

$$y = at^2 + bt + c.$$

This may strike you as somewhat unfair (too easy!), but it turns out to be very effective. In fact, this can be done with any function $y = f(x)$ whatsoever. Let's look at one of the advantages of doing so. If we restrict our attention to the right side of the parabola, we know that the curve is strictly increasing (if $a > 0$) and so has an inverse f^{-1}. We know that the graph of the inverse function is the mirror image of the graph of f about the diagonal line $y = x$. However, in all but the simplest cases, it is not easy, or even possible, to find an explicit, or closed form, expression for the inverse f^{-1}. Thus it is difficult, if not impossible, to construct the graph of the inverse function from an explicit formula. With parametric functions at our disposal, however, this becomes a simple chore. Recall the definition of a function and its inverse function. If $y = f(x)$, then for each value of x, the function determines a single corresponding value for y. The inverse function undoes this in the sense that, for each value of y, f^{-1} returns the value of x that led to y. We can draw the graph of f in the parametric form

$$x = t$$
$$y = f(t)$$

using the parametric mode of the graphing calculator. To produce the graph of the inverse function, then, realize that all we need to do is reverse the roles of x and y. That is, if we set

$$x = f(t)$$
$$y = t,$$

then

$$y = t = f^{-1}(x)$$

and the calculator will draw the graph of the inverse function! Try this out with, say, the right side of a parabola or with an exponential function, where you know what the function and its inverse should look like.

There are many curves that can be represented fairly readily using parametric equations that cannot be represented simply, if at all, with y as a function of x. Suppose that your friend has a reflector attached to the rim of her bicycle tire. As she rides past you at a constant speed, you observe that the path of the reflector is a curve such as the one shown in

Figure 8.35. This curve, showing y as a function of x, is called a **cycloid.** If the radius of the tire is a, then the parametric representation of the cycloid is

$$x = at - a \sin t$$
$$y = a - a \cos t,$$

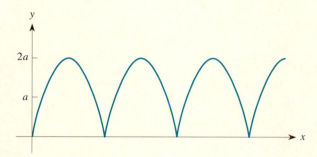

FIGURE 8.35

where the parameter t represents time. These equations are typically derived in calculus; we will only discuss their reasonableness and experiment with them here. Since the variable is time, be sure that you graph all such curves in radian mode.

Let's begin with the expression for the height y of the reflector, as a function of time. The given expression includes a cosine term combined with a vertical shift of a, so that the average height of the reflector will be a inches above the ground; that is, at the level of the center of the rolling circle. The cosine term has an amplitude of a, so the height varies from a minimum of 0 to a maximum of $2a$, which makes sense in terms of the physical phenomenon.

What about the expression $x = at - a \sin t$ for the horizontal distance? Notice that this expression involves a sine term, which oscillates between $-a$ and a, and this term is subtracted from at, which grows linearly. Again, this should make sense. The bicycle wheel is rolling along, so the horizontal distance traveled by the center of the wheel, at, is increasing linearly. Because the reflector is rotating about on the rim of the tire, there must be an oscillatory adjustment to the linear distance covered.

In Exercise 32 of Section 2.3, we raised the question about the shape of a water slide along which a person would slide most rapidly from one point to another; this is called the brachistochrone problem. From physical principles, it makes sense that the curve be decreasing and concave up, so that the person gains the greatest speed at the beginning of the slide. It turns out that the specific curve along which the time needed is a minimum is an upside down cycloid.

Let's consider another application involving a parametric representation for a curve. Most of you have likely seen a spirograph, a toy with

which you can draw intricate shapes by tracing out curves as one plastic wheel rotates
about another plastic wheel. Suppose that
you have a large wheel of radius b and a
smaller wheel of radius a that is rolling
around on the outside of the larger wheel, as
shown in Figure 8.36. A fixed point on the
outer, rolling, circle describes a curve that is
known as an **epicycloid**. A parametric representation of the epicycloid is

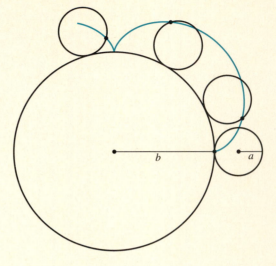

FIGURE 8.36

$$x = (a + b) \cos\left(\frac{at}{b}\right) - a \cos\left(\frac{a + b}{b}\right)t$$

$$y = (a + b) \sin\left(\frac{at}{b}\right) - a \sin\left(\frac{a + b}{b}\right)t$$

These equations are also typically derived in
calculus. For now, let's see what the path of the fixed point on the rolling
circle looks like. As the outer circle rolls on the fixed inner circle, the point
on it moves back and forth, getting closer to and farther from the origin. It
is closest to the origin—at a distance b—at the points where the two circles touch. It is farthest from the origin when the point is at the farthest
possible position on the rolling circle, which is at a distance of $b + 2a$. At
any other time, the point is at an intermediate distance between b and
$b + 2a$.

The actual shapes of epicycloids are often visually surprising and
striking, as shown in Figure 8.37 for $a = 11$, $b = 28$, and t between 0 and
421π. For a much simpler case, if the fixed inner circle has a radius of
$b = 4$ and the rolling circle has a radius of $a = 1$, we get the epicycloid
shown in Figure 8.38 for t between 0 and $8\pi = 25.13$; the same curve thereafter repeats with period 8π in the sense that the identical points are repeatedly traced out. We also superimpose the inner fixed circle to indicate
how the epicycloid is traced out by the fixed point as the outer (unseen)
circle rolls around on the inner circle.

In the exercises for this section we ask you to experiment with the epicycloid and other curves using parametric equations. You will see some surprising shapes if you simply try interesting combinations of functions. The
authors' favorite parametric curve, by the way, is the "snowman" function
whose parametric representation is

$$x = t - \frac{1}{2} \sin 10t$$

$$y = 5 \sin t - \frac{1}{2} \cos 10t.$$

FIGURE 8.37

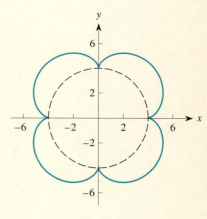

FIGURE 8.38

EXERCISES

1. Consider the parametric representation for the line $x = 2 + 4t$, $y = 5 - 3t$ that we studied in the text.

   a. Find the slope of the line from the table in the text.

   b. Use the slope and a point on the line to write an equation of the line with y as a function of x.

   c. How does the slope of the line relate to the coefficients in the parametric representation?

   d. Eliminate the parameter t algebraically by solving for t from the first equation and substituting into the second.

2. Consider the parametric representation for a line: $x = 7 - 3t$, $y = 4 + 5t$.

   a. Based on the results of part (c) of Exercise 1, what do you expect the slope of this line to be?

   b. Create a table of values for this line and use the entries to sketch the graph of the line.

   c. Find a point-slope form for the equation of the line.

   d. Find an equation for the line by eliminating the parameter t algebraically.

3. Consider the curve given in parametric form $x = t^3 + 1, y = t^2 - 2$.
    a. Create a table of values for this function using $t = -2, -1.5, -1, \ldots,$ 2 and plot the points to construct a rough sketch of the graph.
    b. Draw the curve using parametric mode on your function grapher. How does the result compare to what you did in part (a)?
    c. Eliminate the parameter t algebraically by first solving for t in terms of x.
    d. Graph the function you obtained in part (c) using function mode on your function grapher. How does it compare to your graph in part (b)?

4. Consider the curve with parametric representation $x = t^2 + 1, y = t^2 - 2$.
    a. Create a table of values for this function using $t = -1, 0, 1, \ldots,$ and plot the points to construct a rough sketch of the graph. What surprising result do you get?
    b. Use your function grapher in parametric mode to verify that the result you obtained in part (a) is correct.
    c. In terms of x and y, what are the domain and range for the curve you found in part (a)?
    d. Eliminate the parameter t algebraically and see why you got the shape you did.

5. Use parametric mode on your function grapher to draw the graph of the epicycloid with $a = 1$ and $b = 3$. Repeat this with $b = 4, b = 5$, and $b = 6$ while keeping $a = 1$. Do you see any pattern in the periods of these curves? Do you observe any pattern in the number of loops that you get?

6. Repeat Exercise 5 with $a = 1$ and $b = 6, 8, 10$, and 12. How do the curves you get compare to the ones you have seen before?

7. Repeat Exercise 5 with $a = 2$ and $b = 3, 5, 7$, and 9.

8. Figure 8.35 shows the graph of a cycloid which is the path of a reflector mounted on the rim of a tire of radius a. Determine the coordinates of the points where the curve touches the horizontal axis.

9. A **hypocycloid** is the curve that is generated by a fixed point on a circle of radius a that rolls around the inside of a larger circle of radius b. The parametric equations for a hypocycloid are

    $$x = (b - a) \cos t + a \cos \left(\frac{b - a}{a} \right) t$$

    $$y = (b - a) \sin t - a \sin \left(\frac{b - a}{a} \right) t.$$

    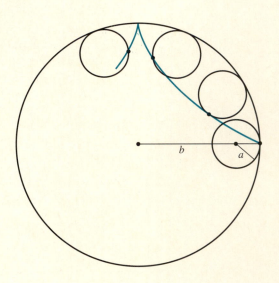

    Use your function grapher to see the shapes that result when
    a. $a = 1, b = 2$
    b. $a = 1, b = 3$
    c. $a = 1, b = 4$
    d. $a = 2, b = 3$

10. Suppose that a bicycle reflector is mounted partway along one of the spokes in a wheel at a distance $b < a$ from the center of the wheel. How should the parametric equations for the cycloid be modified to reflect this new position?

11. In the equations for the cycloid,

$$x = at - a \sin t$$
$$y = a - a \cos t,$$

use the second equation to solve for t and then substitute that into the first equation to eliminate the parameter and so obtain x as a function of y. What does the resulting equation tell you about the path of the reflector?

12. The trigonometric functions $\sin \theta$ and $\cos \theta$ are often referred to as circular functions because they can be defined in terms of the unit circle $x^2 + y^2 = 1$ and because $x = \cos \theta$ and $y = \sin \theta$ are a parametric representation for that circle. Engineers often use a related pair of functions, the hyperbolic cosine and the hyperbolic sine, which are defined as

$$\cosh \theta = \frac{e^\theta + e^{-\theta}}{2} \quad \text{and} \quad \sinh \theta = \frac{e^\theta - e^{-\theta}}{2},$$

where $e = 2.71828 \ldots$. (For instance, if you hang a heavy chain from two fixed points, it will fall into the shape of a hyperbolic cosine.)

a. Use your function grapher to draw the graphs of these two functions.

b. Use the definitions given above to show that $\cosh^2 \theta - \sinh^2 \theta = 1$ for all θ.

c. Explain why the functions $x = \cosh \theta$ and $y = \sinh \theta$ form a parametric representation for the right-hand branch of the hyperbola $x^2 - y^2 = 1$.

d. The motivation for defining the hyperbolic functions in this manner is that the sine and cosine can be written as

$$\cos \theta = \frac{e^{i\theta} + e^{-i\theta}}{2} \quad \text{and} \quad \sin \theta = \frac{e^{i\theta} - e^{-i\theta}}{2i},$$

where $i = \sqrt{-1}$. Use these expressions to show algebraically that $\cos^2 \theta + \sin^2 \theta = 1$.

8.5 *The Average Value of a Function*

We saw in Section 8.3 that the Earth travels around the sun in an elliptical orbit whose equation is

$$\frac{x^2}{92.96^2} + \frac{y^2}{92.95^2} = 1,$$

and that the perihelion distance is 91.38 million miles while the aphelion distance is 94.54 million miles. In this section, we raise a related question: What is the *average* distance of the Earth from the sun in the course of a year? Clearly, the average must be somewhere between the minimum and the maximum distances, but what is it exactly?

To answer this question, we must consider the problem in a deeper way. Rather than concentrating merely on the average distance from the Earth to the sun, let's look at the question of finding the average value of any function, f, on some interval. At this point in your mathematical experience, the notion of *average* makes sense only when we average a finite set of numbers. However, for a smooth curve $f(x)$ on an interval from a to b, there are infinitely many points. The best we can do for an average based on our present knowledge is to select a large number of points on the curve, calculate the mean of their $f(x)$ values, and use this value as an estimate for the average value of the function at all points in the interval. If the points chosen are $(x_1, f(x_1))$, $(x_2, f(x_2))$, $(x_3, f(x_3))$, ..., $(x_n, f(x_n))$, as shown in Figure 8.39, then the average value for the function is approximately the sum $f(x_1) + f(x_2) + \cdots + f(x_n)$ of the values, or heights, of the function divided by the number of values, n. Using *summation notation* (see Appendix C), we can write

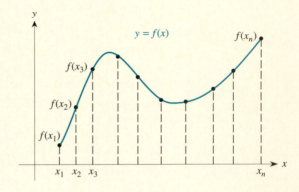

FIGURE 8.39

$$\sum_{k=1}^{n} f(x_k) = f(x_1) + f(x_2) + \cdots + f(x_n)$$

so that the

$$\text{average value} \approx \frac{\sum_{k=1}^{n} f(x_k)}{n}.$$

To be assured that the approximation is reasonably accurate, we need a very large number of points, which suggests that we should use a computer to do the work. How do we decide which points to use? One way to do this is to use a randomly generated set of points. We have the computer randomly generate 1000 points, say, on a given curve, and calculate the average of the values of the function at these points.

Let's see how this actually works in practice. We use such a program with the function $f(x) = \sin x$ on the interval 0 to π with 1000 random points. On one run of the program, the average value of $\sin x$ over the 1000 points was 0.6334. On another run with 1000 different random points, the average value was 0.6368, which is quite close to the first approximation. It turns out that the true average value for the sine function on the interval 0 to π is $2/\pi \approx 0.63662$.

Does this average value for the sine curve make sense? We know that between 0 and π, the sine function assumes values between 0 and 1.

Should the average value be $\frac{1}{2}$? Consider the graph of the sine function shown in Figure 8.40. We have drawn in a horizontal line corresponding to a height of $y = \frac{1}{2}$. We know that the curve rises from $y = 0$ at $x = 0$ to $y = \frac{1}{2}$ at $x = \frac{\pi}{6}$ (or 30°). Then, from $x = \frac{\pi}{6}$ to $x = \frac{\pi}{2}$ (or 90°), the sine curve continues to rise up to a height of $y = 1$; from $x = \frac{\pi}{2}$ to $x = \frac{5\pi}{6}$ (or 150°), the curve falls back down to a height of $y = \frac{1}{2}$; and finally from $x = \frac{5\pi}{6}$ to $x = \pi$ (or 180°), the sine curve continues to fall down to $y = 0$. Thus over the course of the 180° range that we are considering, the sine curve is below the line $y = \frac{1}{2}$ for a total of 60° and above it for a total of 120°. There are considerably more points on the curve above the $y = \frac{1}{2}$ level than there are below it. As a result, we see that the average value for the function on the interval 0 to π probably should be greater than $\frac{1}{2}$. The indicated value of 0.63662 is therefore not unreasonable.

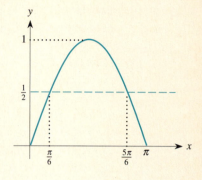

FIGURE 8.40

Think About This

Use a computer simulation to find the average value of the function $f(x) = \sin x$ on the interval 0 to 2π. Does the value you obtain from the simulation match up with what you expect the result should be?

Usually, the greater the number of points we use to estimate the average value of a function, the more accurate the approximation will be. Of course, the fact that the points we are using are random means that, in some instances, we could get a very accurate level of approximation with relatively few points and a lower level of accuracy with more points.

In calculus you will find a way to get a precise average value of a function across an interval for most common functions. However, since these calculus concepts are beyond the scope of this book, we will continue to estimate average values of functions using random simulations.

EXAMPLE

Find the average height above the x-axis of the upper half of the circle $x^2 + y^2 = 1$.

Solution We know that the average height must be between 0 and 1 because the semicircle extends to a maximum height of 1, as shown in Figure 8.41. Might this average value be 0.5? Because of the concavity of the semicircle, there are far more points on the circle that are above a height of 0.5 than there are below it. See Figure 8.41. Therefore, we should expect that the average value will be greater than 0.5. To estimate this average height more accurately, we will use an appropriate computer program. The function here is

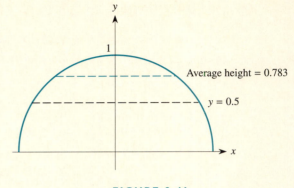

FIGURE 8.41

$$y = f(x) = \sqrt{1 - x^2} = (1 - x^2)^{1/2}.$$

Since we want the average height for the entire semicircle, we use an interval from -1 to 1. On one run with a computer program, using 1000 randomly generated points, the average of the corresponding values of the function turned out to be 0.783. (You will see in calculus that the average height is actually $\frac{\pi}{4} \approx 0.7854$.)

Let's return to the question we raised at the beginning of this section regarding the average distance from the Earth to the sun. In a broader sense, suppose we want to find the average distance from any point $P(x, y)$ on the ellipse

$$\frac{x^2}{a^2} + \frac{y^2}{b^2} = 1$$

to the focus at $F(c, 0)$, as shown in Figure 8.42. First we must find an expression for the distance from an arbitrary point $P(x, y)$ on the ellipse to the focus. Using the distance formula, we obtain

$$D = \sqrt{(x - c)^2 + y^2}.$$

This expression involves both x and y. We can eliminate one of them, say y, by using the fact that x and y satisfy the equation of the ellipse, $\frac{x^2}{a^2} + \frac{y^2}{b^2} = 1$. Solving for y^2, we get

$$y^2 = b^2 \cdot \left(1 - \frac{x^2}{a^2}\right).$$

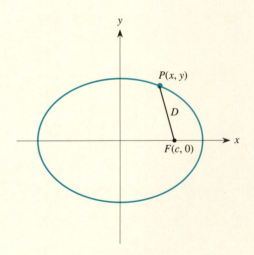

FIGURE 8.42

Substituting this into the expression for the distance, we obtain

$$D = \sqrt{(x^2 - 2cx + c^2) + b^2\left(1 - \frac{x^2}{a^2}\right)}$$

$$= \sqrt{x^2\left(1 - \frac{b^2}{a^2}\right) - 2cx + (c^2 + b^2)}.$$

However, since $a^2 = b^2 + c^2$, we have

$$1 - \frac{b^2}{a^2} = \frac{a^2}{a^2} - \frac{b^2}{a^2}$$

$$= \frac{a^2 - b^2}{a^2}$$

$$= \frac{c^2}{a^2}.$$

Consequently, the above equation reduces to

$$D = \sqrt{x^2\left(\frac{c^2}{a^2}\right) - 2cx + a^2}$$

$$= \sqrt{\frac{c^2x^2 - 2a^2cx + a^4}{a^2}}$$

$$= \sqrt{\frac{(cx - a^2)^2}{a^2}}.$$

Furthermore, for points on the ellipse, we must have $x \leq a$. Also, since c must be positive and $c < a$, it follows that

$$cx < ax \leq a^2,$$

and so

$$cx - a^2 < 0 \qquad \text{or} \qquad a^2 - cx > 0.$$

Therefore, the above expression for D becomes

$$D = \sqrt{\frac{(a^2 - cx)^2}{a^2}}.$$

When we take the square root in this expression for D, we finally obtain

$$D = \frac{a^2 - cx}{a}$$

as the (positive) distance from any given point on the ellipse to the focus.

Now let's investigate the average value for this distance using a computer simulation. We start by considering some simple numbers for a and b to see what happens. Suppose we use $a = 5$ and $b = 4$, so that $c = 3$ (using the fact that $a^2 = b^2 + c^2$), and ask for $n = 2000$ random points between -5 and 5. One typical run of the program then produces an approximation to the average distance equal to 5.00013, which is suggestively close to 5.

Repeated runs of the program with the same set of values invariably produce an average distance very close to 5. Suppose instead we use the ellipse with $a = 10, b = 4$, and $c = \sqrt{a^2 - b^2} = \sqrt{84}$. The results of repeated runs of the program invariably produce approximations to the average distance very close to 10. If you conduct such experiments yourself with other values for a and b, you will discover that the results are always close to the value of a and that b does not seem to have any effect on the average distance.

Let's see why this average distance actually turns out to be precisely equal to the value of a. We previously obtained the expression

$$D = \frac{a^2 - cx}{a} = a - \frac{c}{a}x$$

for the distance of any point on an ellipse to the focus at $F(c, 0)$. Notice that this expression is a *linear* function of x; its graph is a straight line with a negative slope of $-\frac{c}{a}$, as shown in Figure 8.43. Its vertical intercept is at a and it extends from $x = -a$ to $x = a$, the endpoints of the ellipse. For a linear function, however, the concept of average value is very easy to translate into a precise value: The average value for a linear function on an interval is exactly the height from the x-axis to the midpoint of the line segment. For the linear distance function

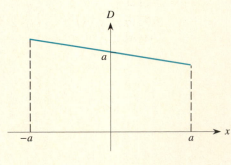

FIGURE 8.43

$$D = a - \frac{c}{a}x,$$

where x ranges from $-a$ to a, the midpoint of the line segment occurs at $x = 0$ and consequently the average height to the line is given by

$$D_{\text{average}} = \frac{a^2 - 0}{a} = a.$$

(Unfortunately, the average value for most other functions does not work out quite so neatly.)

More generally, the average distance from points on any ellipse to a focus is precisely equal to the parameter a representing the length of the *semimajor axis*. In terms of the average distance from the Earth to the sun, this is simply the length of the semimajor axis, which is 92.96 million miles, as we saw in Example 4 of Section 8.3.

There is another interesting interpretation of this fact. Recall that the aphelion distance for the Earth is 94.54 million miles and the perihelion distance is 91.38 million miles. These distances can be expressed as:

$$\text{perihelion} = a - c$$
$$\text{aphelion} = a + c$$

and so their arithmetic average is

$$\frac{1}{2}(\text{perihelion} + \text{aphelion}) = \frac{1}{2}(a - c + a + c) = a;$$

that is, the average distance of the Earth (or any other planet) from the sun is just the average of its perihelion and aphelion distances.

We can give another interpretation to this result based on Figure 8.44. Consider the focus $F_1(c, 0)$ and the point P_1 on the ellipse at a distance d from F_1. Using the symmetry of the ellipse, we know that there is a comparable point P_2 at the same distance d to the other focus $F_2(-c, 0)$. From the geometric definition of the ellipse, we know from Section 8.3 and a little algebra that the sum of the two distances, $|FP_1|$ and $|FP_2|$, must be equal to the constant $2a$ and hence the average of these two distances is just a. Since this can be done for all possible pairs of symmetric points on the ellipse, it is clear that the average distance from F to all points on the ellipse must also be a.

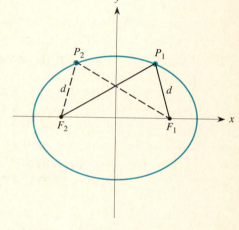

FIGURE 8.44

Notice that the above derivations concerned the average distance from points on the ellipse to a focus. We could also find the average height above the x-axis for points on an ellipse[2]

$$\frac{x^2}{a^2} + \frac{y^2}{b^2} = 1.$$

Suppose we consider the upper half of the ellipse so that the height above the x-axis corresponding to an arbitrary point $P(x, y)$ is simply

$$y = b \cdot \sqrt{1 - \frac{x^2}{a^2}}.$$

To estimate the average value of this height, we again use a computer simulation. As before, suppose we start with $a = 5$ and $b = 4$. A typical run of such a program with x in the interval -5 to 5 and 2000 random points produces an approximate value for the average height of 3.1485, which is very suggestive of π. Additional runs with the same sets of values likewise produce estimates that are relatively close to π, which suggests that the true value likely involves π. Let's continue the investigation with different values of a. If $a = 10$ and $b = 4$, then one set of 2000 randomly generated points leads to an approximation of 3.1314, which is also very close to π.

---

[2]For simplicity we will write a with the x^2-term and b with the y^2-term regardless of whether the major axis of the ellipse is horizontal or vertical.

Similarly, another run using $a = 3$ and $b = 4$ leads to an estimated average height of 3.1388.

Let's consider what happens with different values for b. Suppose we try $a = 3$ and $b = 1$ for comparison. The program now produces, on one sample run, an estimate for the average height of 0.7824, which is approximately $\pi/4$. A comparable run with $a = 3$ and $b = 12$ leads to an estimate of 9.3885, which is close to 3π. Based on these experiences, we might conjecture that the average height above the x-axis of points on an ellipse is independent of the value of a and is given by

$$\text{Average height} = \left(\frac{\pi}{4}\right)b.$$

EXERCISES

1. The number of hours H of daylight in San Diego as a function of the date is approximately

$$H = 12 + 2.4 \sin\left(\frac{2\pi}{365}(t - 80)\right),$$

   where t is the number of days since January 1.

   a. Estimate the average daily number of hours of daylight over the course of one year.
   b. Does your result make sense? Why?

2. During the Apollo moon program, the Apollo spacecraft was first placed into an elliptic orbit about the moon that varied from about 232 to 264 miles above the surface of the moon. The center of the moon was one focus of the ellipse. The radius of the moon is about 1080 miles.

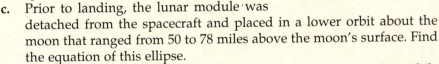
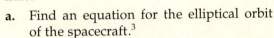

   a. Find an equation for the elliptical orbit of the spacecraft.[3]
   b. What was the average distance of the spacecraft above the moon's surface?
   c. Prior to landing, the lunar module was detached from the spacecraft and placed in a lower orbit about the moon that ranged from 50 to 78 miles above the moon's surface. Find the equation of this ellipse.
   d. Find the average height of the lunar module above the surface while it was moving in its lower orbit.

3. Estimate the average value of $f(x) = \sqrt{x}$ from 0 to 1 if you have an appropriate computer program available.

---

[3]The problem is adapted from *Intermediate Algebra: A Functional Approach* by Shoko Brant and Edward Zeidman, HarperCollins, 1996.

4. Estimate the average value of $f(x) = \sqrt{x}$ from 0 to 4.

5. Estimate the average height of points on the parabolic arch $y = 4 - x^2$ from $x = -2$ to $x = 2$.

6. What is the average height above the x-axis of all points on the circle $x^2 + y^2 = a^2$?

7. In our investigation related to the shape of the Gateway Arch in St. Louis in Section 4.2, we found that its height can be reasonably well modeled by the fourth degree polynomial

$$H(x) = -0.0000000327x^4 - 0.00282x^2 + 644.25,$$

where x is between -315 and 315 feet. Use an appropriate computer program to estimate the average height of the arch.

8. Suppose that for a headache you take two aspirin tablets amounting to 650 mg. After the aspirin has been absorbed into the bloodstream, it is slowly removed by the kidneys as they filter the impurities out of the blood. The level of aspirin remaining in the bloodstream can be well modeled by the function $A(t) = 650(0.976)^t$, where t is measured in minutes. After two hours, or 120 minutes, the effective level of aspirin in the blood is equivalent to 35 mg.

   a. The average of the high and low levels for the aspirin over the 120-minute interval is 342.5 mg. From your knowledge of the behavior of a function such as $A(t)$, do you think the average value for the function over this time interval is higher or lower than 342.5 mg? Explain your answer.

   b. Use the random number generator RAND of your calculator (see Exercise 4 of Section 1.6) to generate 10 random values of t between 0 and 120. Calculate the average of the corresponding values for the function $A(t)$. How does it compare to the average you found in part (a)? Does it support or contradict your answer in the second half of part (a)?

   c. If you have access to a computer or calculator program that estimates the average value of a function over an interval, find a better estimate for the average value of the level of aspirin in the blood over this 120-minute period.

8.6 *The Polar Coordinate System*

As we discussed in Section 8.1, the *polar coordinate system* is based on the idea that every point in the plane must lie on some circle centered at the origin. In this coordinate system, the origin is known as the **pole.** To locate a point P in such a system, we must indicate the radius r of the particular circle on which P lies. That is, does P lie on a circle of radius $r = 3$ or a circle of radius $r = 4$ or a circle of radius $r = 4.2689$? Knowing that P lies on a specific circle still does not locate the point P exactly. We must also specify where the point lies on that circle. Is it at an angle of $30°$ or an angle of $45°$ or an angle of $-63°$ from the horizontal?

To formulate these ideas more precisely, let's develop some appropriate terminology. We first introduce a horizontal axis starting at the pole and pointing to the right. It is called the **polar axis** and serves as a reference. The distance from the pole to the point P, which is equivalent to the radius of a circle centered at the pole, is denoted by the coordinate r. To locate a specific point P on this circle of radius r, we must indicate how far around the circle P lies, starting from the polar axis. We measure this distance around the circle in terms of an angle coordinate θ drawn in a counterclockwise, or positive, direction from the polar axis, as shown in Figure 8.45. Thus, any point in the plane can be located if we know its distance r from the pole (to determine a circle) and the angle θ around this circle. Since the *polar coordinates* of the point P consist of r and θ, we write $P(r, \theta)$. For example, a point that lies 5 units away from the pole at an angle of $60°$, or $\frac{\pi}{3}$ radians, with the polar axis has polar coordinates $P(5, 60°)$ or $P\left(5, \frac{\pi}{3}\right)$. See Figure 8.46. Similarly, the point Q that lies 3 units away from the pole at an angle of $\frac{2\pi}{3}$ or $120°$ has coordinates $P\left(3, \frac{2\pi}{3}\right)$, as shown in Figure 8.46 also.

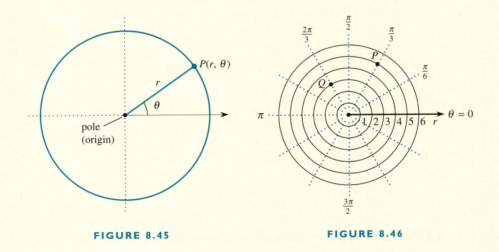

FIGURE 8.45   **FIGURE 8.46**

We can picture the polar coordinates of a point P in the following alternative way. Any point is located on a line passing through the pole, as measured by the angle of inclination θ, which is known as the **polar angle.** The particular location of the point P along that line is determined by its distance, r, from the pole. Thus the point P can be pictured as lying at the intersection of a line through the pole and a circle centered at the pole. This approach has an added geometric advantage. Recall from geometry that the radius drawn to any point on a circle is perpendicular to the tangent line at that point. Therefore the polar coordinates of a point P are determined by the intersection of two curves that we can think of as being "perpendicular" at the point. This is analogous to what we do in rectangular coordinates where the vertical and horizontal lines that determine a point are perpendicular to each other.

Although there are many advantages to working with a polar coordinate system, it does have one disadvantage. In rectangular coordinates, every point has a unique pair of coordinates. However, every point in polar coordinates has more than one address. Consider the point 1 unit to the right of the pole on the polar axis. According to our discussion so far, you might conclude its polar coordinates are $r = 1$ and $\theta = 0$. However, with a little thought, it should be evident that the address for this point could also be $r = 1$ and $\theta = 2\pi$, or $r = 1$ and $\theta = 4\pi$, and so on. Thus there are infinitely many polar coordinate representations of the same point.

In fact, there are still other ways to give the polar address of this point. In general, any angle θ measured from the polar axis in the counterclockwise direction is considered positive; any angle θ measured from the polar axis in the clockwise direction is considered negative. See Figure 8.47. Thus, our point on the polar axis could also be written as $P(1, -2\pi)$, for instance.

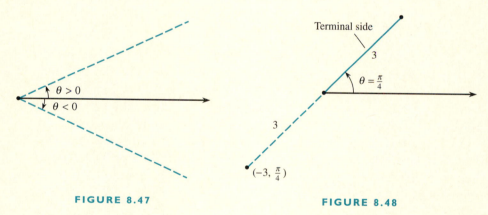

FIGURE 8.47   FIGURE 8.48

Furthermore, we will encounter some situations in the next section where a given angle θ gives rise to a negative value for r. Let's see what it means for r to be negative, since it certainly cannot represent the radius of a circle. If $\theta = \frac{\pi}{4}$ and $r = 3$, then we simply measure off a distance of 3 units along the terminal side of the angle $\theta = \frac{\pi}{4}$. However, if $\theta = \frac{\pi}{4}$ and $r = -3$, how do we locate the corresponding point? One way is to extend the terminal side of the angle backward through the pole, as shown in Figure 8.48, and measure off a distance of 3 units along this extension.

Let's look at this question a little more formally. When we draw any angle θ, it determines a terminal side OP from the pole through some point P (see Figure 8.49). The obvious polar coordinate representation for that point is $P(r, \theta)$, where r is positive since the distance is measured along the terminal side. However, we can also represent that point by considering the angle $\theta + \pi$, corresponding to an additional rotation of π radians or $180°$, and measuring a distance of r from the pole in the opposite direction. In such a case, we think of r as negative and the polar coordinates of the point are $P(-r, \theta + \pi)$. Thus, if a point P is located at $\theta = \pi/4$ and $r = 3$, we can consider the associated angle $\pi/4 + \pi = 5\pi/4$ and assign the coordinates $(-3, 5\pi/4)$ to the point P as well.

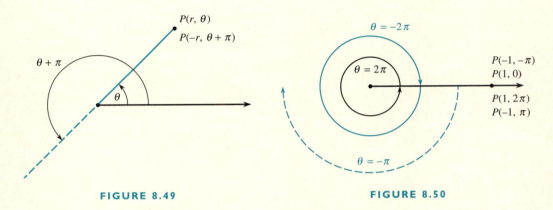

FIGURE 8.49   FIGURE 8.50

With these ideas in mind, we can find even more ways to write our earlier point with $r = 1$ and $\theta = 0$. For example, we can obtain this point when $r = -1$ and $\theta = \pi$ or when $r = -1$ and $\theta = -\pi$. See Figure 8.50. Can you think of any other coordinates for this point when r is negative? when θ is negative?

We therefore see that any given point in the polar coordinate system has an infinite number of pairs of coordinates. Even the pole, where $r = 0$, has infinitely many representations since it can be thought of as corresponding to any possible angle θ.

Transforming Between Polar and Rectangular Coordinates

Often, it is useful to think of the two coordinate systems, polar and rectangular, as being superimposed. In such a case, the pole and the origin are the same point; the polar axis and the positive x-axis coincide. The question then is, how do the coordinates of a point P in one system relate to the coordinates of the same point in the other system? That is, how do we transform the rectangular coordinates of a point into the equivalent polar coordinates and vice versa?

Suppose that we start with a point P whose polar coordinates (r, θ) are given, and we wish to determine the corresponding rectangular coordinates, x and y. From the right triangle shown in Figure 8.51, it is clear that

$$x = r \cdot \cos \theta$$
$$y = r \cdot \sin \theta.$$

For example, if the point P has polar coordinates $r = 5$ and $\theta = \pi/3$, then the corresponding rectangular coordinates x and y are

$$x = r \cdot \cos \theta = 5 \cdot \cos \frac{\pi}{3} = 5\left(\frac{1}{2}\right) = 2.5$$

$$y = r \cdot \sin \theta = 5 \cdot \sin \frac{\pi}{3} = 5 \cdot \left(\frac{\sqrt{3}}{2}\right) = 4.33.$$

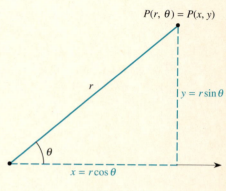

FIGURE 8.51

For the reverse problem, suppose that we start with a point whose rectangular coordinates are $P(x, y)$, as shown in Figure 8.52. We want to find corresponding polar coordinates r and θ. First observe that r is the distance from the pole (origin) to P. The Pythagorean theorem gives

$$r^2 = x^2 + y^2$$

so that

$$r = \pm\sqrt{x^2 + y^2}.$$

From the diagram, we also observe that

$$\tan \theta = \frac{y}{x}$$

so that

$$\theta = \text{Arctan}\,\frac{y}{x}.$$

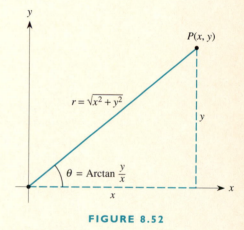

FIGURE 8.52

Given the rectangular coordinates (x, y) of a point, we can find the polar coordinates (r, θ) using

$$r = \pm\sqrt{x^2 + y^2}$$

$$\theta = \text{Arctan}\,\frac{y}{x}.$$

For example, if the rectangular coordinates of a point are $x = 3$ and $y = 4$, then we immediately find that

$$r = \pm\sqrt{x^2 + y^2} = \pm\sqrt{3^2 + 4^2} = \pm\sqrt{25} = \pm5$$

and

$$\theta = \text{Arctan}\,\frac{4}{3} = 0.927 \text{ radian.}$$

Since the point is in the first quadrant, we have $P(r, \theta) = P(5, 0.927)$. There are infinitely many other possibilities, including the representations $P(5, 0.927 + 2\pi)$, $P(5, 0.927 + 4\pi)$, ..., $P(5, 0.927 + 2n\pi)$, for any integer n, either positive or negative. An alternative set of polar coordinates that locate the same point is $(-5, \pi + 0.927)$ or approximately $(-5, 4.069)$, as well as $P(-5, 4.069 + 2n\pi)$ for any integer n. It is extremely important to plot the point in order to decide which r to match with which θ.

Polar coordinates are particularly useful in representing situations where there is a single special point and all other ideas of interest are centered at that point. For instance, in physics, the total mass of a body is often assumed to be at a single point corresponding to the pole. Thus satellites can often be thought of as moving in circular orbits about a planet located at the pole of a polar coordinate system. Similarly, the magnetic field associated with a magnet can be thought of as being centered at the pole of a polar coordinate system and all related phenomena are often best expressed in terms of polar coordinates.

EXERCISES

1. For each of the points P, Q, R, and S shown in the accompanying figure,

    a. write polar coordinates with r and θ both positive;

    b. write polar coordinates with r positive and θ negative;

    c. write polar coordinates with r negative and θ positive;

    d. write polar coordinates with r and θ both negative.

2. A merry-go-round at an amusement park has an inner radius of 9 feet and an outer radius of 26 feet. There are five concentric circles of horses, three feet apart, starting with the innermost circle at a distance of 10 feet from the center. From a reference point at the entrance gate of the ride:

    a. What are the polar coordinates of the horse in the outer or fifth circle that is one-third of the way around the merry-go-round to your right?

    b. What are the polar coordinates of the horse in the second circle that is one-fifth of the way around to your left?

3. Transform the rectangular coordinates into equivalent polar coordinates. Sketch the location of each point in the polar coordinate plane.

    a. $P(4, 4)$ b. $P(-4, 4)$ c. $P(-4, -4)$ d. $P(4, -4)$
    e. $P(3, -4)$ f. $P(-3, 4)$ g. $P(8, 3)$ h. $P(3, 8)$

4. Transform the polar coordinates into equivalent rectangular coordinates. Indicate the location of each point graphically in the polar plane.

   a. $P(5, 0)$ b. $P\left(5, \frac{\pi}{2}\right)$ c. $P\left(-5, \frac{\pi}{2}\right)$

   d. $P(-5, 0)$ e. $P\left(3, \frac{\pi}{3}\right)$ f. $P\left(-3, \frac{-\pi}{3}\right)$

   g. $P\left(2, \frac{3\pi}{2}\right)$ h. $P\left(2, \frac{5\pi}{4}\right)$ i. $P\left(2, -\frac{5\pi}{3}\right)$

5. A satellite is in a circular orbit 22,800 miles above the surface of the Earth about the equator. The radius of the Earth is about 4000 miles. The longitude line running north–south from the north pole to the south pole through Greenwich, England, serves as the 0° reference. Since the circumference of the Earth is about 24,000 miles, each 15° of longitude corresponds to about 1000 miles around the equator.

   a. What are the polar coordinates of the Ugandan capital Kampala, which is 3000 miles east (positive direction) of the Greenwich baseline?
   b. What are the polar coordinates of the capital of Borneo, which is about 7500 miles east of the Greenwich baseline?
   c. What are the polar coordinates of the satellite when it passes over Quito, Ecuador, which is 5200 miles west of the Greenwich baseline?

8.7 *Families of Curves in Polar Coordinates*

In the last section, we introduced the notion of polar coordinates and considered coordinates (r, θ) for individual points in such a system. A far more interesting and useful question is: How do we represent curves and families of curves in polar coordinates?

Recall that, in rectangular coordinates, the curve associated with an equation $y = f(x)$ consists of all points (x, y) whose coordinates satisfy the equation. We adopt the comparable notion when working with polar coordinates with the understanding that it is only necessary that one representation of a point in polar coordinates satisfies the equation.

To begin, it is usually much simpler to think of the angle θ as the independent variable and the distance r from the pole as the dependent variable. Thus for most functions in polar coordinates, we write $r = f(\theta)$ for some set of values of the angle θ. Then for each allowable value of θ, the function determines a corresponding value for r and the pair (r, θ) represents the polar coordinates of a point P in the plane. The totality of all such points determined by the equation constitutes the graph of the function. Note that writing the polar coordinates of a point as (r, θ) reverses our usual notation of writing the independent variable first and the dependent variable second as is done in rectangular coordinates with (x, y).

Let's begin with some particularly simple examples. Consider the equation

$$r = f(\theta) = c,$$

where c is a constant. This says that, no matter what the angle θ is, the distance r from the pole is the constant c. The set of all points that satisfy this condition forms the circle of radius c centered at the pole (see Figure 8.53).

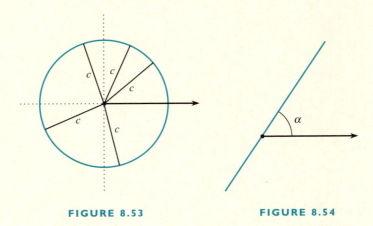

FIGURE 8.53   **FIGURE 8.54**

By way of comparison, consider the equation $\theta = \alpha$, where α is a constant, and where the distance r is not explicitly mentioned. No matter what distance we use, the corresponding point is always located at an angle α as measured from the horizontal polar axis. The set of all such points forms a straight line, inclined at the angle α, which passes through the pole (see Figure 8.54).

Shapes that are far more interesting and intricate than a circle and a line arise from relatively simple polar equations. Our goals are to investigate the types of shapes that arise and to determine some of the underlying patterns in the shapes of various families of polar coordinate curves. In all of the following, we assume that you will use your graphing calculator set in polar mode or that you have access to a polar graphing program for a computer.

Working with polar coordinates often gives us a special advantage over working with rectangular coordinates. Consider the apparently simple curve shown in Figure 8.55. It is known as an Archimedean spiral and its equation in rectangular coordinates is

$$y = x \cdot \tan\left(\sqrt{x^2 + y^2}\right).$$

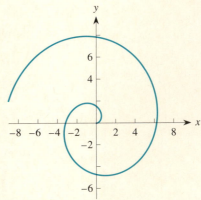

(Recall that a curve such as this does not represent a function.) Now, let's find an equivalent equation in polar coordinates using the transforming equations we derived in the previous section:

$$r^2 = x^2 + y^2$$

$$\theta = \text{Arctan}\,\frac{y}{x}$$

FIGURE 8.55

and
$$x = r \cdot \cos\theta$$
$$y = r \cdot \sin\theta.$$

Upon substituting these expressions into $y = x \cdot \tan\left(\sqrt{x^2 + y^2}\right)$, we get the far simpler and more comprehensible equation $r = f(\theta) = \theta$. Think of this as starting when $\theta = 0$, so $r = 0$ also. As θ increases, so does r. The corresponding points spiral out farther and farther from the pole.

Think About This

What happens to the spiral if $\theta < 0$?

Let's consider the polar function $r = f(\theta) = \cos\theta$, whose graph is shown in Figure 8.56. By eye, the curve appears to be circular, appears to pass through the pole, and appears to be symmetrical about the polar axis. How can we verify these observations? To identify the curve as a circle, we can try to express it in rectangular coordinates where the equation of a circle would be recognizable. While we could attack this head on and attempt to substitute the transforming expressions for r and θ into the equation $r = \cos\theta$, it is much easier to use a little trick. If we multiply both sides of the given equation by r, we obtain

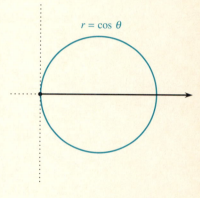

$r = \cos\theta$

FIGURE 8.56

$$r^2 = r \cdot \cos\theta$$

so that this is equivalent to the rectangular equation

$$x^2 + y^2 = x$$

or

$$x^2 - x + y^2 = 0.$$

After completing the square in x, we obtain

$$\left(x - \frac{1}{2}\right)^2 + y^2 = \frac{1}{4},$$

which is the equation of a circle with radius $\frac{1}{2}$ centered at the (rectangular) point $\left(\frac{1}{2}, 0\right)$. Thus as we conjectured, the circle is indeed symmetrical about the horizontal axis and it does pass through the pole.

Think About This

1. Describe the graph of $r = 5\cos\theta$.
2. Describe the graph of $r = a \cdot \cos\theta$ for any multiple a.
3. Describe the graph of $r = a \cdot \sin\theta$ for any multiple a.

The Family of Rose Curves

Let's consider some related curves in polar coordinates. When we graph the equation $r = \cos 2\theta$, we obtain the result shown in Figure 8.57. The graph shown corresponds to angles θ ranging from 0 to 2π; if you extend the values beyond this interval, in either direction, the same points are traced out, so the result is a periodic function with period 2π. If you experiment with your polar function grapher, you will notice that the picture shown is traced out repeatedly if you take a very large range of values for the angle θ. However, you should not just look at the completed shape, but rather consider this and the other polar coordinate curves we treat in a dynamic fashion. How are the particular shapes produced or traced out? Think of the cursor on the calculator or the computer screen as a moving point that traces out the curve, and observe carefully how the curve is generated.

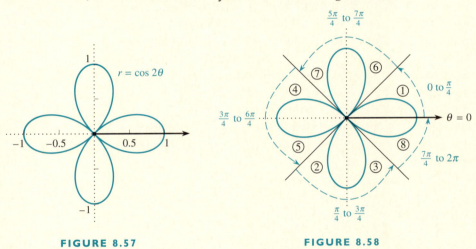

FIGURE 8.57   FIGURE 8.58

Notice that the graph shown in Figure 8.58 for $r = \cos 2\theta$ consists of four loops, each of equal size. (Actually, depending on the calculator or computer graphics package you use, there may be some distortion, so the loops may not appear to be precisely the same size.) To get a better feel for how the particular shape arises, watch carefully as the curve is traced out. Notice that it starts at the far right (corresponding to $\theta = 0$ where $r = 1$) and then loops around (portion ①) until it passes through the pole (corresponding to $\theta = \frac{\pi}{4}$ where $r = \cos 2\left(\frac{\pi}{4}\right) = \cos \frac{\pi}{2} = 0$). It then starts to form a second loop (portion ②) as r takes on negative values. Eventually, it completes the loop (portion ③) before the moving point again passes through the pole, this time at an angle of $\theta = \frac{3\pi}{4}$ so that again $r = 0$. It then begins to form the third loop (portion ④) and completes that loop (portion ⑤) when the moving point passes through the pole, where $\theta = \frac{5\pi}{4}$. It then forms a fourth loop (portions ⑥ and ⑦), for θ between $\frac{5\pi}{4}$ and $\frac{7\pi}{4}$ and finally completes the original loop (portion ⑧) as θ progresses on to 2π. The particular curve is known as a *four leaf rose*.

Think About This

1. What shape is produced if you graph $r = a \cdot \cos 2\theta$ for any multiple a?
2. Describe the graph corresponding to $r = a \cdot \sin 2\theta$ for any multiple a. How does it compare to the graph of the cosine function?

Let us now make a relatively simple change and consider $r = \cos 3\theta$ instead of $r = \cos 2\theta$. The resulting graph is shown in Figure 8.59, but again it is important to observe carefully how the curve is traced out. First of all, notice that the curve now consists of only three loops and that they are traced out for values of θ between 0 and π. For any angles outside the interval $[0, \pi]$, the same points are produced, so the polar curve is periodic with period π, even though the function $f(\theta) = \cos 3\theta$ is periodic with period $2\pi/3$. Notice also that the curve starts when $\theta = 0$ and $r = 1$ to produce the point at the far right. It then forms a half loop and passes through the pole when $\theta = \pi/6$. The lower left full loop is traced out for values of θ between $\pi/6$ and $\pi/2$. The upper left full loop is traced out as θ ranges from $\pi/2$ to $5\pi/6$. The bottom half of the right-hand loop is completed as θ ranges from $5\pi/6$ to π. The full curve is known as a *three leaf rose*.

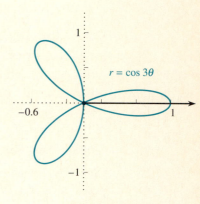

FIGURE 8.59

In general, the family of curves given by $r = \cos n\theta$ or $r = \sin n\theta$ for any positive integer n are called *rose curves*. In Figures 8.60(a) and 8.60(b), we show the graphs corresponding to $r = \cos 4\theta$ and $r = \cos 5\theta$ and note that they contain eight and five loops or *petals* respectively.

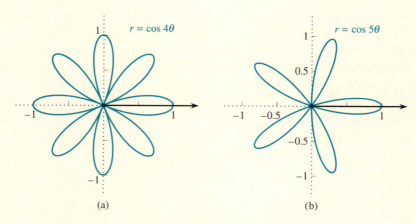

(a)   (b)

FIGURE 8.60

Think About This

1. Investigate some other cases using your polar function grapher until you can devise a rule to predict the number of petals in the rose curve $r = \cos n\theta$ for any integer n. Are there any numbers of petals that cannot occur in this family of rose curves?

2. What can you conclude about the related family of rose curves given by $r = \sin n\theta$?

The Family of Cardioids

Let's consider another family of polar coordinate curves, those given in the form $r = a(1 \pm \cos \theta)$. In Figure 8.61, we show the graph of $r = 1 + \cos \theta$. The heart-shaped appearance of this curve suggested the name given to it, a *cardioid*. Notice how the curve is traced out. Starting with $\theta = 0$ and $r = 2$, the curve begins at the point at the far right. As θ increases to $\frac{\pi}{2}$, the curve arches upward. It then bends downward and eventually inward to hit the pole when θ is π; the resulting point is called a *cusp*. As the angle θ increases from π to 2π, the curve traces out the mirror image of the upper half of the cardioid; this cardioid is symmetric about the polar axis. Those of you who have read Section 7.9 on chaos have seen that the primary central portion of the Mandelbrot set is a cardioid.

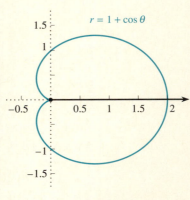

FIGURE 8.61

Think About This

1. a. What is the effect of a multiple a on the shape of the curve

$$r = a(1 + \cos \theta)?$$

   b. Describe the graph of the related equation $r = 1 - \cos \theta$. How does it compare with the cardioid $r = 1 + \cos \theta$?

2. Sketch some graphs of the related equations $r = 1 \pm \sin \theta$. Can you identify an axis of symmetry for them?

3. Suppose you combine the ideas on the rose curves and the cardioids to consider the class of polar equations of the form $r = 1 \pm \cos n\theta$ for different positive integers, n. Can you determine any patterns regarding their shapes?

The Family of Limaçons

An extension of the cardioid known as the *limaçon* is defined by the equations $r = a \pm b \cdot \cos \theta$ and $r = a \pm b \cdot \sin \theta$. In particular, we consider two cases: $a > b$ and $a < b$. When $a = b$, either of the two equations reduces to that of a cardioid. Figure 8.62 shows the graph of $r = 3 + 4 \cos \theta$. Notice how the curve starts at the far right, where $\theta = 0$ and $r = 7$. The curve then traces around the upper arch and eventually bends inward to pass through the pole. After passing through the pole, the curve traces out the small inner loop and then passes through the pole again. It then traces out the large outer loop below the polar axis, which is a mirror image of the large loop above the polar axis. The resulting curve is called a *limaçon with a loop*. (It comes from the Greek word *limax*, for snail, since the first half of the curve traced out from $\theta = 0$ to $\theta = \pi$ resembles a snail-like shape.) For $\theta > 2\pi$, the curve precisely repeats this behavior.

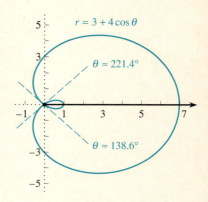

FIGURE 8.62

One question that we might want to ask is: What is the angle at which the curve passes through the pole? Actually, there are two such angles, one in the "second quadrant" and the other in the "third quadrant." To find these angles, we use the fact that the pole corresponds to $r = 0$. Therefore, if we set $r = 0$ in the equation, we get

$$3 + 4 \cos \theta = 0$$

$$\cos \theta = -\frac{3}{4}.$$

Thus the angle at which the limaçon passes through the pole must satisfy

$$\theta = \text{Arccos}\left(-\frac{3}{4}\right) = 2.419 \text{ radians or } 138.59°.$$

We use the symmetry of the cosine function to find the second solution at $\theta = 3.864$ radians or $221.41°$. These values certainly agree with the visual estimates you can make by looking at Figure 8.62.

Figure 8.63 shows the graph of the polar curve $r = 5 + 4 \cos \theta$. This curve is known as a *limaçon without a loop* or a *dimpled limaçon*. It is similar in appearance to a cardioid, but it does not pass through the pole.

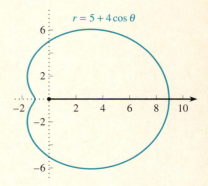

FIGURE 8.63

Think About This

1. Can you account for the fact that the curve $r = 5 + 4 \cos \theta$ never passes through the pole (where $r = 0$) and so never produces a loop?
2. Can you devise any criteria based on the values of a and b in $r = a + b \cos \theta$ so that you can know if there is a loop? Make sure that you graph a variety of different limaçons using your polar function grapher to collect enough information so that you are correct.
3. Describe the shape of limaçons given by $r = a - b \cos \theta$.
4. What happens in the related family of limaçons given by $r = a \pm b \sin \theta$?

We urge you to experiment with the curves generated by polar coordinate equations. You can get some incredibly striking effects by just creating strange combinations of different functions. For instance, the graph of the polar function

$$r = \sin^5 \theta + 8 \sin \theta \cos^3 \theta$$

is the butterfly-shape shown in Figure 8.64. We suggest you explore some family of polar functions in a systematic way, say $r = \sin^n \theta$ for various integers n. You may very well discover some fascinating new patterns and add some new items to the literature of mathematics.

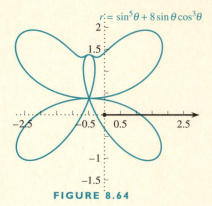
$r = \sin^5 \theta + 8 \sin \theta \cos^3 \theta$

FIGURE 8.64

EXERCISES

Graph the polar curves in Exercises 1–11 using your polar function grapher. For each, use a variety of different intervals for the angle θ until you obtain a "good" picture of the graph.

1. $r = \dfrac{4 \sin^2 \theta}{\cos \theta}$ (*Cissoid of Diocles*)

2. $r = \dfrac{1}{\sin \theta} - 2$ (*Conchoid of Nicomedes*)

3. $r = 4 \sin \theta \cos^2 \theta$ (*Bifolium*)   4. $r = 5\left(4 \cos \theta - \dfrac{1}{\cos \theta}\right)$ (*Trisectrix*)

5. $r = \dfrac{3 \sin \theta}{\theta}$ (*Cochleoid*)   6. $r = \dfrac{a}{\sqrt{\theta}}$ (*Lituus*)

7. $r = \dfrac{8}{\sin 2\theta}$ (*Cruciform*)   8. $r = \dfrac{10}{3 + 2 \cos \theta}$ (*Ellipse*)

9. $r^2 = 4 \cos 2\theta$ (*Lemniscate of Bernoulli*)
   (*Caution:* Be careful to restrict your attention to values of θ that cause the right-hand side to be positive.)

10. $r^3 = 4 \cos 3\theta$ (*Generalized Lemniscate*)
    (*Caution:* Some programs and calculators will not be able to evaluate the cube root of a negative term.)

11. $r = \dfrac{4}{\sin \theta}$

12–22.   Repeat Exercises 1–11 by changing some of the terms. What happens to the shape you produced if you use different values for the coefficients? What happens if you interchange sines and cosines? What happens if you change the multiples of θ? Keep a record of what you do and of your findings.

23. Consider the family of "hybrid rose curves" given by $r = \cos\left(\frac{a}{b}\theta\right)$ for any rational number $\frac{a}{b}$.[4]

    a. By experimenting with different combinations of a and b, can you determine any rules for predicting the number of (overlapping) loops that will result?

    b. Can you determine any rules for predicting the angle θ needed to trace out one complete petal of this curve?

    c. Can you determine any rules for predicting the total angle θ needed to trace out the entire curve? (*Hint:* Consider different cases, depending on whether a and b are odd or even.)

24. Consider the family of polar curves given by $r = \sin^n \theta$.

    a. After graphing the curves corresponding to $n = 1$ and $n = 2$, what shape do you expect if $n = 3$? if $n = 4$?

    b. Can you account for the fact that the shapes are not what you expected?

    c. Can you determine any pattern for the number of loops that will occur corresponding to any value of n?

    d. What range of angles corresponds to a complete curve? Do the same conclusions apply to $r = \cos^n \theta$?

25. Consider the family of generalized lemniscates given by $r^2 = \cos n\theta$. Can you find any pattern for the number and location of the loops that will result for any n? (Be very careful when you try to graph these curves; you must take into account ranges of angles for which the function is well-defined.)

26. Consider the family of generalized lemniscates given by $r^n = \cos n\theta$. Can you determine any pattern for the number and location of the loops that will result for any n?

27. Consider the family of generalized limaçons given by $r = a + b \cos n\theta$. Can you find any pattern for the number and location of the loops that will result for any n?

---

[4]These curves were studied in detail by a student, Kenneth Gordon, in an article *Investigating the Petals of Hybrid Roses*, Mathematics and Computer Education, vol. 26, (1992), 66–73.

CHAPTER SUMMARY

In this chapter, you have learned the following:

- What a coordinate system is.
- How to find the distance between points in the plane.
- How to find the midpoint of a line segment.
- How to find a point at any given distance along the line through two points.
- The parametric equations of a line.
- The equation of a circle.
- The equation of an ellipse, including finding its center, vertices, and foci.
- The reflection property of an ellipse and its applications.
- The equation of a hyperbola, including finding its center, vertices, and foci.
- The equation of a parabola, including finding its vertex, focus, and directrix.
- The reflection property of a parabola and its applications.
- The parametric representation for curves in the plane.
- How to estimate the average value of a function on an interval.
- What the polar coordinate system is and how to transform back and forth between polar and rectangular coordinates.
- The behavior of families of curves in polar coordinates.

REVIEW EXERCISES

1. Find an equation of the ellipse shown in each graph.

a.

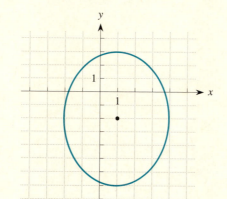

b.

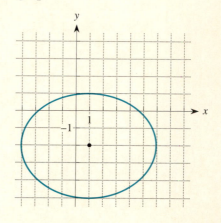

2. Find an equation of the hyperbola shown in each graph.

a.

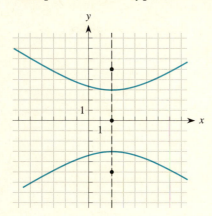

b.

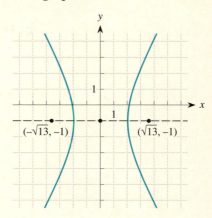

3. Find an equation of the parabola shown in each graph.

a.

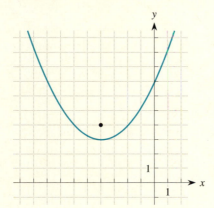

b.

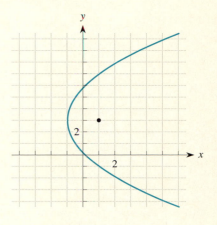

4. Identify the conic whose equation is $xy = 5$ and sketch the curve.

In Exercises 5–9, determine the equation for the standard form of the conic section. Identify the conic and sketch the curve.

5. $x^2 - 6x + y - 34 = 0$

6. $x^2 + y^2 - 8x + 6y + 9 = 0$

7. $2x^2 + 3y^2 + 20x - 12y + 28 = 0$

8. $3x^2 - 4y^2 - 6x - 24y - 45 = 0$

9. $3y^2 - 2x^2 - 12y + 12x - 24 = 0$

10. Wherever applicable, give the focus (foci), vertex (vertices), and center for each of the conics in Exercise 5–9.

11. Find the equation of the ellipse with foci $(-8, 0)$ and $(8, 0)$ and with vertices $(-10, 0)$ and $(10, 0)$.

12. Find the equation of the ellipse centered at $(-6, 3)$ with one focus at $(0, 3)$ and minor axis 4 units long.

13. Find the equation of the hyperbola centered at $(2, 3)$ with one focus at $(2, 7)$ and the corresponding vertex at $(2, 6)$.

14. Find the equation of the hyperbola that has vertices $(0, \pm 4)$ and passes through the point $(6, \sqrt{80})$.

15. Find the equation of the set of points $P(x, y)$ in the plane such that the sum of the distance from $(12, 0)$ to P and the distance from $(-12, 0)$ to P is 30.

16. An ellipse passes through the point $P\left(\frac{\sqrt{3}}{2}, 2\right)$ and has foci at $(-3, 0)$ and $(3, 0)$. Use the geometric definition of an ellipse to find the equation of this ellipse.

17. The ceiling of a whispering gallery is built so that the highest point of the structure is 16 feet above the floor. The floor has vertices 40 feet apart. Where along the axes should each person stand to be able to get the "whispering effect"? Ignore the height of the two people.

18. A lithotripter is a medical device used by doctors to break up kidney stones by bombarding them with intense bursts of sound waves using the reflection property of an ellipse. The device is situated in such a way that the sound waves emanate from one focus, reflect off an elliptic-shaped bowl, and come together to strike the kidney stone at the other focus. The distance between the two foci is 23 cm and the distance from the source focus to the vertex on the elliptic reflector bowl is 3 cm. Find the equation of an elliptic cross section of the lithotripter bowl.

19. When a raw chicken is taken from the kitchen table at a temperature of 70°F and placed into an oven kept at a constant 350°F to cook, the temperature of the chicken can be well modeled by the function $T(t) = 350 - 280(0.992)^t$. The chicken is typically removed from the oven when its internal temperature reaches 180°F, which takes about 60 minutes.

    a. The average of the high and low temperatures for the chicken over the 60-minute interval is 125°F. From your knowledge of the behavior of a function such as $T(t)$, do you think the average value for the function over this time interval is higher or lower than 125°F? Explain your answer.

    b. Use the random number generator RAND of your calculator (see Exercise 4 of Section 1.6) to generate 10 random values of t between 0 and 60. Calculate the average of the corresponding values for the function T. How does it compare to the average you found in part (a)? Does it support or contradict your answer in the second half of part (a)?

    c. If you have access to a computer or calculator program that estimates the average value of a function over an interval, find a better estimate for the average value of the temperature of the chicken while cooking.

20. Let $x = 4 - t$, $y = 2 + 3t$. Graph the points (x, y) for $t = -2, -1, 0, 1$, and 2. Find the function determined by the parametric equations.

21. Sketch the parametric curve given by

$$x = 3t, \qquad y = t^2 + 1 \quad \text{for } -2 \le t \le 2.$$

22. Let $x = t^2 + 3$, $y = t^3 - 1$.

    a. Graph the curve for $-4 \le t \le 4$.

    b. Eliminate t and write an expression for the curve in x and y.

    c. At what value of x is $y = 0$?

23. Sketch the curve

$$x = 1 - \log t, \qquad y = \log t \quad \text{for } 1 \le t \le 10.$$

a. Eliminate the parameter to find the expression for y as a function of x.
b. What is the largest possible domain for this function?

24. Graph the equations

$$x = \sin 2t, \quad y = \cos \tfrac{1}{2}t \quad \text{for } -2\pi \le t \le 2\pi.$$

25. Compare the graph in Exercise 24 to the graphs of

a. $x = \sin 4t, \quad y = \cos 2t$
b. $x = \sin 6t, \quad y = \cos 2t$
c. $x = \sin 6t, \quad y = \cos t$

Determine the period of each of the graphs in parts (a)–(c).

26. Use appropriate trig identities to eliminate t and write the following expression in terms of x and y

$$x = \cos 2t, \qquad y = \sin t.$$

27. A hypocycloid is the curve traced out by a point on a circle of radius a, rolling on the inside of a larger circle of radius b. Its parametric equations are

$$x = (b - a) \cos t + a \cos \left(\frac{b - a}{a} \right) t$$

$$y = (b - a) \sin t - a \sin \left(\frac{b - a}{a} \right) t$$

Suppose $a = 1$ and $b = 3$. Sketch the curve and show the orientation. How many revolutions will the small circle make before it reaches its starting point?

28. Transform each point given in rectangular coordinates into an equivalent point in polar coordinates.

a. $P(3, 3)$ **b.** $P(-1, 3)$ **c.** $P(4, -1)$ **d.** $P(0, 6)$.

29. Transform each point from polar coordinates to rectangular coordinates.

a. $P\left(3, \frac{\pi}{3}\right)$ **b.** $P\left(3, \frac{\pi}{4}\right)$ **c.** $P\left(4, \frac{3\pi}{2}\right)$

d. $P\left(4, \frac{5\pi}{4}\right)$ **e.** $P\left(5, \frac{5\pi}{6}\right)$ **f.** $P(5, 2)$

30. Using polar coordinates, sketch the curve

$$r = \frac{1}{1 + \cos \theta}.$$

Convert to rectangular coordinates and find the equation of the conic.

31.   The polar equation of a well-known family of curves is

$$r = \frac{1}{\sqrt{\dfrac{\cos^2 \theta}{a^2} + \dfrac{\sin^2 \theta}{b^2}}}$$

What are these curves?

In Exercises 32–35, compare the graphs of each set of equations.

32. $r = \cos \theta$, $r = \cos 2\theta$,   and $r = \cos 4\theta$.

33. $r = \cos \theta$, $r = \cos 3\theta$,   and $r = \cos 5\theta$.

34. $r = \sin \theta$, $r = \sin 2\theta$,   and $r = \sin 4\theta$.

35. $r = \sin \theta$, $r = \sin 3\theta$,   and $r = \sin 5\theta$.

APPENDICES

A: *Introduction to Trigonometry*

A.1   The Tangent of an Angle

Suppose your math instructor has assigned you the task of calculating the height of a tall flagpole in the middle of your campus. The direct approach would be to climb to the top, release a string until the bottom reaches the ground, and then measure the length of string. However, this is not the simplest method and you likely would be either unwilling or unable to do it. So, you would have to come up with some less physical approach.

Assume that at the time you go out to the flagpole, you notice that the pole is casting a shadow that is 66 feet long. Let's see how we can use this extra piece of information to determine the height of the pole. Suppose you enlist the aid of your friend Ron, who is exactly six feet tall. Have him stand in the shadow cast by the pole in such a way that the tip of his shadow falls exactly on the same spot, *A*, as the tip of the shadow of the flagpole, as shown in Figure A.1. Suppose you measure the length of his shadow as $8\frac{1}{4}$ feet. The two triangles shown, *ABC* and *ADE*, are similar (see Appendix C) because the angles are the same and so the corresponding sides are proportional. Therefore,

$$\frac{\text{Ron's height}}{\text{Length of his shadow}} = \frac{\text{Height of pole}}{\text{Length of pole's shadow}}$$

$$\frac{6}{8.25} = \frac{\text{Height of pole}}{66}.$$

Thus, we can calculate the height of the pole as $\frac{6 \times 66}{8.25} = 48$ feet.

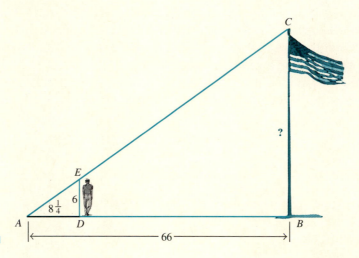

FIGURE A.1

A-1

Is this correct? You can check it out by enlisting the aid of another friend, Sue, who is five feet tall. Have her stand in the appropriate spot so that the tip of her shadow precisely matches the end of the pole's shadow. Suppose that the length of her shadow turns out to be $6\frac{7}{8}$ feet. Again, this leads to another right triangle, *AFG*, that is similar to the previous two, as shown in Figure A.2. Using the fact that corresponding sides are proportional, we get

$$\frac{\text{Sue's height}}{\text{Length of her shadow}} = \frac{\text{Height of pole}}{\text{Length of pole's shadow}}$$

$$\frac{5}{6\frac{7}{8}} = \frac{\text{Height of pole}}{66}.$$

Again, we find that the height of the pole is $(5 \times 66)/6\frac{7}{8} = 48$ feet.

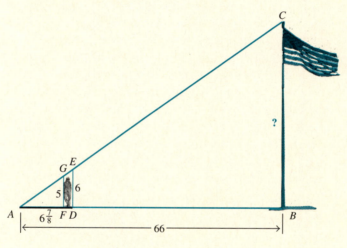

FIGURE A.2

Let's look at this situation from a slightly more sophisticated point of view to get a better idea of what is happening. In each of the triangles shown in Figure A.3, the various lengths are different, but the angles in similar positions all remain the same size. Consider the angle *CAB* (which is the same as angle *EAD* and angle *GAF*.) We call this the *angle of inclination* and denote it by the Greek letter θ (theta). For the right triangles shown in Figure A.3, we can measure this angle and find it to be $\theta = 36°$. Thus, in *any* right triangle where the angle of inclination is 36°, the ratio of the vertical height to the horizontal distance or width will always be the same, in this case,

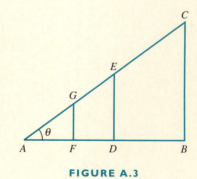

FIGURE A.3

$$\frac{\text{height}}{\text{width}} = \frac{6}{8.25} = \frac{5}{6\frac{7}{8}} = 0.727.$$

Of course, if the angle θ changes, then there is a different configuration of height and width and their ratio will also be different. That is, the ratio of height to width depends only on the measure of the angle of inclination θ; this ratio is a function of that angle. We call this function the **tangent of the angle**, the **tangent ratio**, or the **tangent function** and write it as

$$\tan \theta = \frac{\text{height}}{\text{width}}.$$

Use your calculator, in degree mode, to verify that $\tan 36° = 0.7265$.

Since we are concerned exclusively with right triangles here, the angle θ must be between $0°$ and $90°$ and so, for our purposes, the domain of the tangent function consists of all angles $0° < \theta < 90°$. (Later we will see how this can be extended to a larger domain.) Further, we can have a right triangle in a different orientation, as shown in Figure A.4, so the words height and width may not be quite appropriate. The standard terminology is to call the two sides of the triangle that flank the right angle the **opposite** and the **adjacent** sides with respect to an angle θ. Consequently

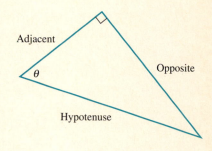

FIGURE A.4

$$\tan \theta = \frac{\text{opposite}}{\text{adjacent}}.$$

The third side of the triangle, which is opposite the right angle, is called the **hypotenuse.**

For example, recall from geometry that in a 45–45–90 right triangle, the two sides flanking the hypotenuse are equal and, by the Pythagorean theorem,

$$c^2 = a^2 + a^2 = 2a^2,$$

so $c = \sqrt{2}\,a$. That is, the hypotenuse must be $\sqrt{2}$ times the length of either side. See Figure A.5. In this triangle with angle $\theta = 45°$ and sides a, a, and $\sqrt{2}\,a$, the ratio of the side opposite θ to the side adjacent θ is clearly 1; using a calculator, you also can verify that $\tan 45° = 1$.

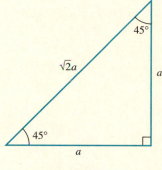

FIGURE A.5

Similarly, in a 30–60–90 right triangle, recall from geometry that the side opposite the 30° angle is one-half the hypotenuse. In such a triangle, suppose that the side opposite the 30° angle has length a and the hypotenuse has length $2a$, as shown in Figure A.6. We find the length of the third side from the Pythagorean theorem. Because $a^2 + b^2 = c^2$, we have $b^2 = c^2 - a^2$, and so

$$b^2 = (2a)^2 - a^2 = 4a^2 - a^2 = 3a^2$$
$$b = \sqrt{3}a.$$

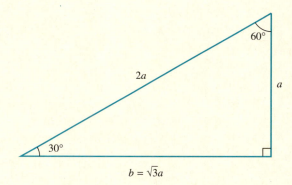

FIGURE A.6

Consequently, for an angle of 30°, the ratio of the opposite side to the adjacent side is

$$\tan 30° = \frac{a}{\sqrt{3}a} = \frac{1}{\sqrt{3}} = 0.577.$$

Alternatively, using a calculator, we find tan 30° = 0.577. Similarly, to find the tangent of 60°, we note that the side opposite the 60° angle in the triangle is now $\sqrt{3}a$ and the side adjacent to it is now a, so that

$$\tan 60° = \frac{\sqrt{3}a}{a} = \sqrt{3} = 1.732.$$

For any angle θ between 0° and 90°, you can use a calculator to obtain a value for tan θ. For instance,

$$\tan 10° = 0.176,$$
$$\tan 20° = 0.364,$$
$$\tan 50° = 1.192,$$
$$\tan 80° = 5.671.$$

Notice that as θ increases toward 90°, the value of tan θ also increases; the tangent is an increasing function of θ. Does this make sense? Picture walking toward the 556-foot high Washington Monument while keeping your eye fixed on the top of the monument, as shown in Figure A.7. The vertical height remains the same while the horizontal distance gets smaller and smaller. The nearer you get to the monument, the larger the angle of inclination and the larger the ratio of the fixed vertical height to the diminishing horizontal distance. By the time your eye is practically touching the side of the monument and the angle is virtually 90°, the value for the tangent function has gotten very large indeed. The tangent function is *not defined* for $\theta = 90°$.

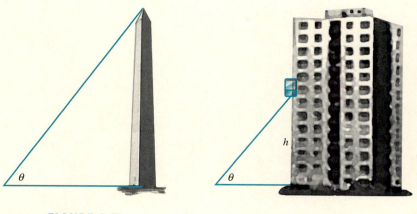

FIGURE A.7   FIGURE A.8

What about the tangent of 0°? Suppose that you are standing across the street from a glass elevator descending along the outside of a tall building, as shown in Figure A.8. Now the horizontal distance is fixed, the vertical height is decreasing, and the angle is decreasing toward 0°. Therefore the value of the tangent function is likewise diminishing because it is the ratio of the diminishing vertical height and the fixed horizontal distance. Clearly, tan 0° is 0. We can therefore conclude that the domain of the tangent function can be extended to $0° \leq \theta < 90°$.

Let's consider the values for the tangent function and investigate their growth pattern. Using a calculator, we obtain the following values.

θ	0°	10°	20°	30°	40°	50°	60°	70°	80°
tan θ	0	0.176	0.364	0.577	0.839	1.192	1.732	2.747	5.671

Notice that as the angle θ increases from 0° to 10° to 20°, and so on, the change in the function clearly is not constant. If the tangent function were linear, then we would expect, for instance, tan 40° to be twice tan 20°, and this is not the case. In fact, the tangent function grows much more rapidly than a linear function, particularly for angles considerably larger than 0°. Actually, the tangent function is growing ever more quickly, so the function is concave up. How does the growth pattern compare to that of an exponential function? If you examine the successive ratios of the values of tan θ, you will find that they are not constant, but rather are increasing considerably. In fact, the tangent function grows extremely rapidly near $\theta = 90°$ since $\theta = 90°$ is a vertical asymptote for the function. You might want to look at its graph with your function grapher for angles between 0° and somewhat less than 90°. We examine the properties of this function in considerably more detail in Chapter 7.

Suppose we have a right triangle in which we can measure one of the angles other than the 90° angle using a protractor. We can find the tangent of that angle with a calculator. Then, if we know the length of either the adjacent side or the opposite side, we can easily find the length of the other without involving Ron, Sue, or anyone else to solve the kind of problem we used to begin this discussion.

E X A M P L E

While hiking through the mountains, you find yourself at the edge of a deep gorge and you wonder how far across it is to the other side. There is a tree rooted on your side at the edge of the gorge. From a point 15 feet up in the tree, you find that the angle of depression (measured down from the horizontal at eye level) to the opposite edge of the gorge is 22°. How far is it across the gorge?

Solution The height (15 feet) to the point in the tree and the unknown distance D across the gorge form the two sides of a right triangle, as shown in Figure A.9. Notice that the 22° angle of depression is not an angle of the triangle. However, it does determine the measures of the triangle's angles θ and ϕ (the Greek letter phi). The angle $\theta = 68°$ because it is the complement of 22°. The angle ϕ is 22° because $\phi = 180° - 68° - 90° = 22°$ (there are 180° in any triangle). We therefore have

$$\tan \theta = \tan 68° = 2.475.$$

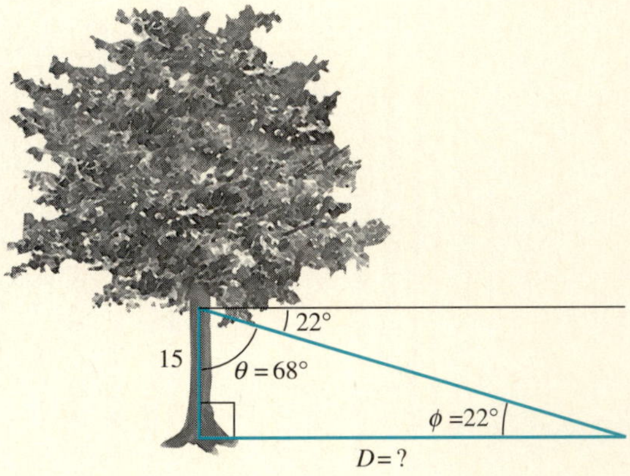

As a consequence,

$$\tan \theta = \frac{D}{15} = 2.475,$$

$$D = 15 \times 2.475 = 37.125.$$

Therefore the gorge is about 37 feet across.

Note that if we worked with the angle ϕ instead, we would obtain the same result:

$$\tan \phi = \tan 22° = \frac{15}{D}$$

$$D = \frac{15}{0.404} \approx 37.13.$$

FIGURE A.9

We often are faced with the problem of determining an angle θ in a right triangle when two of the sides are known. For example, suppose that the two sides of a right triangle are $a = 20$ and $b = 13$ (see Figure A.10) so that $\tan \theta = 0.65$ and we want to find θ. We know from the given table of values that $\tan 30° = 0.577$ and $\tan 40° = 0.839$. Since the values for the tangent are strictly increasing, we would obviously expect that θ is between 30° and 40°. We certainly can improve on these very rough estimates by trial and error. For instance, using the calculator, we might find that $\tan 35° = 0.700$ (too high), $\tan 32° = 0.625$ (too low), $\tan 34° = 0.6745$ (slightly too high), and so forth.

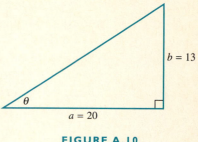

FIGURE A.10

A far more effective method is to use the inverse of the tangent function (which we develop in detail in Chapter 7). On the calculator, simply press either "2nd" or "INV" followed by "TAN" and then the known tangent value. For this example, INV TAN 0.65 returns 33.024. This means that 33.024° is the angle whose tangent value is 0.65. That is, $\tan 33.024° = 0.65$.

In general, problems in right angle trigonometry typically involve knowing a small amount of information about a right triangle and using that information intelligently to determine values for the other parts (either the sides or the angles) of the triangle. In fact, there are only a limited number of possibilities. We list all of these cases (based on the right triangle in Figure A.11) in the following table. We leave the last column for you to complete. Decide on the appropriate strategy for finding each of the missing pieces based on the information given or previously determined.

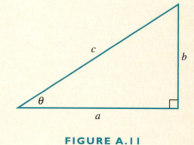

FIGURE A.11

Given	Objective	Strategy
a and b	Find c.	
	Find θ.	
a and c	Find b.	
	Find θ.	
b and c	Find a.	
	Find θ.	
a and θ	Find b.	
	Find c.	
b and θ	Find a.	
	Find c.	
c and θ	Find a.	Cannot be done simply using the tangent of θ.
	Find b.	Cannot be done simply using the tangent of θ.

We will examine the last case, which cannot be solved using the tangent of an angle, in the next section.

Finally, whenever you face any problem involving a triangle, your first step should be to draw a simple representation of the situation in order to identify the different parts of the triangle and see how they are related to one another. Your drawing will help you determine which strategy, if any, to use to solve for the remaining parts of the triangle.

EXERCISES

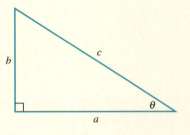

For Exercises 1–5, refer to the accompanying figure. Given the following information, find all other parts of the triangle.

1. $\theta = 52°$ and $b = 12$
2. $\theta = 16°$ and $a = 12$
3. $c = 15$ and $a = 6$
4. $c = 30$ and $b = 18$
5. $a = 72$ and $b = 47$

6. The shadow of a flagpole is 50 feet long. A line of sight from the tip of the shadow to the tip of the pole makes an angle of 28° with the ground. How high is the pole?

7. You want to find the distance across a fast-flowing river. You find two trees that are directly across the river from one another at points *A* and *B*. You then measure off a distance of 32 feet to another tree at point *C* on the edge of the river on your side. From the tree at *C*, you find that the angle *ACB* is 56°. Find the distance across the river.

8. The line of sight from the top of a lighthouse to a Jet-Ski makes an angle of depression of 5° with the horizontal. The lighthouse is 50 feet high. How far is the Jet-Ski from the base of the lighthouse? How far is the straight line distance from the top of the lighthouse to the Jet-Ski?

9. A searchlight on the ground is aimed at the sky at an angle of 20° from the ground. It casts a bright spot on the bottom of a cloud. An observer moves to a point on the ground directly under the bright spot. This observer then measures the distance to the searchlight and finds it to be 3000 feet. How high is the cloud?

10. A wheelchair ramp is to be built from ground level to a platform that is 9 feet above the ground. The angle of inclination with the ground is required to be no greater than 11°. What is the shortest length for a ramp that meets this requirement? How far is the start of the ramp from the base of the platform?

11. Jill is standing at the top of a vertical cliff. Jack is standing 25 feet away from the foot of the cliff and estimates that the angle of elevation θ from his position to Jill's is 40°. How high is the cliff?

12. Suppose Jack's measurement in Exercise 11 of the distance to the cliff is off by one foot. How much of a difference does this make in the calculated height of the cliff? (*Hint*: Recalculate your answer to Exercise 11 two ways, once with a distance of 24 feet and then with a distance of 26 feet.)

13. Suppose Jack's estimate of the angle of elevation is off by one degree in Exercise 11. How much of a difference does this make in the calculated height of the cliff?

14. Suppose Jill, at the top of the cliff, wants to find the horizontal distance from the foot of the cliff to where Jack is standing without climbing down and measuring it directly. She drops a rock at the end of a long measuring tape down the cliff and finds that the height of the cliff is 75 ft. Next, she measures the angle of depression from her position to Jack's to be 70°. How far is Jack from the foot of the cliff?

A.2 The Sine and Cosine of an Angle

Suppose you are flying a kite at the end of 400 feet of string and are curious about how high the kite is. How can you find its height? If you consider Figure A.12, you will see that the length of string is simply the hypotenuse of the large right triangle shown. You can certainly measure the angle of inclination θ that the kite string makes with the horizontal: say $\theta = 37°$. However, if you think back, you'll realize that this is precisely the last case in the strategy table given in the previous section where we pointed out that the tangent function is of little help. We must come up with a different scheme to determine the height y.

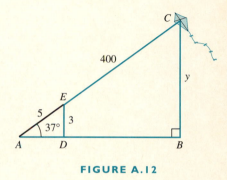

FIGURE A.12

Using a yardstick, you could measure off a length of 5 feet along the kite string and measure the height from the horizontal to that point on the string; say you get 3 feet. As seen in Figure A.12, you have a pair of similar right triangles ABC and ADE, so you know that corresponding sides are proportional. Consequently,

$$\frac{\text{Height of kite}}{\text{Length of hypotenuse}} = \frac{\text{Height to point on string}}{\text{Length of string to that point}}$$

$$\frac{y}{400} = \frac{3}{5}.$$

Therefore the height of the kite is $y = 400 \times \frac{3}{5} = 240$ feet above the "horizontal." (If, in fact, you hold the kite string chest-high, say four feet above the ground, then the kite is 240 feet above your hand. So the kite is 244 feet above ground level.)

The Sine of an Angle   The key to solving this problem was to construct the ratio of the height and the hypotenuse of the right triangle. As in our discussion on the tangent of an angle, for any given angle θ we can construct infinitely many right triangles that are all similar, so all of them have the same value for this ratio. Because the ratio changes as the angle changes, this ratio is a function of the angle. We define this ratio to be the **sine of the angle** or the **sine function** and write

$$\sin \theta = \frac{\text{opposite}}{\text{hypotenuse}}.$$

As with the tangent function, realize that *opposite* refers to the side opposite the angle θ regardless of the orientation of the right triangle. It is essential that you think of the opposite side in terms of the angle θ and not as the side of a triangle that is in some particular location, such as the vertical position.

Since we are concerned only with right triangles for now, the domain of the sine function consists of angles between $0°$ and $90°$. The following comments apply only to this situation. In the next section of this Appendix and in Chapter 7, we will consider cases where this restriction is lifted, and so more interesting and useful behavior patterns will emerge for the sine function.

As with the tangent of an angle, you can get the values for the sine of any angle in a right triangle using your calculator. For instance,

$$\sin 10° = 0.174, \quad \sin 20° = 0.342, \quad \sin 30° = 0.5,$$
$$\sin 40° = 0.643, \quad \text{and} \quad \sin 75° = 0.966.$$

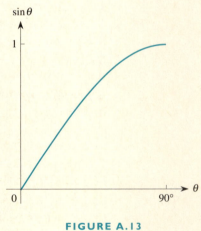

Notice that the values for the sine function are increasing as the angle θ increases from $0°$ to $90°$. Further, these values follow a pattern that is concave down, so the sine function grows more rapidly for small angles and less rapidly as angles get closer to $90°$. In Figure A.13 we show a graph of the sine function for θ between $0°$ and $90°$. (We discuss the sine's behavior for angles outside this interval in Section 7.1.)

FIGURE A.13

Let's look again at some of the special angles, notably $\theta = 0°$, $30°$, $45°$, $60°$, and $90°$. To begin, think about what happens in a right triangle as the angle shrinks down to $0°$ for a fixed hypotenuse, c. (Picture the kite nosediving toward the ground at the end of the taut string.) The length of the opposite side also shrinks down to zero and so

$$\sin 0° = \frac{0}{c} = 0.$$

What about $\sin 90°$? Again, for a fixed hypotenuse, c, think about what happens in a right triangle as the angle increases to $90°$. (Although improbable, picture the kite moving directly over your head so that the height of the kite becomes equal to the length of the string.) The length of the opposite side in the triangle grows until it approaches the length of the hypotenuse, so

$$\sin 90° = \frac{c}{c} = 1.$$

Check these two facts on your calculator.

Now let's look at the other special angles. As seen in Figure A.14, when $\theta = 45°$, we have a right triangle with two angles equal to $45°$ and the two corresponding sides of equal length, say a. By the Pythagorean theorem, the length of the hypotenuse is $\sqrt{2}a$, as we saw in the previous section. So the ratio of the opposite side and the hypotenuse is

$$\sin 45° = \frac{a}{\sqrt{2}a} = \frac{1}{\sqrt{2}} = 0.707.$$

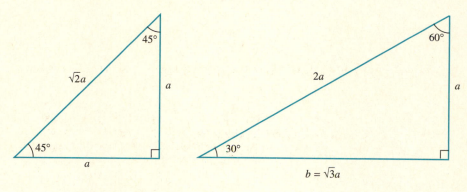

FIGURE A.14   FIGURE A.15

Similarly, as seen in Figure A.15, when $\theta = 30°$, the remaining angle is $60°$. Further, the length of the side opposite the $30°$ angle is half the length of the hypotenuse. If the hypotenuse has length $2a$, then the opposite side has length a. Consequently, we see that

$$\sin 30° = \frac{\text{opposite}}{\text{hypotenuse}} = \frac{a}{2a} = 0.5.$$

By the Pythagorean theorem, if the remaining side has length b, then

$$b^2 = (2a)^2 - a^2 = 4a^2 - a^2 = 3a^2,$$

so $b = \sqrt{3}a$. We therefore have

$$\sin 60° = \frac{\sqrt{3}a}{2a} = \frac{\sqrt{3}}{2} = 0.866,$$

as we found previously using a calculator.

We summarize these facts in the following table.

θ	0°	30°	45°	60°	90°
$\sin \theta$	0	0.5	$0.707 = \frac{1}{\sqrt{2}}$	$0.866 = \frac{\sqrt{3}}{2}$	1

Note that the values for the sine of any angle in a right triangle must always be between 0 and 1. This is true because the sine is the ratio of the opposite side and the hypotenuse, and in any right triangle the hypotenuse is always the longest side.

E X A M P L E   1

A highway through the mountains has a stretch that drops at a steep grade of 5°. If you drive a distance of 12 miles along this road, how much of a vertical descent do you make?

Solution   To help us "see" the situation, we "straighten out" all curves in the road and draw a picture (Figure A.16), which is not drawn to scale. Note that a 5° grade also can be viewed as a 5° angle of descent. We know that the length of the hypotenuse is 12 miles. Therefore,

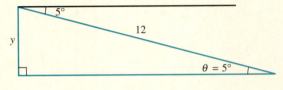

FIGURE A.16

$$\sin 5° = \frac{y}{12}$$

$$y = 12 \cdot \sin 5° = 1.05.$$

Consequently, while driving along this stretch of the highway, the road drops 1.05 miles or about 5544 feet.

Often, we are faced with the problem of determining an angle given the value of the sine of that angle. For instance, if the hypotenuse of a right triangle is 20 and the side opposite the angle θ is 15, as shown in Figure A.17, then $\sin \theta = 0.75$. What is the angle θ? We could find it by trial and error (we know that $\sin 45° = 0.707$ and $\sin 60° = 0.866$, so we might try 50°, and so on). But a far more effective approach is to use the inverse sine function (which is discussed in detail in Chapter 7). With your calculator, simply press either "2nd" or "INV" followed by "SIN", and then the known sine value, say 0.75, and the calculator returns 48.590. To verify that this is the correct angle, we check that

$$\sin (48.59°) = 0.749996.$$

In the previous section, you were asked to complete a table outlining strategies that would allow you to solve for all the parts of a right triangle given various combinations of sides and angles. In all but one of those cases, you could

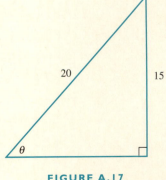

FIGURE A.17

determine all the other parts of the triangle using the tangent of an angle. Now we ask you to complete the table a second time by deciding on appropriate strategies to determine the parts of a right triangle using the sine of an angle instead of the tangent. Refer to Figure A.18 to see the parts of the right triangle.

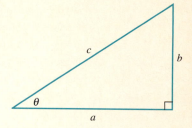

FIGURE A.18

Given	Objective	Strategy
a and b	Find c.	
	Find θ.	
a and c	Find b.	
	Find θ.	
b and c	Find a.	
	Find θ.	
a and θ	Find b.	
	Find c.	
b and θ	Find a.	
	Find c.	
c and θ	Find a.	
	Find b.	

Realize that, in practice, it doesn't matter which function you apply as long as you use a correct strategy; so for most of the cases, there are a variety of different approaches that will give the correct answers. Incidentally, note that with the sine function, you are able to solve for all the parts of a right triangle in all six cases.

The Cosine of an Angle   So far, we have looked at two possible ratios among the sides of a right triangle:

$$\tan \theta = \frac{\text{opposite}}{\text{adjacent}} \qquad \sin \theta = \frac{\text{opposite}}{\text{hypotenuse}}$$

There is one other ratio that is also very useful. We now define a third trigonometric function, the **cosine of an angle:**

$$\cos \theta = \frac{\text{adjacent}}{\text{hypotenuse}}.$$

As with the sine and the tangent, you can get the values for the cosine of any angle in a right triangle using your calculator. For instance,

$$\cos 10° = 0.985, \quad \cos 20° = 0.940, \quad \cos 30° = 0.866,$$
$$\cos 40° = 0.766, \quad \text{and} \quad \cos 50° = 0.643.$$

Notice that these values are decreasing and the pattern is concave down, so that the values decrease more slowly for small angles and more rapidly as angles get closer to 90°. We show a graph of the cosine function for θ between 0° and 90° in Figure A.19.

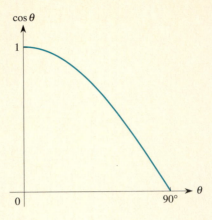

FIGURE A.19

E X A M P L E   2

To get onto a straight water slide at an amusement park, you must climb a flight of steps 60 feet high. The slide itself is inclined downward at a 42° angle. How long is the actual slide?

Solution From Figure A.20, we see that the angle in the right triangle is 48° and the adjacent side is 60 feet long. Therefore, to find the length L of the slide, we use

$$\cos 48° = \frac{\text{adjacent}}{\text{hypotenuse}} = \frac{60}{L}.$$

Thus

$$L = \frac{60}{\cos 48°} = 89.67.$$

So the slide is about 90 feet long.

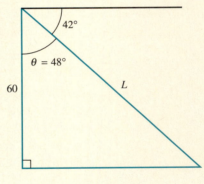

FIGURE A.20

Think About This

Can you solve the problem of Example 2 using the sine function instead? the tangent function?

    Again, let's consider the special angles, $\theta = 0°$, 30°, 45°, 60°, and 90°. To begin, what is cos 0°? Think about a right triangle in which the hypotenuse remains constant and the angle θ shrinks to 0°. (Picture again the kite as it nosedives toward the ground on a windy day so that the string remains taut.) The hypotenuse gets closer and closer to the adjacent side, so that

$$\cos 0° = \frac{\text{adjacent}}{\text{hypotenuse}} = 1.$$

Similarly, think about a right triangle in which the hypotenuse remains fixed and the angle approaches 90°. (Picture the kite moving directly over your head.) The adjacent side gets closer to 0 and so

$$\cos 90° = 0.$$

Next, let's look at the other special angles. As seen in Figure A.21, when $\theta = 45°$,

$$\cos 45° = \frac{\text{adjacent}}{\text{hypotenuse}} = \frac{a}{\sqrt{2}a} = \frac{1}{\sqrt{2}} = 0.707.$$

This is the same value we found for the sine of 45°. Can you explain why this is the case?

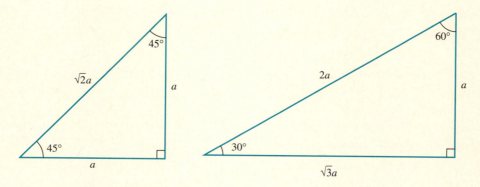

FIGURE A.21   FIGURE A.22

Also, when $\theta = 30°$, we see from Figure A.22 that

$$\cos 30° = \frac{\sqrt{3}a}{2a} = \frac{\sqrt{3}}{2} = 0.866,$$

which is the identical value as sin 60°. Can you explain why? Similarly,

$$\cos 60° = \frac{a}{2a} = \frac{1}{2},$$

which is the same as sin 30°. Why?

We summarize these key values for the cosine function as follows:

θ	0°	30°	45°	60°	90°
$\cos \theta$	1	$0.866 = \frac{\sqrt{3}}{2}$	$0.707 = \frac{1}{\sqrt{2}}$	0.5	1

Relationships among Trigonometric Functions If you compare the values in the preceding table to the corresponding values for the sine function, you will notice that the values for 30° and 60° have been reversed. This is no coincidence. In general, for any angle θ,

$$\cos \theta = \sin (90° - \theta)$$

and

$$\sin \theta = \cos (90° - \theta).$$

Can you see why these two relationships are true? Consider the right triangle shown in Figure A.23. The side a opposite angle θ is the side adjacent to the angle $90° - \theta$. Similarly, the side b adjacent to angle θ is the side opposite angle $90° - \theta$. That is, their roles are reversed depending on which angle, θ or $90° - \theta$, you consider. What happens when you write out the ratios, using a and b, for the sine and cosine for both angles?

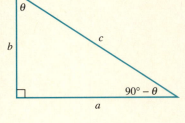

FIGURE A.23

One reason why the trig functions are so useful is that there are many interrelationships among them. The two formulas above are examples of such. But there are others that are far more important. Suppose we look at the right triangle in Figure A.23. We start with the definitions of the sine and cosine, which give

$$\sin \theta = \frac{a}{c} \quad \text{and} \quad \cos \theta = \frac{b}{c}.$$

If we multiply both sides of these equations by c, we obtain

$$a = c \sin \theta \quad \text{and} \quad b = c \cos \theta.$$

Now consider the tangent function, which is the ratio of the opposite and adjacent sides, so that

$$\tan \theta = \frac{a}{b} = \frac{c \sin \theta}{c \cos \theta}.$$

We therefore have, for any angle θ whatsoever,

$$\tan \theta = \frac{\sin \theta}{\cos \theta} \qquad \text{if } \cos \theta \neq 0.$$

Next consider the Pythagorean theorem applied to the right triangle in Figure A.23, so that

$$a^2 + b^2 = c^2.$$

When we substitute $a = c \sin \theta$ and $b = c \cos \theta$, we have

$$(c \sin \theta)^2 + (c \cos \theta)^2 = c^2$$
$$c^2 (\sin \theta)^2 + c^2 (\cos \theta)^2 = c^2.$$

We divide both sides by c^2 to obtain

$$(\sin \theta)^2 + (\cos \theta)^2 = 1,$$

which holds for any angle θ. For convenience, it is customary to write

$$(\sin \theta)^2 = \sin^2 \theta \quad \text{and} \quad (\cos \theta)^2 = \cos^2 \theta$$

and we have the *Pythagorean identity*

$$\sin^2 \theta + \cos^2 \theta = 1.$$

Check this on your calculator using different values for θ. Be careful to enter the expressions as (SIN x)$\wedge$2 and (COS x)$\wedge$2.

To summarize the definitions and special relationships among the three trigonometric functions:

$$\sin \theta = \frac{\text{opposite}}{\text{hypotenuse}}$$

$$\cos \theta = \frac{\text{adjacent}}{\text{hypotenuse}}$$

$$\tan \theta = \frac{\text{opposite}}{\text{adjacent}}$$

$$\tan \theta = \frac{\sin \theta}{\cos \theta}$$

$$\sin^2 \theta + \cos^2 \theta = 1$$

E X A M P L E   3

Suppose that the SIN and TAN keys on your calculator are broken. You can use your COS key to find that cos 20° = 0.940. Determine the values for sin 20° and tan 20°.

Solution   We illustrate three different approaches to solving this problem.

Method 1   Using the Pythagorean relationship between the sine and the cosine,

$$\sin^2 \theta + \cos^2 \theta = 1,$$

we find that

$$\sin^2 \theta = 1 - \cos^2 \theta$$

for any angle θ. Therefore, when $\theta = 20°$,

$$\sin^2 20° = 1 - \cos^2 20° = 1 - 0.8830 = 0.1170$$

and so, when we take the square root of both sides, we find

$$\sin 20° = \sqrt{0.1170} = 0.342.$$

Further, we have

$$\tan 20° = \frac{\sin 20°}{\cos 20°} = \frac{0.342}{0.940} = 0.364.$$

Method 2   From Figure A.24, we see that $\sin 20° = \dfrac{b}{c}$. How-ever, it is also true that $\cos 70° = \dfrac{b}{c}$ and since we can find cos 70° = 0.342, we know that sin 20° = 0.342 also. Knowing sin 20° and cos 20°, we now can find tan 20° = 0.364, as we did in Method 1.

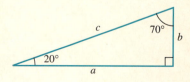

FIGURE A.24

Method 3   We know that

$$\cos 20° = 0.940 = \frac{\text{adjacent}}{\text{hypotenuse}},$$

so from the triangle in Figure A.22, the ratio $\frac{a}{c}$ must be 0.940. Thus we can assume, for instance, that $a = 940$ and $c = 1000$. (There are infinitely many other possibilities; another is $a = 0.940$ and $c = 1$.) Consequently, using the Pythagorean theorem, we can find the third side b.

$$b^2 = c^2 - a^2 = 1000^2 - 940^2 = 116{,}400$$
$$b = \sqrt{116{,}400} = 341.17.$$

As a result, we have

$$\sin 20° = \frac{b}{c} = \frac{341.17}{1000} = 0.341.$$

$$\tan 20° = \frac{b}{a} = \frac{341.17}{940} = 0.363.$$

Notice that both of these values differ slightly from the results in Methods 1 and 2 due to rounding.

EXERCISES

For Exercises 1–6, refer to the accompanying figure. Given the following information, find all other parts of the triangle.

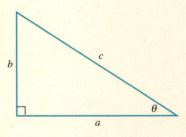

1. $\theta = 52°$ and $b = 15$
2. $\theta = 16°$ and $c = 12$
3. $c = 22$ and $a = 16$
4. $c = 30$ and $b = 8$
5. $a = 12$ and $b = 9$
6. $a = 42$ and $\theta = 72°$

7. A road up a hill is inclined at 11° to the horizontal. A driver starts driving up this hill and, by checking the odometer, discovers that the steep portion of the road extends for $\frac{3}{4}$ of a mile. How much has the car gained in altitude?

8. With its radar, an aircraft spots another aircraft 10,000 ft away at an angle of depression of 15°. Find the horizontal distance from one aircraft to the other.

9. As a pendulum of length 21 inches swings back and forth, the maximum angle it makes from the vertical is $\theta = 18°$. What is the greatest height that the end of the pendulum reaches compared to its lowest height when it passes the vertical?

10. From takeoff, an airplane reaches a height of two miles (10,560 feet) in the process of covering 20 miles horizontally.

    a. Find the average angle of ascent of the airplane as it climbs.

    b. Do you think that the actual path upward of the airplane is a straight line, or is the path curved? If you believe it is curved, is it concave up or concave down?

    c. If the airplane were to climb along a straight-line path, find the distance it would travel as it goes from the ground to the two-mile height. Do you think the distance that the airplane actually travels is greater than or less than the distance you calculated? Why?

11. When the space shuttle comes in for a landing at Cape Canaveral, its descent to the ground for the final 10,000 feet of height is at an angle of 19° with the horizontal.

    a. What actual distance does the shuttle traverse along this final glide path?

    b. How far from touchdown should the shuttle be when it passes the 10,000-foot altitude?

12. Suppose that the COS and TAN keys on your calculator are broken. You can use your SIN key to find that, for some angle θ, $\sin \theta = 0.3$. Determine the values for $\cos \theta$ and $\tan \theta$. What is the angle θ?

13. You will be hammering a three-inch-long nail into a piece of wood that is two inches thick. Find the steepest angle at which you can hammer the nail all the way into the wood without having it come out the opposite side.

14. The cranberry sauce to go with your holiday turkey comes out of a can and has a diameter of 3 inches. When you slice the roll of cranberry sauce at an angle, most of the slices will be an ellipse with a minor axis of 3 inches. Suppose that you slice the roll at an angle of 27° to the vertical. Find the length of the major axis of each elliptical slice.

A.3   The Sine, Cosine, and Tangent in General

So far, we have considered the trigonometric functions only for angles in a right triangle; therefore, all angles have been between 0° and 90°. However, we often encounter situations where we need to consider angles that are larger than 90° and a natural question to ask is: How do we adapt the ideas from the previous two sections to apply to such cases? Before we discuss this, suppose you choose such an angle, say $\theta = 125°$, and check what happens when you use your calculator. You will get

$$\sin 125° = 0.819 \qquad \cos 125° = -0.574 \qquad \tan 125° = -1.428.$$

Let's see why.

Suppose we have any angle between 90° and 180°; picture it with one side, its *initial side*, along the positive *x*-axis and the other side, its *terminal side*, in the second quadrant. See Figure A.25. This terminal side forms the hypotenuse of a right triangle which we construct by drawing a vertical line from the terminal side to the *x*-axis. Suppose that the hypotenuse of this right triangle is h, the height is y, and, because the horizontal side extends in the negative direction, we indicate the base as $-x$. The angle ϕ in this right triangle is just the supplement of the angle θ, since $\theta + \phi = 180°$ (and so $\theta = 180° - \phi$). We define the trigonometric functions for the angle θ in terms of the comparable "directed" values for the angle ϕ. Therefore,

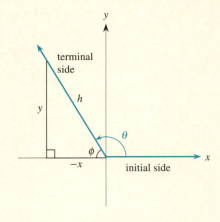

FIGURE A.25

$$\sin \theta = \frac{\text{opposite}}{\text{hypotenuse}} = \frac{y}{h}$$

$$\cos \theta = \frac{\text{adjacent}}{\text{hypotenuse}} = \frac{-x}{h} = -\frac{x}{h}$$

$$\tan \theta = \frac{\text{opposite}}{\text{adjacent}} = \frac{y}{-x} = -\frac{y}{x}.$$

With these definitions we see that for any angle θ between 90° and 180° with terminal side in the second quadrant, the cosine and tangent are negative, while the sine is positive. We saw examples of this above with sin 125° = 0.819, cos 125° = −0.573, and tan 125° = −1.428.

What about an angle θ between 180° and 270° whose terminal side is in the third quadrant, as shown in Figure A.26? Again, we construct a right triangle by drawing a vertical line from the terminal side to the *x*-axis, and so determine an angle ϕ. Notice that both the *x*- and *y*-values are negative and that $\theta = 180° + \phi$. As before, we define the trigonometric functions for this angle θ in terms of the appropriate "directed" lengths in the right triangle with angle ϕ, so that

$$\sin \theta = \frac{\text{opposite}}{\text{hypotenuse}} = \frac{-y}{h} = -\frac{y}{h}$$

$$\cos \theta = \frac{\text{adjacent}}{\text{hypotenuse}} = \frac{-x}{h} = -\frac{x}{h}$$

$$\tan \theta = \frac{\text{opposite}}{\text{adjacent}} = \frac{-y}{-x} = \frac{y}{x}.$$

For any angle between 180° and 270° exclusive, the sine and the cosine are negative and the tangent is positive. For instance, suppose $\theta = 211°$, so $\phi = 31°$. Use your calculator to find the values for sin 211°, cos 211°, and tan 211°. How do they compare with sin 31°, cos 31°, and tan 31°?

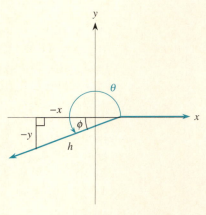

FIGURE A.26

Next, consider an angle θ between 270° and 360° whose terminal side is in the fourth quadrant, as shown in Figure A.27. Once more, we construct a right triangle and define each of the trigonometric functions in terms of the angle ϕ in that triangle. Notice that the y-value is negative and that $\theta + \phi = 360°$, so $\theta = 360° - \phi$. Also,

$$\sin \theta = \frac{\text{opposite}}{\text{hypotenuse}} = \frac{-y}{h} = -\frac{y}{h}$$

$$\cos \theta = \frac{\text{adjacent}}{\text{hypotenuse}} = \frac{x}{h}$$

$$\tan \theta = \frac{\text{opposite}}{\text{adjacent}} = \frac{-y}{x} = -\frac{y}{x}.$$

Therefore for any angle between 270° and 360° exclusive, the cosine is positive and the sine and the tangent are negative.

What happens if θ is greater than 360°, say 410°? As shown in Figure A.28, we can construct such an angle by looping around a full 360° and then an additional 50°; essentially, this is equivalent to an angle of $\phi = 50°$ in the first quadrant. Thus, if you check using your calculator, you will find that

$$\sin 410° = 0.766 = \sin 50°$$

$$\cos 410° = 0.643 = \cos 50°$$

$$\tan 410° = 1.192 = \tan 50°.$$

In a totally similar way, if $\theta = 775°$, then we make two full rotations (accounting for $2 \times 360° = 720°$), which leaves an angle $\phi = 55°$, and so

$$\sin 775° = 0.819 = \sin 55°$$

$$\cos 775° = 0.574 = \cos 55°$$

$$\tan 775° = 1.428 = \tan 55°.$$

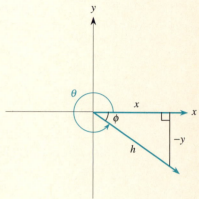

FIGURE A.27

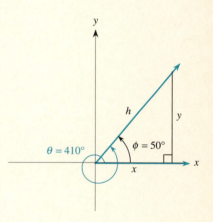

FIGURE A.28

It is easy to see that the values for the three trig functions repeat exactly every 360°. We say that these functions are **periodic** because their behavior repeats. The smallest interval over which they repeat is called the **period.** The periods of the sine and cosine functions are both 360°. In general, for any angle θ,

$$\sin (\theta + 360°) = \sin \theta$$

$$\cos (\theta + 360°) = \cos \theta.$$

On the other hand, the period of the tangent function is 180° because its values repeat every 180°. Thus, for any angle θ,

$$\tan (\theta + 180°) = \tan \theta.$$

Check this on your calculator.

Finally consider a negative angle, say $\theta = -30°$, drawn in the clockwise direction as shown in Figure A.29. This is equivalent to a positive angle of 330° since both angles have the same terminal side. We therefore have

$$\sin \theta = \frac{-y}{h} = -\frac{y}{h}$$

$$\cos \theta = \frac{x}{h}$$

$$\tan \theta = \frac{-y}{x} = -\frac{y}{x}.$$

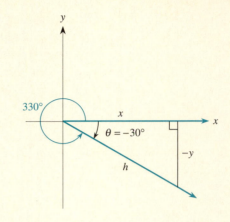

Let's summarize the behavior pattern for the sine function. For θ between 0° and 90°, $\sin \theta$ increases from 0 to 1. For θ between 90° and 180°, $\sin \theta$ decreases from 1 back down to 0. For θ between 180° and 270°, $\sin \theta$ continues to decrease from 0 down to -1. For θ between 270° and 360°, $\sin \theta$ increases from -1 up to 0. Thus, we see that the sine function has a maximum value of 1 and a minimum value of -1 and this oscillatory pattern will continue indefinitely in both directions (for $\theta > 360°$ and for $\theta < 0°$). Use your function grapher with θ between $-500°$ and 500°, say, to see this pattern.

Think About This

Give a similar summary for the behavior of the cosine function.

In Figure A.30, we summarize the information about the signs of the three trigonometric functions based on the quadrant containing the terminal side.

You can visualize the behavior of the trig functions by looking at their graphs. In Figure A.31(a), we show the graph of the function $y = \sin \theta$ for θ between 0° and 360° and how the graph relates to the signs of $\sin \theta$ in the four quadrants shown in Figure A.30. In Figure A.31(b), we expand the graph of $y = \sin \theta$ to show its behavior between $-360°$ and 720°. Notice that this portion of the curve consists of three full repetitions of the *basic sine curve* that occurs between 0° and 360°, which is one full period of the function. Also, the curve oscillates between a minimum height of $y = -1$ (when $\theta = \ldots, -90°, 270°, 630°, \ldots$) and a maximum height of

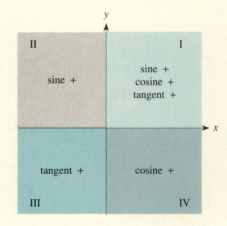

$y = 1$ (when $\theta = \ldots, -270°, 90°, 450°, \ldots$). Starting at $\theta = 0°$, notice how the values of $\sin \theta$ increase from $y = 0$ to $y = 1$ for θ from 0° to 90°; then decrease from $y = 1$ through $y = 0$ and on to $y = -1$ for θ from 90° to 180° and then on to 270°; and finally increase from $y = -1$ to $y = 0$ for θ between 270° and 360°. In addition, the sine curve is concave down between $\theta = 0°$ and $\theta = 180°$ and is concave up from $\theta = 180°$ to 360°.

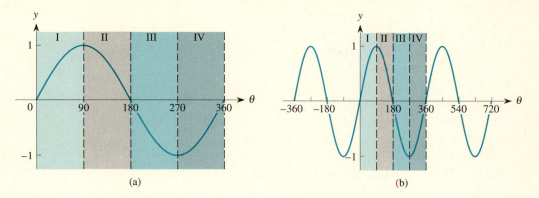

FIGURE A.31

You also should be very careful about distinguishing between the information shown in Figure A.30 regarding the signs of the sine function in different quadrants and what happens in the graph of the sine function in Figure A.31(a). The quadrants referred to in Figure A.30 are based on a coordinate system with y versus x and values for the angle θ are measured as you rotate the terminal side of the angle. These are not the same quadrants that you see in Figure A.31(a) because that graph shows y as a function of θ, so θ is measured horizontally. In particular, angles in the first quadrant in Figure A.30 correspond to the portion of the θ-axis in Figure A.31(a) between $\theta = 0°$ and $\theta = 90°$; the second quadrant in Figure A.30 corresponds to the portion of the θ-axis between $\theta = 90°$ and $\theta = 180°$ in Figure A.31(a); and so forth. We have marked this in Figures A.31(a) and (b) using Roman numerals and corresponding shadings for the different quadrants to help make the point. Be sure that you understand the subtle differences here before going on.

In Figure A.32, we show the graph of the cosine function from $\theta = -360°$ to $\theta = 720°$. Use the graph to answer the following questions: Where is the cosine function increasing? Where is it decreasing? What are its maximum and minimum values? Where do they occur? Where is the cosine concave up? Where is it concave down? Also, be sure that you see how the information in Figure A.30 on the sign of the cosine function in the different quadrants relates to the behavior you see in the graph of the cosine function.

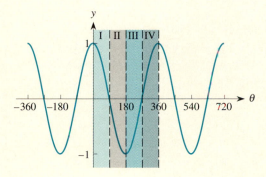

FIGURE A.32

The Law of Sines Everything we have done so far with the trigonometric functions has been developed in terms of angles in right triangles. In fact, these functions can be used with any triangle. Consider the triangle ABC shown in Figure A.33 where the sides opposite the angles A, B, and C are denoted respectively by a, b, and c. As drawn, all three angles are acute; that is, each is less than 90°. We will consider the case later where one angle is greater than 90°.

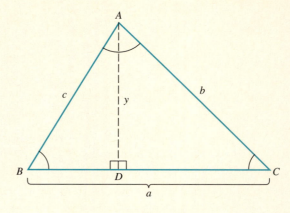

FIGURE A.33

Suppose we drop a perpendicular from the vertex at angle A down to a point D on the base a. That perpendicular line AD, whose length we call y, produces two right triangles. In triangle ABD on the left, we have

$$\sin B = \frac{y}{c} \quad \text{or} \quad y = c \sin B.$$

Similarly, in triangle ACD on the right, we have

$$\sin C = \frac{y}{b} \quad \text{or} \quad y = b \sin C.$$

Since these two expressions for y must be equal, we find

$$y = c \sin B = b \sin C,$$

and so

$$\frac{\sin B}{b} = \frac{\sin C}{c}.$$

However, we could just as easily have drawn a perpendicular from the vertex at angle B to the opposite side b. In that case, using an identical line of reasoning, we would find that

$$\frac{\sin A}{a} = \frac{\sin C}{c}.$$

Together, these results yield

$$\frac{\sin A}{a} = \frac{\sin B}{b} = \frac{\sin C}{c}$$

for any triangle with three acute angles.

What about a triangle with an angle greater than 90°? Consider the one shown in Figure A.34. We can still drop a perpendicular of length y from the vertex at angle A to point D on an extension of side a as shown. This line forms two right triangles. Clearly, in the large right triangle ACD,

$$\sin C = \frac{y}{b} \quad \text{or} \quad y = b \sin C.$$

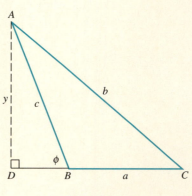

FIGURE A.34

For sin B, we determine the trig functions using the angle ϕ in the smaller right triangle ABD. We thus find that

$$\sin B = \frac{y}{c} \quad \text{or} \quad y = c \sin B.$$

Consequently,

$$y = c \sin B = b \sin C,$$

so that again we have

$$\frac{\sin B}{b} = \frac{\sin C}{c}.$$

We can similarly drop a perpendicular either from the vertex at angle B onto side b or from the vertex at angle C onto an extension of side c and obtain a similar relationship involving $\frac{\sin A}{a}$.

Therefore for *any* triangle, we have

$$\frac{\sin A}{a} = \frac{\sin B}{b} = \frac{\sin C}{c}.$$

This is known as the **law of sines.** It can be used to find all the remaining sides and angles in any triangle if appropriate combinations of sides and angles are known that include at least one angle and the side opposite it. We illustrate its use in the following example.

E X A M P L E

The Federal Communications Commission (FCC) is attempting to locate a pirate radio station by *triangulation*. The FCC set up two monitoring stations 30 miles apart on an east–west line and took simultaneous readings on the direction of the radio signal. The western-most monitor measured the signal as coming from a direction 42° north of east; the other monitor measured the signal as coming from a point 56° north of west. Where is the pirate station located?

Solution The information recorded determines the triangle shown in Figure A.35. The two monitoring stations are located 30 miles apart at the points A and B. The signal directions determine the angles of 42° and 56°. The pirate is located at point C. We see that the angle at C must be 82° (= 180° − 42° − 56°).

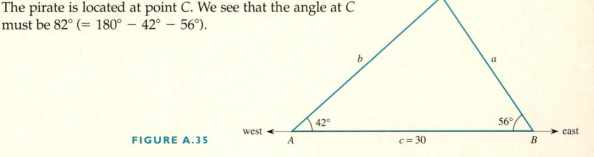

FIGURE A.35

We now apply the law of sines to find the lengths of sides a and b. Using angles $A = 42°$ and $C = 82°$, we find

$$\frac{\sin A}{a} = \frac{\sin C}{c}$$

$$\frac{\sin 42°}{a} = \frac{\sin 82°}{30}$$

$$a = \frac{30 \sin 42°}{\sin 82°} \approx 20.27.$$

Similarly, to find b, we apply the law of sines using the angles B and C to get

$$\frac{\sin B}{b} = \frac{\sin C}{c}$$

$$\frac{\sin 56°}{b} = \frac{\sin 82°}{30}$$

$$b = \frac{30 \sin 56°}{\sin 82°} \approx 25.12.$$

Therefore the pirate station is located at a distance of 25.12 miles from station A in a direction of 42° toward the northeast and at a distance of 20.27 miles from station B in a direction of 56° toward the northwest. The point C is determined precisely by these two facts and the pirate station can be shut down by the FCC.

In this example, we used the law of sines when two angles and one side of a triangle are known. The law of sines can also be used when two sides (say a and b) and the angle opposite one of them (either A or B) are known. However, depending on the sizes of the two known sides, it is possible to obtain either a unique answer or two distinct configurations for the triangle. This *ambiguous case* occurs when you try to find the angle from its sine. Recall that there will be two angles, one less than 90° and the other greater than 90°, that both possess the same sine value. We ask you to explore possible ambiguous cases in the exercises at the end of this section.

There is another possible complication when using the law of sines if you know two sides and the angle opposite one of them. If in the midst of such a set of calculations, you obtain a sine value that is greater than 1, it indicates that the values you are working with could not come from a real triangle. Again, you will encounter such a case in the exercises.

EXERCISES

For Exercises 1–6, refer to the notation for the sides and angles in the accompanying figure. Given the following information, find all other parts of the triangle.

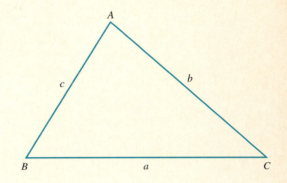

1. $A = 26°$, $B = 63°$, $b = 12$

2. $A = 47°$, $C = 72°$, $c = 60$

3. $A = 35°$, $B = 65°$, $c = 24$

4. $A = 40°$, $a = 10$, $b = 6$ (*Hint:* Is it possible to have two different values for B?)

5. $A = 40°$, $a = 10$, $b = 12$ (*Hint:* Is it possible to have two different values for B?)

6. $A = 40°$, $a = 10$, $b = 18$

7. Two ships at sea are 50 miles apart on a north–south line when they both receive an SOS signal from a third ship in trouble. One ship receives the SOS from a direction of 41° north of east. The other ship receives the signal from a direction of 54° south of east. Where is the third ship?

8. You want to find the distance across a fast-flowing river. You pick two large trees, at points A and B, that are 35 feet apart along the edge of the river on your side. You then spot another tree on the opposite side of the river at point C. The angle CAB at point A is 43°; the angle CBA at point B is 52°. Find the distance across the river.

9. In Exercises 4, 5, and 6, you saw three cases where the law of sines works very differently because of the relative sizes of sides a and b. Based on those results, explain the following statements.

    a. Given a value for angle A, there will always be one triangle whenever $b < a$.

    b. There will be two different possible triangles whenever b is somewhat larger than a.

    c. There will be no triangle whenever b is much larger than a.

B: *Arithmetic of Complex Numbers*

Complex numbers were originally introduced to allow people to solve quadratic equations such as $x^2 + 4 = 0$. This equation is equivalent to $x^2 = -4$, which has no solution among the real numbers. When we try to solve this equation by taking square roots, we obtain

$$x = \pm\sqrt{-4}.$$

To work with numbers such as these that involve the square root of a negative number, we introduce the **imaginary number** i.

$$i = \sqrt{-1}.$$

In terms of i, the roots of the equation $x^2 + 4 = 0$ are therefore

$$x = \pm\sqrt{-4} = \pm\sqrt{4}\sqrt{-1} = \pm 2i,$$

which is equivalent to $x = 2i$ and $x = -2i$.

Similarly, we can find the roots of $x^2 - 2x + 10 = 0$ using the quadratic formula:

$$x = \frac{(-2) \pm \sqrt{(-2)^2 - 4(1)(10)}}{2(1)}$$

$$= \frac{2 \pm \sqrt{4 - 40}}{2}$$

$$= \frac{2 \pm \sqrt{-36}}{2}$$

$$= \frac{2 \pm 6i}{2}$$

$$= 1 \pm 3i.$$

So the roots are $x = 1 + 3i$ and $x = 1 - 3i$.

Each of these roots is called a *complex number*. In general, we write a **complex number** in the form $z = a + bi$, where a and b are both real numbers. Thus, z is composed of a real number, a, and an imaginary number, bi. We call a the *real part* of z and bi the *imaginary part* of z.

In the first of the two examples above, the roots were $x = 2i$ and $x = -2i$. In the second example, the roots were $x = 1 + 3i$ and $x = 1 - 3i$. As in both of these examples, complex numbers typically arise in pairs of the form $a + bi$ and $a - bi$, called *complex conjugates*.

Since $i = \sqrt{-1}$, it immediately follows that

$$i^2 = \left(\sqrt{-1}\right)^2 = -1;$$

that is,

$$i^2 = -1.$$

From this we can find other powers of i.

$$i^3 = (i^2)(i) = (-1)(i) = -i$$
$$i^4 = (i^2)(i^2) = (-1)(-1) = 1$$
$$i^5 = (i^4)(i) = (1)(i) = i,$$

and so on. All higher powers simply cycle through these four values: $i, -1, -i,$ and 1.

The arithmetic of complex numbers, for the most part, is quite straight-forward. Consider the two complex numbers $z = 5 + 4i$ and $w = 3 - 11i$.

Addition   To add complex numbers, we add the real parts and separately add the imaginary parts. For example,

$$z + w = (5 + 4i) + (3 - 11i) = 8 - 7i.$$

In symbols,

> If $z = a + bi$ and $w = c + di$, then
>
> $$z + w = (a + c) + (b + d)i.$$

Subtraction   To subtract complex numbers, we subtract the real parts and separately subtract the imaginary parts. For our numbers z and w,

$$z - w = (5 + 4i) - (3 - 11i)$$
$$= (5 - 3) + (4i - (-11i)) = 2 + 15i.$$

In symbols,

> If $z = a + bi$ and $w = c + di$, then
>
> $$z - w = (a - c) + (b - d)i.$$

Multiplication   To multiply complex numbers, we multiply them alge-braically, use the fact that $i^2 = -1$ to simplify any power of i, and collect like terms consisting of the real part and the imaginary part. For example,

$$z \times w = (5 + 4i)(3 - 11i)$$
$$= (5)(3) + (5)(-11i) + (4i)(3) + (4i)(-11i)$$
$$= 15 - 55i + 12i - 44i^2$$
$$= 15 - 43i + 44 \qquad \text{since } i^2 = -1$$
$$= 59 - 43i.$$

In general,

> If $z = a + bi$ and $w = c + di$, then
>
> $$z \times w = (ac - bd) + (ad + bc)i.$$

Notice that, in the particular case where z and w are complex conjugates, say $z = 6 + 8i$ and $w = 6 - 8i$, we have

$$z \times w = (6 + 8i)(6 - 8i)$$
$$= 6(6) + 6(-8i) + (8i)(6) - (8i)(8i)$$
$$= 36 - 48i + 48i - 64i^2$$
$$= 36 - 64(-1) = 100.$$

In general,

> If $z = a + bi$ and $w = a - bi$ are complex conjugates, then
>
> $$z \times w = a^2 + b^2.$$

Consequently, the product of complex conjugates is always a real number.

Division of complex numbers is somewhat harder, but the ideas we develop in Section 7.8 provide a simple way to do it.

EXERCISES

Simplify each expression.

1. i^{23} **2.** i^{72} **3.** i^{58} **4.** i^{45}

Perform each of the following operations. Write each answer in the form $a + bi$, where a and b are real numbers.

5. $(8 + i) + (-6 + 3i)$ **6.** $(5 - 2i) + (7 + 6i)$

7. $(10 - i) - (-1 + i)$ **8.** $(4 + 5i) - 8i$

9. $(1 - 3i)(2 + i)$ **10.** $(5 + 6i)^2$

11. $(15 + 2i)(15 - 2i)$ **12.** $(7 - 4i)(7 + 4i)$

Find the roots of each equation.

13. $x^2 + 25 = 0$ **14.** $4x^2 + 9 = 0$

15. $5x^2 + 2x + 1 = 0$ **16.** $x^2 - 4x + 7 = 0$

C: *Some Mathematical Moments to Remember*

Absolute Value   Absolute value is used to transform any number, either positive or negative, into the corresponding positive value. We write the absolute value of a number a as $|a|$. Thus,

$$|5| = 5 \quad \text{and} \quad |-6| = 6.$$

In general, for any number $x \geq 0, |x| = x$ and for any number $x < 0$, $|x| = -x$. We can write this as

$$|x| = \begin{cases} x & \text{if } x \geq 0 \\ -x & \text{if } x < 0. \end{cases}$$

Factorial Notation $n!$ Expressions of the form $4 \times 3 \times 2 \times 1$ involving a product of consecutive positive integers that includes 1 arise so often in mathematics that a special symbol is given to them. We write such products as $n!$ (and read it as n *factorial*). For example,

$$3! = (3)(2)(1) \quad\quad = 6$$
$$4! = (4)(3)(2)(1) \quad = 24$$
$$5! = (5)(4)(3)(2)(1) = 120,$$

and so forth. In general, for any positive integer n,

$$n! = n(n - 1)(n - 2) \cdots (3)(2)(1).$$

For completeness, we define $0! = 1$.

Summation or Sigma Notation   *Summation notation* is used as a shorthand for writing expressions such as

$$1 + 2 + 3 + \cdots + 100,$$
$$1 + 2 + 2^2 + 2^3 + \cdots + 2^{50}$$
$$1 + \frac{1}{2} + \frac{1}{3} + \cdots + \frac{1}{n}.$$

We use the Greek letter sigma (Σ) to stand for *summation* and an *index of summation*, say k, to indicate which integers are intended. Thus, to add all the integers between $k = 1$ and $k = 100$, we write $\sum_{k=1}^{100}$ because

$$\sum_{k=1}^{100} k = 1 + 2 + 3 + \cdots + 100.$$

To add all the integers between 25 and 60,

$$\sum_{k=25}^{60} k = 25 + 26 + 27 + \cdots + 60.$$

Similarly, we can write the sum of the powers of 2 from $k = 0$ to $k = 50$ as

$$\sum_{k=0}^{50} 2^k = 1 + 2 + 2^2 + 2^3 + \cdots + 2^{50}.$$

Also, to add the *reciprocals* of all the integers k between $k = 1$ and some unspecified upper limit $k = n$, we write

$$\sum_{k=1}^{n} \frac{1}{k} = 1 + \frac{1}{2} + \frac{1}{3} + \frac{1}{4} + \cdots + \frac{1}{n}.$$

Notice that the letter we use for the index of summation is immaterial; we could equivalently write

$$\sum_{i=1}^{n} \frac{1}{i} = 1 + \frac{1}{2} + \frac{1}{3} + \frac{1}{4} + \cdots + \frac{1}{n} \quad \text{or} \quad \sum_{j=1}^{n} \frac{1}{j} = 1 + \frac{1}{2} + \frac{1}{3} + \frac{1}{4} + \cdots + \frac{1}{n}.$$

The numerical result of these summations will be the same regardless of the letter we use as the index.

Similar Triangles   Two triangles are similar if all three angles in one triangle are the same as all three angles in the other. The other key fact about similar triangles is:

Corresponding sides of similar triangles are proportional.

The right triangles ABC and ADE shown in Figure A.36 are similar because the angle θ is common to both, the angle ϕ is the same in both, and both triangles have a right angle. Therefore, we have a variety of ratios that are equal, including

$$\frac{AB}{AD} = \frac{BC}{DE} = \frac{AC}{AE} \quad \text{and} \quad \frac{AB}{AC} = \frac{AD}{AE} \quad \text{and} \quad \frac{BC}{AC} = \frac{DE}{AE}.$$

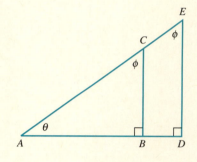

FIGURE A.36

Distance Between Points in the Plane   The distance from a point $A(x_0, y_0)$ to a point $B(x_1, y_1)$ is given by the *distance formula*

$$|AB| = \sqrt{(x_1 - x_0)^2 + (y_1 - y_0)^2}.$$

It is based on the Pythagorean theorem, as illustrated in Figure A.37. The distance $|AB|$ is the hypotenuse of the right triangle formed by a base of $x_1 - x_0$, which is the horizontal change, and a height of $y_1 - y_0$, which is the vertical change.

For instance, the distance between the two points $A(2, 5)$ and $B(6, 8)$ is

$$\begin{aligned} |AB| &= \sqrt{(6 - 2)^2 + (8 - 5)^2} \\ &= \sqrt{16 + 9} \\ &= \sqrt{25} = 5. \end{aligned}$$

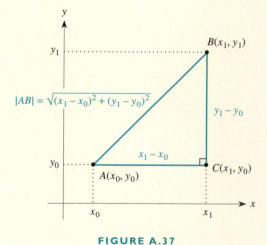

FIGURE A.37

The Equation of a Circle   The equation of the circle with radius r and center at (x_0, y_0) is

$$(x - x_0)^2 + (y - y_0)^2 = r^2.$$

For instance, the equation of the circle with radius 7 and center at $P(2, -5)$ is

$$(x - 2)^2 + (y + 5)^2 = 7^2 = 49.$$

The Equation of an Ellipse   The equation of the ellipse with center at (x_0, y_0) having its horizontal major (or longer) axis of length $2a$ and its minor (or shorter) axis of length $2b$ is

$$\frac{(x - x_0)^2}{a^2} + \frac{(y - y_0)^2}{b^2} = 1.$$

See Figure A.38.

For instance, the equation of the ellipse with center at $P(2, -5)$, whose major axis has length 12 and whose minor axis has length 8 is

$$\frac{(x - 2)^2}{6^2} + \frac{(y + 5)^2}{4^2} = 1$$

or

$$\frac{(x - 2)^2}{36} + \frac{(y + 5)^2}{16} = 1.$$

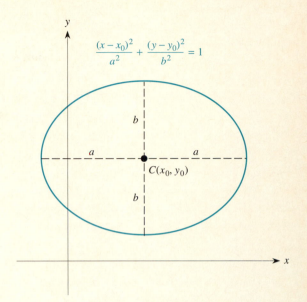

FIGURE A.38

Completing the Square The two quadratic functions $y = x^2 - 6x + 13$ and $y = (x - 3)^2 + 4$ are algebraically equivalent, so they represent the same function and their graphs are the same parabola. While the first representation is more common, the second gives more information about the behavior of the corresponding parabola: It is shifted 3 units to the right and 4 units up compared to the basic parabola $y = x^2$. A process for transforming the first expression into the second is called *completing the square*. We illustrate it as follows:

- Start with $y = x^2 - 6x + 13$.
- Take one-half of the coefficient of x: $\frac{1}{2}(-6) = -3$.
- Square this number to get 9.
- Add and immediately *subtract* this number 9 (so that we are actually adding 0 and thus still have the equivalent of the original expression):

$$y = (x^2 - 6x + 9) - 9 + 13 = (x^2 - 6x + 9) + 4.$$

- Recognize that the first three terms, $(x^2 - 6x + 9)$, form a perfect square—it is the square of $(x - 3)$, so we have

$$y = (x - 3)^2 + 4.$$

Notice that the -3 in this expression is the same as half of the original coefficient of x.

As another example, if $y = x^2 + 10x - 11$, we have

$$y = [x^2 + 10x \quad] - 11$$
$$= [(x^2 + 10x + 5^2) - 5^2] - 11 \qquad \text{since half of 10 is 5, and } 5^2 = 25$$
$$= (x + 5)^2 - 25 - 11$$
$$= (x + 5)^2 - 36.$$

Thus, the corresponding parabola is obtained by shifting $y = x^2$ to the left by 5 and down by 36.

If the original quadratic expression has a leading coefficient other than 1, it is first necessary to factor out that coefficient. For example,

$$2x^2 + 16x + 12 = 2[(x^2 + 8x \quad) + 6]$$
$$= 2[(x^2 + 8x + 4^2) - 4^2 + 6] \qquad \text{since half of 8 is 4}$$
$$= 2[(x + 4)^2 - 10]$$
$$= 2(x + 4)^2 - 20.$$

The corresponding parabola is obtained by doubling $y = x^2$, and then shifting it 4 units to the left and 20 units down.

Symmetry   The notion of *symmetry* arises throughout mathematics in a variety of ways. One is to describe the behavior of functions and other geometric objects when one portion is a mirror image of another portion. We typically describe a curve (whether or not it represents a function) as being *symmetric about a line,* or *symmetric with respect to an axis,* or occasionally *symmetric with respect to the origin.* Consider the ellipse shown in Figure A.39(a). It is symmetric with respect to the x-axis because the lower half is the mirror image of the upper half. Similarly, as shown in Figure A.39(b), the ellipse is symmetric with respect to the y-axis because the left half is the mirror image of the right half. The ellipse also is symmetric with respect to the origin since, for any point P on the ellipse, you can find the mirror image P' through the origin on the ellipse as well, as shown in Figure A.39(c).

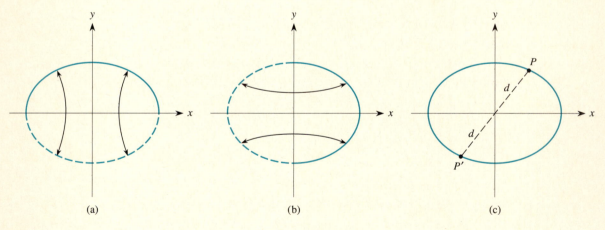

(a)   (b)   (c)

FIGURE A.39

On the other hand, the curve shown in Figure A.40 is not symmetric about the x-axis nor is it symmetric about the y-axis, since the two portions of the curve are not mirror images of each other about either axis.

of the curve are not mirror images of each other about either axis. However, the curve is *symmetric* with respect to the origin; if any point $P(a, b)$ is on the curve, then so is the point $P'(-a, -b)$, as shown in Figure A.40. This is equivalent to one portion of the curve being rotated through an angle of 180° to produce the other portion.

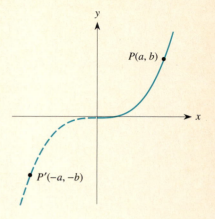

FIGURE A.40

We summarize the key information about symmetry as follows:

1. If a curve is symmetric about the x-axis and a point $P(a, b)$ lies on the curve, then the point $P'(a, -b)$ also lies on the curve. See Figure A.41(a).
2. If a curve is symmetric about the y-axis and a point $P(a, b)$ lies on the curve, then the point $P'(-a, b)$ also lies on the curve. See Figure A.41(b).
3. If a curve is symmetric about the origin and a point $P(a, b)$ lies on the curve, then the point $P'(-a, -b)$ also lies on the curve. See Figure A.40.

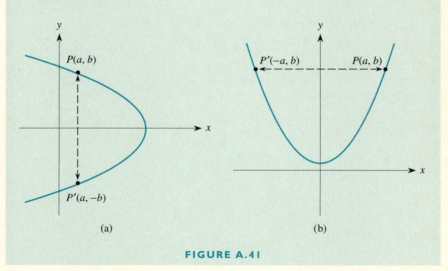

FIGURE A.41

EXERCISES

Write each number without the absolute value sign.

1. $|0|$ **2.** $\left|\dfrac{2}{3}\right|$ **3.** $|-\pi|$ **4.** $|9+12|$

5. $|6-10|$ **6.** $|-7-3|$ **7.** $|-5|-|4|$ **8.** $|-5|-|-4|$

9. Calculate the value of $y=|x|$ for $x=-4,-3,-2,\ldots,4$. Plot the points and then connect them. How would you describe the graph of $y=|x|$?

10. Repeat Exercise 9 using $y=|x+3|$ for $x=-7,-6,-5,\ldots,1$. How does the graph $y=|x+3|$ compare to the graph of $y=|x|$?

Evaluate each expression.

11. $6!$ **12.** $1!$ **13.** $2!\,0!$

14. $3!\,2!$ **15.** $\dfrac{5!}{3!}$

Write out the terms in each summation and calculate the sum.

16. $\displaystyle\sum_{k=1}^{4}(2k-3)$ **17.** $\displaystyle\sum_{j=1}^{4}\dfrac{1}{j}$ **18.** $\displaystyle\sum_{k=1}^{6}k^2$

19. $\displaystyle\sum_{i=0}^{3}5(2)^i$ **20.** $\displaystyle\sum_{j=3}^{7}(4j+1)$ **21.** $\displaystyle\sum_{i=15}^{25}8$

22. In the accompanying figure, triangle ABC is similar to triangle ADE.

    a. Find DE.
    b. Find AE.

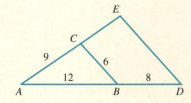

Find the distance between points A and B.

23. $A(4,3)$ $B(7,9)$ **24.** $A(-2,-3)$ $B(0,-1)$

25. $A(7,-2)$ $B(4,-6)$

Complete the square to identify the center and radius of the following circles.

26. $x^2+y^2-2x-10y-55=0$

27. $x^2+y^2+8x-6y-3=-11$

28. Write an equation for the ellipse at the right.

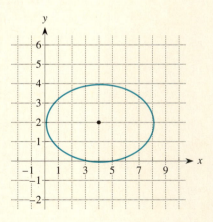

APPENDIX D: *1995 World Population Data*

	Population mid-1995 (millions)	Birth Rate per 1000	Death Rate per 1000	Natural Increase (annual %)	Doubling Time in Years at Current Rate	Projected Population (millions) 2010	Projected Population (millions) 2025	Life Expectancy at Birth
WORLD	**5,702**	**24**	**9**	**1.5**	**45**	**7024**	**8312**	**66**
NORTH AMERICA	**293**	**15**	**9**	**0.7**	**105**	**334**	**375**	**76**
Canada	29.6	14	7	0.7	102	33.6	36.6	78
United States	263.2	15	9	0.7	105	300.4	338.3	76
LATIN AMERICA and CARIBBEAN	**481**	**26**	**7**	**1.9**	**36**	**601**	**706**	**69**
CENTRAL AMERICA	**126**	**29**	**5**	**2.3**	**30**	**163**	**196**	**71**
Belize	0.2	38	5	3.3	21	0.3	0.4	68
Costa Rica	3.3	26	4	2.2	32	4.4	5.5	76
El Salvador	5.9	32	6	2.6	27	7.6	9.4	68
Guatemala	10.6	39	8	3.1	22	15.8	21.7	65
Honduras	5.5	34	6	2.8	25	7.6	9.7	68
Mexico	93.7	27	5	2.2	34	117.7	136.6	72
Nicaragua	4.4	33	6	2.7	26	6.7	9.1	65
Panama	2.6	29	8	2.1	33	3.3	3.8	72
CARIBBEAN	**36**	**23**	**8**	**1.5**	**46**	**43**	**50**	**70**
Cuba	11.2	14	7	0.7	102	12.3	12.9	75
Dominican Republic	7.8	27	6	2.1	32	9.7	11.2	70
Haiti	7.2	35	12	2.3	30	9.8	13.1	57
Jamaica	2.4	25	6	2.0	35	2.8	3.3	74
Puerto Rico	3.7	18	8	1.0	67	4.1	4.6	74
Trinidad and Tobago	1.3	17	7	1.1	64	1.6	1.8	71
SOUTH AMERICA	**319**	**25**	**7**	**1.8**	**38**	**395**	**460**	**68**
Argentina	34.6	21	8	1.3	55	40.8	46.1	71
Bolivia	7.4	36	10	2.6	27	10.2	13.1	60
Brazil	157.8	25	8	1.7	41	194.4	224.6	66
Chile	14.3	22	6	1.7	41	17.3	20.1	72
Colombia	37.7	24	6	1.8	39	46.1	53.0	69
Ecuador	11.5	28	6	2.2	31	14.9	17.8	69
Guyana	0.8	25	7	1.8	39	1.0	1.1	65
Paraguay	5.0	33	6	2.8	25	7.0	9.0	70
Peru	24.0	29	7	2.1	33	30.3	35.9	66
Suriname	0.4	25	6	2.0	36	0.5	0.6	70
Uruguay	3.2	17	10	0.7	102	3.5	3.7	73
Venezuela	21.8	30	5	2.6	27	28.7	34.8	72
EUROPE	**729**	**11**	**12**	**−0.1**	**—**	**743**	**743**	**73**
NORTHERN EUROPE	**94**	**13**	**11**	**0.2**	**443**	**97**	**99**	**76**
Denmark	5.2	13	12	0.1	770	5.3	5.3	75
Estonia	1.5	9	14	−0.5	—	1.4	1.4	70
Finland	5.1	13	10	0.3	227	5.2	5.2	76
Iceland	0.3	17	7	1.1	64	0.3	0.3	79
Ireland	3.6	14	9	0.5	139	3.5	3.5	75
Latvia	2.5	10	15	−0.5	—	2.4	2.4	68
Lithuania	3.7	13	12	0.0	6931	3.8	3.9	71
Norway	4.3	14	11	0.3	224	4.7	5.0	77

	Population mid-1995 (millions)	Birth Rate per 1000	Death Rate per 1000	Natural Increase (annual %)	Doubling Time in Years at Current Rate	Projected Population (millions) 2010	2025	Life Expectancy at Birth
Sweden	8.9	13	12	0.1	990	9.2	9.6	78
United Kingdom	58.6	13	11	0.2	385	61.0	62.1	76
WESTERN EUROPE	**181**	**11**	**10**	**0.1**	**741**	**187**	**184**	**77**
Austria	8.1	12	10	0.1	533	8.3	8.3	77
Belgium	10.2	12	11	0.1	578	10.4	10.5	77
France	58.1	12	9	0.3	217	61.7	63.6	78
Germany	81.7	10	11	−0.1	—	81.2	76.1	76
Liechtenstein	0.03	12	6	0.6	108	0.03	0.04	—
Luxembourg	0.4	13	10	0.4	193	0.4	0.4	76
Netherlands	15.5	13	9	0.4	182	16.9	17.6	77
Switzerland	7.0	12	9	0.3	224	7.6	7.5	78
EASTERN EUROPE	**310**	**10**	**14**	**−0.3**	**—**	**315**	**320**	**68**
Belarus	10.3	11	13	−0.2	—	10.9	11.3	69
Bulgaria	8.5	10	13	−0.3	—	7.9	7.5	71
Czech Republic	10.4	12	11	0.0	2,310	10.5	10.7	73
Hungary	10.2	12	14	−0.3	—	9.9	9.3	69
Moldova	4.3	15	12	0.4	193	4.8	5.1	68
Poland	38.6	12	10	0.2	301	40.2	41.7	72
Romania	22.7	11	12	−0.1	—	22.2	21.6	70
Russia	147.5	9	16	−0.6	—	149.5	153.1	65
Slovakia	5.4	14	10	0.4	178	5.7	6.0	71
Ukraine	52.0	11	14	−0.4	—	53.0	54.0	69
SOUTHERN EUROPE	**144**	**11**	**9**	**0.1**	**516**	**144**	**139**	**76**
Albania	3.5	23	5	1.8	39	4.1	4.7	72
Bosnia-Herzegovina	3.5	14	7	0.7	95	4.4	4.5	72
Croatia	4.5	10	11	−0.1	—	4.4	4.2	70
Greece	10.5	10	9	0.0	1,733	10.2	10.0	77
Italy	57.7	9	10	−0.0	—	56.5	52.8	77
Macedonia	2.1	16	8	0.8	85	2.3	2.5	72
Malta	0.4	14	7	0.7	102	0.4	0.4	75
Portugal	9.9	12	11	0.1	866	9.9	9.8	75
San Marino	0.03	10	6	0.4	169	0.03	0.03	76
Slovenia	2.0	10	10	0.1	1,386	2.0	1.9	73
Spain	39.1	10	9	0.1	578	39.0	37.1	77
Yugoslavia	10.8	13	10	0.3	204	11.1	11.5	72
OCEANIA	**28**	**19**	**8**	**1.2**	**60**	**34**	**39**	**73**
Australia	18.0	15	7	0.8	91	20.8	23.1	78
New Zealand	3.5	16	8	0.9	81	4.1	4.5	76
AFRICA	**720**	**41**	**13**	**2.8**	**24**	**1,069**	**1,510**	**55**
NORTHERN AFRICA	**162**	**32**	**8**	**2.4**	**29**	**219**	**279**	**64**
Algeria	28.4	30	6	2.4	29	38.0	47.2	67
Egypt	61.9	30	8	2.3	31	80.7	97.9	64
Libya	5.2	42	8	3.4	21	8.9	14.4	63
Morocco	29.2	28	6	2.2	32	38.4	47.4	69
Sudan	28.1	41	12	3.0	23	41.5	58.4	55
Tunisia	8.9	25	6	1.9	36	11.2	13.3	68
WESTERN AFRICA	**199**	**45**	**14**	**3.1**	**22**	**311**	**467**	**53**
Benin	5.4	49	18	3.1	22	8.3	12.3	48
Burkina Faso	10.4	47	19	2.8	24	14.5	21.6	48
Cape Verde	0.4	36	7	2.8	25	0.6	0.7	65
Ivory Coast	14.3	50	15	3.5	20	23.1	36.8	51

	Population mid-1995 (millions)	Birth Rate per 1000	Death Rate per 1000	Natural Increase (annual %)	Doubling Time in Years at Current Rate	Projected Population (millions) 2010	2025	Life Expectancy at Birth
Gambia	1.1	48	21	2.7	26	1.5	2.1	45
Ghana	17.5	42	12	3.0	23	26.6	38.0	56
Guinea	6.5	44	19	2.4	29	9.3	12.9	44
Guinea-Bissau	1.1	43	21	2.1	32	1.5	2.0	44
Liberia	3.0	47	14	3.3	21	4.8	7.2	55
Mali	9.4	51	20	3.2	22	15.0	23.7	47
Mauritania	2.3	40	14	2.5	27	3.3	4.4	52
Niger	9.2	53	19	3.4	21	14.8	22.4	47
Nigeria	101.2	43	12	3.1	22	162.0	246.0	56
Senegal	8.3	43	16	2.7	26	12.2	16.9	49
Sierra Leone	4.5	46	19	2.7	26	6.4	8.7	46
Togo	4.4	47	11	3.6	19	7.4	11.7	58
EASTERN AFRICA	**226**	**46**	**15**	**3.0**	**23**	**345**	**491**	**50**
Burundi	6.4	46	16	3.0	23	9.5	13.5	50
Comoros	0.5	46	11	3.6	20	0.9	1.4	58
Djibouti	0.6	38	16	2.2	32	0.8	1.1	48
Eritrea	3.5	42	16	2.6	27	5.2	7.0	—
Ethiopia	56.0	46	16	3.1	23	90.0	129.7	50
Kenya	28.3	45	12	3.3	21	43.6	63.6	56
Madagascar	14.8	44	12	3.2	22	23.3	34.4	57
Malawi	9.7	47	20	2.7	25	14.7	21.3	45
Mauritius	1.1	21	7	1.5	47	1.3	1.5	69
Mozambique	17.4	45	19	2.7	26	26.9	38.3	46
Reunion	0.7	23	6	1.8	40	0.8	0.9	73
Rwanda	7.8	40	17	2.3	30	10.4	12.8	46
Somalia	9.3	50	19	3.2	22	14.5	21.3	47
Tanzania	28.5	45	15	3.0	23	42.8	58.6	49
Uganda	21.3	52	19	3.0	21	32.3	48.1	45
Zambia	9.1	47	17	3.1	23	13.0	17.0	48
Zimbabwe	11.3	39	12	2.7	26	15.3	19.6	54
MIDDLE AFRICA	**83**	**46**	**16**	**2.9**	**24**	**127**	**191**	**51**
Angola	11.5	47	20	2.7	26	17.6	24.7	46
Cameroon	13.5	40	11	2.9	24	21.2	32.6	58
Central African Republic	3.2	42	22	2.0	34	3.9	5.2	41
Chad	6.4	44	18	2.6	27	9.3	12.9	48
Congo	2.5	40	17	2.3	31	3.2	4.2	46
Equatorial Guinea	0.4	40	14	2.6	27	0.6	0.9	53
Gabon	1.3	37	16	2.2	32	1.9	2.7	54
Sao Tome & Principe	0.1	35	9	2.6	27	0.2	0.2	64
Zaire	44.1	48	16	3.2	22	69.1	107.6	48
SOUTHERN AFRICA	**50**	**31**	**8**	**2.3**	**30**	**67**	**83**	**65**
Botswana	1.5	31	7	2.3	30	2.2	3.0	64
Lesotho	2.1	31	12	1.9	36	3.0	4.2	61
Namibia	1.5	37	10	2.7	26	2.2	3.0	59
South Africa	43.5	31	8	2.3	30	57.5	70.1	66
Swaziland	1.0	43	11	3.2	22	1.6	2.5	57
ASIA	**3,451**	**24**	**8**	**1.7**	**42**	**4,242**	**4,939**	**65**
WESTERN ASIA	**168**	**31**	**7**	**2.4**	**29**	**242**	**329**	**67**
Armenia	3.7	16	7	0.8	83	4.2	4.3	71
Azerbaijan	7.3	23	7	1.6	43	9.0	10.3	71
Bahrain	0.6	29	4	2.5	28	0.8	1.1	74

	Population mid-1995 (millions)	Birth Rate per 1000	Death Rate per 1000	Natural Increase (annual %)	Doubling Time in Years at Current Rate	Projected Population (millions)		Life Expectancy at Birth
						2010	2025	
Cyprus	0.7	17	8	0.9	76	0.8	0.9	77
Gaza	0.9	52	6	4.6	15	1.8	2.8	69
Georgia	5.4	12	10	0.2	462	5.7	6.0	73
Iraq	20.6	43	7	3.7	19	34.5	52.6	66
Israel	5.5	21	6	1.5	47	6.9	8.0	77
Jordan	4.1	38	4	3.3	21	6.2	8.3	72
Kuwait	1.5	25	2	2.2	31	2.5	3.6	75
Lebanon	3.7	25	5	2.0	34	5.0	6.1	75
Oman	2.2	53	4	4.9	14	3.7	6.0	71
Qatar	0.5	19	2	1.8	39	0.6	0.7	73
Saudi Arabia	18.5	36	4	3.2	22	30.0	48.2	70
Syria	14.7	41	6	3.5	20	23.6	33.5	66
Turkey	61.4	23	7	1.6	44	79.2	95.6	67
United Arab Emirates	1.9	23	4	1.9	36	2.5	3.0	72
West Bank	1.9	41	7	3.4	20	2.7	3.8	68
Yemen	13.2	50	14	3.6	19	21.9	34.5	52
SOUTH CENTRAL ASIA	**1,355**	**31**	**10**	**2.1**	**33**	**1,772**	**2,138**	**60**
Afghanistan	18.4	50	22	2.8	24	31.1	41.4	43
Bangladesh	119.2	36	12	2.4	29	160.8	194.1	55
Bhutan	0.8	39	15	2.3	30	1.1	1.5	51
India	930.6	29	9	1.9	36	1,182.7	1384.6	60
Iran	61.3	36	7	2.9	24	83.7	106.1	67
Kazakhstan	16.9	19	9	0.9	74	18.4	20.5	69
Kyrgyzstan	4.4	26	8	1.8	38	5.6	7.0	68
Maldives	0.3	43	7	3.6	19	0.4	0.6	65
Nepal	22.6	38	14	2.4	29	32.2	43.3	54
Pakistan	129.7	39	10	2.9	24	187.7	251.8	61
Sri Lanka	18.2	21	6	1.5	46	21.0	24.0	73
Tajikistan	5.8	33	9	2.4	29	9.2	13.1	70
Turkmenistan	4.5	33	8	2.5	28	5.9	7.9	66
Uzbekistan	22.7	31	7	2.5	28	31.9	42.3	69
SOUTHEAST ASIA	**485**	**26**	**8**	**1.9**	**37**	**601**	**704**	**64**
Brunei	0.3	27	3	2.4	29	0.4	0.4	74
Cambodia	10.6	44	16	2.8	25	15.7	22.8	50
Indonesia	198.4	24	8	1.6	43	240.6	276.5	63
Laos	4.8	42	14	2.8	25	7.2	9.8	52
Malaysia	19.9	29	5	2.4	29	27.5	34.5	71
Myanmar (Burma)	44.8	28	9	1.9	36	57.3	69.3	60
Phillipines	68.4	30	9	2.1	33	87.2	102.7	65
Singapore	3.0	17	5	1.2	56	3.6	4.0	74
Thailand	60.2	20	6	1.4	48	68.7	75.4	70
Viet Nam	75.0	30	7	2.3	30	92.5	108.1	65
EAST ASIA	**1,442**	**17**	**6**	**1.0**	**66**	**1,628**	**1,768**	**70**
China	1,218.8	18	6	1.1	62	1,385.5	1,522.8	69
Hong Kong	6.0	12	5	0.7	99	6.4	6.3	78
Japan	125.2	10	7	0.3	277	130.4	125.8	79
Korea, North	23.5	23	6	1.8	40	28.5	32.1	70
Korea, South	44.9	15	6	1.0	72	49.7	50.8	72
Macao	0.4	16	4	1.2	57	0.5	0.6	79
Mongolia	2.3	22	8	1.4	51	3.0	3.6	64
Taiwan	21.2	16	5	1.0	67	24.0	25.5	74

SELECTED ANSWERS

CHAPTER 1

Section 1.1

2. a. (ii)   **b.** (iii)   **c.** (i)   **d.** (vi)   **e.** (iv)
   f. (v)

7. a. y is a function of x.   **b.** y is not a function of x.

Section 1.2

2.

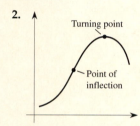

6. a, b, c, d, e, h, i, and **j** are periodic; **f** and **g** are not periodic.

Section 1.3

1. a, c, e, f, h, i, k, and **l** are functions; **b, d, g, j,** and **m** are not functions.

3. $f(P) = 170 + 6P$

5. a. $f(100) = 35$ **b.** $f(150) = 47.5$, $f(200) = 60$,
   $f(500) = 135$

   c. The variable m represents the number of miles greater than or equal to 100. A reasonable domain is $100 \le m \le 1560$ (based on driving for 24 hours at 65 mph). The corresponding range (cost values) is $35 \le c \le 400$.

8. $F(0) = -\frac{1}{4}$, $F(1) = -\frac{1}{3}$, $F(3) = \frac{1}{5}$, $F(4) = \frac{1}{12}$,

   $F(5) = \frac{1}{21}$. When $t = 2$, the denominator is 0, and we cannot divide by 0. Yes, $t = -2$ should also be skipped. The domain of F is the set of all real numbers except 2 and -2.

10. $g(4) = 6$, $g(16) = 20$, $g(25) = 30$, $g(100) = 110$. There are no values of s that produce a negative output for the function g. The range of g must be a subset of the nonnegative real numbers. The domain of g is the set of all nonnegative real numbers since we cannot take the square root of a negative number.

12. $f(3) = -1$, $f(4) = -5$, $f(5) = -1$, $f(3.5) = -3.625$,
   $f(4.5) = -4.375$

Section 1.4

2. One possible answer is: $f(-3) = 20$, $f(-2) = 12$,

$f(-1) = 6$, $f(0) = 2$, $f(1) = 0$, $f(2) = 0$, $f(3) = 2$,
$f(4) = 6$, $f(5) = 12$; $f\left(\frac{1}{2}\right) = \frac{3}{4}$, $f\left(\frac{3}{2}\right) = -\frac{1}{4}$,

$f\left(\frac{-5}{2}\right) = \frac{63}{4}$

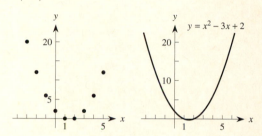

4. $h(-4) = -31$, $h(-3) = -1$, $h(-2) = 11$, $h(-1) = 11$,
   $h(0) = 5$, $h(1) = -1$, $h(2) = -1$, $h(3) = 11$, $h(4) = 41$

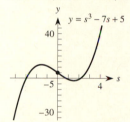

Section 1.5

5. a. 38 mph   **b.** 49 mph   **c.** 58 mph
   d. 69 mph   **e.** 150 feet

Section 1.6

1. 11% of the M&M's are red.

3. The area is about 18.768 square units.

8. The area is about 4.2 square units.

Review Exercises

1. independent variable: the depth of the tumor
   dependent variable: amount of radiation

2. One possible scenario is as follows: After the visiting team's first batter grounded out, their second batter hit a home run to put his team ahead by two. However, the next two batters struck out. The lead hitter for the home team struck out, but the second hitter hit a double. The next batter hit a sacrifice bunt to advance the runner on second to third base. The next batter hit a double that scored the runner on third. Alas, with a runner on second to possibly tie the score, the last batter struck out, ending the game.

3.

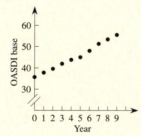

4.

5. a. function
      b. not a function

6. Overall, the zoos with larger budgets have greater attendance.

7. The biggest annual increases were $3,300 from 1989 to 1990 and $3,000 from 1988 to 1989. The smallest increase occurred between 1987 and 1988 when the OASDI base rose by $1,200.

8. $f(0) = 1, f(1) = 2, f(1.1) = 2.43, f(1.01) = 2.0403,$
$f(-3) = 34, f(a) = 3a^2 - 2a + 1$

9. a. $81,800   **b.** $23,203   **c.** $72,661 to $107,262

10.

$f(t) = 1.04t^{0.49}$

    a. 5.6 years
    b. 3.9 to 24.8 years
    c. increasing, concave down
    d. 232 days
    e. 16.2 years

11. a. domain: all real numbers; range: all real numbers
    b. domain: all real numbers; range: all real numbers greater than or equal to $-\dfrac{89}{16}$

11. c. domain: all real numbers greater than or equal to -5; range: all nonnegative real numbers
    d. domain: all real numbers less than or equal to -4 or greater than or equal to 4; range: all nonnegative real numbers
    e. domain: all real numbers except -3 and 3; range: all real numbers greater than 1 or less than or equal to $-\dfrac{4}{9}$
    f. domain: all real numbers; range: all real numbers greater than $-\dfrac{4}{9}$ and less than 1

12. a.

weight	0–1	1–2	2–3	3–4	4–5
postage (cents)	32	55	78	101	124

    b.

13. a. The track is: increasing and concave up from A to B, increasing and concave down from B to C, decreasing and concave down from C to D, decreasing and concave up from D to E, increasing and concave up from E to F. **b.** The car's speed is: decreasing at an increasing rate from A to B, decreasing at a decreasing rate from B to C, increasing at an increasing rate from C to D, increasing at a decreasing rate from D to E, decreasing at an increasing rate from E to F.

CHAPTER 2

Section 2.2

1. a. (iii)   **b.** (i)   **c.** (v)   **d.** (vi)   **e.** (iv)
    f. (ii)

3. a. $y = 7x - 9$ **b.** $y = -2$
    c. $y = -0.626x + 7.164$

5. $y = 0.057x + 1.329$

7. a.

    b. 1.8
    c. $F = 1.8C + 32$
    d. 86°
    e. 37°C
    f. $-40°$

9. a. $DJ_1 = 60H + 120, DJ_2 = 75H + 100$

9. b.

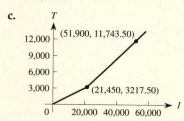

11. a. $T = 0.28(I - 21{,}450) + 3217.50$;
    domain is $21{,}450 \le I \le 51{,}900$ and
    range is $3217.50 \le T \le 11{,}743.50$

c.

16. a. $3N + 2G = 30$ **b.** See graph.
    c. Assuming N is the independent variable,
    domain is $0 \le N \le 10$ and range is $0 \le G \le 15$.
    d. $3N + 2G = 60$
    e. $3N + G = 30$
    f. $6N + 2G = 30$

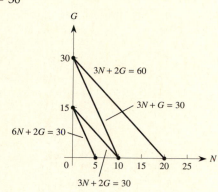

17. a.   **b.**

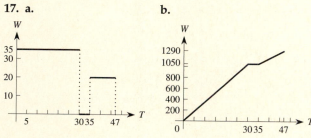

    c. $W = 35T$ for $0 \le T \le 30$
        $W = 1050$ for $30 \le T \le 35$
        $W = 20(T - 35) + 1050$ for $35 \le T \le 47$

19. a. $y = 5x - 26$ **b.** $y = -\frac{1}{5}x + \frac{26}{5}$

21. 10, 24

Section 2.3

5. a. $-1 < x < 0, 2 < x < 3$
    b. $-2.5 < x < -1, 0 < x < 2$
    c. near $x = 0$
    d. near $x = -1$ and near $x = 2$
    e. between $x = 2$ and $x = 2.5$
    f. between $x = -2.5$ and $x = -2$
    g. $-2.5 < x < 0, 0.5 < x < 2.5$
    h. $0 < x < 0.5, 2.5 < x < 3$
    i. near $x = 0, x = 0.5, x = 2.5$
    j. approximately at x between 1.0 and 1.5 and also
      between 2.5 and 3.0

6. a. A   **b.** C   **c.** B and D   **d.** B and C   **e.** D

9. a. about 2 units of time   **11.** 3.5%

15. a. (ii)   **b.** (iii)   **c.** (i)

19. a. $g(x)$ **b.** $f(x)$ **c.** $h(x)$ **21.** Finest; 1992

23. a. $\overline{PQ}, \overline{PR}, \overline{QR}$ **b.** $\overline{QR}, \overline{PR}, \overline{PQ}$ **25.** \$37,507.13

Section 2.4

1. The power functions are **a, d, e, h, j,** and **k**; the exponential functions are **b, c, g,** and **n**; the functions that
are neither power nor exponential are **f, i, l,** and **m**.

2. a. $h(x)$ **b.** $f(x)$ **c.** $g(x)$

5. $f(x)$ is quadratic, $g(x)$ is exponential, and $h(x)$ is cubic.

7. a. 54 mph   **b.** 38 mph   **c.** 76 mph
    d. 83 ft   **e.** 167 ft

11. a. $x^3 \to \infty$ as $x \to \infty$; $x^3 \to -\infty$ as $x \to -\infty$
    b. $-x^3 \to -\infty$ as $x \to \infty$; $-x^3 \to \infty$ as $x \to -\infty$
    c. $x^{1/3} \to \infty$ as $x \to \infty$; $x^{1/3} \to -\infty$ as $x \to -\infty$
    d. $-x^{1/3} \to -\infty$ as $x \to \infty$; $-x^{1/3} \to \infty$ as $x \to -\infty$
    e. $x^{-3} \to 0$ as $x \to \infty$; $x^{-3} \to 0$ as $x \to -\infty$
    f. $x^{-3} \to \infty$ as $x \to 0$ from the right;
      $x^{-3} \to -\infty$ as $x \to 0$ from the left

13. $D = \sqrt{H^2 + 7920H}$ **15.** 46.7 mi   **17.** 5,102,000 mi^2

21. slope of $\overline{PQ} = 1$, slope of $\overline{QR} = 3$, and slope of $\overline{PR} = 2$

25. a. x^7 **b.** a^{-2} **c.** r^4 **d.** z^{21}

Section 2.5

2. a. $\log x^6$ **b.** $\log (xy)$ **c.** $\log x^{3/2}$ **d.** $2 \log \left(\frac{x}{y}\right)$
    e. x^2 **f.** x^2

5. b. 37.81 million   **c.** about 50 years

7. in 11.83 years, or in 2007

9. a. 23.45, 17.61, 14.21, 11.90, and 10.24 years

11. in 38.55 years, or in 2033

13. 100 times; 1000 times

15. a. $x \approx 1.2323$ **b.** $x \approx 14.2067$ **c.** $x \approx 13.0244$
    d. $x \approx 23.3183$ **e.** $x \approx 9.191538$ **f.** $x \approx 2.02235$

Section 2.7

2. a. $x \approx 0.5$ **b.** $x \approx 4$ **c.** $x \approx 5$ **d.** $x \approx -2$

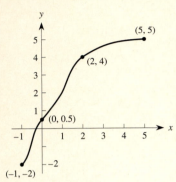

3. a. domain = $\{0, 1, 2, 3, 4, 5\}$;
range = $\{1.12, 1.44, 1.84, 2.05, 2.48, 2.94\}$

b.

x	1.12	1.44	1.84	2.05	2.48	2.94
$f(x)$	5	4	3	2	1	0

domain of f^{-1} = $\{1.12, 1.44, 1.84, 2.05, 2.48, 2.94\}$;
range of f^{-1} = $\{0, 1, 2, 3, 4, 5\}$

7. $p^{-1}(t) = 58.7084 \log t$

9. a. $c \approx 1.11134$ **b.** $t \approx 16.2376$
c. $k \approx 9.230896$ **d.** $m \approx 1.0007$

Review Exercises

1. a. exponential; base $a < 1$
b. power function; power p is between 0 and 1
c. exponential; $a > 1$

2. $F(t) = 5(1.193483192)^t$; growth factor = 1.193483192

3. a. $P(t) = 1.5 + 0.1t$; 1985

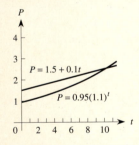

b. $P(t) = 950,000(1.1)^t$; during 1987
c. $t = 10.26$; during 1990

4. a. 17.227 years   **b.** 122 years from now

5. $F(180) = 50.5°$

6. One possible answer is $I(y) = 2.3y - 5.3$;
$I(26) = 54.5$ billions

7. One possible answer is $P(t) = -1.4t + 35$; The slope of the line, -1.4, represents the average drop in the percentage of median income per year that mortgage payments represent.

8. 19.4%; during 2002.

9. $G(16) = 18.20$; We can expect the percentage to drop below 18% sometime in 1996.

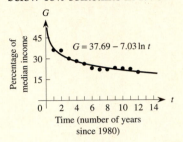

10. a. $F(x) = 8(0.72492)^x$ **b.** $G(x) = -\dfrac{13}{3}x + 8$
c. $H(x) = 2.72222(1.60357)^x$

11. a. $c = 1.05450$ **b.** $x = 8.82747$ **c.** $x = 63,096$
d. $a = 56,623$ **e.** $a = 1.69381$ **f.** $w = 5.77503$
g. $x = 23.1016$

12. a.

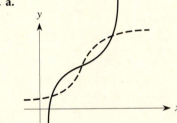

b.

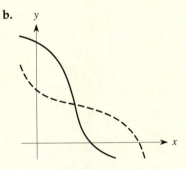

c. An inverse function does not exist since the function is not monotonic.

13. a. The domain of f is $t > 2$; $f^{-1}(t) = \dfrac{100^t + 4}{2}$

b. The domain of g is all real values of x;
$g^{-1}(x) = (x - 6)^{1/3}$

14. $F(x) = 4(1.5)^x$

15. a. $g(x)$ **b.** $h(x)$ **c.** $f(x)$

16. a. $f(t) = 0.2(1.05511)^t$ **b.** ice cream
c. in 2095   **d.** a little over $280

CHAPTER 3

Section 3.2

1. $C = 0.9T - 4$; $r = 0.8741$

2. r is 0.9897, which exceeds the critical value for a sample of size 6, namely, 0.811; Time = 0.231 × Speed − 4.06; about 6.3 seconds (for 45 mph) and 16.7 seconds (for 90 mph)

7. a. (iii)   **b.** (i)   **c.** (ii)

12. a. $N = 300 + 0.015T$ **b.** 20,000 years before 1900

13. a. $C = 1.32 + 1.08(t − 1)$

b. The slope is the rate per additional minute; the intercept is the charge for the initial minute.

c. $28.32   **d.** $C = 0.924 + 0.756(t − 1)$; $19.82

Section 3.3

1. a. the best linear fit: $h = 1.3577m + 21.655$, with $r = 0.9337$; the best exponential fit: $h = 24.0806(1.03045)^m$, with $r = 0.931315$; the best power fit: $h = 20.0558m^{0.324327}$, with $r = 0.903$

b. 93.6 hours (linear); 118.1 hours (exponential); 73 hours (power)

c. The linear fit seems the most reasonable; the exponential seems the least reasonable

3. yes, population = $3.000317(1.33031)^t$, where t = no. of decades since 1780 and with $r = 0.99892$

11. a. impossible   **b.** impossible   **c.** possible

d. impossible   **e.** impossible   **f.** possible

g. impossible   **h.** impossible   **i.** possible

12. a. $y = 2.25(12.1423)^x$ **13. a.** $y = 2.25x^{1.0843}$

b. $y = 0.05(6.08135)^x$ **b.** $y = 0.05x^{0.7840}$

c. $y = 7.12(0.0533)^x$ **c.** $y = 7.12x^{-1.2733}$

Section 3.4

10. a. possible   **b.** impossible   **c.** impossible

d. possible   **e.** impossible   **f.** impossible

g. possible   **h.** impossible   **i.** impossible

11. Craig is furthest ahead when the pursuit car's speed equals his car's speed.

12. a. 3 seconds   **b.** 103 mph   **c.** 6 seconds   **d.** 528 ft

Section 3.5

1. * Let $Y = 1, 2, \ldots, 10$

(corresponds to 1900, 1910, . . . , 1990)

•Let $Y = 1900, 1910, \ldots, 1990$

Function	G(in 2000)	Year for G = 2,000,000	
* $G = 124.27Y − 254$	1,113,000	2071	
$r = 0.962$			
* $G = 23.89(1.521)^Y$	2,408,000	1995	doubling time is 16.5 years
$r = 0.983$			
* $G = 18.19(Y^{1.706})$	1,088,000	2047	
$r = 0.969$			
• $G = 55570 \log Y$ − 182332	1,106,000	2075	
$r = 0.961$			

10. 3.34 seconds

13. a. increasing slope: $\overline{PQ}, \overline{PR}, \overline{QR}$

b. increasing steepness: $\overline{QR}, \overline{PR}, \overline{PQ}$

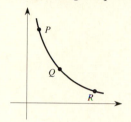

Section 3.6

2. The best fit is $T = 193.82565(0.96937)^t$ with $r = -0.99178$

Review Exercises

1. The best linear fit: $A = 0.111B + 0.25811$; the best exponential fit: $A = 0.50077(1.080835)^B$; the best power fit: $A = 0.167755B^{0.875746}$; the best logarithmic fit: $A = -1.256558 + 2.829569 \log B$

2. The linear model gives an excellent fit.

3. Letting l represent longevity and g represent gestation, the best exponential fit is $l = 7.40311(1.002367)^g$, $r = 0.502$; the best linear fit is $l = 0.023408g + 8.79526$, $r = 0.480$; the best logarithmic fit is $l = -7.063717 + 9.55164 \log g$, $r = 0.612$; the best power function fit is $l = 1.361457g^{0.438243}$, $r = 0.669$; power function is best.

4. a. Letting t represent the number of years since 1900, the best exponential fit is $N = 0.004506(1.10818)^t$ and the best power function fit is $N = (8.98165 \times 10^{-14})t^{7.5081}$; the exponential model seems to give the best fit.

b. 86,373 jail sentences in 1996

5. $P = 0.09 + 0.23w$, where w is the weight in ounces.

6. a. $C = 36.0126t − 2451.04$, where t represents the number of years since 1900.   **b.** 1993

7. a. $N = 0.633684(1.026615)^t$, where t represents the number of years since 1900, is a good fit.

c. 2001

8. a. $6.27 \le y \le 13.77$ **b.** $x \approx 19.5884$

c. $x = 10^{(y-3.77)/2.5}$

9. a. $H = 10.3N + 210$

b. slope = height of each additional story. **c.** 10.3 feet

10.

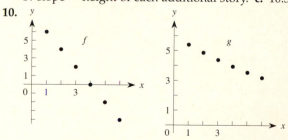

10.

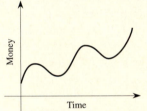

f is linear,

g is exponential, and

h is logarithmic.

CHAPTER 4

Section 4.1

1. Polynomial of degree at least three with a positive leading coefficient.

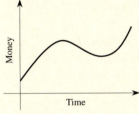

2. Polynomial of degree at least five with a positive leading coefficient.

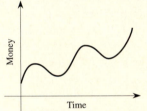

3. a. (i) degree 3 (ii) degree 4 (iii) degree 4 (iv) degree 5
   b. (i) positive (ii) positive (iii) negative (iv) negative

5.

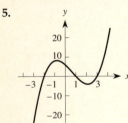

6.

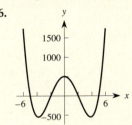

7.

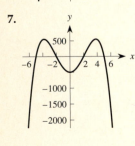

8.

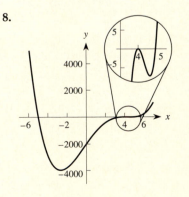

9. a. 6

   b. real roots: 2, 1, $-\frac{1}{2}$, and -3; complex roots: $-1 + i$ and $-1 - i$.

   c. as $x \to \infty$, $P(x) \to \infty$; as $x \to -\infty$, $P(x) \to \infty$ **d.** 5; 4

11. a. (i) increasing on $-2 < x < 3$; decreasing elsewhere
      (ii) concave up for $x < 1$; concave down for $x > 1$
      (iii) $y = -x^3 + x^2 + 16x - 16$

   b. (i) increasing on $-1 < x < 1$ and $x > 3$; decreasing elsewhere   (ii) concave down for $0 < x < 2$; concave up elsewhere   (iii) $y = x^4 - 4x^3 - 4x^2 + 16x$

15. $P(x) = -5x^2 + 10x$

17. a. $f(x) = x^2 - 4x - 12$ **b.** $(2, -16)$

   c. $f(x) = -\frac{5}{4}x^2 + 5x + 15$ **d.** $f(x) = \frac{5}{4}x^2 - 5x - 15$

19. a. The object starts at height 0   **b.** $\frac{2v_0}{g}$ seconds

   c. $\frac{v_0}{g}$ seconds   **d.** $\frac{v_0^2}{2g}$ feet

21. a. $(x + 3)(x + 4)$ **b.** $(x - 5)(x + 1)$

   c. $x(x + 6)(x - 6)$ **d.** $x(x - 3)(x - 1)$

   e. $x(x + 5)^2$ **f.** $x(x + 5)(x - 4)$

Section 4.2

(Many different answers are possible.)

2. $-0.0064x^2 + 4.165x + 22.185$; $(-3.27 \times 10^{-8})x^4 + (4.25 \times 10^{-5})x^3 - 0.02356x^2 + 6.3267x - 18.733$

4. men: $-28.638x^3 - 409.96x^2 + 8006.08x + 12069.12$; women: $-5.305x^3 + 20.21x^2 + 955.64x + 1018.83$

Section 4.3

1. 50%

3. The discriminant becomes $k^2(b^2 - 4ac)$. If $k \neq 0$, we get the same roots.

5. a.

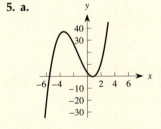

b.

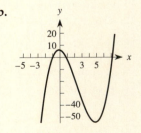

c.

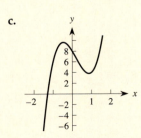

d.

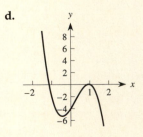

5. e.

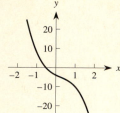

7. a. slope = 1.73   **b.** (1.73, 3)   **c.** (−1.73, 3)

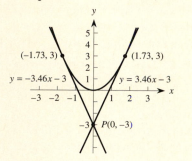

Section 4.4

1. a. $\dfrac{56}{5}$ **b.** $\dfrac{54}{5}$ **c.** $\dfrac{11}{5}$ **d.** 55   **e.** $-\dfrac{17}{5}$

f. $\dfrac{1}{11}$ **g.** 29   **h.** 5   **i.** $3x - 4 + \dfrac{1}{x}$

j. $3x - 4 - \dfrac{1}{x}$ **k.** $3 - \dfrac{4}{x}$ **l.** $3x^2 - 4x$

m. $\dfrac{3}{x} - 4$ **n.** $\dfrac{1}{3x - 4}$ **o.** $9x - 16$ **p.** x

3. a. $10^5 + \log 5$ **b.** $10^5 - \log 5$ **c.** $10^5 \cdot \log 5$

d. $\dfrac{10^5}{\log 5}$ **e.** $10^{\log 5} = 5$ **f.** $\log 10^5 = 5$

g. $10^{10^5} = 10^{100,000}$ **h.** $\log (\log 5)$

i. $10^x + \log x$ **j.** $10^x - \log x$ **k.** $10^x \cdot \log x$

l. $\dfrac{10^x}{\log x}$ **m.** x **n.** x **o.** 10^{10^x}

p. $\log (\log x)$

5. a. (iv)   **b.** (v)   **c.** (vi)   **d.** (iii)   **e.** (ii)

f. (i)

7. a.

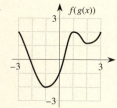

b.

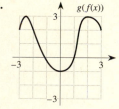

9. $G(x) = x + 5$ and $F(x) = x^4$ is one such pair.

11. $G(x) = \log x$ and $F(x) = x + 3$ is one such pair.

13. a. $y = m(x - 5) + 12$

    b. $y = m(x - x_0) + y_0$; the point-slope form.

15. a.

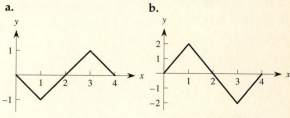

b.

c.

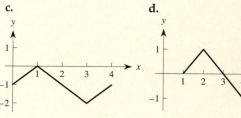

d.

e.

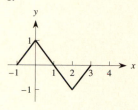

f.

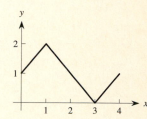

17. a. $x^2 - 3x + 4 + h$ **b.** $x^2 + 2xh + h^2 - 3x - 3h + 4$

    c. $2xh - 3h + h^2$ **d.** $2x - 3 + h$

21.

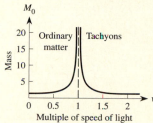

23. a.

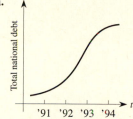

b. The single point of inflection corresponds to the maximum annual debt.

27. Only in the unlikely event that $ad + b = bc + d$.

Section 4.5

1. a. $x \approx 0.35$　　**b.** $x \approx 1.53$　　**c.** $x \approx 1.88$

3. $(1.37, 2.58)$ and $(9.94, 982.29)$

5. $(1.168, 0.068)$

6. $-1.88 < x < 0.35$ and $x > 1.53$

7. $-1.596 < x < -0.183$ and $1.516 < x < 2.263$

9. $x \approx 0.3487$; the interval we obtain for the bisection method is of length $\frac{1}{8}$. The secant method gives an interval of length $\frac{121}{347}$; the secant method.

Section 4.6

1. $f(x) = 3x^2 - 5x, g(x) = x^2 + 4x + 1$, and $k(x) = 2x^2 - 4x + 3$

3. $q(n) = \frac{1}{6}n^3 - \frac{1}{2}n^2 + \frac{1}{3}n$

5. Using the points $(0, 15)$, $(2100, 546)$, and $(-2100, 546)$, $F(x) = \frac{59}{490,000}x^2 + 15$

7. 5525; 42,925　　**9.** 22,022; 3,635,060

13. $\Delta^2 y_0 = \Delta y_1 - \Delta y_0 = (y_2 - y_1) - (y_1 - y_0) = y_2 - 2y_1 + y_0; \Delta^2 y_n = \Delta y_{n+1} - \Delta y_n = (y_{n+2} - y_{n+1}) - (y_{n+1} - y_n) = y_{n+2} - 2y_{n+1} + y_n$

15. $p(n) = \frac{1}{4}n^4 + \frac{1}{2}n^3 + \frac{1}{4}n^2 = \frac{n^2(n+1)^2}{4}$

19. $L(x) = 1 - 0.0083333x - 0.47222222x^2$;

$L(0.1) = 0.99444445$	$\cos(0.1) = 0.99500417$
$L(0.25) = 0.96840279$	$\cos(0.25) = 0.96891242$
$L(0.4) = 0.92111112$	$\cos(0.4) = 0.92106099$
$L(0.75) = 0.728125$	$\cos(0.75) = 0.73168887$

Review Exercises

1.

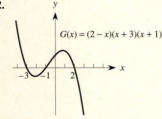

$f(x) = (x + 3)(x - 2)(x - 4)$

2.

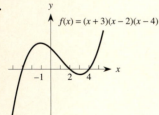

$G(x) = (2 - x)(x + 3)(x + 1)$

3.

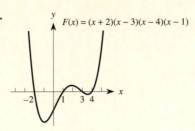

$F(x) = (x + 2)(x - 3)(x - 4)(x - 1)$

4.

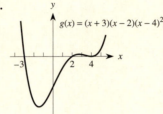

$g(x) = (x + 3)(x - 2)(x - 4)^2$

5. $P(x) = (x - 2)(x + 3)$; the roots are $x = 2$ and $x = -3$.

6. $Q(x) = (x + 5)(2x - 1)$; the roots are $x = -5$ and $x = \frac{1}{2}$.

7. $R(x) = x(x - 2)(x - 1)$; the roots are $x = 0, x = 1$ and $x = 2$

9. $x = 11.39$

10. $x = 0, x = -2.66$

11. a.

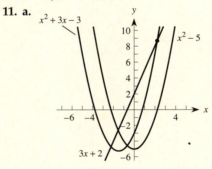

$x^2 + 3x - 3$　　$x^2 - 5$　　$3x + 2$

12. a. Zeros at $x = -2, x = 2$; turning point at $x = 0$; no vertical asymptotes; end behavior as x approaches ∞ is 1 and end behavior as x approaches $-\infty$ is 1.
b. Zeros at $x = -2, x = 2$; turning point at $x = 0$; vertical asymptotes $x = -3, x = 3$; end behavior as x approaches ∞ is 1 and end behavior as x approaches $-\infty$ is 1.
c. No zeros; turning point at $x = 0$; no vertical asymptotes; end behavior as x approaches ∞ is 1 and end behavior as x approaches $-\infty$ is 1.
d. No zeros; turning point at $x = 0$; vertical asymptotes $x = -3, x = 3$; end behavior as x approaches ∞ is 1 and end behavior as x approaches $-\infty$ is 1.

13. a.

(i)

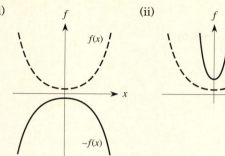

(ii)

(ii)

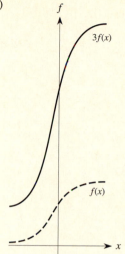

(iii)

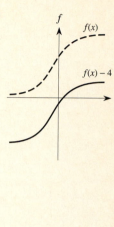

(iii)

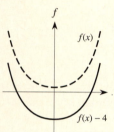

(iv)

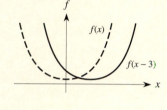

(iv)

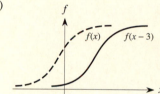

(v)

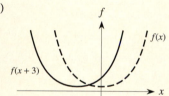

(v)

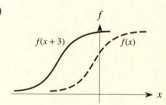

(vi)

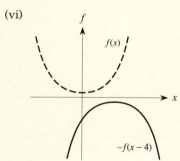

(vi)

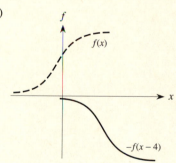

13. b. (i)

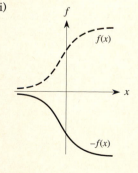

13. c. (i)

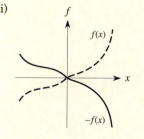

13. c.

(ii)　　　　　　　　　　(iii)

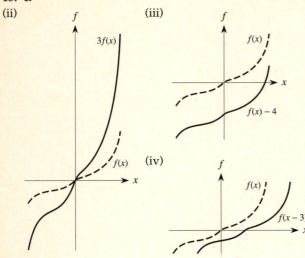

(iv)

(v)　　　　　　　　　　(vi)

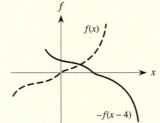

14. a. 19.4　　**b.** 723　　**c.** $\dfrac{6}{7}$　　**d.** $-\dfrac{1}{4}$　　**e.** 7.6

　f. 47.5　　**g.** $\dfrac{3(x^2 + 2)}{(x + 2)^2}$

　h. $8x^4 + 8x^2 + 3$　　**i.** $\dfrac{2x^2}{2x^2 + 3}$　　**j.** $\dfrac{-1}{x + 1}$

　k. $\dfrac{2x^3 - 2x^2 + x - 1}{x + 2}$　　**l.** $\dfrac{2x^3 + 4x^2 + x + 2}{x - 1}$

15. a. $f(g(0)) = 2, f(g(1)) = 2, f(g(2)) = 3, f(g(3)) = 0$
　b. $g(f(0)) = 2, g(f(1)) = 2, g(f(2)) = 3, g(f(3)) = 1$
　c. $f(0) + g(0) = 3, f(1) + g(1) = 2, f(2) + g(2) = 5,$
　　$f(3) + g(3) = 3$
　d. $\dfrac{f(0)}{g(0)} = 2, \dfrac{f(1)}{g(1)}$ is not defined, $\dfrac{f(2)}{g(2)} = \dfrac{3}{2}, \dfrac{f(3)}{g(3)} = 0$

16. a. $f(g(1)) = 3, f(g(2)) = 3, f(g(3)) = 1, f(g(4)) = 1$
　b. $g(f(1)) = 3, g(f(2)) = 1, g(f(3)) = 1, g(f(4)) = 3$
　c. $f(1) + g(1) = 4, f(2) + g(2) = 5, f(3) + g(3) = 4,$
　　$f(4) + g(4) = 2$
　d. $\dfrac{f(1)}{g(1)} = \dfrac{1}{3}, \dfrac{f(2)}{g(2)} = \dfrac{3}{2}, \dfrac{f(3)}{g(3)} = 3, \dfrac{f(4)}{g(4)} = 1$

17. $0.106262 < t < 3.29374$　　**18.** 15150

19. $Q(x) = 3x^2 - 2x + 5$　　**20.** $C(x) = x^3 - 2x^2 - 4$

CHAPTER 5

Section 5.1

1. {4, 8, 12, 16, 20, 24}　　**3.** {0.5, 1, 1.5, 2, 2.5, 3}

5. $\{-9, -2, 17, 54, 115, 206\}$

7. $\left\{\dfrac{2}{3}, \dfrac{4}{9}, \dfrac{8}{27}, \dfrac{16}{81}, \dfrac{32}{243}, \dfrac{64}{729}\right\}$

9. $\left\{1, \dfrac{1}{2}, \dfrac{1}{3}, \dfrac{1}{4}, \dfrac{1}{5}, \dfrac{1}{6}\right\}$

11. {0.8, 0.96, 0.992, 0.9984, 0.99968, 0.999936}

13. diverges　　**15.** diverges　　**17.** diverges

19. converges to 0　　**21.** converges to 0

23. converges to 1　　**25.** strictly increasing, no concavity

27. strictly increasing, no concavity

29. strictly increasing, concave up

31. strictly decreasing, concave up

33. strictly decreasing, concave up

35. strictly increasing, concave down

37. $13, 15; a_n = 2n + 1$; diverges.

39. $6, 3; a_n = 192(0.5)^{n-1}$; converges to 0

41. $\dfrac{5}{7}, \dfrac{6}{8}; a_n = \dfrac{n}{n + 2}$; converges to 1

43. $e_{1000} \approx 2.7169239, e_{1,000,000} \approx 2.7182804,$
　　$e_{10,000,000} \approx 2.7182817, \; e_{100,000,000} \approx 2.7182818;$
　　this sequence coverges to the number e.

45. g_n has limiting value e^2; h_n has limiting value e^5.

47. No. The terms increase without bound as n becomes
　　larger, and so they approach ∞.

Section 5.2

2. $x_n = 4^n$; diverges　　**4.** $x_n = 20(1.5)^n$; diverges

6. $y_n = 64(0.75)^n$; converges to 0

8. $w_n = 5$; converges to 5

10. {100, 150, 175, 187.5, . . . }; increasing and concave
　　down

12. {10, 8, 12, 4, 20, . . . }; neither decreasing nor increasing,
　　neither concave up nor concave down.

13. $100(1.06)^{10} = \$179.08$.

15. 64% increase, and the doubling time is roughly 35
　　years; 109% increase, and doubling time is 23 years.

18. \$124,157,667,678.78

21. If $x_0 = 1, x_1 = \sqrt{2}, x_2 = \sqrt{3}, x_3 = \sqrt{4}, x_4 = \sqrt{5}, \ldots$
　　$x_n = \sqrt{n + 1}$. If $x_0 = 10$, then $x_1 = \sqrt{101}, x_2 = \sqrt{102},$
　　$x_3 = \sqrt{103}, x_4 = \sqrt{104}, \ldots, x_n = \sqrt{n + 100}.$

25. The process described in Exercise 23 as a difference
　　equation is $x_{n+2} = \dfrac{x_{n+1} + 1}{x_n}$, with $x_1 = a$ and $x_2 = b$.

Section 5.3

1. {10, 10, 20, 30, 50, 80, 130, 210, . . . }; each value is ten
　　times greater.

3. 1,020,585,563 people; 37 years; in 2085.

5. For Israel, $I_n = 5.4 (1.015)^n$ and for Jordan, $J_n = 4.2 (1.033)^n$; Jordan's population will equal Israel's population 14.3 years after 1994.

6. a. $P_{n+1} = (1 + a)P_n$ **b.** $P_{n+1} = (1 + a)P_n - 100$

   c. $P_{n+1} = (1 + a)P_n - 0.4P_n$

   d. $P_{n+1} = (1 + a)P_n - b\sqrt{P_n}$

   e. $P_{n+1} = (1 + a)P_n - 0.4(P_n) + 500$

8. a. $P_{n+1} = 1.06P_n - 10,000, P_0 = 80,000$

   b. Juanita will deplete her account in the 13th year.

   c. $P_{n+1} = 1.06P_n - 0.2P_n, P_o = 80,000$

   d. She will be down to $105 in 45 years and $P_n < 0.01$ for $n \geq 111$.

   e. 6% of $80,000 = $4800.

15. first differences: $0, k, 4k, 9k, 16k, \ldots$;

   second differences: $k, 3k, 5k, 7k, 9k, \ldots$;

   third differences: $2k, 2k, 2k, 2k, \ldots$; the solution is a cubic polynomial.

Section 5.4

1. 5.5 mL after 12 hours, 1.9 mL after 24 hours; it takes 31.1 hours for the level to drop below 1 mL and 82.7 hours to drop below 0.01 mL.

3. 91 mL, 84.25 mL, 79.188 mL, $\ldots$; if the rate of elimination and the repeated dosage does not change, the limiting value does not change.

8. Yes, limiting value $L = \dfrac{B}{1 - a}$ varies directly with B.

10. Roughly 36.92 mg; 2.911 hours; 652 mg

14. $r_3 < r_1 < r_2$

Section 5.5

2. $\{3, 3.06, 3.11, 3.17, 3.23, 3.29, 3.35, 3.41, 3.47, 3.53, 3.60, \ldots\}; L = 40$

3. $\{5, 5.08, 5.15, 5.23, 5.3, 5.38, 5.46, 5.54, 5.62, 5.7, 5.78, \ldots\}; L = 20$

5. $\{1, 2, 3, 5, 8, 13, 21, 34, 55, 89, \ldots\}$; the values appear to increase indefinitely with approximately exponential growth.

7. $\{1, 5, 6, 11, 17, 28, 45, 73, 118, 191, \ldots\}$; the values appear to increase indefinitely with approximately exponential growth.

10. $P_0 = 1, P_1 = 4, P_2 = 16, P_3 = 64, P_4 = 252, P_5 = 964,$ and $P_6 = 3159$; everybody will have heard the rumor during the 7th hour.

12. a. $P_{n+1} = 1.4P_n - 0.0002P_n^2$

   b. $P_{n+1} = 1.4P_n - 0.0002P_n^2 - 120$

   c. $P_{n+1} = 1.1P_n - 0.0002P_n^2$

   d. $P_{n+1} = 1.4P_n - 0.0002P_n^2 - k\sqrt{P_n}$

   e. $P_{n+1} = P_n - 0.0002P_n^2 + 75$

14. For $a = 0.055$ and $b = 0.00002$, L is 2750; for $a = 0.06$ and $b = 0.00002$, L is 3000.

For $a = 0.065$ and $b = 0.00002$, L is 3250; for $a = 0.045$ and $b = 0.00002$, L is 2250.

From the formula $L = \dfrac{a}{b}$, we see that L is directly proportional to a.

Section 5.6

1. 3.32 half-lives, or 96.3 years

3. Sometime after 2137 B.C.

5. About 780 years old, dating to the 13th century

Section 5.7

1. 1.999023438, 1.999999046, 1.999999999; the partial sums are approaching 2.

3. 4.57050327, 4.95388314, 4.99504824; the partial sums are approaching 5.

5. 170.9952, 9973.7702, 575,251.1777; the partial sums do not converge.

7. A total of 29,997 cases from 1950 through 1990; no more than 4 new cases between 1990 and 2000

9. 46.6 billion kilowatt-hours

11. 478,969 metric tons of rice

13. The total distance the ball travels is 42 ft.

15. The ball at the 80% bounce travels 20 feet farther than at the 75% bounce.

17. a. Growth is very much faster than the growth of a geometric sequence.

   b. $x_n = (n - 1)! \, x_1$

   c. $x_n = 5^n(n - 1)! \, x_1$

   d. $x_n = r^n(n - 1)! \, x_1$

   e. $x_n = (r + b)(2r + b)(3r + b) \ldots [(n - 1)r + b]x_1$

19. $x_n = \left[\dfrac{1}{(n - 1)!}\right]x_1$ converges very quickly to 0;

$x_n = \left(\dfrac{1}{r}\right)^n x_0$ is a geometric sequence that converges to 0, but not as quickly.

Section 5.8

2. 36.3 minutes

4. a difference of 16 minutes

6. If the body temperature was 98.6° at the time of death, then about 4.57 hours elapsed, which sets the time of death slightly before 4:30 A.M.

8. Temperature $= 70.2247(0.987948)^{\text{Time}}$

12. $r_3 < r_1 < r_2$

13. $t_1 > t_2$

Section 5.9

1. $\{4, 4, 4, 4, 4, \ldots\}; \Delta x_n = 4$

3. $\left\{\dfrac{1}{2}, \dfrac{1}{2}, \dfrac{1}{2}, \dfrac{1}{2}, \dfrac{1}{2}, \ldots\right\}; \Delta x_n = \dfrac{1}{2}$

5. $\{18, 28, 38, 48, 58, \ldots\}; \Delta x_n = 10n + 8$

7. $\{6, 18, 54, 162, 486, \ldots\}; \Delta a_n = 2 \cdot 3^n$

15. 85

18. $x_{101} < x_{100}$

20. 22, 27, 37, 41, 33

Review Exercises

1. a. $\{5, 11, 17, 23, 29, \ldots\}$

   b. $\left\{3, \dfrac{9}{2}, 9, \dfrac{81}{4}, \dfrac{243}{5}, \ldots\right\}$

   c. $\{0.7, 0.91, 0.973, 0.9919, 0.99757, \ldots\}$

2. a. increasing and divergent

   b. increasing and divergent

   c. increasing, converges to 1

   d. increasing and divergent

   e. decreasing, converges to $\dfrac{1}{7}$

   f. increasing, converges to 7

3. a. $3^n, n \geq 0.$ **b.** $\left(\dfrac{1}{4}\right)^n, n > 0.$

   c. $(2n + 1)^2, n \geq 0$ **d.** $40 \cdot 3^n, n \geq 0.$

4. about 23.5 years

5. 254,000,000 people; 284,000,000 people; 62 years

6. 1167 students; 44 years

7. 5 hours and 39 minutes; 8 hours and 24 minutes

8. a. $\{2, 10, 18, 26, 34, \ldots\}$ **b.** $\{12, 4, -4, -12, -20, \ldots\}$

   c. $\left\{5, \dfrac{5}{3}, \dfrac{5}{9}, \dfrac{5}{27}, \dfrac{5}{81}, \ldots\right\}$ **d.** $\{10, 11, 8, 17, -10\}$

9. a. $\{2, 7, 9, 16, 25, \ldots\}$ **b.** $\{3, 7, 10, 17, 27, \ldots\}$

10. 167 mg

11. 21 mg

12. yes, 150 units

13. 15,000; at year $n = 22$.

14. $b = 0.000017; u_{n+1} = 1.2u_n - 0.000017u_n^2, u_0 = 1000$

15. a. $P_{n+1} = 1.3P_n - 0.0004P_n^2$

   b. $P_{n+1} = 1.3P_n - 0.0004P_n^2 - 2000$

   c. $P_{n+1} = 1.3P_n - 0.0004P_n^2 + 400$

   d. $P_{n+1} = 1.3P_n - 0.0004P_n^2 - 0.10P_n$

16. a. $M_{n+1} = 0.85M_n, M_0 = 100\%$; about 0.29%

   b. $M_{n+1} = 1.1M_n + 2000, M_0 = \$50,000; \$104,009$

   c. $P_{n+1} = P_n - 20,000, P_0 = \$2,000,000$ and
$T_{n+1} = 1.1T_n, T_0 = \$2,000,000;$
$P_3 = \$1,940,000, T_3 = \$2,662,000$

CHAPTER 6

Section 6.1

1. first order, linear, homogeneous difference equation.

3. first order, linear, nonhomogeneous difference equation.

5. first order, linear, nonhomogeneous difference equation.

7. second order, linear, nonhomogeneous difference equation.

9. first order, nonlinear, nonhomogeneous difference equation.

11. $\{1, 1, 0.7, 0.50667, 0.388, 0.312, 0.26302, 0.22321,$ $0.19531, 0.17361, 0.1562\}$ As n increases, the terms approach 0.

13. a. $\{1, -1, -1, -1, -1, \ldots\}$; it becomes constant after the first term.

   b. $\{1.5, 0.25, -1.938, 1.7539, 1.0762, -0.8418,$ $-1.291, -0.3324, -1.889, 1.5701, 0.4653, \ldots\};$ it doesn't seem to converge but values are between -2 and 2.

   c. The solution begins $\{1.9, 1.61, 0.5921, -1.64941,$ $0.720578, -1.48076, 0.192670, -1.96287, 1.85289,$ $1.4332, 0.0540736, \ldots\}$; it doesn't seem to converge but values are between -2 and 2.

17. $x_n = x_0 + n^3$. The first values of the solution for the choices $x_0 = -10, 0, 10,$ and 20 are:

n	$x_0 = -10$	$x_0 = 0$	$x_0 = 10$	$x_0 = 20$
1	-9	1	11	21
2	-2	8	18	28
3	17	27	37	47
4	54	64	74	84
5	115	125	135	145
6	206	216	226	236
7	333	343	353	363
8	502	512	522	532
9	719	729	739	749
10	990	1000	1010	1020

18. b. The term $C2^n$ is the general solution to the homogeneous equation $x_{n+1} = 2x_n$; the term $6(3^n)$ is the nonhomogeneous term of the equation $x_{n+1} - 2x_n = 6(3^n)$.

20. b. The term $C3^n$ is the general solution of the homogeneous equation $x_{n+1} = 3x_n$; the term $2n^2 + 2n - 3$ is a polynomial in n of the same degree as the term $-4n^2 + 8n + 6$ in the difference equation $x_{n+1} = 3x_n - 4n^2 + 8n + 6$.

24. a. 1   **b.** $\left(\dfrac{1}{2}, \dfrac{1}{4}\right)$; (1,1)

Section 6.2

1. $x_n = C \cdot (1.5)^n$

3. $x_n = C \cdot (-2)^n + \left(\dfrac{20}{7}\right)5^n$

5. $z_n = C \cdot 5^n - 6 \cdot 3^n$ 7. $x_n = C \cdot 2^n - 10 \cdot 4^n$

9. $z_n = C \cdot 3^n - 6n - 4$

11. $x_n = C \cdot 4^n + 3n\,4^n$

13. $x_n = -\dfrac{13}{7}(-2)^n + \dfrac{20}{7}5^n$ if $x_0 = 1$; $x_n = -\dfrac{6}{7}(-2)^n +$

$\dfrac{20}{7}5^n$ if $x_0 = 2$

15. $z_n = 7 \cdot 5^n - 6 \cdot 3^n$ if $x_0 = 1$; $z_n = 8 \cdot 5^n - 6 \cdot 3^n$ if $z_0 = 2$

17. $x_n = 11 \cdot 2^n - 10 \cdot 4^n$ if $x_0 = 1$; $x_n = 12 \cdot 2^n - 10 \cdot 4^n$ if $x_0 = 2$

19. $z_n = 5 \cdot 3^n - 6n - 4$ if $z_0 = 1$; $z_n = 6 \cdot 3^n - 6n - 4$ if $z_0 = 2$

21. **a.** $x_n = C\left(\dfrac{1}{4}\right)^n + 72\left(\dfrac{1}{3}\right)^n$

    b. Taking $C = 1$, we get $x_{15} = 0.000005018735$, $x_{20} = 0.0000000206503$, and $x_{25} = 0.00000000008497784$. The terms of the sequence approach 0.

    c. Taking $C = 100$, we get $x_{15} = 0.0000511094$, $x_{20} = 0.0000000207403$, and $x_{25} = 0.0000000000850658$. The terms of the sequence approach 0.

Section 6.3

1. **a.** $D_{n+1} = 0.75D_n + 32$

   b. $D_n = 128 - 96(0.75)^n$

   c. The maintenance level is doubled, to 128 mL.

3. $B_{18} = \$27{,}671.23$

5. $D = \$3{,}274.50$

7. $B_n = (1.05)^n (121{,}000) - 5{,}000n - 120{,}000.$ $B_{18} = \$81{,}201$

9. $B_{40} = \$330{,}095.36$

11. $B_{n+1} = 1.007B_n + 1 + 0.25n$ $B_n = (1.007)^n (5492.9) - 35.714n - 5245.90$ In the year 2000 the population will be $B_{10} = 287.70$.

13. **a.** positive, less than 1

    b. $K_n = (1 - m)^n K_0 - 100(1 - m)^n + 100$

    c.
    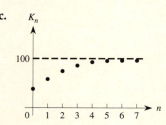

Section 6.4

1. $P_n = -1.2314n^2 + 5.2837n + 24.71$

3. Setting the origin at the lowest point of the cable, the best quadratic fit is
   $$h = (1.05 \times 10^{-4})\,d^2 + 0.00357d + 20.678.$$

5. $x_n = 2n^2 - 12n + 7$

7. The first differences are constant.

Section 6.5

3. **a.** payments: \$350.39, \$354.91, \$359.39
       total amounts: \$10,511.70, \$10,647.30, \$10,781.72

5. **a.** for \$8000, payment is \$276.77; for \$9000, payment is \$311.37; for \$12,000, payment is \$415.16

7. $p = \$632.07$, or about \$33 less than the monthly payments at 7%. The total cost of the loan is \$227,545.

9. \$4850.37

Section 6.6

1. $C_{100} = 4{,}050$, $C_{1000} = 499{,}500$, and $C_{10{,}000} = 49{,}995{,}000$.

3. The formula for the bubble sort exceeds the formula developed for the simple insertion sort for all integer values of n greater than 10.

5. The following is the table of differences for the data on the time in seconds needed to bubble sort a list of n quantities.

n	size	T_n	ΔT_n	$\Delta^2 T_n$
1	200	0.12		
2	400	0.51	0.39	
3	600	1.26	0.75	0.36
4	800	2.16	0.90	0.15
5	1000	3.36	1.2	0.30
6	1200	4.83	1.47	0.27
7	1400	6.56	1.73	0.26
8	1600	8.61	2.05	0.32
9	1800	10.77	2.16	0.11
10	2000	13.38	2.61	0.45

The best linear model for the first differences is $\Delta T_n = 0.265667n + 0.145$ with a correlation coefficient $r = 0.997$. This corresponds to the first order linear difference equation $T_{n+1} = T_n + 0.265667n + 0.145$ with solution $T_n = 0.132833n^2 + 0.012167n - 0.025$, where n is the number of increments of 200 objects to be sorted.

15. R needs to be between 0.15 and 0.3 ppm. If $R = 0.25$ ppm, then the concentration approaches $\dfrac{5}{3} \approx 1.67$ ppm.

Section 6.8

1. Predictions for the various models are displayed in the following table. The choice of which best fits the situation is left to the reader.

Age	1	1.5	2	2.5	3	3.5
Vocabulary	50	200	350	600	880	1200
Linear	−139	141	420	700	980	1260
Exponential	140	199	282	398	564	797
Power	72	171	316	511	755	1052
Logistic	50	117	267	584	1153	1830

4	4.5	5	5.5	6
1450	1800	2150	2425	2750
1539	1819	2099	2379	2658
1128	1596	2258	3194	4519
1401	1803	2261	2774	3343
2064	2001	2025	2017	2020

5. The logistic transformed values seem to be 1 less than the ratio values. Observing the logistic transformation quotient, $\dfrac{\Delta P_n}{P_n} = \dfrac{P_{n+1} - P_n}{P_n} = \dfrac{P_{n+1}}{P_n} - 1$.

Section 6.9

1. $x_0 = 0.2$: {0.2, 0.512, 0.79954, 0.51288, 0.79947, 0.51302, 0.79946, 0.51304, 0.79946, . . .}.
   $x_0 = 0.8$: {0.8, 0.512, 0.79954, 0.51288, 0.79947, 0.51302, 0.79946, 0.51304, 0.79946, . . .}.

3. 0.842154 and 0.451963

Review Exercises

1. **a.** $L = -2$, $L = 0$, and $L = 3$.
   b. The first few terms of each solution starting with the initial conditions $x_0 = -3, -2, \ldots , 4$ are as follows:

$n\backslash x_0$	−3	−2	−1	0	1	2
1	−21	−2	3	0	−5	−6
2	−9597	−2	3	0	−125	−222
3	-8.8×10^{11}	−2	3	0	−1,968,125	-1.1×10^7
4	-6.9×10^{35}	−2	3	0	-7.6×10^{18}	-1.3×10^{21}
5	$\to -\infty$	−2	3	0	$\to -\infty$	$\to -\infty$

$n\backslash x_0$	3	4
1	3	28
2	3	21,028
3	3	9.3×10^{12}
4	3	8.0×10^{38}
5	3	$\to \infty$

2. $123,002.97

3. $173,290.39

4. $755.64; $P_{n+1} = P_n + 250(1.0075)^n$

5. approximately $19,727.40

6. $D_{n+1} = (0.92)^n D_n$, $D_0 = 100$, $D_3 = 77.87$; after 28 years

7. $D_n = 1250 - 1150(0.92)^n$; about 426.15 pounds; limiting value = 1250 pounds

8. maintenance level = 150 units

9. $d_n = (d_0 - 40)(0.8)^n + 40(1.3)^n$

10. **a.** $\dfrac{3}{2}, \dfrac{5}{3}, \dfrac{8}{5}, \dfrac{13}{8}, \dfrac{21}{13}, \dfrac{34}{21}, \dfrac{55}{34}, \dfrac{89}{55}, \dfrac{144}{89}, \dfrac{233}{144}$

   b. $\dfrac{1 + \sqrt{5}}{2}$ **d.** $x_{n+1} = 1 + \dfrac{1}{x_n}$

   e. $\dfrac{1 + \sqrt{5}}{2}, \dfrac{1 - \sqrt{5}}{2}$

CHAPTER 7

Section 7.1

1. The length of Janis's fingernails is a periodic function with a period of one week.

3. **a.** $\dfrac{\pi}{12}$ radians   **b.** $\dfrac{5\pi}{12}$ radians   **c.** $\dfrac{2\pi}{3}$ radians
   d. $\dfrac{5\pi}{6}$ radians   **e.** $\dfrac{5\pi}{4}$ radians   **f.** $\dfrac{7\pi}{4}$ radians

5. **a.** $\dfrac{5}{2}$ **b.** $\dfrac{5}{2}\sqrt{2}$ **c.** $\dfrac{5}{2}\sqrt{3}$ **d.** $\dfrac{5}{2}\sqrt{3}$
   e. −1.2941   **f.** 2.26995   **g.** $\dfrac{5}{2}\sqrt{2}$ **h.** $\dfrac{5}{2}\sqrt{3}$
   i. 1.29410   **j.** $-\dfrac{5}{2}$ **k.** −4.24261   **l.** −2.31448

8. The sine curve is simply the cosine curve shifted to the right by 90°, or $\dfrac{\pi}{2}$ radians. The graph of $\cos\left(x - \dfrac{\pi}{2}\right)$ will be identical to the graph of $\sin x$.

10. **a.** $\dfrac{\pi^2}{8} \approx 1.233701$ **b.** $\dfrac{\pi}{4} \approx 0.785398$
    c. The average of the over- and under-estimates, which is 1.009549, would be reasonable.
    d. About 2 square units.

Section 7.2

1. **a.** periodic; period = 2
   b. periodic; period = 2
   c. periodic; period = 2
   d. periodic; period = 1
   e. periodic; period = 2
   f. not periodic   **g.** not periodic
   h. A constant function can be considered periodic, but a definite period cannot be determined.
   i. periodic; period = 2   **j.** periodic; period = 20

2. March 1 has 11.19 hours of daylight. May 12 has 13.87 hours of light, and July 4 has 14.33.

4. $H(t) = 12 - 8.3 \sin\left(\dfrac{2\pi}{365}(t - 80)\right)$

5. $H(t) = 12 - 2.4 \sin\left(\dfrac{2\pi}{365}(t - 80)\right)$

7. $W(t) = 8 + 4 \sin\left(\dfrac{15\pi}{128}\left(t - \dfrac{94}{15}\right)\right)$

10. a. $2 \sin\left(\dfrac{\pi}{40}t\right)$ **b.** $7.5 \sin\left(\dfrac{\pi}{150}t\right)$

14. Because the sine function is periodic there are infinitely many correct formulas for each graph.

a. $y = 2 \sin\dfrac{x}{3}$ **b.** $y = 5 \sin 2x$

c. $y = 4 \cos 4x$ **d.** $y = -3 \sin\dfrac{x}{5}$

e. $y = 1.5 \cos\dfrac{x}{4}$ **f.** $y = 2 - \cos 2x$

g. $y = -5 \cos 2x$ **h.** $y = 10 + 6 \sin 8x$

i. $y = -3 + 3 \sin 4x$ **j.** $y = 3 + 3 \cos 4x$

k. $y = \sin\left(\dfrac{\pi}{2}x\right)$ **l.** $y = 4 \cos \pi x$

15. $f_1(x) = \sin x$ $f_2(x) = 2 + \sin x$
$f_3(x) = 2 \sin x$ $f_4(x) = \sin 2x$
$f_5(x) = \sin\dfrac{x}{2}$ $f_6(x) = 1 - \sin x$

21. a. not periodic   **b.** not periodic
c. periodic with period π
d. periodic with period π
e. not periodic
f. periodic with period π
g. periodic with period π
h. periodic with period 2π
i. periodic with period π

Section 7.3

4. not an identity   **6.** an identity
8. an identity   **10.** not an identity
12. not an identity   **14.** not an identity
16. $\cos^6 x = \dfrac{5}{16} + \dfrac{7}{16}\cos 2x + \dfrac{3}{16}\cos 4x + \dfrac{1}{16}\cos 2x \cos 4x$
18. $\cos 3x = 4 \cos^3 x - 3 \cos x$
20. $\cos 5x = 16 \cos^5 x - 20 \cos^3 x + 5 \cos x$. There are many equivalent forms.
24. a. $\sin(x + \pi) = -\sin x$
b. $\cos\left(x + \dfrac{\pi}{2}\right) = -\sin x$
26. $c = 2.41$
29. Barnard's Star: 0.3476×10^{14}
Sirius: 0.4771×10^{14}
Epsilon-Eridani: 0.6291×10^{14}

Section 7.4

1. a. no solutions
b. $0.411517 \pm 2\pi n$ and $2.730076 \pm 2\pi n$
c. no solutions
d. $0.848062 \pm 2\pi n$ and $2.293531 \pm 2\pi n$
e. $-0.848062 \pm 2\pi n$ and $-2.293531 \pm 2\pi n$
f. $0.32175 \pm \pi n$ and $1.24905 \pm \pi n$
g. $1.10715 \pm \pi n$ and $2.03444 \pm \pi n$
h. $0.841068 \pm 2\pi n$ and $-0.841068 \pm 2\pi n$
i. $\dfrac{\pi}{2} \pm \pi n, 0.304693 \pm 2\pi n,$ and $2.83690 \pm 2\pi n$

2. Eleven hours of daylight on day #55 (Feb. 24) and again on day #287 (Oct. 14). Ten hours of daylight on January 23 and November 16. San Diego always has more than 9 hours of daylight.
9. $C = 0.64350$ radians.

Section 7.5

3. $\tan 2x = \dfrac{2 \tan x}{1 - \tan^2 x}$
4. an identity
6. an identity
8. an identity
10. an identity
12. not an identity
14. The maximum angle is about $19° \ 30'$ when the distance is 65 meters.
16. a. $0.896055 \pm n\pi$ **b.** $0.67474 \pm n\pi$ **c.** $0.67474 \pm n\pi$

Section 7.6

5. $g(x) = \begin{cases} 1 & 0 \leq x \leq \pi \\ -1 & \pi \leq x \leq 2\pi \end{cases}$

8. $f(x) = \begin{cases} \dfrac{2}{\pi}x & \text{for } 0 \leq x \leq \dfrac{\pi}{2} \\ -\dfrac{2}{\pi}(x - \pi) & \text{for } \dfrac{\pi}{2} < x \leq \dfrac{3\pi}{2} \\ \dfrac{2}{\pi}(x - 2\pi) & \text{for } \dfrac{3\pi}{2} < x \leq 2\pi \end{cases}$

10. One possible formula is

$$T(t) = 55 - 30 \cos\left(\dfrac{2\pi}{365}(t - 32)\right),$$

where time t is measured in days from midnight of New Year's Eve.

Section 7.7

1. $T_2(0) = 1,$ $T_2(0.1) = 0.995,$
$T_2(0.2) = 0.98,$ $T_2(0.3) = 0.955,$ $T_2(0.4) = 0.92,$
$T_2(0.5) = 0.875,$ $T_2(0.6) = 0.82$

5. The quadratic best fit is $y = -0.496024x^2 + 0.999957$

x	0.05	0.15	0.25	0.40	0.50
$\cos x$	0.998750	0.988771	0.968912	0.921061	0.877583
$T_2(x)$	0.998750	0.988750	0.968750	0.920000	0.875000
y	0.998717	0.988796	0.968955	0.920593	0.875951

9. $P_5(x) = -x - \dfrac{(-x)^3}{6} + \dfrac{(-x)^5}{120} = -x + \dfrac{x^3}{6} - \dfrac{x^5}{120}$

11. a. $P_5(x) = 2x - \dfrac{4}{3}x^3 + \dfrac{4}{15}x^5$

    b. $P_3(x) \cdot P_4(x) = x - \dfrac{2}{3}x^3 + \dfrac{1}{8}x^5 - \dfrac{1}{144}x^7$

Section 7.8

1. $|z| = 5$, $\theta = -\text{Arctan}\,\dfrac{3}{4}$

3. $|z| = 13$, $\theta = -\text{Arctan}\,\dfrac{5}{12}$

5. $|z| = 73.4302$, $\theta = -0.512389$

7. $|z| = 8.60233$, $\theta = 2.19104$

9. $|z| = 8.18535$, $\theta = 3.3548$

11. $13\left[\cos\left(\text{Arctan}\,\dfrac{12}{5}\right) + i\sin\left(\text{Arctan}\,\dfrac{12}{5}\right)\right]$

13. $25\left[\cos\left(\text{Arctan}\,\dfrac{4}{3}\right) + i\sin\left(\text{Arctan}\,\dfrac{4}{3}\right)\right]$

15. $8.544[\cos(-0.358771) + i\sin(-0.358771)]$

17. $4.12311\,[\cos(0.755969) + i\sin(0.755969)]$

19. $z^2 = 7 - 24i$

21. $z^2 = 119 - 120i$

23. $z^2 = 25\left[\cos\left(-2\,\text{Arctan}\,\dfrac{3}{4}\right) + i\sin\left(-2\,\text{Arctan}\,\dfrac{3}{4}\right)\right]$

25. $z^2 = 169\left[\cos\left(-2\,\text{Arctan}\,\dfrac{5}{12}\right) + i\sin\left(-2\,\text{Arctan}\,\dfrac{5}{12}\right)\right]$

27. $z^2 = 5392\,[\cos(-1.02478) + i\sin(-1.02478)]$

29. $z^2 = 74\,[\cos(4.38208) + i\sin(4.38208)]$

31. $z^2 = 67\,[\cos(6.7096) + i\sin(6.7096)]$

33. $z^3 = -2035 - 828i$

35. $z^3 = -14625 + 5500i$

37. $z^3 = 2197\left[\cos\left(3\,\text{Arctan}\,\dfrac{12}{5}\right) + i\sin\left(3\,\text{Arctan}\,\dfrac{12}{5}\right)\right]$

39. $z^3 = 15625\left[\cos\left(3\,\text{Arctan}\,\dfrac{4}{3}\right) + i\sin\left(3\,\text{Arctan}\,\dfrac{4}{3}\right)\right]$

41. $z^3 = 623.7\,[\cos(-1.076312) + i\sin(-1.076312)]$

43. $z^3 = 70.0928\,[\cos(2.267908) + i\sin(2.267908)]$

45. $z^0 = 1$, $z = 1 + 2i$, $z^2 = -3 + 4i$,
     $z^3 = -11 - 2i$, $z^4 = -7 - 24i$

Section 7.9

2. a. The limiting value will be real whenever $C < \dfrac{1}{4}$.

    b. {0.5, 0.35, 0.2225, 0.14950625, 0.12235212,
        0.11497004, 0.11321811, 0.11281834, 0.11272798,
        0.1127076, 0.112703, 0.112702, 0.112702, . . . }

6. Every graph of the form $y = x^3 + C$ intersects the line $y = x$.

Review Exercises

1. $\cos x = \sin\left(x + \dfrac{\pi}{2}\right)$, and $5.70 + \dfrac{\pi}{2} \approx 7.2708$.

2. $\theta = 120°, 420°, 480°, \ldots$; $\theta = -240°, -300°, -600°, -660°, \ldots$

3. a. $\theta = 315°, 405°, 675° \ldots$; $\theta = -45°, -315°, -405°, -675° \ldots$

    b. $\theta = \dfrac{7\pi}{4}, \dfrac{9\pi}{4}, \dfrac{15\pi}{4} \ldots$; $\theta = \dfrac{-\pi}{4}, \dfrac{-7\pi}{4}, \dfrac{-9\pi}{4}, \dfrac{-15\pi}{4} \ldots$

4. a. 3142 miles   **b.** 4189 miles

    c. 10472 miles   **d.** 1047 miles

5. a. 360π radians   **b.** 720π feet

6. Numerical answers will vary depending on the diameter D. With D measured in inches, the tire makes $\dfrac{5280 \times 12}{\pi D}$ revolutions per minute.

7. a.

	vertical shift	amplitude	frequency	period	phase shift
i.	325	10	$\dfrac{2\pi}{9}$	9	0
ii.	63	3	$\dfrac{2\pi}{25}$	25	0
iii.	71	2	$\dfrac{2\pi}{15}$	15	0
iv.	80	13	$\dfrac{\pi}{12}$	24	15
v.	38	8	$\dfrac{\pi}{12}$	24	5
vi.	100	25	$\dfrac{\pi}{36}$	72	0
vii.	100	25	$\dfrac{2\pi}{97}$	97	0
viii.	145	40	$\dfrac{2\pi}{83}$	83	0

8.

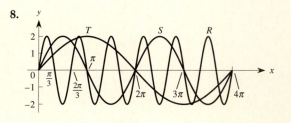

9. a. frequency $\frac{3}{4}$, period $\frac{8\pi}{3}$, amplitude 2, and phase shift 0

b. frequency $\frac{3}{4}$, period $\frac{8\pi}{3}$, amplitude 2, and phase shift $-\frac{4\pi}{3}$

c. frequency π, period 2, amplitude 2, and phase shift $-\frac{3}{4\pi}$

d. frequency $\frac{3\pi}{4}$, period $\frac{8}{3}$, amplitude 2, and phase shift $\frac{1}{\pi}$

10. a. $x = \pm\frac{4\pi}{9} + \frac{8\pi n}{3}$

b. $x = \pm\frac{4\pi}{9} + \frac{8\pi n}{3}$

c. $x = \frac{2}{3} - \frac{3}{4\pi} + 2n$ or $x = \frac{4}{3} - \frac{3}{4\pi} + 2n$

d. $x = \frac{8}{9} + \frac{1}{\pi} + \frac{8n}{3}$ or $x = \frac{16}{9} + \frac{1}{\pi} + \frac{8n}{3}$

11. a.

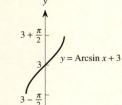

b.

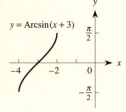

c.

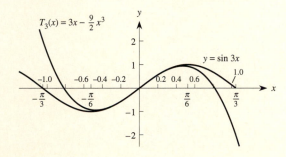

12. a. $\theta = -\text{Arctan}\,\frac{3}{2} + \pi n$ **b.** $\theta = \frac{1}{2}\,\text{Arctan}\,2 + \frac{\pi}{2}\,n$

c. $\theta = \frac{1}{3}\,\text{Arctan}\,7 + \frac{\pi}{3}\,n$

13. a. $x = 4.45522$ **b.** $x = 1.070796$

14. With $T_3(x) = 3x - \frac{9}{2}x^3$, we have $T_3(0.2) = 0.564$.

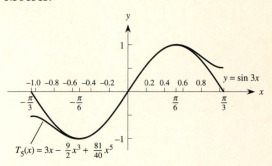

15. With $T_5(x) = 3x - \frac{9}{2}x^3 + \frac{81}{40}x^5$, we have $T_5(0.2) = 0.564648$.

CHAPTER 8

Section 8.2

1. 5

3. 5

5. $\sqrt{20} \approx 4.47$

7. $(3.5, 6)$

9. $\left(\frac{-8}{3}, 2\right)$

11. $(x - 5)^2 + (y - 2)^2 = 25$

13. $(x - 6)^2 + (y - 4)^2 = 16$

15. $(x - 6)^2 + (y - 4)^2 = 16$; $(x - 6)^2 + (y - 9)^2 = 41$

17. $(x + 2)^2 + (y + 3)^2 = 25$; center $(-2, -3)$, radius $= 5$

19. $(x + 5)^2 + (y - 2)^2 = 100$; center $(-5, 2)$, radius $= 10$

21. $(x - 1)^2 + (y + 3)^2 = 4$; center $(1, -3)$, radius $= 2$

25. $(5, 0)$ and $(-3, 4)$

Section 8.3

1. a. point on the right   **b.** point on the right
   c. point on the left

3. The orbit will decay and the satellite spiral in.

5. Venus' orbit is the most circular and Pluto's the least circular.

7. $\frac{x^2}{3352.41} + \frac{y^2}{3210.4} = 1$

9. a. $\frac{x^2}{96,100} + \frac{y^2}{65,792.25} = 1$ **b.** 348.182 ft

   c. 402.358 ft

11. 5.0625 in.

13. $a - c =$ distance at perihelion; $a + c =$ distance at aphelion

15. $\frac{(x + 1)^2}{16} + \frac{(y + 1)^2}{4} = 1$; ellipse with center at $(-1, -1)$, major axis parallel to x-axis, and $a = 4$, $b = 2$, so $c = \sqrt{12}$.

17. $\dfrac{(x+1)^2}{16} - \dfrac{(y+1)^2}{4} = 1$; hyperbola with center at

$(-1, -1)$, axis parallel to the x-axis and $a = 4$, $b = 2$,
so $c = \sqrt{20}$.

19. $\dfrac{(x-2)^2}{\left(\dfrac{38}{4}\right)} - \dfrac{(y+1)^2}{\left(\dfrac{38}{9}\right)} = 1$; hyperbola with center at

$(2, -1)$, axis parallel to the x-axis, and $a = \dfrac{\sqrt{38}}{2}$,

$b = \dfrac{\sqrt{38}}{3}$, so $c = \dfrac{\sqrt{494}}{6}$.

21. $(x-3)^2 + (y+2)^2 = \dfrac{9}{4}$; circle with center at $(3, -2)$

and radius $= \dfrac{3}{2}$.

23. $\dfrac{(y+2)^2}{\left(\dfrac{1}{2}\right)^2} - \dfrac{(x+1)^2}{\left(\dfrac{1}{3}\right)^2} = 1$; hyperbola with center at

$(-1, -2)$, axis parallel to the y-axis, and $a = \dfrac{1}{2}$,

$b = \dfrac{1}{3}$.

Section 8.4

1. a. Slope $= -\dfrac{3}{4}$ **b.** $y = \dfrac{-3}{4}x + \dfrac{13}{2}$

3. a.

t	-2	-1.5	-1	-0.5	0
x	-7	-2.375	0	0.875	1
y	2	0.25	-1	-1.75	-2

0.5	1	1.5	2
1.125	2	4.375	9
-1.75	-1	0.25	2

(b), (d)

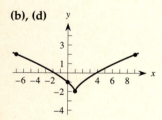

c. $y = (x-1)^{2/3} - 2$

5. $a = 1, b = 3$; period $= 6\pi$

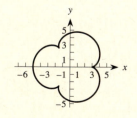

$a = 1, b = 4$; period $= 8\pi$

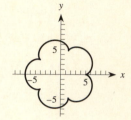

$a = 1, b = 5$; period $= 10\pi$

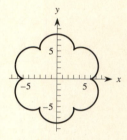

$a = 1, b = 6$; period $= 12\pi$

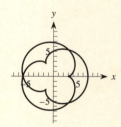

If the period is $n\pi$, then the pattern for the number of loops appears to be $\dfrac{n}{2}$.

7. $a = 2, b = 3$; period $= 6\pi$

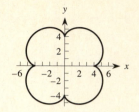

Section 8.5

1. 12 hours

3. about $\dfrac{2}{3}$.

5. about $\dfrac{8}{3}$.

7. about 486.5 ft.

Section 8.6

1. a. $P\left(5, \dfrac{\pi}{6}\right)$ $Q\left(3, \dfrac{2\pi}{3}\right)$ $R\left(6, \dfrac{7\pi}{6}\right)$ $S\left(2, \dfrac{5\pi}{3}\right)$

b. $P\left(5, \dfrac{-11\pi}{6}\right)$ $Q\left(3, \dfrac{-4\pi}{3}\right)$ $R\left(6, \dfrac{-5\pi}{6}\right)$ $S\left(2, \dfrac{-\pi}{3}\right)$

c. $P\left(-5, \dfrac{7\pi}{6}\right)$ $Q\left(-3, \dfrac{5\pi}{3}\right)$ $R\left(-6, \dfrac{\pi}{6}\right)$ $S\left(-2, \dfrac{2\pi}{3}\right)$

d. $P\left(-5, \frac{-5\pi}{6}\right)$ $Q\left(-3, \frac{-\pi}{3}\right)$ $R\left(-6, \frac{-11\pi}{6}\right)$

$S\left(-2, \frac{-4\pi}{3}\right)$

3. a. $r = \sqrt{32}$ and $\theta = \frac{\pi}{4}$

b. $r = \sqrt{32}$ and $\theta = \frac{3\pi}{4}$

c. $r = \sqrt{32}$ and $\theta = \frac{5\pi}{4}$

d. $r = \sqrt{32}$ and $\theta = \frac{-\pi}{4}$

e. $r = 5$ and $\theta = \text{Arctan}\frac{-4}{3}$

f. $r = 5$ and $\theta = \text{Arctan}\left(\frac{-4}{3}\right) + \pi$

g. $r = \sqrt{73}$ and $\theta = \text{Arctan}\frac{3}{8}$

h. $r = \sqrt{73}$ and $\theta = \text{Arctan}\frac{8}{3}$

5. a. $\left(4000, \frac{\pi}{4}\right)$ **b.** $\left(4000, \frac{5\pi}{8}\right)$ **c.** $\left(26800, \frac{-13\pi}{30}\right)$

Section 8.7

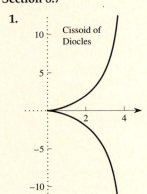

1. Cissoid of Diocles

3. Bifolium

5. Cochleoid

7. Cruciform

9. Lemniscate of Bernoulli

11. $r = \dfrac{4}{\sin\theta}$

Review Exercises

1. a. $\frac{(x-1)^2}{9} + \frac{(y+2)^2}{25} = 1$ **b.** $\frac{(x-1)^2}{25} + \frac{(y+2)^2}{9} = 1$

2. a. $\frac{y^2}{9} - \frac{(x-2)^2}{16} = 1$ **b.** $\frac{x^2}{4} - \frac{(y+1)^2}{9} = 1$

3. a. $y = \frac{(x+4)^2}{4} + 3$ **b.** $x = \frac{(y-3)^2}{8} - 1$

4. $xy = 5$ is a hyperbola with axis on the line $y = x$.

5. $y - 43 = -(x-3)^2$ is a vertical parabola, with vertex at $(3, 43)$, opening downward.

6. $(x-4)^2 + (y+3)^2 = 16$ is a circle of radius 4 and center at $(4, -3)$.

7. $\frac{(x+5)^2}{17} + \frac{(y-2)^2}{\left(\frac{34}{3}\right)} = 1$ is an ellipse with center at

$(-5, 2)$, major axis parallel to the x-axis and $a = \sqrt{17}$,

$b = \sqrt{\frac{34}{3}}$.

8. $\frac{(x-1)^2}{4} - \frac{(y+3)^2}{3} = 1$ is a hyperbola with center

$(1, -3)$ whose axis is parallel to the x-axis.

9. $\frac{(y-2)^2}{6} - \frac{(x-3)^2}{9} = 1$ is a hyperbola with center

$(3, 2)$ whose axis is parallel to the y-axis.

10. Exercise 5: Focus at $(3, 42.75)$, vertex at $(3, 43)$
Exercise 6: Center at $(4, -3)$, radius $= 4$
Exercise 7: Foci at $(-2.61952, 2)$, $(-7.38048, 2)$
Vertices at $(-0.876894, 2)$, $(-9.12311, 2)$, $(-5, 5.3665)$, $(-5, -1.3665)$
Center at $(-5, 2)$
Exercise 8: Foci at $(3.64575, -3)$, $(-1.64575, -3)$
Vertices at $(3, -3)$, $(-1, -3)$
Center at $(1, -3)$
Exercise 9: Foci at $(3, 5.87298)$, $(3, -1.87298)$
Vertices at $(3, 4.44949)$, $(3, -0.44949)$
Center at $(3, 2)$

11. $\frac{x^2}{100} + \frac{y^2}{36} = 1$

12. $\frac{(x+6)^2}{52} + \frac{(y-3)^2}{16} = 1$

13. $\frac{(y-3)^2}{9} - \frac{(x-2)^2}{7} = 1$

14. $\frac{y^2}{16} - \frac{x^2}{9} = 1$ **15.** $\frac{x^2}{225} + \frac{y^2}{81} = 1$

17. 12 ft from the point below the highest point, along the axis of the vertices.

20. The points are $(6, -4)$, $(5, -1)$, $(4, 2)$, $(3, 5)$, $(2, 8)$; the function is $y = -3x + 14$.

21. Same as parabola $y = \frac{x^2}{9} + 1$, between $x = -6$ and $x = 6$.

22. a.

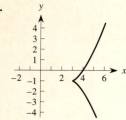

b. $x = (y + 1)^{2/3} + 3$

c. $x = 4$

23. a. Same as the graph of $y = 1 - x$, for $0 \leq x \leq 1$.

26. $x - 1 = -2y^2$

28. a. $\left(\sqrt{18}, \dfrac{\pi}{4}\right)$ **b.** $\left(\sqrt{10}, 1.89255\right)$

c. $\left(\sqrt{17}, -0.244979\right)$ **d.** $\left(6, \dfrac{\pi}{2}\right)$

29. a. $\left(1.5, 1.5\sqrt{3}\right)$ **b.** $\left(\dfrac{3\sqrt{2}}{2}, \dfrac{3\sqrt{2}}{2}\right)$

c. $(0, -4)$ **d.** $\left(-2\sqrt{2}, -2\sqrt{2}\right)$

e. $\left(-2.5\sqrt{3}, 2.5\right)$ **f.** $(-2.08073, 4.54649)$

APPENDIX A

Section A.1

1. $a = 9.3754$, $c = 15.2282$

2. $b = 3.4409$, $c = 12.4836$

3. $b = 13.748$, $\theta = 66.42°$

4. $a = 24$, $\theta = 36.87°$

5. $c = 85.9826$, $\theta = 33.14°$

6. The pole is 26.6 feet high.

7. The width of the river is 47.4 feet.

8. The distance from the base of the tower to the Jet-Ski is 571.5 feet. The distance from the top of the tower to the Jet-Ski is 573.7 feet.

9. The height of the cloud is 1092 feet.

10. The length of the ramp is 47.2 feet. It starts 46.3 feet from the base of the platform.

11. The cliff is just short of 21 feet high.

12. The estimates of the height of the cliff range from 20.14 to 21.82 feet.

13. The estimates of the height of the cliff range from 20.24 to 21.73 feet.

14. Jack is 27.3 feet from the base of the cliff.

Section A.2

1. $c = 19$, $a = 11.7$ **2.** $a = 11.5$, $b = 3.3$

3. $b = 15.1$, $\theta = 43.342°$ **4.** $\theta = 15.47°$, $a = 28.9$

5. $c = 15$, $\theta = 36.87°$ **6.** $c = 135.915$, $b = 129.263$

7. 0.1431 miles, or 756 feet   **8.** 9659 feet

9. 1.03 inches   **10. a.** 5.71°   **c.** 20.0998 miles

11. a. 30,715.5 feet   **b.** 29,042.1 feet, or 5.5 miles

12. $\cos \theta = 0.95$, $\tan \theta = 0.33$; $\theta \approx 18°$ **13.** 41.81°

14. $\dfrac{3}{\cos 27°} \approx 3.37$ inches

Section A.3

1. $C = 91°$, $a = 5.90$, and $c = 13.47$

2. $B = 61°$, $a = 46.14$, and $b = 55.18$

3. $C = 80°$, $a = 13.98$, and $b = 22.09$

4. $B = 22.69°$, $C = 117.31$, and $c = 13.82$

5. $B = 50.47°$, $C = 89.53°$, and $c = 15.56$; or $B = 129.53°$, $C = 10.47°$, and $c = 2.83$

6. The given conditions do not correspond to any triangle.

7. There are various ways to describe the location of the third ship. The ship is 37.9 miles in the direction of 54° south of east from the northernmost ship and 29.5 miles in the direction of 41° north of east from the southernmost ship.

8. 18.8 feet.

9. a. Exercise 4 represents the case $b < a$, where $b = 6$ and $a = 10$. The following diagram shows that only one triangle can be formed: $AC = 6$ is less than the radius $BC = 10$, which intersects the ray AB exactly once.

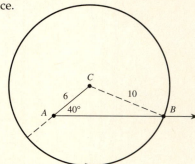

b. Exercise 5 represents the case $b > a$, where $b = 12$ and $a = 10$ (12 is somewhat larger than 10). The following diagram shows that two different triangles can be formed: $AC = 12$ is greater than the radius $BC = 10$, but BC is short enough to intersect the ray AB at two different points.

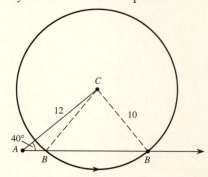

c. Exercise 6 represents the case $b > a$, where $b = 18$ and $a = 10$ (18 is much larger than 10). The following diagram shows that no triangle can be formed: $AC = 18$ is greater than the radius, but it exceeds the radius too much so that side BC cannot intersect the ray AB.

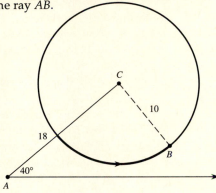

9. A "V-shaped" curve that is the same graph as the line $y = x$ in which the part of the line corresponding to negative values of x is reflected to be above the x-axis.

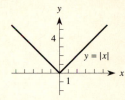

10. The graph of $y = |x + 3|$ is the same as the graph of $y = |x|$ that has been shifted to the left 3 units.

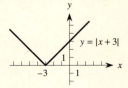

APPENDIX B

1. $-i$

2. 1

3. -1

4. i

5. $2 + 4i$

6. $12 + 4i$

7. $11 - 2i$

8. $4 - 3i$

9. $5 - 5i$

10. $-11 + 60i$

11. 229

12. 65

13. $5i, -5i$

14. $\frac{3}{2}i, -\frac{3}{2}i$

15. $-\frac{1}{5} + \frac{2}{5}i, -\frac{1}{5} - \frac{2}{5}i$

16. $2 + \sqrt{3}\,i, 2 - \sqrt{3}\,i$

APPENDIX C

1. 0

2. $\frac{2}{3}$

3. π

4. 21

5. 4

6. 10

7. 1

8. 1

11. 720

12. 1

13. 2

14. 12

15. 20

16. 8

17. $\frac{25}{12}$

18. 91

19. 75

20. 105

21. 88

22. $DE = 10, AE = 15$

23. $3\sqrt{5}$

24. $2\sqrt{2}$

25. 5

26. $C(1, 5), r = 9$

27. $C(-4, 3), r = \sqrt{17}$

28. $\dfrac{(x - 4)^2}{16} + \dfrac{(y - 2)^2}{4} = 1$

INDEX

Absolute value, A-30
AIDS, spread of, 57, 84, 194–195
Amortization, 386
Amplitude, 438–439
Analytic geometry, 517–523
Angle of inclination, A-2
Aphelion, 531, 558–559
Arccos function, 468–470
Archimedean spiral, 568–569
Arcsin function, 464–468
   properties of, 467
Arctan function, 473–474, A-7
Area of circle, 74
Asymptote
   horizontal, 64
   vertical, 80, 88
Attractor, 414
Average value of function, 553–560
Axis, 22

Bacterial culture model, 298
Base
   changing, 94–96
   e, 94–96
   exponential function, 61
   logarithm, 86
Best fit
   curve, 121
   line, 52–54, 122
Bifurcation diagram, 420
Bisection method, 224–230
Black thread method, 53
Bubble sort, 391–396
Budget constraint, 54
Butterfly effect, 418

Cannonball problem, 240
Carbon dating, 309–310
Cardioid, 508, 572
Cartesian coordinate system, 22, 515
   and polar coordinates, 564–566
Challenger shuttle, 119–122, 172–174
Chaos, 415–421
   and complex numbers, 503–510
   Julia set, 504–507

Mandelbrot set, 508–510
Circle
   area of, 74
   as conic section, 525
   equation of, 521–523, A-32
   parametric equations, 545
   polar coordinates, 568, 569
   unit, 34–35, 425, 504–505
Circular function, 430
Cobweb diagram, 415
Coefficient, leading, 180–181
Communications satellite, 84–85
Complete graph, 189
Complete solution of difference equation, 354
Completing the square, A-33–A-34
Complex number, 183, 496–502
   arithmetic of, A-27–A-30
   complex plane, 496–497
   conjugate pair, 184, 501–502
   modulus, 497
   powers of, 498–501
   and sequences, 504–510
   trignometric form, 497–498
Complex root, 183–184, 199–206
Composite function, 213–214
Compound interest, 258–259
Concavity, 10–11
Conic section, 525–540
   ellipse, 525–534
   general equation, 539–540
   hyperbola, 534–537
   parabola, 537–539
Conjugate pair, 184, 501–502
Constant of proportionality, 42
Contaminants model, 273–276
Continuous function, 224
Continuous variable, 49
Convergence, 227
Convergent sequence, 253–254
Cooling model, 322–323
Coordinate system
   Cartesian/rectangular, 22, 515
   complex plane, 496–497
   polar, 516, 561–574
Correlation coefficient, 126–127, 158, 166
   critical values of, 128–130
Cosecant function, 475

Cosine function, 425–430, A-13–A-15, A-19–A-26
   approximating with polynomials, 481–492
   graph of, 429, A-23
   identities. *See* Trigonometric identities
   inverse, 468–470
   law of cosines, 460–462
   period, 429
Cotangent function, 475
Cricket chirping, 24–25
   line of best fit, 52–54
   point–slope form, 46–47
Critical values for correlation coefficient, 128–130
Cubic polynomial, 184–186
   difference patterns, 234, 237–239
   fitting to data, 194–195
   leading coefficient, 186
   roots of, 184–186, 203–206
   and sine function, 483–488
Curve. *See* Graph
Curve fitting. *See* Fitting to data; Nonlinear regression
Cusp, 572
Cycloid, 549

Data, 2
   difference patterns, 232–233, 236–237
   discrete, 49
   fitting line to, 52–54
   linear, 51
   sampling, 127–129
   table of, 4–5, 23–24
   See also Linear regression; Nonlinear regression
Daylight model, 436–437, 443, 464–465
Dead body model, 324
Decay factor, 65
Decay rate, 65
Decreasing function, 9–11
Degree of polynomial, 180
DeMoivre's theorem, 501
Dependent variable, 16–17, 22, 47, 66
Depreciation, straight-line, 58
Descartes, Rene, 517
Diagonal shift, 215–216

Difference
   of functions, 208–209
   of logarithms, 89–90
Difference equation, 257–276, 341–351
   and algebraic equation, 344–345
   amortization model, 86
   arbitrary constant, 344, 346
   closed form solution, 271–272, 362
   constant coefficient, 343
   and differences, 334–336
   drug model, 280–288, 364–366
   equilibrium solution, 349–350
   exponential, 268–271, 334
   Fibonacci model, 268–270
   general solution, 285–290, 344, 346
   graph of solution, 346–349
   homogenous, 343
   limiting value, 283, 288, 290
   linear, 342–343
   logistic growth, 295–303, 334–336
   Newton's Laws of Cooling and Heating, 321–327
   nonhomogeneous, 343
   nonlinear, 342, 348–349
   particular solution, 354
   radioactive decay, 306–310
   regression, 374–377
   solution field, 347–349
   solving, 284–285, 343–350
   specific solution, 344
   See also Iteration; Sequence
Difference equation, first order, 272, 341, 354–371
   complete solution, 354–356
   and financing, 385–389
   fitting to data, 374–382, 405–410
   general/particular solution, 354
   geometric sequence, 367–369
   method of undetermined coefficients, 357–360
   nonhomogeneous term, 355, 361–362
   nonlinear, 379–382
   and sorting methods, 391–399
   stochastic, 403
   stock market model, 400–404
   sum of integers, 369–371
   trial solution, 357, 359–360
Difference equation, second order, 272, 341–342
Difference operator, 331
Difference pattern, 232–233, 236–237
   logistic growth, 334–336
   and sequences, 329–334
Directrix of parabola, 537
Discontinuous function, 224–225
Discrete dynamical system, 257
Discriminant, 184
Distance formula, 517, A-32

Divergent sequence, 253
Domain of function, 17–19, 24
   inverse, 109–110
   split, 478
Double angle identity, 456–457
Double root, 187
Doubling time, 62, 65–66
Dow Jones average, 8, 71, 170, 189
   modeling, 400–404
Drug model, 279–288
   eliminating medication, 279–281
   and geometric sequence, 316–317
   maintenance level, 283, 288, 364–366
   repeated dosage, 281–285

e, base, 94–96
Earthquakes, 92–93
Ellipse, 525–534
   average value, 556–560
   foci of, 526
   major/minor axes, 527
   parametric equations, 545–547
   reflection property of, 532–534
   standard form equations, 528–529, A-32–A-33
   symmetry, A-34
   vertices, 527
Epicycloid, 550–551
Equation
   circle, 521–523, A-32
   conic section, 539–540
   ellipse, 528–529, A-32–A-33
   hyperbola, 536
   line, 45
   parabola, 538
   parametric, 520, 543–551
   quadratic, 182–184
   roots of, 224–230
   systems of linear equations, 234, 247
   See also Difference equation
Equilibrium solution, 349–350
Even function
   power, 76–77
   trignometric, 456
Exponent, 66–67
   fractional, 77–79
   negative, 79
Exponential function, 60–68, 100
   asymptote, 64
   base of, 61
   converting bases, 94–96
   decaying, 62–64, 103–104
   doubling time, 65–66
   fitting to data, 135–139, 148, 158–160, 194
   formula, 64–65
   graph of, 61, 64
   growth/decay factor, 61, 65, 101–104

   growth/decay rate, 65
   half–life, 65–66
   identities, 91
   inverse, 107, 109
   logistic growth, 301
   and sinusoid, 445–446
   vertical shift, 217–219
   See also Logarithmic function; Power function
Exponential growth, 61
   difference equations, 268–271
   and differences, 334
   population, 135–139
Exponential sequence, 256–257
Extrapolation, 31, 50
   exponential function, 62
   linear regression, 125

Factor, polynomial, 183
   and roots, 183, 185
Factorial notation, A-31
Falling body, 5, 26
   nonlinear regression, 167–168
Federal debt, 158–160, 223
Fiber–optic cable
   exponential function, 63–64
   half–life of signal, 66
Fibonacci model, 268–270
Fibonacci sequence, 264–266
Financing a car, 385–389
Finite sequence, 252–253, 314
Fitting to data, 148–149
   difference equation, 374–382, 405–410
   exponential function, 135–139
   line, 52–54
   logarithmic function, 147–148, 164–165
   logistic function, 405–410
   polynomial, 194–197
   power function, 139–146, 148
   quadratic function, 374–382
   residuals, 153–160
   sinusoidal function, 445, 450–452
   See also Nonlinear regression
Fixed cost, 44
Focus
   ellipse, 526
   hyperbola, 534
   parabola, 537
Fourier approximation, 479
Fractal, 510
Frequency, 440–442
Function, 1–3, 9–13
   average value, 553–560
   circular, 430
   composite, 213–214
   concave up/down, 10–11
   continuous/discontinuous, 224–225

and data tables, 4–5
difference of, 208–209
domain and range, 17–19, 24, 109–110
even/odd, 76–77, 456
graphs of, 3–6, 22–23, 26–27
increasing/decreasing, 9–11
inverse, 106–113
limiting, 212
maximum/minimum of, 9
and model, 30
multiple of, 216–217
periodic, 12–13, 428
point of inflection, 11
product/quotient, 209–211
rational, 210–213
shifting of, 214–216
sinusoidal, 437–446
split domain, 478
stretching of, 216
sum of, 4, 208–209
transcendental, 244
transformed, 154
turning point, 9
See also Exponential function; Inverse
    function; Linear function;
    Logarithmic function; Polynomial
    function; Power function; Sequence;
    Trignometric function
Future value, 72

Galileo, 5
Galois, Evariste, 224
General solution, 344
Geometric sequence, 256–257, 312–319
  behavior of, 313
  and difference equations, 367–369
  finite/infinite, 314, 315
  sum of, 314–315, 367–369
Geometry, analytic, 517–523
Graph, 3–6, 22–23, 26–27
  complete, 189
  cosine function, 429, A-23
  cubic polynomial, 184
  exponential function, 61, 64
  inverse function, 108, 111–112
  linear function, 42–44
  logarithmic function, 87–89
  logistic curve, 294
  polynomial function, 179
  power function, 75–81
  quadratic polynomial, 181, 182
  sine function, 427, A-22–A-23
  symmetry, 74, 89, A-34–A-35
  tangent function, 471
Gravitation, Inverse Square Law of, 74
Growth factor, 61, 65
Growth rate, 65

Half angle identity, 458–459
Half-life
  exponential decay, 65–66
  radioactive, 307–308
Heating model, 217–218, 327
Homogeneous difference equation, 343
Horizontal asymptote, 64
Horizontal axis, 22
Horizontal line test, 110
Horizontal shift, 215
Hours of daylight model. *See* Daylight
  model
Hyperbola, 525, 534–537
  branches of, 534
  standard form equations, 536
Hypotenuse, A-3

i, 496
Identity, logarithmic–exponential, 91
  See also Trigonometric identity
Imaginary number, 496
Implicit linear function, 54–55
Income tax model, 42, 58
Increasing function, 9–11
Independent variable, 16–17, 22, 47, 66
Infinite sequence, 252
Inflection point, 11
Inhibited growth. *See* Logistic growth
Initial condition, 344, 354
Initial side of angle, A-20
Interpolation, 31
  linear regression, 125
  polynomials, 241–245
Interval notation, 200
Inverse function, 106–113
  cosine, 468–470
  exponential, 107, 109
  graph of, 108, 111–112
  horizontal line test, 110
  logarithmic, 107
  monotonic, 110, 111
  multiple roots, 107
  nonlinear regression, 137
  parametric equations, 548
  power, 107, 109
  restricted domain, 109–110
  solving, 108–109
  trigonometric, 464–470, 473–474, A-7
Inverse Square Law of Gravitation, 74
Iteration, 412–421
  attractor, 414
  bifurcation diagram, 420
  butterfly effect, 418
  chaos, 415–421
  cobweb diagram, 415
  iterates, 413
  limit of sequence, 413–414

period-doubling, 417
random behavior, 418
sine function, 414–415
  See also Difference equation; Sequence

Julia set, 504–507

Kepler, Johannes, 140, 144

LORAN system, 516
Law of cosines, 460–462
Law of sines, A-24–A-26
Leading coefficient, 180–181, 186
Least squares line, 124–125
Limaçon, 572–573
Limiting function, 212
Line
  of best fit, 52–54, 122
  least squares, 124–125
  parallel and perpendicular, 55
  parametric form, 543–544
  point of intersection, 55
  point-slope form, 45–47
  polar coordinates, 568
  and sine function, 483
  slope-intercept form, 45
  slope of, 42–44, 232
Linear correlation coefficient, 126–127, 158,
  166
  critical values of, 128–130
Linear equations, system of, 234, 237
Linear function, 41–55, 100
  continuous, 49
  data, 51
  defined, 48
  extrapolation, 50
  graph of, 42–44
  implicit, 54–55
  line of best fit, 52–54
  point-slope form, 45–47
  slope-intercept form, 45
Linear regression, 122–130
  and calculators, 124–125
  correlation coefficient, 126–130
  data sampling, 127–129
  extrapolation/interpolation, 125
  least squares/regression line, 124–125
  residuals, 124
  See also Nonlinear regression
Logarithm, 86
  changing bases, 94–96
  natural, 94–96
  nonlinear regression, 137–139, 148
  properties of, 89–90

Logarithmic function, 85–96, 101
    applications of, 92–93
    asymptote, 88
    base 10, 86
    fitting to data, 147–148, 164–165
    graph of, 87–89
    identities, 91
    inverse, 107
    and natural logarithm, 94–96
    rate of growth, 104–105
    See also Exponential function; Power
        function
Logistic growth, 293–303
    difference equation, 295–303, 347–348
    and differences, 334–336
    and exponential growth, 301
    fitting to data, 405–410
    graph of, 294
    inhibiting constant, 293–294, 303
    iteration, 415–420
    limit to, 294, 296, 303
    maximum sustainable population, 294,
        296
    point of inflection, 295, 297
    and quadratic function, 297
    rabbit population model, 293–298
    rate of, 293, 303
    spread of technology, 301–303
    U.S. population, 407–410
    world population, 298–299

Maintenance level of drug. *See* drug model
Mandelbrot set, 508–510
Mathematical model, 28–31
    See also Difference equation
Maximum of function, 9
Medication model. *See* Drug model
Method of undetermined coefficients,
    357–360
Midpoint formula, 518–519
Minimum of function, 9
Model, mathematical, 28–31
Modulus, 497
Monotonic function, 110, 111
Monte Carlo method, 32–36

$n!$, A-31
Natural logarithm, 94–96
Newton's Laws, 5
    Cooling, 320–326
    Heating, 326–327
    Motion, 167, 222–223
Nonhomogeneous difference equation, 343
Nonlinear regression, 135–149
    calculator, 163–164, 377
    correlation coefficient, 158, 166

difference equations, 374–377
    exponential functions, 135–139, 148,
        158–160, 194
    falling body, 167–168
    inverse function, 137
    logarithmic functions, 147–148, 164–165
    logistic curve, 405–410
    planetary motion, 140–144
    power functions, 139–146, 148
    residuals, 140, 153–160
    See also Linear regression
Normal distribution, 404

Odd-even identity, 455–456
Odd function
    power, 75–77
    trigonometric, 456
Order of difference equation, 272, 341
Origin of coordinate system, 22
Oscillation, periodic, 425
Outlier, 157

Parabola, 180–184
    axis of symmetry, 538
    as conic section, 525, 537–539
    directrix, 537
    focus, 537
    iteration, 418–420
    parametric equations, 544–545
    reflection property, 538
    standard form equations, 538
    vertex, 181
Parallax, 461
Parallel line, 55
Parametric equation, 520, 543–551
    and calculators, 546–547
    circle, 545
    cycloid, 549
    ellipse, 545–547
    epicycloid, 550–551
    inverse function, 548
    line, 543–544
    parabola, 544–545
Particular solution of difference equation,
    354
Pascal's Triangle, 333
Perihelion, 531, 558–559
Period, 425, A-21
    sinusoid, 441–442
Periodic function, 12–13, 425, 428, A-21
    See also Trigonometric function
Perpendicular line, 55
Perturbation, 206
Phase shift, 442–443
pH value, 92

Pi, 33–35
Planetary motion, 140–144
    average value, 553–554, 558–559
    ellipse, 531–532
Point of inflection, 11
    logistic curve, 295, 297
Point of intersection, 55
Point–slope form of line, 45–47
Polar coordinate system, 516, 561–574
    Archimedean spiral, 568–569
    cardioid, 572
    circle, 568, 569
    limaçon, 572–573
    line, 568
    polar angle/axis, 562
    pole, 561
    and rectangular coordinates, 564–566
    rose curve, 570–572
Polynomial function, 179–206
    approximating sine and cosine, 481–492
    coefficients of, 180
    cubic, 184–186
    of degree n, 186–189
    degree of, 180
    difference patterns, 232–233, 236–237
    end behavior, 187–188
    fitting to data, 194–197
    graph of, 179
    interpolating, 241–245
    leading coefficient, 180–181
    and power function, 180, 188
    quadratic, 180–184
    quartic, 197
    roots of, 199–206, 224–230
Population growth, 25
    and difference equations, 260–261
    and differences, 334–336
    doubling time, 62
    exponential, 60–62, 135–139, 270–271
    prediction, 85–86, 90–91, 410
    and sequences, 251, 256–257
    See also Logistic growth
Power function, 74–81, 100
    asymptote, 80
    even/odd, 75–77
    fitting to data, 139–146, 148
    with fractional powers, 77–79
    graph of, 75–81
    inverse, 107, 109
    with negative powers, 79–81, 103–104
    and polynomials, 180, 188
    rate of growth, 101–105
    root, 107, 109
    See also Exponential function;
        Logarithmic function
Present value, 72
Principal values, 467
Probability, 34, 404

Product
    of functions, 209–210
    logarithm of, 89
Pythagorean identity, 430, A-16

Quadrant, 22
Quadratic equation, 182
Quadratic formula, 182
Quadratic polynomial, 180–184
    difference patterns, 233–234, 236–237
    fitting to data, 374–382
    graph of, 181, 182
    interpolating, 242–245
    linear factors, 183
    roots of, 182–184, 199–202
Quartic polynomial, 197
Quicksort, 397–399
Quotient
    of functions, 210–211
    logarithm of, 89–90

Rabbit population growth, 264–270
    logistic model, 293–298
Radian measure, 430–432
Radioactive decay, 306–311
    exponential pattern, 310–311
    half–life, 307–308
Range of function, 17–19, 24
Rational function, 210–213
Rectangular coordinate system, 22, 515
    and polar coordinates, 564–566
Recursion equation. See Difference equation
Reflection identity, 455–456
Reflection property of ellipse, 532–534
Reflection property of parabola, 538
Regression. See Linear regression;
        Nonlinear regression
Relativity, 222, 383
Residual, 124, 140, 153–160
    pattern, 156–157
    plot, 155
    transformed/detransformed function,
        154
Richter scale, 92–93
Root
    complex, 183–184, 199–206
    cubic equation, 184–186, 203–206
    double, 187
    and factors, 183, 185
    finding, 224–230
    multiple, 107
    power function, 107, 109
    quadratic equation, 182–184, 199–202
Rose curve, 570–572
Rumor model, 299–301

Savings plan model, 366–367
Scatterplot, 121
Secant function, 475
Second difference, 233, 331
Second order difference equation, 272,
        341–342
Sequence, 251–262
    and complex numbers, 504–510
    convergent/divergent, 253–254
    difference operator, 331
    and difference patterns, 329–334
    difference table, 332
    elements/terms of, 251
    Fibonacci, 264–266
    finite/infinite, 252–253
    general/nth term of, 252
    geometric/exponential, 256–257,
        312–319
    limit of, 253
    sum of, 255
    See also Difference equation; Iteration
Shifting and stretching functions, 214–219
Sigma notation, A-31–A-32
Similar triangles, A-32
Simple insertion sort, 396–397
Simulation, 33
    computer, 555, 557–560
Sine function, 425–428, A-9–A-13, A-19–
        A-26
    approximating with polynomials,
        481–492
    behavior, 432–433
    and calculator, 427
    graph of, 427, A-22–A-23
    inverse, 464–468
    law of sines, A-24–A-26
    period, 428
Sinusoidal function, 437–446
    amplitude, 438–439
    approximating with polynomials,
        477–479
    and exponential decay, 445–446
    fitting to data, 445, 450–452
    frequency, 440–442
    period, 441–442
    phase shift, 442–443
    vertical shift, 438
Slope-intercept form of line, 45
Slope of line, 42–44
    differences, 232
    parallel and perpendicular lines, 55
Solution field of difference equation,
        347–349
Sorting methods, 391–399
    bubble sort, 391–396
    quicksort, 397–399
    simple insertion sort, 396–397
Specific solution, 344

Sphere, volume of, 74
Spiral, Archimedean, 568–569
Split domain function, 478
Spring model, 444–446
Square, completing the, A-33–A-34
Square wave, 477–478
Stochastic difference equation, 403
Stock market model, 400–404
Straight–line depreciation, 58
Strictly increasing/decreasing function, 10
Sum
    functions, 208–209
    integers, 235–237, 369–371
    logarithms, 89
    sequence, 255, 314–315, 367–369
    squares of integers, 237–241
Sum and difference identity, 457
Summation notation, 235, A-31–A-32
Symmetry, 74, 89, A-34–A-35
Systems of linear equations, 234, 237

Table, data, 4–5, 23–24
Tangent function, 471–474, A-1–A-8,
        A-19–A-26
    graph of, 471
    inverse, 473–474, A-7
Taylor polynomial approximation, 487, 492
Temperature conversion, 106
Terminal side of angle, A-20
Third difference, 332
Tide model, 443–444, 448, 470, 481
Torricelli's Law, 379–382
Transcendental function, 244
Transforming functions, 208–219
    residuals, 154
Trial solution of difference equation, 357,
        359–360
Triangles, similar, A-32
Triangulation, A-25–A-26
Tribbles, 278–279
Trigonometric function, 425–510, A-19–A-26
    approximating, 477–492
    and chaos, 503–510
    and complex numbers, 496–502
    cosine, 414–415, 425–433, A-9–A-15
    even/odd, 456
    inverse, 464–470, 473–474
    radian measure, 430–432
    sine, 414–415, 425–433, A-9–A-15
    sinusoidal, 437–446
    tangent, 471–474
Trigonometric identity, 455–460, A-15–A-18
    approximating sine and cosine, 484–485
    double angle, 456–457
    half angle, 458–459
    law of cosines, 460–462

Trigonometric identity (cont.)
    odd-even, 455–456
    Pythagorean, 430, A-16
    reflection, 455–456
    sum and difference, 457
    tangent, 472–473
Trigonometry, A-1–A-26
    cosine, A-13–A-15
    graphs, A-22–A-23
    initial/terminal side of angle, A-20
    law of sines, A-24–A-26
    period, A-21
    sine, A-9–A-13
    tangent, A-1–A-8
Turning point of function, 9

Unit circle, 34–35, 426
    and complex numbers, 504–505
U.S. population
    exponential growth, 135–139, 270–271
    logistic growth, 407–410
    and sequences, 251

Variable
    continuous, 49
    dependent/independent, 16–17, 22, 47, 66
Variable cost, 44
Vertex
    ellipse, 527
    parabola, 181

Vertical asymptote, 80, 88
Vertical intercept, 43
Vertical line test, 27
Vertical shift, 215, 217–219
    sinusoidal, 438

World population. *See* Population growth

y–intercept, 43

Zero of a function. *See* Root